PROTEIN BIOCHEMISTRY, SYNTHESIS, STRUCTURE AND CELLULAR FUNCTIONS

ALPHA-FETOPROTEIN

FUNCTIONS AND CLINICAL APPLICATIONS

PROTEIN BIOCHEMISTRY, SYNTHESIS, STRUCTURE AND CELLULAR FUNCTIONS

Additional books in this series can be found on Nova's website
under the Series tab.

Additional e-books in this series can be found on Nova's website
under the eBooks tab.

PROTEIN BIOCHEMISTRY, SYNTHESIS, STRUCTURE AND CELLULAR FUNCTIONS

ALPHA-FETOPROTEIN

FUNCTIONS AND CLINICAL APPLICATIONS

NISHA LAKHI
AND
MICHAEL MORETTI
EDITORS

nova publishers
New York

Additional color graphics may be available in the e-book version of this book.

Library of Congress Cataloging-in-Publication Data

Names: Lakhi, Nisha, editor. | Moretti, Michael, editor.
Title: Alpha-fetoprotein : functions and clinical application / editors, Nisha Lakhi and Michael Moretti (Department of Department of Obstetrics and Gynaecology, Richmond University Medical Centre, Staten Island, New York, USA).
Description: Hauppauge, New York : Nova Science Publisher's, Inc., [2016] | Series: Protein biochemistry, synthesis, structure and cellular functions. | Includes index.
Identifiers: LCCN 2016007435 (print) | LCCN 2016017035 (ebook) | ISBN 9781634848756 (hardcover) | ISBN 9781634849005 (ebook) | ISBN 9781634849005 ()
Subjects: LCSH: Alpha fetoproteins.
Classification: LCC QP552.A439 A47 2016 (print) | LCC QP552.A439 (ebook) | DDC 616.99/40756--dc23
LC record available at https://lccn.loc.gov/2016007435

Published by Nova Science Publishers, Inc. † New York

To Dr. G.J. Mizejewski

For his inspiration, mentorship, and guidance as I edited my first book, and for devoting his career to studying and writing about AFP.

I am sure there is still much more to come!

To My Mom and Dad

For their countless sacrifices and love, without which, I would not be who I am today.

I love you both *ad infinitum*

CONTENTS

FOREWORD
AN OVERVIEW OF ALPHA-FETOPROTEIN RESEARCH THROUGHOUT THE DECADES

Gerald J. Mizejewski, PhD

The following discourse is not a history of AFP (see Chapter by Vladimer Jurisic), but rather an overview as observed through the eyes of the author (GJM) who has researched AFP since 1970. Thus, this "Forward Section" of the book reflects a personalized lifelong journey of the author's pursuit of knowledge in understanding the biology and chemistry of AFP. The background for this discussion began in 1937 when the Russian Virologist/immunologist Dr. Lev Zilber discovered the virus responsible for causing Eastern Russian tic-borne Encephalitis. As a reward for this major discovery, Dr. Zilber was commissioned and founded the Microbiology Virus Institute in Moscow to study cancer-specific antigens. In 1951, a young scientist named Dr. Garry Abelev joined Zilber's laboratory team to research cancer antigens. Interest in linking serum fetal proteins to cancer was first peaked in 1956 when C. Bergstrand and B. Czar discovered an unknown protein in the sera of human fetuses by the use of paper electrophoresis.

The pre-AFP decade of the 1950s proved to be a banner period for visualizing the antigen-antibody reaction as an opague zone of precipitation in agar and gels. In 1953, O. Ouchterlony developed an agar double immunodiffusion method; later that same year P. Grabar and J. Williams adopted this concept to fit an immunoelectrophoresis procedure. G. Bertil and B. Bjorklund further advanced the method of immunoprecipitation in agar and gels in 1954. Dr. Zilber himself demonstrated in 1958 a hepatoma specific antigen in mice by agar immunodiffusion not knowing that the same protein was present in fetuses. By 1959, the procedures of electro- immunodiffusion, radial immunodiffusion, autoradiographic double diffusion, and radioimmunoassay (R. Farr) were already in use. In the1959-1960 period, Dr. R. Yallow reported using radioimmunoassay (RIA) to measure insulin in human sera. Thus, the decade of the 1960s would produce the immunoassays necessary to detect and quantitate proteins in very low concentrations in the sera and biological fluids of both man and animals.

It was in the decade of the 1960s that AFP was discovered and research was initiated to pursue further knowledge concerning this tumor-associated fetal protein. It was the first time that a protein was found that possessed the dual role of a fetal protein and a tumor-associated protein, i.e, an oncofetal protein that re-appeared in the adult. In 1963, Dr. Garry Abelev pursued Zilber's earlier efforts to study cancer specific and cancer-associated antigens and discovered a fetal protein (AFP) in the sera of hepatoma-bearing mice. Thus, Abelev is considered the uncontested discoverer of AFP and the father of tumor markers. Two years later in 1965, Dr. Yuri Tatarinov reported a similar fetal protein in the sera of human adults bearing hepatomas (hepatocellular carcinomas). In that same year, J. Masopust and L. Kotal labeled the new protein as "fetoprotein," and in 1966 David Gitlin further coined the term alpha-fetoprotein" due to its electrophoretic migration pattern. Thus, AFP became the first true tumor-specific biomarker reported in the world's medical literature. Since that decade, AFP has become the "gold standard" as a specific cancer biomarker for hepatomas and as a tumor-associated marker for many other cancers. Although various authors have written obituaries for AFP as a clinical cancer biomarker, none of these critical commentaries have withstood the test of time.

The decade of the 1970s proved to be an era of isolation, purification and characterization of AFP from hepatoma extracts/serum and from fetal tissues. Scientists from this era of research included E. Ruoslahti, S. Nishi, H. Pilko, L. Purves, H. Watabe, M. Seppala, L. Savu, M. Gorin, S. Sell, E. Alpert, E. Zimmerman, J. Masopust, V. Zizkovsky, and T. Tamaoki. A multitude of new techniques and methods were introduced to facilitate protein purification. Such methods encompassed Sephadex and Sepharose chromatography, isoelectric focusing, lectin/mannosyl and immunoaffinity chromatography, preparative electrophoresis, and high-pressure liquid chromatography. RIAs for both rat and human AFP were developed by E. Ruoslahti, P. Gold, L. Purves, S. Sell and H. Hirai. New techniques to characterize proteins encompassed isoelectric pH focusing, molecular weight measurements, mass spectroscopy, sedimentation rates, extinction coefficients, and ultracentrifugation. These techniques provided purified forms of proteins(AFP) in which to study ligand-binding, in vivo transport, and immunoregulation. Outstanding researchers in these areas included E. Nunez, J. Uriel, G. Vallette, C. Aussel, J. Trojan, C. Esteban, F. Herve, R. Masseyeff, R. Hassoux, S. Sell and C. Benassayag. This decade also experienced the use of AFP in prenatal screening for fetal spinal cord disorders such as neural tube defects. The scientists responsible for developing such testing were D. Brock, R. Sutcliff, A. Leek, B. Crandall and H. Cuckle.

The decade of the 1980s demonstrated the ability of scientists to determine the sequence of the AFP gene, its promoters and mRNA, together with the translated amino acid sequences of its polypeptide chain. Thus, the AFP gene was cloned, its mRNA was identified, and the protein amino acid sequence was determined. The names emerging from such studies included S. Tilghman, J. Chou, J. Lemire, N. Fausto, M. Gorin, N. Kioussis, T. Morinaga, A. Dugaiczyk, A. Luft, F. Lorscheider, and P. Pucci. This also was a period in which the study of AFP was pursued regarding immunoregulation which included topics such as immunosuppression, immunoenhancement, innunotolerance, and autoimmunity. Researchers linked to these studies were R. Murgita, T. Tomasi, C. Lu, H. Wigzell, Y. Suzuki, N. VanOers, D. Hopper, E. Zimmerman, R. Sharony, and M. Gerschwin. Also, prenatal screening biomarkers employing AFP, HCG, and estriol were combined into a "Triple test" for use in Down syndrome screening. The triple test was developed by clinical scientists such as N. Wald, J. Canick, H. Cuckle and I. Merkatz. The use of antibodies to study AFP physiology emerged with scientists such as GJM, S. Nishi, J. Masopust, R. Hassoux, P. Strop, E. Engvall, S. Sell and H. Shepard. Physiological studies of AFP were also reported in which endocrine, neuronal, reproductive growth regulations were elucidated.

The decade of the 1990s focused on results and findings regarding the biological activities of AFP receptors and binding proteins, platelet aggregation, growth regulation, apoptosis, conformational transformations, placental transfer, and molten globular formations. Other functions included were cell proliferation/multiplication/differentiation, cell migration and adherence, and cell-to-cell contact. Scientists associated with such activities were GJM, E. Nunez, J. Uriel, R. Moro, M. Geuskins, M. Lauderoute, J. Torres, J. Navel, J. Laborda, J. Trojan, C. Esteban, M. Alava, M. Villecampa, E. Sarcione, Y. Suzuki, T. Weizmann, and M. Longnecker. This was a period in which two different AFP receptors were isolated, purified, but only partially characterized; thus, such studies were pursued by the production of antibodies to the AFP receptor. It was during this era that the prenatal "Quad test" was developed for more proficient Down Syndrome screening. The quad screening test was comprised of analytes which included AFP, HCG, estriol, and dimeric inhibin developed by researchers such as N. Wald, J. Cuckle, J. Haddow, P. Benn, J. Macri, J. Canick, and G. Knight. It was also a time in which AFP-derived peptides were developed and their biological activities determined by GJM, A. Terentiev, and N. Moldogazieva. Finally, it was a period in which AFP was found to be involved in cell cycle regulation.

The decade of the 2000s again demonstrated numerous advances in the use of AFP in signal transduction pathways, and transcriptional regulation. Such progress introduced topics such as post-translational (epigenetic) expression, zinc fingers, homeoboxes, gene promoters, postnatal expression, cell cycle regulation (cyclin-A, Cyclin-E), and hepatocyte regeneration/proliferation. Additional studies included binding to chemokine and retinoic acid receptors, and the use of recombinant third domain AFP as a vector for drug delivery. Such studies involved a host of researchers including N. Lazarevich, A. Atemezem, M. Yamamoto, G. Posypanova, S. Severin, G. Li, E. Dudich, L. Semenkova, J. Li, S. Zhou, T. Korpela, L. Guo, M. Irony-Tur-Sinai, E. Nizri and J. Bartha. As in the 1990s decade, the autoimmune connection to AFP was extended to involve neuroinflammatory diseases such as multiple sclerosis, rheumatoid arthritis, myasthenia gravis, system lupus, and ataxia telangiectasia. In the field of Down Syndrome prenatal testing, a new test screen for first trimester screening was developed which included HCG, PAPP-A, nuchal translucency, and ultrasound markers. The investigators responsible for these later advancements included N. Wald, L. Platt, L. Lugaff, K. Nicolaides, H. Cuckle, J. Canick and G. Knight.

The decade of 2010 (although not completed) shows that AFP research progressed to new heights taking advantage of new and novel technologies. Such advanced technologies involved signal transduction, transcriptional regulation, gene knock-out and knock-in, and confocal microscopy. New transcription related factors inlcuded caspase 3 and 9 enzyme interactions, tumor apoptosis, metastasis, retinoic acid signaling, Trail Death Receptor, PTEN, PI3K/AKT, GADD153, FN-14, AKT/mTOR and the CXCR4 chemokine receptor. Other techniques included microarray analysis, RNA interference, GST pulldown assays, knockdown using SiRNA-transfected cells, co-immuno-precipitation, overexpression of AFP, and chromatin immunoprecipitation assays. The investigators associated with these AFP studies included G. Li, Y. Lin, M. Li, S. Wang, C. Li, M. Zhu, L. Semenkova, E. Dudich, and C. Zhang. The direct molecular interactions of AFP with receptors and transcription factors were proven by: (a) co-localization protein-to-protein interaction studies; and (b) fluorescence resonance energy transfer (FRET) analysis. Such protein interactions can modulate intra-cellular signaling cascades leading to regulation of transcription factors with initiation of cell growth, proliferation, and metastasis.

The decade of 2020 and beyond can only be considered as a prospectus for future AFP research. AFP may well be employed as a key component in vaccines directed against hepatocellular carcinomas (HCC) employing AFP-pulsed dendritic cells, cross-linking of AFP to Heat Shock Protein (HSP-72), engineered cytotoxic T-cells with AFP-specific TCR genes, AFP-DNA prime adenovirus boost, and a host of others. New potential diagnostic biomarkers like circulating micro RNA-150 and micro RNA-301 assayed together with AFP will be utilized for HCC and hepatitis-B virus related infections. Serum AFP will also be used as a surrogate biomarker for liver disease progression of Hepatitis-C. AFP measurements together with high-powered microwave ablation and radiofrequency ablation may be employed following transcatheter arterial chemo-embolization. Together with AFP measurements, serum level of chemokine CCL4 and CCL5 in cirrhotic patients should improve the detection and diagnosis of HCC. AFP might be more frequently used to monitor patients awaiting liver transplantation. In the research field, more detailed mapping of ligands such as ion channels and various receptor binding sites on the AFP third domain (and other domains) will be achieved. In prenatal screening, AFP together with other serum and placental biomarkers will be employed to predict pre-term birth, pre-eclampsia, and other adverse outcome and pregnancy complications.

Postscript: The author (GJM) spend his entire Ph.D. research career of 45 years in pursuit of elucidating the biological, physiological, and biochemical activities of human and animal alpha-fetoprotein. However, the end is not yet in sight.

Summary of AFP Research through the Decades

1950's – Birth of Immunoassay: RIA, RID, ID, CIEP and others.

1960's – AFP discovery: mouse hepatomas, human fetal sera tumor-associated Fetal Proteins

1970's – AFP isolation, purification, and characterization; prenatal screening for neural tube defects developed.

1980's – AFP gene cloned DNA/RNA, and amino acid sequence determined; triple test developed for Down Syndrome developed.

1990's – Immunoregulation, biological activities endocrine interactions, quad test for Down Syndrome advanced.

2000's – Cell cycle regulation, zinc fingers, autoimmune diseases, signal transduction pathways, and transcriptional regulation was studied.

2010's – Gene knockout, Gene knock-in; inhibitory RNA utilized.

2015 plus – Micro RNA biomarker and AFP used for HCC (hepatomas)vaccines: and microarray analysis utilized for AFP associated liver disease states.

PREFACE

Although it has been sixty years since Bergstrand and Czar first discovered an unknown protein fraction in fetal serum using paper electrophoresis, Alpha-fetoprotein (AFP) continues to capture the attention of scientists and clinicians worldwide. It should be of no surprise however, as not many other molecules have such a diverse and widespread appeal.

In addition to being necessary for normal fetal development and modulating the immune tolerance needed to prevent maternal rejection of her fetus, AFP has several other clinical applications during pregnancy. For over three decades, due to the notable work of Drs. Wald, Cuckle, and Merkaz among others, AFP is now routinely used to screen for fetal chromosomal aneuploidies and neural tube defects. New screening strategies that incorporate AFP along with other modalities and serum markers may improve the ability to detect pregnancies destined for adverse maternal or fetal outcomes, including preeclampsia, preterm birth, and other placental complications.

AFP also has utility in the diagnosis and treatment monitoring of several other disease conditions. In the discipline of oncology, AFP is used for the diagnosis, management, and surveillance of several malignant conditions, including hepatocellular carcinoma, testicular and gynecological germ cell malignancies, as well as certain pediatric tumors. AFP may also be a useful biomarker for assessing risks associated with viral hepatitis.

It has been demonstrated that AFP receptors are present on several human cancer cell lines, making it possible to use AFP, or its recombinant fragments, for drug delivery and tumor targeting. In addition, AFP has been engineered to develop cancer vaccines and T cell immunotherapies directed against overcoming the host tolerance to tumor associated antigens. The anti-estrogenic properties of AFP have been utilized to synthesize an orally active peptide that has been shown to suppress tumor growth of estrogen receptor positive breast cancers.

Clearly AFP has many applications across several disciplines that only further study will elucidate. The aim of this book is to bring together some of the great research and clinical insights that currently surround AFP, and to pay homage to the many esteemed scientists before us, who dedicated their life to the study of AFP. The target audience for this book is research scientists as well as clinicians. I am certain it will be an inspiration for future research and clinical utilization of AFP.

This book is divided into five sections. In Section One, the history of AFP is recollected, and the role of AFP in normal and pathological conditions is explored. Section Two and Three focus on the utility of AFP during pregnancy. In Section Two, the use of routine maternal serum AFP screening for detection of chromosomal aneuploidies and neural tube defects in normal pregnancies and pregnancies with pre-existing disease is discussed. The potential use of AFP for the detection of preeclampsia is also reviewed. Section Three explores the pathogenesis and data linking unexplained elevations of AFP to adverse obstetrical outcomes. Section Four presents chapters that discuss AFP ligand binding, receptor targeting, and drug delivery strategies. Finally Section Five serves to explain the regulatory roles of AFP, its use in immunotherapy, and in the development of cancer vaccines.

Serving as the editor of this book has been an extremely rewarding experience. I want to thank all my contributors, who spent a tremendous amount of time in developing and writing chapters for this book. It has truly been an honor to work with such esteemed scientists and clinicians, many of whom have devoted significant portions of their lifetime to the study of AFP. My journey has been truly rewarding, and I want to express my sincere gratitude to all those involved in the project.

My recent research interest involves maternal serum AFP levels, its relationship to oxidative stress during pregnancy, and prediction of subsequent adverse obstetrical outcomes. This book has inspired me to continue both basic science and clinical research in this area. I hope that others researches are inspired equally.

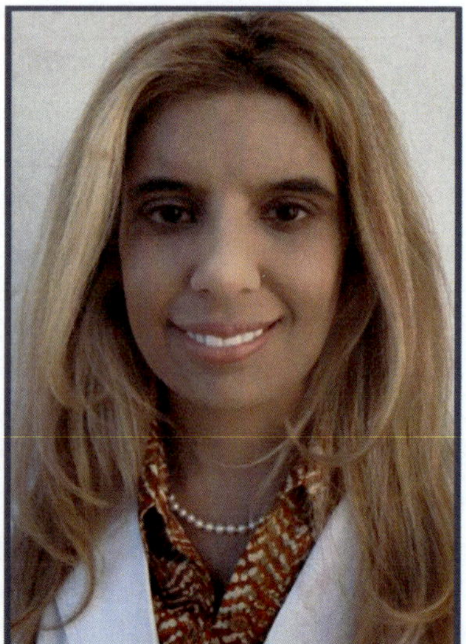

Nisha Lakhi, MD

SECTION 1: ALPHA-FETOPROTEIN HISTORY AND NORMAL AND PATHOLOGICAL DEVELOPMENT

In: Alpha-Fetoprotein: Functions and Clinical Applications
Editors: Nisha Lakhi and Michael Moretti
ISBN: 978-1-63484-875-6

Chapter 1

THE HISTORY OF ALPHA-FETOPROTEIN DISCOVERY

Vladimir Jurišić**, MD, PhD, and Dina Nikulina, Dr Sci
University of Kragujevac, Faculty of Medical Sciences, Serbia
Department of Biochemistry, State Medical Academy, Astrakhan, Russia

ABSTRACT

One of the first tumor markers to be described, and is still used in clinical practice today is Alpha-fetoprotein (AFP). AFP represents the most prominent oncobiomarker; widely used for the early diagnosis of hepatocellular carcinoma, monitoring of the tumor progression, presence of metastasis, and cancer prognosis. Until today, this protein is of great interest for study and investigation. This chapter presents historically important dates and discoveries related to the AFP.

INTRODUCTION

Bearing in mind that cancer is one of the leading causes of death in the world, new methods and techniques to improve early detection and treatment of cancer are constantly being investigated. In this sense, seeking and identifying molecules that could indicate the early onset of cancer and serve as a tumor biomarkers is an imperative [1-7]. The aim is to identify biomarkers that indicate a change within the body, even when the tumor is very small or at an early stage, and that are useful in treatment monitoring, as well as influencing clinical response [5, 8]. To achieve all this with a single cancer biomarker is very difficult. However, contemporary possibilities are much better today than a several decades ago due to the application of new technologies [9-11].

One of the first described tumor markers, which is still used for the diagnosis of hepatocellular carcinoma is alpha-fetoprotein (AFP). Until now, AFP represents the most prominent oncobiomarker; widely used for the early diagnosis of hepatocellular carcinoma,

* Corresponding author. Vladimir Jurišić: vdvd@lycos.com.

monitoring of tumor progression and presence of metastasis, as well as for determining cancer prognosis and efficacy of antitumor therapeutic measures [12-19].

The History of Alpha-Fetoprotein Discovery

The First Ideas

The first successful attempt to identify a specific protein, in the extracts and biological fluids of the human fetus and animals other than adults, was undertaken in the fourth decade of 20th century. At that time there were no sensitive methods for the definitive identification and quantification of individual proteins, molecules or ultrastuctures, as now, in the era of modern medicine [20 -22]. A modified method, similar to that used for the detection of anaphylaxis in guinea pigs, was used. In 1956, Swedish researchers, Bergstrand and Czar, who worked at Karolinska Hospital in Stockholm, used paper electrophoresis of serum proteins to discover new human protein fraction with the mobility of the alpha-1-globulin that did not exist in the serum of adults [23, 24]. This finding was not unexpected, as Pedersen Kai (1944) previously revealed by using ultracentrifugation and ammonium sulphate for calf blood proteins absent in adult animals, a new protein called fetuin [23, 25]. This protein was shown to electrophoretically belong to alpha-globulin fraction.

During gestation, marked changes in plasma proteins in various species have been reported [26]. Some of these plasma proteins have shown electrophoretic variations during fetal development. While exploring the developmental changes in fetal plasma proteins of mice, three electrophoretically separable transferrins (Tr 1-3) and five tufetoproteins (Fpl-5)' were described, at day 14.5 of development [23]. Mouse fetoprotein showed changes in concentration of its electrophoretic variants in the course of the development. Mixtures of the fetoprotein containing Fpl-3, Fpl-5, and Fp5 were purified from day 12.5 amniotic fluid, day 15.5 amniotic fluid, and day 18.5 plasma, respectively. The molecular weights of these purified protein mixtures, determined by sodium dodecyl sulfate-polyacrylamide gel electrophoresis, were similar (about 70 000). Antibody produced against Fpl-3 immunologically reacted with Fpl-5 and Fp5. Analyses of amino acids and carbohydrates indicated that they were similar in these AFP mixtures, except in the number of sialic acid residues [23, 25, 26].

The Main Scientific Researchers Engaged in Investigation on AFP from Russia

The discovery of alpha fetoprotein in literature during the sixties was mostly associated with research groups of *Garry I. Abelev* and *Yuri S. Tatarinov*, from Russia.

Garry Abelev was engaged in research at N.F. Gamaleya Research Institute of Epidemiology and Microbiology as well as in N.N. Blokhin Research Oncological Center in Moscow, Russia, from 1959 to 2008. Abelev, with his colleagues, firstly detected a specific antigen from a chemically induced hepatoma in an animal model in 1959 [27-29]. Using immunological methods, this antigen was also detected in the amniotic fluid, tissue extracts of the liver, and in the blood serum of mouse embryos in 1962 [27]. It became clear that hepatocytes produce and secrete into the blood an embriospecific alpha-globulin. In the same year, at the VIII World Congress of Cancer, which was held in Moscow, Russia, this data was discussed, and subsequently published a year later [30, 31].

Apart this, Abelev described a method for antigen determination that combined the resolving power of immunodiffusion tests with the sensitivity of radioimmunodiffusion test [32].

This method of electrophoresis precipitation (EPAG techniques) is based on concentrating the antigen by electrophoresis in a discontinuous system of buffers and then detecting it by immunodiffusion in the same gel plate [32]. Sensitivity of the method is raised by highly diluted antisera for antigen precipitation, and it was in direct proportion to the dilution used. Different variants of the method have been also developed, mainly for the exact determintion of AFP in various systems [31-33].

In the first years of research of AFP, investigation by Abelev and his colleagues was mainly focused on experimental model with mice [30]. The following idea was that this protein could be used for monitoring during the chemically induced tumor response and possibly for the estimation of cell proliferation during tissue regeneration. He focused on the obtained results on cell proliferation in normal and pathological conditions [29]. Tests of proliferation were supported by the use of radioactive C14. At that time, Abelev announced that he believed that this protein could not be used in clinical diagnosis, because a lot of murine hepatocellular carcinoma failed to produce this protein [28].

Figure 1. Astrakhan Medical Institute, Russia (1950-1960).

The big step into the clinical significance of AFP for diagnosis of cancer were made by biochemist, *Yuri S. Tatarinov*, who firstly worked at Astrakhan State Medical Institute at Department of Biochemistry (Figure 1), and then at N.I. Pirogov Russian National Research Medical University, in Moscow. Therefore, on the basis of his works, Yuri Tatarinov (Figure 2a, and 2b) was the first, or among the first, who came to the opposite conclusion in comparison to Abelev, and made connection of this protein with its clinical application [17, 26, 34]. His work was based on the idea that liver disease varies the antigenic specificity of certain serum proteins, including albumin, and that on this basis, it was possible to create immunodiagnostic methods for the identification of various diseases [35].

(A)

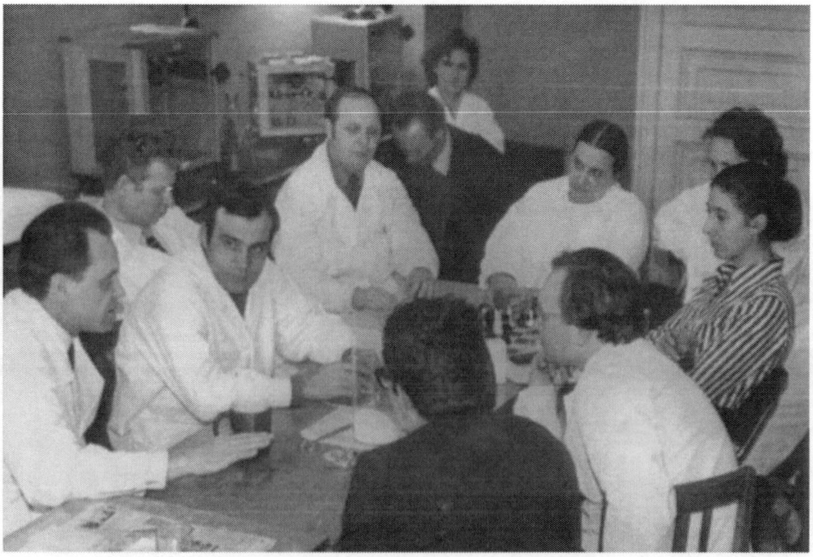

(B)

Figure 2. Yuri S Tatarinov (A) (03.09.1928 – 27.01.2012) and his associates (B), Astrakhan, Russia, 1967 (from the archives of Department of Biochemistry, Astrakhan State Medical Academy, Russia).

МИНИСТЕРСТВО ЗДРАВООХРАНЕНИЯ РСФСР
ПЕРВЫЙ МОСКОВСКИЙ ОРДЕНА ЛЕНИНА МЕДИЦИНСКИЙ ИНСТИТУТ ИМЕНИ И. М. СЕЧЕНОВА

На правах рукописи

Ю. С. ТАТАРИНОВ

ИММУНОХИМИЧЕСКИЕ ИССЛЕДОВАНИЯ СЫВОРОТОЧНЫХ БЕЛКОВ ЧЕЛОВЕКА

Автореферат диссертации на соискание ученой степени доктора медицинских наук

Астрахань
1965

(A)

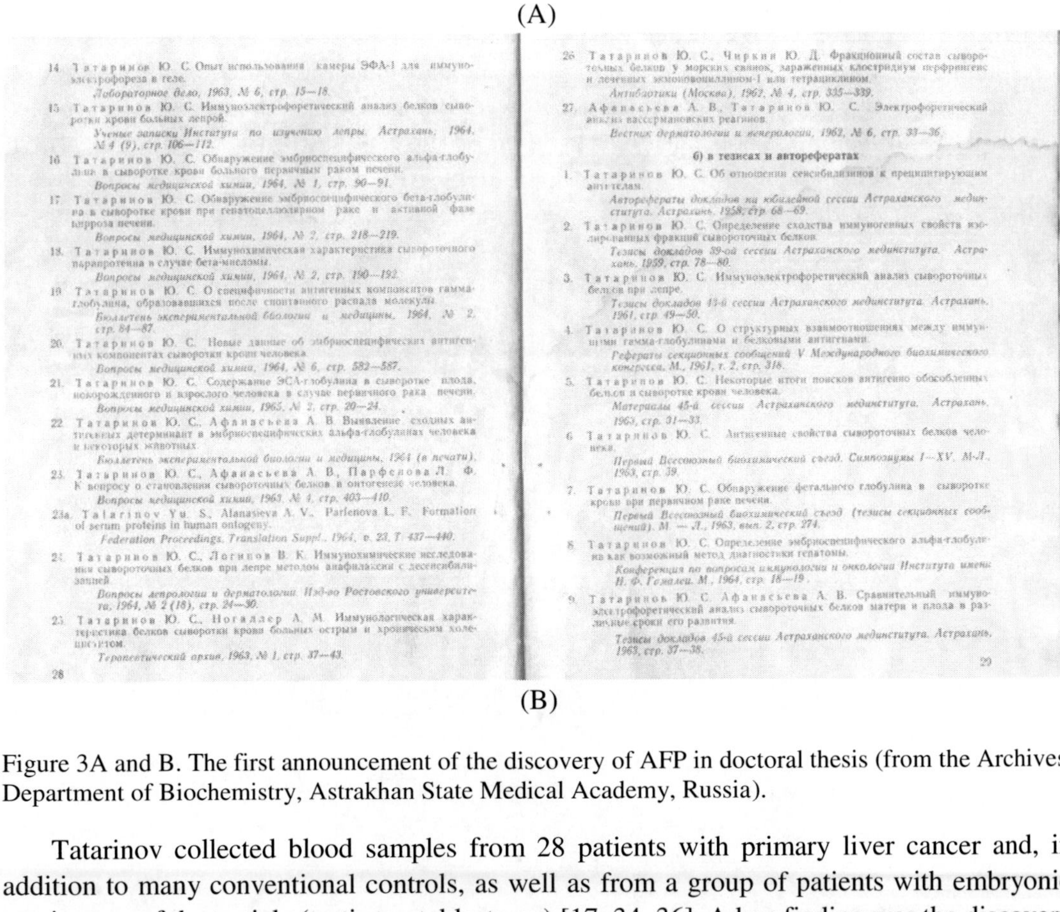
14. Татаринов Ю. С. Опыт использования камеры ЭФА-1 для иммуноэлектрофореза в геле.
Лабораторное дело, 1963, № 6, стр. 15—18.
15. Татаринов Ю. С. Иммуноэлектрофоретический анализ белков сыворотки крови больных лепрой.
Ученые записки Института по изучению лепры. Астрахань, 1964, № 4 (9), стр. 106—112.
16. Татаринов Ю. С. Обнаружение эмбриоспецифического альфа-глобулина в сыворотке крови больного первичным раком печени.
Вопросы медицинской химии, 1964, № 1, стр. 90—91.
17. Татаринов Ю. С. Обнаружение эмбриоспецифического бета-глобулина в сыворотке крови при гепатоцеллюлярном раке и активной фазе цирроза печени.
Вопросы медицинской химии, 1964, № 2, стр. 218—219.
18. Татаринов Ю. С. Иммунохимическая характеристика сывороточного парапротеина в случае бета-миеломы.
Вопросы медицинской химии, 1964, № 2, стр. 190—192.
19. Татаринов Ю. С. О специфичности антигенных компонентов гамма-глобулина, образовавшихся после спонтанного распада молекулы.
Бюллетень экспериментальной биологии и медицины, 1964, № 2, стр. 84—87.
20. Татаринов Ю. С. Новые данные об эмбриоспецифических антигенных компонентах сыворотки крови человека.
Вопросы медицинской химии, 1964, № 6, стр. 582—587.
21. Татаринов Ю. С. Содержание ЭСА-глобулина в сыворотке плода, новорожденного и взрослого человека в случае первичного рака печени.
Вопросы медицинской химии, 1965, № 2, стр. 20—24.
22. Татаринов Ю. С., Афанасьева А. В. Выявление сходных антигенных детерминант в эмбриоспецифических альфа-глобулинах человека и некоторых животных.
Бюллетень экспериментальной биологии и медицины, 1964 (в печати).
23. Татаринов Ю. С., Афанасьева А. В., Парфенова Л. Ф. К вопросу о становлении сывороточных белков в онтогенезе человека.
Вопросы медицинской химии, 1963, № 4, стр. 403—410.
23a. Tatarinov Yu. S., Afanasieva A. V., Parfenova L. F. Formation of serum proteins in human ontogeny.
Federation Proceedings. Translation Suppl., 1964, v. 23, T. 437—440.
24. Татаринов Ю. С., Логинов В. К. Иммунохимические исследования сывороточных белков при лепре методом анафилаксии с десенсибилизацией.
Вопросы лепрологии и дерматологии. Изд-во Ростовского университета, 1964, № 2 (18), стр. 24—30.
25. Татаринов Ю. С., Ногаллер А. М. Иммунологическая характеристика белков сыворотки крови больных острым и хроническим холециститом.
Терапевтический архив, 1963, № 1, стр. 37—43.

28

26. Татаринов Ю. С., Чиркин Ю. Д. Фракционный состав сывороточных белков у морских свинок, зараженных клостридиум перфрингенс и леченных экмоновоциллином-1 или тетрациклином.
Антибиотики (Москва), 1962, № 4, стр. 335—339.
27. Афанасьева А. В., Татаринов Ю. С. Электрофоретический анализ вассермановских реагинов.
Вестник дерматологии и венерологии, 1962, № 6, стр. 33—36.

б) в тезисах и авторефератах

1. Татаринов Ю. С. Об отношении сенсибилизинов к преципитирующим антителам.
Авторефераты докладов на юбилейной сессии Астраханского мединститута. Астрахань, 1958, стр. 68—69.
2. Татаринов Ю. С. Определение сходства иммуногенных свойств изолированных фракций сывороточных белков.
Тезисы докладов 39-ой сессии Астраханского мединститута. Астрахань, 1959, стр. 78—80.
3. Татаринов Ю. С. Иммуноэлектрофоретический анализ сывороточных белков при лепре.
Тезисы докладов 43-й сессии Астраханского мединститута. Астрахань, 1961, стр. 49—50.
4. Татаринов Ю. С. О структурных взаимоотношениях между иммунными гамма-глобулинами и белковыми антигенами.
Рефераты секционных сообщений V Международного биохимического конгресса, М., 1961, т. 2, стр. 318.
5. Татаринов Ю. С. Некоторые итоги поисков антигенно обособленных белков в сыворотке крови человека.
Материалы 45-й сессии Астраханского мединститута. Астрахань, 1963, стр. 31—33.
6. Татаринов Ю. С. Антигенные свойства сывороточных белков человека.
Первый Всесоюзный биохимический съезд. Симпозиумы I—XV. М.-Л., 1963, стр. 39.
7. Татаринов Ю. С. Обнаружение фетального глобулина в сыворотке крови при первичном раке печени.
Первый Всесоюзный биохимический съезд (тезисы секционных сообщений). М. — Л., 1963, вып. 2, стр. 274.
8. Татаринов Ю. С. Определение эмбриоспецифического альфа-глобулина как возможный метод диагностики гепатомы.
Конференция по вопросам иммунологии и онкологии Института имени Н. Ф. Гамалеи. М., 1964, стр. 18—19.
9. Татаринов Ю. С., Афанасьева А. В. Сравнительный иммуноэлектрофоретический анализ сывороточных белков матери и плода в различные сроки его развития.
Тезисы докладов 45-й сессии Астраханского мединститута. Астрахань, 1963, стр. 37—38.

29

(B)

Figure 3A and B. The first announcement of the discovery of AFP in doctoral thesis (from the Archives Department of Biochemistry, Astrakhan State Medical Academy, Russia).

Tatarinov collected blood samples from 28 patients with primary liver cancer and, in addition to many conventional controls, as well as from a group of patients with embryonic carcinomas of the testicle (testis teratoblastoma) [17, 34, 36]. A key finding was the discovery that in patients with tumors of other localizations except liver and testicle, AFT could not be found. Definitely, he made a great contribution to the discovery of this protein in human patients with hepatocellular carcinoma [17, 18, 34-39].

Tatarinov also reported the presence of AFP in the sera of women after spontaneous abortion, suggesting that under these circumstances, AFP can traverse the placenta in the direction of fetus to mother [38]. All findings were generally collected and first published in his doctoral dissertation in 1965, in Astrakhan, Russia (Figure 3a and Figure 3b).

On a personal initiative by Tatarinov and Abelev, the first international expedition to various regions of Africa including Uganda, Kenya and Mozambique was organized in 1968. At that time, a high incidence of hepatocellular carcinoma, much higher than in Europe and USA, was in this region. Their aim was to analyze and establish the significance of the detected AFP proteins in these patients. For the purpose, all the equipment necessary for analysis was brought from Russia. Indeed, the first announcements of increased AFP values associated with the geographic distribution of the occurrence of hepatocellular carcinoma arose from this expedition [37].

Abelev with colleagues also reported important data regarding techniques for measurement of AFP [31]. Among 200 serum samples of normal human donors, AFP was clearly detected by EPAG technique in approximately half of the cases (50%), whereas in the rest of samples, doubtful and unstable results were observed near the sensitivity limit of the method. In these cases, AFP concentration was about 3 ng/ml, and identical concentrations corresponded to measurements made with radioimmunodiffusion test. Interestingly, in all 20 examined cases of cirrhosis of the liver, AFP was demonstrated, although all the samples were negative in radioimmunodiffusion test. Based on the findings, it became very clear that the liver cancer cells produced and secreted into the blood serum, an embryo-specific alpha-globulin, that was initially named embryonal protein in mice, and alpha globulin in humans.

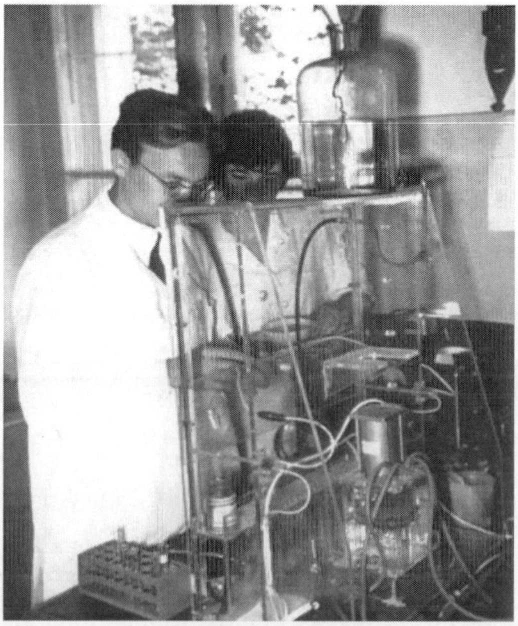

Figure 4. Aleksander Terentijev, first postgraduate student of Tatarinov, during the experiment about AFP, Astrakhan, Russia, 1997 (from the archives of Department of Biochemistry, Astrakhan State Medical Academy, Russia).

From 1963 to 2005, Yuri Tatarinov continued to work in Moscow and investigated carbohydrate heterogeneity, lectin-binding and estrogen-binding ability of human AFP [38] that corresponding to functional activity of AFP [39].

The first post-graduate student of Professor Tatarinov was Alexander A. Terentiev. He initially worked at Astrakhan and later continued his work in Moscow, until today (Figure 4). He used the method of immune-precipitation, which allowed distinguishing AFP from fetuin, another embryo-specific protein, and investigated the functional activity of AFP and its regulation [41-48]. Later, in 1970, this antigen was named "alpha-fetoprotein" (AFP) and has been long recognized to be the first oncodevelopmental biomarker. Confirmation of this discovery was registered in "Synthesis of Embryo-specific Proteins and the Resumption of their Production in Adult Animals and Humans," in 1971 by G.I. Abelev, Yu.S. Tatarinov and others (Diploma for the discovery, № 90). (Figure 5)

Many years later, in 1980, W. Fishman and H. Hirai, two of the founders of the International Society for Oncobiology (ISOBM), confirmed the significance of this discovery noting in the conference proceedings books that: ... "Although the idea of the relationship between embryogenesis and carcinogenesis has been discussed in the last 100 years, the discovery of the synthesis of AFP in 1963 by Abelev and Tatarinov in fetal liver in experimental hepatoma and hepatocellular cancer in humans was the beginning of a new era in oncology and medicine"... .

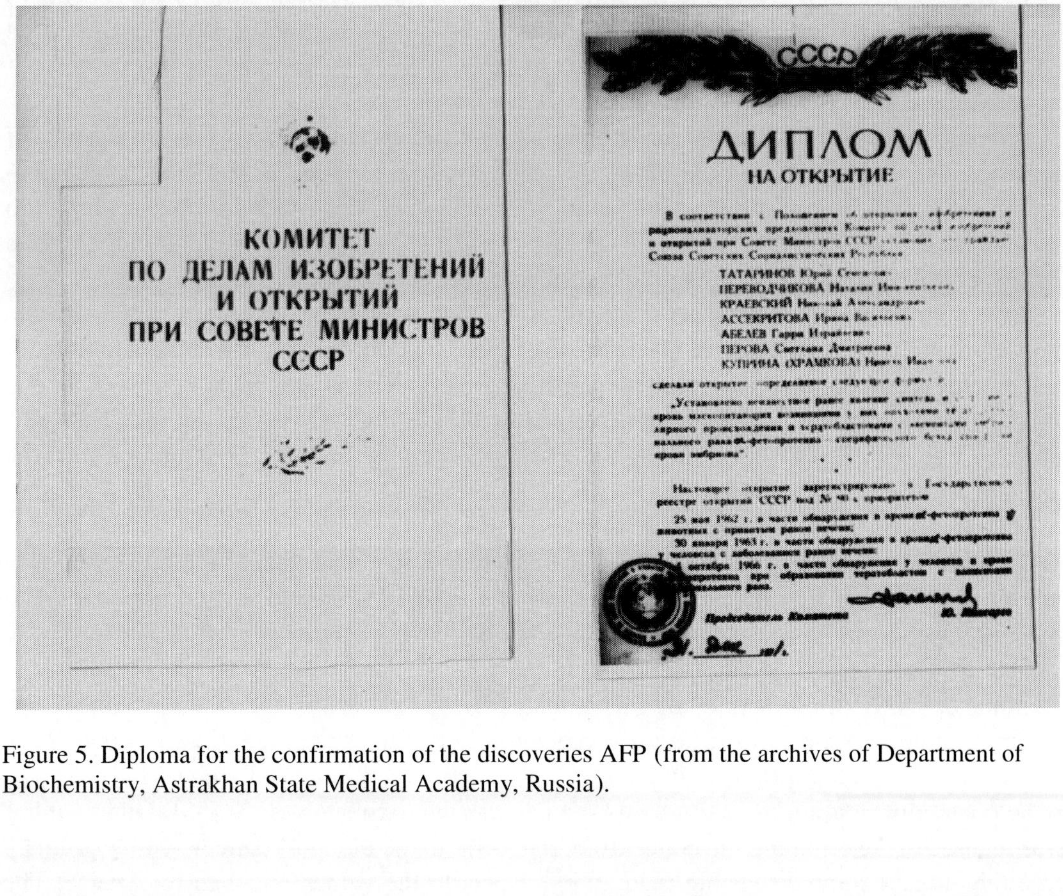

Figure 5. Diploma for the confirmation of the discoveries AFP (from the archives of Department of Biochemistry, Astrakhan State Medical Academy, Russia).

Investigation about AFP in the World

Abelev participated at the Congress of the United States and his lecture at Columbia University, New York, in 1967, held immediately after the first announcements about AFP, launched a great interest in AFP investigation in United States.

David Gitlin from University of Pittsburgh School of Medicine and the Children's Hospital of Pittsburgh, also worked on the investigation of AFP characteristics similar to Abelev [49, 50]. Almost at the same time in 1966, using electrophoretic mobility of serum fetoprotein, he reported that AFP can migrate between albumin and the alpha globulins fractions. He also reported that AFP could be detected in the sera of all embryos and fetuses. Gitlin compared the concentration of AFP to the concentration of albumin and gamma globulin [52]. He estimated that the sedimentation coefficient of this protein is approximately 5.0 S by ultra centrifugation of fetal serum in a sucrose density gradient. He also observed that by gel filtration of fetal serum on Sephadex G-200, that this protein was eluted between transferrin and albumin, the molecular weights of which are 69,000 and 90,000, respectively [51].

Regarding the of synthesis of AFP in the human embryo, it was found that the serum concentration of AFP progressively increased from 6.6 weeks to reach a maximal concentration at approximately 13 weeks, and that the concentration then decreased rapidly to reach levels of less than 2% of the maximum by 34 weeks' gestation. Gitlin with colleagues also reported in 1972 that cultures of liver, developing yolk sac and gastrointestinal tract consistently yielded relatively large amounts of labeled AFP, but in none of the cultures of lung, thymus, pancreas, skeletal muscle, amnion, or chorion AFP was detected [50].

Independently from Tarantiev in Russia, Kithier and colleagues, working in Faculty of Pediatrics, at Charles University in Prague, Czechoslovakia, reported in 1968 that fetal alpha-globulin of bovine serum differs from fetuin [52]. This was determined after working with mammalian fetus serum for several years, and by using rabbit antisera against the serum proteins of bovine fetuses and using radio-diffusion methods.

Development of new highly sensitive methods, such as enzyme-linked immunoassay and radio-immunoassay (RIA), has allowed for the determination of very low concentrations of AFP in the blood serum of normal human adults [53]. RIA is the most sensitive among immunochemical methods and permits detection of as little AFP as 1 ng/ml. This new approach was with about 1000-fold higher sensitivity than the immune-precipitation method. Ruoslahti and M. Seppalato discovered AFP at concentrations as low as 5–7 ng/ml in adult human blood serum [54, 55, 56].

Since incidence of gastric cancer is four times higher in Japan than in the UK, interest in exploring the clinical significance of AFP was great in this region and several important data where obtained from Japanese researchers. Shinzo Nishi from Department of Biochemistry, at School of Medicine, Hokkaido University, Sapporo, Japan, worked on purification and isolation of AFP from serum of hepatoma patients or from the fetus [58-61]. An immunochemical method was used for purification in his investigation. He also reported the content of the peptide portion, estimated from the recovery of analyzed amino acids, as well as the content of non-nitrogenous sugars, which was approximately 97% [58]. Other studies have focused on determining the concentration of AFP in the blood of patients with extremely large tumors and comparisons are made with respect to the values obtained by Abelev. The use of sensitive technique of radioimmunoassay, enabled measuring AFP in series of yolk sac tumors (a kind of teratocarcinoma) in respect to size of the tumors, clinical characteristics including fluctuation with presence necrosis, surgical removal and recurrence [61]. Investigation demonstrated that rate of positivity and the levels of AFP are particularly high in hepatocellular carcinomas and teratocarcinomas, and that the levels of AFP in these Japanese patients were extremely high, usually higher than 10,000 ng/ml, and corresponded to tumor size [61].

Table 1. A table with a chronological important dates related to the investigation of AFP (Modified from A. Terentiev)

Authors	Affiliation	City and Country	Year of publication
Pedersen KO	The University of Upsala, Sweden	Upsala, Sweden	1944
Bergstrand C.G. and Czar B	Karolinska Hospital	Stockholm, Sweden	1956
Abelev G.I	N.F. Gamaleya Research Institute of Epidemiology and Microbiology,	Moskow, Rusia	1963-2008
Abelev G.I	N.N. Blokhin Research, Oncological Center	Moskow, Rusia	
Tatarinov Yu.S	Astrakhan State Medical Institute,	Astrakhan, Russia	1963-2005
Ernest F. Zimmerman	University of Cincinnati	Cincinnati, USA	1967
Gitlin D. and Boesman M	University of Pittsburgh School of Medicine	Pittsburg, USA	1967
K. Kithier	Charles University, Prague	Prague Czechoslovakia	1968
Uriel J	Institut de Recherches Scientifiques sur le Cancer	Villejuif, France	1972-1976
Glisić Lj and Perisić V	Medical Faculty, University of Belgrade	Belgrade, Yugoslavia	1997-1981
Terentiev A.A. and Moldogazieva N.T	N.I. Pirogov Russian National Research Medical University	Moskow, Russia	1997-2012
Dugaiczyk A	University of California	Riverside, California, USA	1982-1996
Deutsch HF	Department of Biomolecular Chemistry, Laboratory Medicine and Medicine, University of Wisconsin	Madison, USA	1988-2000
H. Taga and K. Taketa	Okayama University Medical School	Okayama, Japan	1989–1998
Uversky V.N	Institute of Protein Research	Moskow Russia	1995-1999
Mizejewski G.J. and Jacobson H.I.	Wadsworth Center, New York State Department of Health	New York, USA	1983-1995
Dudich I.V. and Dudich E.I	Institute of Immunological Engineering	Moskow Russia	1997-2012
Behboudi S. and Alisa A	Institute of Hepatology University College London	London, UK	2005-2010
Szpirer J. and Szpirer C	Institut de Biologie et de Médecine Moléculaires	Bruxelles, Belgium	2002-2006
Li G	Peking University Health Science Center	Beijing, China	2002-2011
Severin S.E. and Severin E.S	Moscow Research Institute of Medical Ecology	Moskow, Russia	2000-2012

In Serbia, clinical studies on the value of AFP in primary and secondary liver tumors as well as the high prevalence of infection with hepatitis B were initially conducted and published from 1979 to 1981. The data indicated the problems in the diagnosis of tumors and suggest the use of immunological methods for detection of AFP [62, 63].

The most important findings about the estrogen-binding activity of AFP were obtained by the French researchers including Uriel J working in Villejuif, France in period from 1972 to 1976 [64, 65, 66]. Estrogen-binding activity of AFP is of crucial importance because it may represent a significant regulatory mechanism during embryonic development, probably protecting some organs from the effects of maternal estrogens [67, 68].

The studies of the AFP gene structure and its regulation will undoubtedly lead to a better understanding of the normal physiological functions of this protein and may provide insight into the molecular basis of oncogenic development and malignant transformation [69-77]. The first analysis of the structure AFP have emerged between the seventies and eighties, mostly based on experimental and comparative model [78, 79]. Analyses of human and rat AFP during the seventies suggest microheterogeneity in several fractions, mostly based on results obtained using polyacrylamide gel and electrophoresis methods, separating according to both size and charge [80]. More recently, the same has been done by new techniques including mass spectrometric investigation [81-83].

Chemically AFP is a glycoprotein with molecular weight of about 68-73 kDa and carbohydrate content of 3%-5% [84, 85]. Primary structure of full length human AFP molecule is represented by 609 amino acid residues from which a signal peptide of 18 amino acid residues is removed during processing to yield the mature protein of 591 residues [83, 85-90].

Regarding structure and functional activity of AFP, the data from the past decades have shown that total AFP is a collection of heterogeneous glycoproteins consisting of three different glycoforms [91, 92]. The total AFP can be separated into three fractions, AFP-L1 to AFP-L3, based on its reactivity to Lens culinaris agglutinin (LCA) on affinity electrophoresis [91-93]. AFP-L1 does not react with LCA. The L1 fraction of the total AFP is present in chronic hepatitis and liver cirrhosis, and constitutes a major fraction of total AFP in the non-malignant liver diseases. AFP-L3 has LCA-binding activity, and has an additional a 1–6 fucose residue attaching at reducing terminus of N-acetylglucosamine. It appears to be produced only by cancer cells [92]. AFP-L2 is mostly derived from yolk sac tumors and also could be detected in maternal serum during pregnancy. The clinical investigation has suggested that AFP-L3 is a marker of biologic malignancy of HCC. Liver cancer cells with expression of AFP-L3 glycoform have a tendency for early vascular invasion and intrahepatic metastasis. Early diagnosis of hepatocellular carcinoma has been difficult, if not impossible, in a background of presence cirrhosis. Screening program exists in some hospitals in North America or other countires for chronic cirrhotic patients by evaluation of abdominal ultrasonography and by measurement of serum AFP. It seems that the measurement of AFP-L3 fraction could be most appropriate for screening program in these patients. Laboratory assay for AFP-L3 used to be based on lectin affinity electrophoresis and Western blotting. These assays were tedious with long turnaround times. However, an automated assay for measuring the AFP-L3 has been developed and introduced for clinical use in Japan [94, 95].

Molecular Analyses of AFP in Last Period

Results of proteomic analysis of gene expression in tumor tissues of HCC patients demonstrated that different molecular markers can be used for assessment of liver cancerogenesis [79]. The global analysis of several genes in AFP producing cancer cell lines showed a correlation between high blood level of AFP and overexpression of specific genes that encode growth factors as well as other key molecules involved in cancerogenesis [22 42, 79, 95, 96].

Structural and functional mapping of AFP performed with the use of experimental and computer based technologies represents a new approach for determination of AFP functionality. To date, a large number of linear structural motifs have been revealed in AFP that demonstrate its multimodular and polyfunctional character. Generation of the computer-based model of a three dimensional structure of AFP was performed with perspectives to be used in study of mechanisms of ligand- or receptor binding. Using 'MODELLER' program, Terentiev generated a 3D model of human AFP on the basis of homology with human serum albumin and Vitamin D–binding protein. After this, subsequent molecular binding of a synthetic estrogen analog to the model structure and molecular dynamics simulation studies were performed [69]. The model was constructed to determine the U shaped structural cavity in which a putative estrogen-binding site may be localized. Results using this computer program indicated good quality of the model and the stability of extended region of four alpha helical structures that contains putative hormone-binding residues [95]. Data suggested two types of interactions between amino acid residues of human AFP and vitamin D-binding protein that were involved: hydrogen bonding residues as well as and hydrophobic interactions with participation of L138, M448, and M548 residues [22]. A suggestion was made that binding and immobilization of the hormone using a long spacer provides delivery of the estrogen molecule to the binding site and, thereby, facilitates interaction between AFP and the hormone [39].

Conclusion

This chapter presents historically important dates and discoveries related to the AFP in order to provide current and future scientists and researchers with new ideas for further scientific work. Moreover, it highlights the importance of research and development of new techniques and ideas that could improve the diagnosis of cancer as well as other diseases.

Acknowledgments

Writing this chapter was supported by grants from the Ministry of Science and Technology of the Republic of Serbia: grant number 175056 and Russian Foundation for Basic Research, grant 14-04-01075/15. Authors thank Jasmina Obradović for help in resolving references.

[illegible]

using anion exchange chromatography and sulfated peptide introduced antibody. *J* [illegible]

[79] Ruoslahti E, Terry WD. (1976) Alpha-fetoprotein and serum albumin show sequence homology. *Nature*, 260, 804-805.

[80] [illegible] (1988) Microheterogeneity of human alpha-fetoprotein. [illegible] 2-14.

[81] Morinaga T, Sakai M, Wegmann TG, Tamaoki T. (1983) Primary structure of human alpha-fetoprotein and its mRNA. *Proc Natl Acad Sci U S A*, 80, 4604-4608.

[82] Pucci P, Siciliano R, Malorni A, Marino G, Tecce MF, Ceccarini C, Terrana B. (1991) Human alpha-fetoprotein primary structure: a mass spectrometric study. *Biochemistry*, 30, 5061-5066.

MOLECULAR ANALYSES OF AFP IN LAST PERIOD

Results of proteomic analysis of gene expression in tumor tissues of HCC patients demonstrated that different molecular markers can be used for assessment of liver cancerogenesis [79]. The global analysis of several genes in AFP producing cancer cell lines showed a correlation between high blood level of AFP and overexpression of specific genes that encode growth factors as well as other key molecules involved in cancerogenesis [22 42, 79, 95, 96].

Structural and functional mapping of AFP performed with the use of experimental and computer based technologies represents a new approach for determination of AFP functionality. To date, a large number of linear structural motifs have been revealed in AFP that demonstrate its multimodular and polyfunctional character. Generation of the computer-based model of a three dimensional structure of AFP was performed with perspectives to be used in study of mechanisms of ligand- or receptor binding. Using 'MODELLER' program, Terentiev generated a 3D model of human AFP on the basis of homology with human serum albumin and Vitamin D–binding protein. After this, subsequent molecular binding of a synthetic estrogen analog to the model structure and molecular dynamics simulation studies were performed [69]. The model was constructed to determine the U shaped structural cavity in which a putative estrogen-binding site may be localized. Results using this computer program indicated good quality of the model and the stability of extended region of four alpha helical structures that contains putative hormone-binding residues [95]. Data suggested two types of interactions between amino acid residues of human AFP and vitamin D-binding protein that were involved: hydrogen bonding residues as well as and hydrophobic interactions with participation of L138, M448, and M548 residues [22]. A suggestion was made that binding and immobilization of the hormone using a long spacer provides delivery of the estrogen molecule to the binding site and, thereby, facilitates interaction between AFP and the hormone [39].

CONCLUSION

This chapter presents historically important dates and discoveries related to the AFP in order to provide current and future scientists and researchers with new ideas for further scientific work. Moreover, it highlights the importance of research and development of new techniques and ideas that could improve the diagnosis of cancer as well as other diseases.

ACKNOWLEDGMENTS

Writing this chapter was supported by grants from the Ministry of Science and Technology of the Republic of Serbia: grant number 175056 and Russian Foundation for Basic Research, grant 14-04-01075/15. Authors thank Jasmina Obradović for help in resolving references.

References

[1] Jurisic V, Radenkovic S, Konjevic G. (2015) The Actual Role of LDH as Tumor Marker, Biochemical and Clinical Aspects. *Adv Exp Med Biol,* 867,115-24.

[2] Jurisic V. (2003) Estimation of cell membrane alteration after drug treatment by LDH release. *Blood*, 101 (7), 2894.

[3] Nikulina D, Terentyev A, Galimzyanov K, Jurišić V. (2015) Fifty years of discovery of alpha-fetoprotein as the first tumor marker. *Srp Arh Celok Lek*, 143, 100-104.

[4] Petrović M, Bukumirić Z, Zdravković V, Mitrović S, Atkinson HD, Jurišić V. (2014) The prognostic significance of the circulating neuroendocrine markers chromogranin A, pro-gastrin-releasing peptide, and neuron-specific enolase in patients with small-cell lung cancer. *Med Oncol*, 31, 823.

[5] Radenkovic S, Milosevic Z, Konjevic G, Karadzic K, Rovcanin B, Buta M, Gopcevic K, Jurisic V. (2103) Lactate dehydrogenase, catalase, and superoxide dismutase in tumor tissue of breast cancer patients in respect to mammographic findings. *Cell Biochem Biophys*, 66, 287-295.

[6] Stankovic S, Konjevic G, Gopcevic K, Jovic V, Inic M, Jurisic V. (2010). Activity of MMP-2 and MMP-9 in sera of breast cancer patients. *Pathol Res Pract*, 206, 241-247.

[7] Jurisić V, Konjević G, Banićević B, Duricić B, Spuzić I. (2002) Different alterations in lactate dehydrogenase activity and profile of peripheral blood mononuclear cells in Hodgkin's and non-Hodgkin's lymphomas. *Eur J Haematol,* 64, 259-266.

[8] Petrović M, Bukumirić Z, Zdravković V, Mitrović S, Atkinson HD, Jurišić V. (2014) The prognostic significance of the circulating neuroendocrine markers chromogranin A, progastrin releasing peptide, and neuron-specific enolase in patients with small-cell lung cancer. *Med Oncol*, 31, 823.

[9] Nikulina DM, Kriventsev Iu A, Bisalieva RA, Lapeko SV. (2009) Novel immunoassay for laboratory assessment of the erythron state. *Klin Lab Diagn*, 12, 27-30.

[10] Carreres G, Dauphinee MJ, Eisele LE, MacColl R, Mizejewski GJ. (2002). Anti-prostate cancer and anti-breast cancer activities of two peptides derived from alpha-fetoprotein. *Anticancer Res*, 22, 2817-2820.

[11] Debruyne EN, Delange JR. (2008) Diagnosing and monitoring hepatocellular carcinoma with alpha-fetoprotein: new aspects and applications. *Clin Chim Acta*, 395,19–26.

[12] Terentiev AA, Moldogazieva NT. Alpha-fetoprotein: a renaissance.(2013) *Tumor Biol*,34, 2075-2091.

[13] Godovannyi AV, Vorontsov EA, Gukasova NV, Pozdnyakova NV, Vasilenko EA, Yabbarov NG, Dubovik EG, Severin SE, Severin ES, Ghuchev NV. (2011) Targeted delivery of paclitaxel-loaded recombinant alpha-fetoprotein fragment-conjugated nanoparticles to tumor cells. *Dokl Biochem Biophys*, 439, 158–160.

[14] Aoyagi Y, Saitoh A, Suzuki Y, Igarashi K, Oguro M, Yokota T, Mori S, Suda T, Isemura M, Asakura H. (1993) Fucosylation index of alpha-fetoprotein, a possible aid in the early recognition of hepatocellular carcinoma in patients with cirrhosis. *Hepatology,* 17, 50-52.

[15] Seta N, Gayno S, Jezequel-Guer M, Toueg ML, Erlinger S, Durand G. (1997) Sequential study of serum glycoprotein fucosylation in acute hepatitis. *J Hepatol*, 26, 265-271.
[16] Tatarinov, YS. (1963) Content of embryo-specific alpha-globulin in fetal and neonatal sera and sera from adult humans with primary carcinoma of the liver. *Fed Proc (Trans. Suppl.)*, 25, 344-346.
[17] Tatarinov YS. (1963). Detection of embryospecific alpha-globulin in serum of patients with primary liver cancer. In *I-st All-Union Biochem Cong Abstr Book*, Moscow-Leningrad, 2, 274.
[18] Debruyne EN, Delange JR. (2008). Diagnosing and monitoring hepatocellular carcinoma with alpha-fetoprotein: new aspects and applications. *Clin Chim Acta*, 395, 19–26.
[19] Ruiz-Gutierrez V, Moreno R, Moreda W, Copado MA, Rodriguez-Burgoz A. (2001) Detection of squalene in alpha-fetoprotein and fetal serum albumin from bovine. *J Prot Chem*, 20, 19-23.
[20] Valette G, Vranckx R., Martin M-E, Benassayag C, Nunez EA. (1989). Conformational changes in rodent and human alpha-fetoprotein: influence of fatty acids. *Biochim Biophys Acta*, 997, 302-312.
[21] Festin SM, Bennett JA, Fletcher PW, Jacobson HI, Shaye DD, Andersen TT. (1999) The recombinant third domain of human alpha-fetoprotein retains the antiestrothophic activity found in the full-length molecule. *Biochim Biophys Acta*, 1427, 307-314.
[22] Marrero JA, Feng Z, Wang Y, Nguyen MH, Befeler AS, Roberts LR, et al. (2009). Alpha-fetoprotein, des-gamma carboxyprothrombin, and lectin-bound alpha-fetoprotein in early hepatocellular carcinoma. *Gastroenterology*, 137, 110–18.
[23] Bergstrand CG, Czar B. (1956) Demonstration of a new protein fraction in serum from the human fetus. *Scan J Clin Lab Invest*, 8, 174–176.
[24] Bergstrand CG, Czar B. (1957) Paper electrophoretic study of human fetal serum proteins with demonstration of a new protein. *Scand. J. Clin. Lab. Invest*, 9, 277.
[25] Pedersen, KO. (1944) Fetuin, a New Globulin Isolated from Serum. *Nature*, 154: 575.
[26] Tatarinov YuS. (1965) Content of embryo-specific alpha-globulin in the blood serum of human fetus, newborns and adults. *Vorp Med Khim,* 11, 20-24.
[27] Abelev GI, Avenirova ZA, Tsvetkov VS. (1960) The elution and purification of organ specific antigen of the liver. *Vopr Onkol,* 6, 43-9.
[28] Nartsissov NV, Abelev GI. (1959) Antibody formation in primary rat sarcomas induced with carcinogens. *Neoplasma,* 6, 353-60.
[29] Abelev GI. (1971) Alpha-fetoprotein in ontogenesis and its association with malignant tumors. *Adv Cancer Res*, 14, 295-358.
[30] Abelev GI, Perova SD, Khramkova NI, Postnikova ZA, Irlin IS. (1963). Production of embryonic alpha-globulin by the transplantable mouse hepatomas. *Transplantation*, 1, 174–180.
[31]. Abelev GI. (1986) Production of embryonal serum alpha-globulin by hepatomas: review of experimental and clinical data. *Cancer Res*, 28, 1344-50.
[32] Abelev GI, Perova SD, Sokolenko AA. (1976) High sensitivity radioimmunodiffusion tests for tumor-associated antigens. *Annals of The New York Academy Of Sciences,* 276, 1, 75–90.
[33] Abelev GI. (1983). Alpha-fetoprotein: the genesis. *Oncodev Biol Med*, 4:371-381.

[34] Tatarinov YS. (1963) Detection of embryospecific alpha-globulin in serum of patients with primary liver cancer. In *I-st All-Union Biochem Cong Abstr Book,* Moscow-Leningrad, 2, 274.

[35] Tatarinov, YS (1964). Detection of embryo-specific alpha-globulin in the blood sera of patients with primary liver tumour. *Vorp Med Khim*, 10, 90-91.

[36] Tatarinov YS. (1965) Content of embryo-specific alpha-globulin in the blood serum of human fetus, newborns and adults. *Vorp Med Khim,* 11, 20–24.

[37] Elgort DA, Abelev GI, O'Conor GT. (1972) Dependence of the specificity of the serologic test for primary liver cancer in different areas of the world on sensitivity of the method used for detecting alpha-fetoprotein. *Int J Cancer,* 10, 331-7.

[38] Tatarinov YS, Nikulina DM, Phalaleeva DM, Kozlyaeva GM. (1976) Human trophoblastic beta-1-globulin and chorionepithelioma. In: *W. Fichman and S. Sell /Eds/ Oncodevelipmental gene Expression*, Academic Press N, York, 463-466.

[39] Tatarinov YuS, Terentiev AA, Moldogazieva NT, Tagirova AK. (1991) Human alpha-fetoprotein and its purification by chromatography on immobilized estrogens. *Tumor Biol*, 12, 125-130.

[40] Terentiev AA. (1969) Comparative study of fetuin and bovine embryo-specific alpha-globulin. *Bull Exp Biol Med,* 7, 42-45.

[41] Terentiev AA. (1968) Comparative studies of physico-chemical properties of human, bovine and sheep embryo-specific alpha-globulins. *Proc ASMU*, Astrakhan, 16, 336–341.

[42] Terentiev AA, Moldogazieva NT, Tagirova AK, Tatarinov YuS. (1988) Affinity chromatography of human alpha-fetoprotein on immobilized estrogens. *Byul Eksp Biol Med*, 4, 422-424.

[43] Terentiev AA, Moldogazieva NT. (2006) Structural and functional mapping of alpha-fetoprotein. *Biochemistry (Mosc)*, 71, 120–132.

[44] Krivonosov SK, Zorin NA, Ionova NK, Terentiev AA, Tatarinov Yu.S. (1985) Interaction of pregnancy-associated proteins with hemagglutinin and concanavalin A. *Byul Eksp Biol Med*, 5, 568-71.

[45] Zorin NA, Krivonosov SK, Zorina RM, Terentiev AA., Ionova NK. (1986) Competitive inhibition of interactions between phytohemaglutinin A and blood serum glycoproteins by carbohydrates. *Vopr Med Khim*. 32, 96-98.

[46] Terentiev AA, Moldogazieva NT, Tagirova AK, Tatarinov YuS. (1990) Competitive affinity chromatography of human alpha-fetoprotein on immobilized diethylstilbestrol. *Byul Eksp Biol Med*, 5, 438-440.

[47] Terentiev AA, Moldogazieva NT, Tagirova AK, Tatarinov YuS. (1992) Affinity chromatography of rodent alpha-fetoprotein on immobilized diethylstilbestrol. *Byul Eksp Biol Med,* 6, 588-590.

[48] Terentiev AA. (2004) Structural prerequisition for participation of alpha-fetoprotein and transforming growth factor (TGF-beta) in cell cycle regulation. *Med Immunol*, 6, 256.

[49] Gitlin D, Boesman M. (1967).\ Sites of serum alpha-fetoprotein synthesis in the human and in the rat. *J Clin Invest*, 46, 1010-1016.

[50] Gitlin D, Perricelli A, Gitlin GM. (1972) Synthesis of α-fetoprotein by liver, yolk sac, and gastrointestinal tract of the human conceptus. *Cancer Res,* 32, 979-82.

[51] Gitlin D, Boesman, M. Serum a-Fetoprotein, Albumin, and G-Globulin in the Human Conceptus. (1966) *J. Clin. Invest*, 45:1826-1838.

[51] Kithier K, Masopust J, Rádl J. (1968) Fetal alpha-globulin of bone serum differing from fetuin. *Biochim Biophys Acta*, 160, 135-137.

[52] Ruoslahti E, Seppala M. (1971) Studies of carcino-fetal proteins. III. Development of radio-immunoassay for alpha-fetoprotein. Demonstration of alpha-fetoprotein in serum of healthy human adults. *Intern J Cancer*, 8, 374–383.

[53] Ruoslahti E, Seppala M. (1979) Alpha-Fetoprotein in cancer and fetal development. *Adv Cancer Res*. 29, 275–346.

[54] Ruoslahti E, Seppala M. Studies of carcino-fetal proteins III. (1971). Development of radio-immunoassay for alpha-fetoprotein. Demonstration of alpha-fetoprotein in serum of healthy human adults. *Intern J Cancer*, 8, 374–383.

[55] Ishii M. (1973). Radioimmunoassay of alpha-fetoprotein. *Gann Monogr Cancer Res*, 14, 89-98.

[56] Ruoslahti E, Seppala M. (1972). α-Foetoprotein in normal human serum. *Nature*, 235, 161–162.

[57] Taketa K, Okada S, Win N et al. (2002). Evaluation of tumor markers for the detection of hepatocelular carcinoma in Yangon General Hospital, Myanmar. *Acta Medica Okayama,* 56, 317-320.

[58] Nishi S, Watabe H, Hirai H. (1967) Production of antibody to homologous alpha-fetoprotein in rabbits, rats and horses by immunization with human alpha-fetoprotein. *J Immunol*, 109, 957–960.

[59] Nishi S, Watabe H, Hirai H. (1973) Species cross-reaction of alpha-fetoprotein and break down of the tolerance to alpha-fetoprotein by immunization with heterologous alpha-fetoprotein. *Tumor Res*, 8, 17-22.

[60] Nishi S, Matsue H, Yoshida H, Yamamoto R, Sakai M. (1991) Localization of estrogen-binding site of alpha-fetoprotein in chimeric human-rat proteins. *Proc Natl Acad Sci USA*, 88, 3102-3105.

[61] Nishi S. (1979) Isolation and Characterization of a human fetal α-globulin from the sera of fetuses and a hepatoma patients. *Cancer Research* 30, 2507-2513.

[62] Perisić-Savić M, Perisić V, Sinćić M, Janev A, Nastić-Mirić D. (1981) Immunologic and immuno-histochemical determination of alpha l-fetoprotein and HBs antigens in hepatocellular carcinoma. *Srp Arh Celok Lek*, 109, 775-782.

[63] Glisić Lj, Perisić V, Lemberger J, Pastrakuljić N, Novaković R. (1979) Levels of carcinoembryonic antigen and alpha fetoproteins in primary and secondary liver carcinoma. *Srp Arh Celok Lek,* 107, 117-123.

[64] Uriel J, de Nechaut B, Dupiers M. (1972) Estrogen-binding properties of rat, mouse and man fetospecific serum protein. Demonstration by im muno-autoradiographic methods. *Biochem Biophys Res Commun*, 46, 1175-1180.

[65] Uriel J, Bouillon D, Dupiers M. (1975) Affinity chromatography of human, rat and mouse alpha-fetoprotein on estradiol-sepharose adsorbents. *FEBS Lett*, 53, 305-308.

[66] Uriel J, Bouillon D, Aussel C, Dupiers M. (1976) Alpha-fetoprotein: the major high affinity estrogen binder in rat uterine cytosols. *Proc Natl Acad Sci U S A*, 75, 1452–1456.

[67] Milligan SR, Khan O, Nash M. (1998) Competitive binding of xenobiotic oestrogens to rat alpha-fetoprotein and sex steroid binding proteins in human and rainbow trout (Oncorhynchus mykiss) plasma. *J Comp Endocrinol,* 112, 89-95.

[68] Soloff MS, Swartz SK, Pearlmutter F, Kithier K. (1972) Binding of 17-β-estradiol by variants of alpha-fetoprotein in rat amniotic fluid. *Biochim Biophys Acta*, 427, 644-651.

[69] Terentiev AA., Moldogazieva NT, Levtsova OV, Maximenko D, Borozdenko D, Shaitan KV. (2012) Modeling of three-dimensional structure of human alpha-fetoprotein complexed with diethylstilbestrol: docking and molecular dynamics simulation study. *J Bioinform Comput Biol*, 10, 1241012.

[70] Peck AB, Murgita RA, Wigzell H. (1982) Cellular and genetic restrictions in the immunoregulatory activity of α-fetoprotein. III. Role of the MLC-stimulating cell of α–fetoprotein induced suppression of T cell-mediated cytotoxicity. *J Immunol,* 128, 1134-1140.

[71] Luft AJ, Loscheider FL. (1983) Structural analysis of human and bovine α-fetoprotein by electron microscopy, image processing and circular dichroism. *Biochemistry*, 22, 5971-5978.

[72] Barker ME. (1988) Evolution of alpha-fetoprotein: sequence comparisons among AFP species and with albumin species. *Tumor Biol*, 9, 123-136.

[73] Gorin MB, Cooper DL, Eiferman F, van de Rijn P, Tilghman SM. (1981) The evolution of alpha-fetoprotein and albumin: 1. A comparison of the primary amino acid sequences of mammalian alpha-fetoprotein and albumin. *J Biol Chem,* 256:1954-1959.

[74] Beatie WG, Dugaiczyk A. (1982) Structure and evolution of human alpha-fetoprotein deduced from partial sequence of cloned cDNA. Gene, 20, 415–22.

[75] Uversky VN, Kirkitadze MD, Narizhneva NV, Potekhin SA, Tomashevsky AYu. (1995) Structural properties of alpha-fetoprotein from human cord serum: the protein molecule at low pH possesses all the properties of the molten globule. *FEBS Letts,* 364, 165-167.

[76] Song Y-H, Naumova AK, Liebhaber SA, Cooke NE. (1999) Physical and meiotic mapping of the region of human chromosome 4q11-q13 encompassing the vitamin D-binding protein DBP/Gc-globulin and albumin multigene cluster. *Genome Res,* 9, 581–587.

[77] Law S, Tamoaki T, Kreuzaler F, Dugaiczyk A. (1980) Molecular cloning of DNA omplementary to a mouse α-fetoprotein mRNA sequence. *Gene*, 10, 53-61.

[78] Yamagata Y, Katoh H, Nakamura K, Tanaka T, Satomura S, Matsuura S. (1988) Determination of alpha fetoprotein concentration based on liquid phase binding assay using anion exchange chromatography and sulfated peptide introduced antibody. *J Immunol Methods*, 212:161–8.

[79] Ruoslahti E., Terry WD. (1976) Alpha-fetoprotein and serum albumin show sequence homology. *Nature*, 260, 804–805.

[80] Breborowich J. (1988) Microheteroheneity of human alpha-fetoprotein. *Tumor Biol*, 9, 3-14.

[81] Morinaga T, Sakai M, Wegmann TG, Namaoki T. (1983) Primary structure of human alpha-fetoprotein and its mRNA. *Proc Natl Acad Sci U S A*, 80, 4604-4608.

[82] Pucci P, Siliciano R, Malorni A, Marino G, Tecce MF, Ceccarini C, Terrana B. (1991) Human alpha-fetoprotein primary structure: a mass spectrometric study. *Biochemistry*, 30, 5061-5066.

[83] Dudich I, Tokhtamysheva N, Semenkova L, Dudich E, Hellman Y, Korpela T. (1999) Isolation and structural and functional characterization of two stable peptic fragments of human alpha-fetoprotein. *Biochemistry,* 38, 10406-10414.

[84] Deutsch HF. (1991) Chemistry and biology of alpha-fetoprotein. *Adv Cancer Res*, 56, 253-312.

[85] Taketa K. (1998) Characterization of sugar chain structures of human AFP by lectin affinity electrophoresis. *Electrophoresis*, 19, 1774-1779.

[86] He XM, Carter D. (1992) Atomic structure and chemistry of human serum albumin. *Nature*, 358, 209-215.

[87] Lutsenko SV, Feldman NB, Finakova GV, Gukasova NV, Petukhov SV, Posypanova GA, Scryabin KG, Severin SE. (2000) Study of antitumor activity of alpha-fetoprotein and epidermal growth factor conjugates in vitro and in vivo. *Tumor Biol*, 21, 367–374.

[88] Zizkovsky V, Strop P, Korcakova J, Havranova M, Mikes F. (1983) Fluorescence spectroscopy, fluorescence polarization and circular dichroism in studies of pH-dependent changes in the alpha-fetoprotein molecule. *Ann N Y Acad Sci*, 417, 49-56.

[89] Mizejewski GJ. (2001) Alpha-fetoprotein structure and function: relevance to isoforms, epitopes and conformational variants. *Exp Biol Med*, 226, 377-408.

[90] Mizejewski GJ. (1995) The phylogeny of alpha-fetoprotein in vertebrates: Survey of biochemical and physiological data. *Crit Rev Eucaryot Gene Expr*, 5, 281-316.

[91] Li D, Mallory T, Satomura S. (2001) AFP-L3: a new generation of tumor marker for hepatocellular carcinoma. *Clinica Chimica Acta* 313, 15–19.

[92] Marrero JA, Feng Z, Wang Y, NguyenMH, Befeler AS, Roberts LR, et al. (2009) Alpha-fetoprotein, des-gamma carboxyprothrombin, and lectin-bound alpha-fetoprotein in early hepatocellular carcinoma. *Gastroenterology*, 137, 110–8.

[93] Kumada T, Nakano S, Takeda I, et al. (1999) Clinical utility of lens culinaris agglutinin-reactive alpha-fetoprotein in small hepa-tocellular carcinoma: special reference to imaging diagnosis. *J Hepatol*, 30, 125–130.

[94] Uversky VN, Narizhneva NV, Ivanova TV, Kirkitadze MD, Tomashevsky AYu. (1997) Ligand-free form of human alpha-fetoprotein: evidence for the molten globule state. *FEBS Letts*, 410, 280-284.

[95] Uversky VN, Narizhneva NV, Ivanova TV, Tomashevski AYu. (1997) Rigidity of human alpha-fetoprotein tertiary structure is under ligand control. *Biochemistry*, 36, 13638-13645.

In: Alpha-Fetoprotein: Functions and Clinical Applications ISBN: 978-1-63484-875-6
Editors: Nisha Lakhi and Michael Moretti

Chapter 2

Alpha-Fetoprotein in Development, after Liver Injury, during Chemical Hepatocarcinogenesis and in liver Cancer: How the Study of Alpha-Fetoprotein Led to the Concept of Cancer Stem Cells

Stewart Sell*[*]*, MD

Senior Research Physician, Wadsworth Laboratory, New York State Dept. of Health.
Wadsworth Center, Empire State Plaza, Albany, NY, US

Abstract

Study of the production of Alpha-fetoprotein (AFP) by the fetus during gestation, by the newborn after birth, by adult rats and mice after liver injury, during growth of transplanted hepatocellular carcinoma (HCC) and during chemical hepatocarcinogenesis led to the identification of a population of small "oval" cells that are not easily seen in normal livers but expand greatly during chemical carcinogenesis. Although tissue stem cells had been shown to give rise to cancers in other tissues, the liver was considered the exception to the rule. Because the liver responds to most injury protocols by proliferation of mature hepatocytes that replace the injured cells, most investigators concluded that the liver did not contain stem cells as many other organs did. However, during carcinogenic regimens that inhibit proliferation of mature hepatocytes, the oval cells proliferate and differentiate into mature liver cells as well as give rise to HCC. Systematic analysis of the newly arising oval cell population convincingly demonstrates origin of HCC from putative liver stem cells.

The author knows of no conflicts of interest.

[*] Corresponding Author: stewart.sell@health.

INTRODUCTION

I became interested in investigation the relationship of the production of alpha-fetoprotein (AFP) to the early changes during the induction of liver cancer (hepatomas) by chemical carcinogens in 1964 when I joined the Department of Pathology at the University of Pittsburgh [1]. AFP was first recognized as a fetal serum protein by Bergstrand and Czar in 1957 using paper electrophoresis [2] and further characterized by immunoelectrophoresis [3, 4]. Then Garri Abelev and his group noted production of an "embryonal α-globulin" in serum of mice bearing transplantable hepatomas in 1963 using double-diffusion-in agar [5]. This was extended to the serum of human patients with primary hepatocellular cancer by Tatarinov the following year [6] (Figure 1). For a more detailed summary of this work see ref. [7].

The major research program at the Pathology Department at U. Pittsburgh was chemical hepatocarcinogenesis and the nature of lesions induced by chemical carcinogens (so-called "premalignant lesions") by Emmanuel Farber and his colleagues [for reviews see 8, 9]. What could knowledge of the relationship of AFP production to the preneoplastic changes during chemical hepatocarcinogenesis tell us about the cellular origin of liver cancer? The first step to address this question was to develop a sensitive radioimmunoassay for AFP.

Figure 1. Garri Abelev and Yuri Tatarinov in 1986.

Radioimmunoassay for AFP

Immunodiffusion methods have maximum sensitivities between 0.5 and 1.0 μg/ml and are not reproducible at these levels. On the other hand radioimmunoassay (RIA) sensitivity can be reproducible at 0.001 μg/ml. Fortunately as a medical student at Pitt in 1957, we were taught and actually carried out RIAs under the direction of Richard Farr [10]. This was 3 years before the RIA for insulin won the Nobel Prize for Yalow and Berson [11]. The first step was to identify AFP and prepare specific antibody to it. Figure 2 shows the identification of AFP by immunoelectrophoresis [12]. Then pure AFP was isolated and a radioimmunoassay accurate to 1 ng was developed [13, 14]. Figures 3 and 4. Figure 4 shows the results of replicate assays for normal adult rat sera, demonstrating the reproducibility of the RIA at very low levels, 0.01 μg/ml. The RIA was then used to measure AFP accurately in rats during pregnancy, after 70% partial hepatectomy, chemical injury (galactosamine, carbon tetrachloride), during growth of a transplanted tumor (Morris hepatoma 7777) and after exposure to a chemical carcinogen (N-2-acetlyaminofluorenie (AAF) Figure 5 [summarized in ref 15].

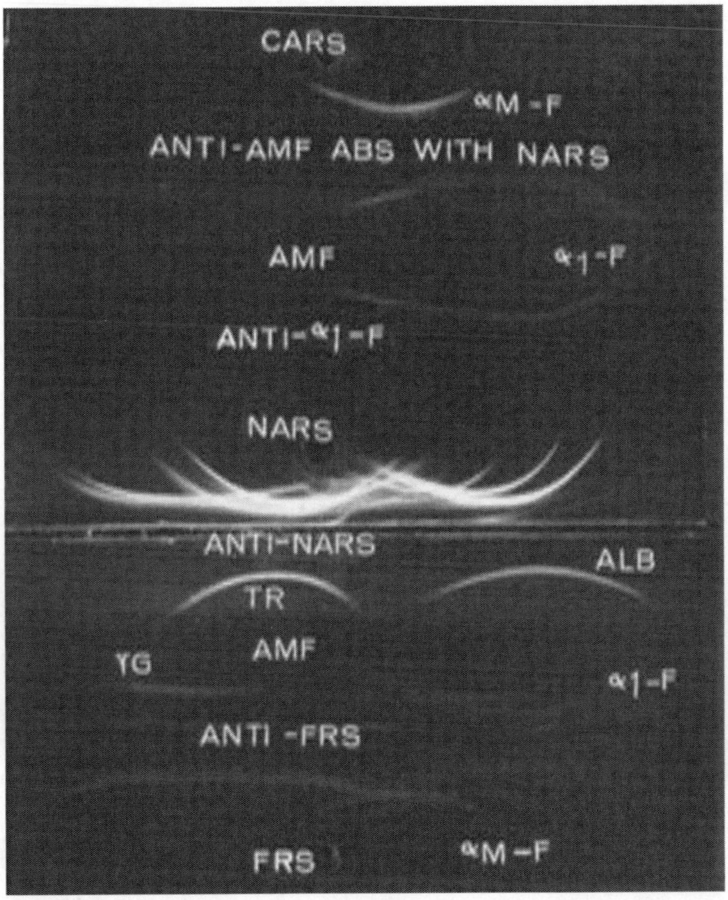

Figure 2. Immunoeletrophoretic identification of AFP. CARS; croton oil treated rat srerum (contains increased levels of α- 2 macroglobulin (αM-F); α1-F, alpha-1 fetoprotein; AMF, amniotic fluid (contains alpha fetoprotein); Anti-AMF, anti-amniotic fluid; Anti-α1-F, antiserum to rat alpha-fetoprotein. NARS, normal adult rat serum; ALB albumin; Anti-NARS, antiserum to NARS; FRS fetal rat serum; TR transferrin. The anti-α1-F only reacts with α1-F, not with αM-F, TR, albumin or other serum proteins.

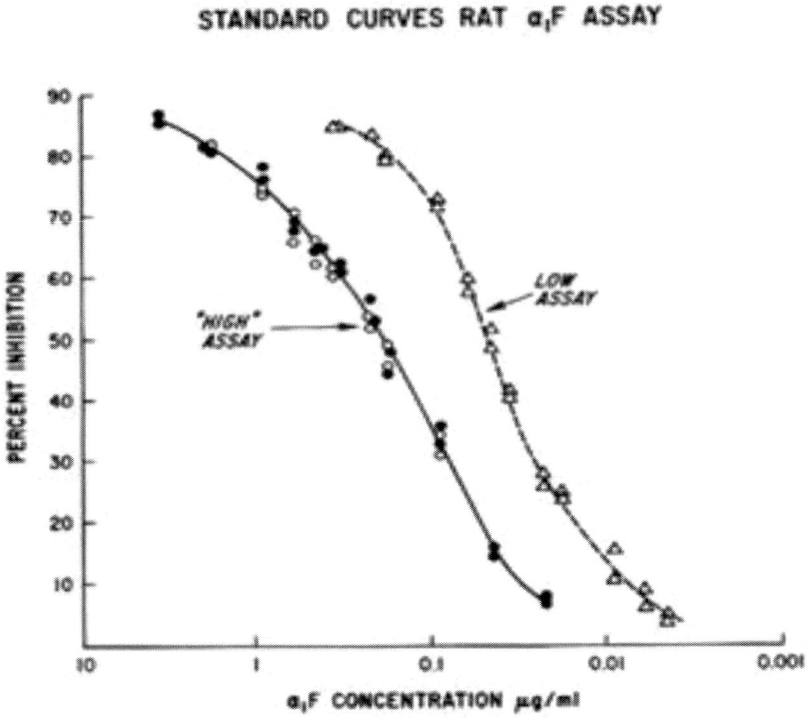

Figure 3. Standard curves for rat RIA. The sensitivity can be increased 10 fold by bias incubation of the unlabeled unknown with the antiserum dilution for 20 hours before the radiolabeled α1-F is added.

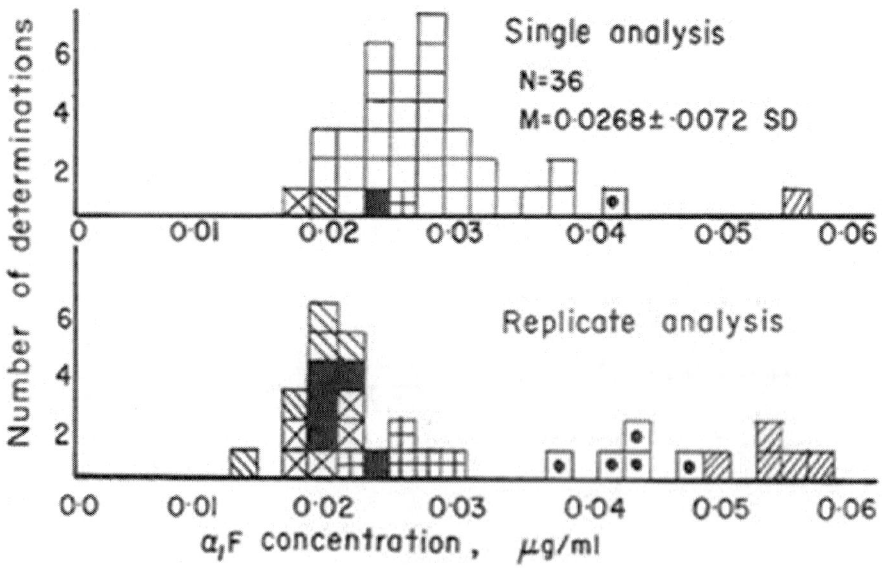

Figure 4. Results of repetitive assays of the same serum samples from normal mice. The squares on the top show the results of assay for normal adult rat sera. The squares with symbols show the results of replicate assays of the same sample.

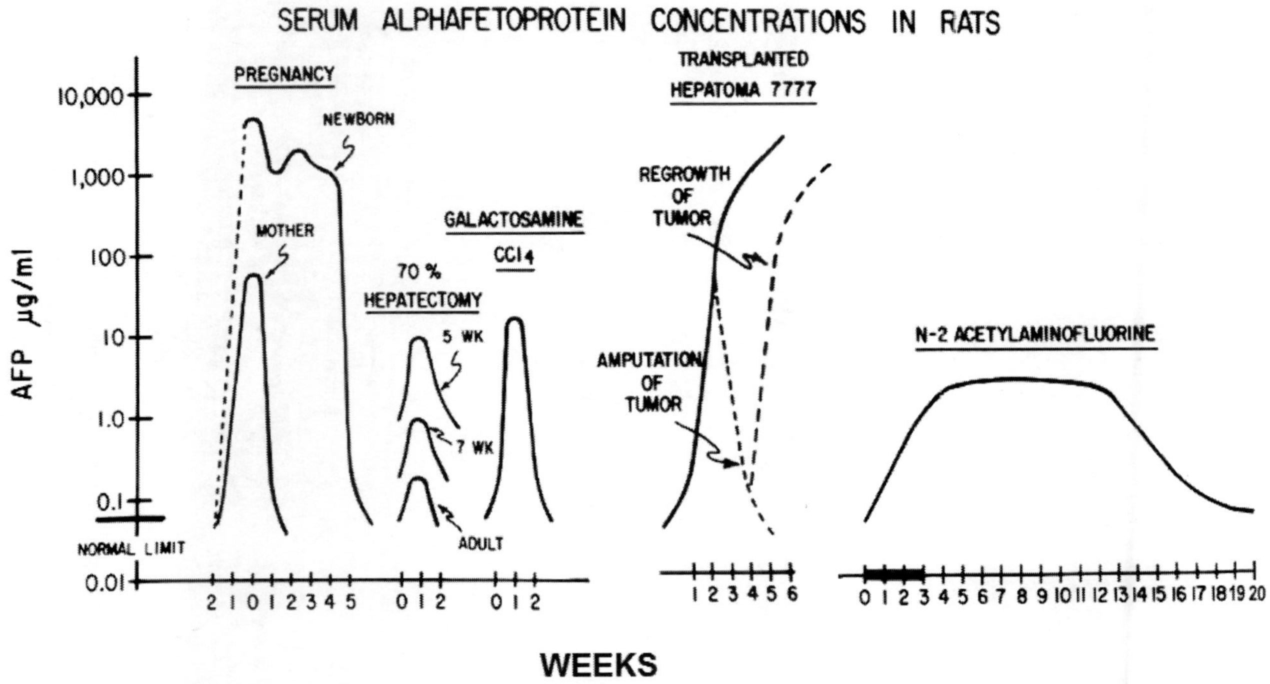

Figure 5. Serum concentrations of AFP in pregnancy, after liver injury, during growth of a transplantable hepatoma, and during exposure to AAF.

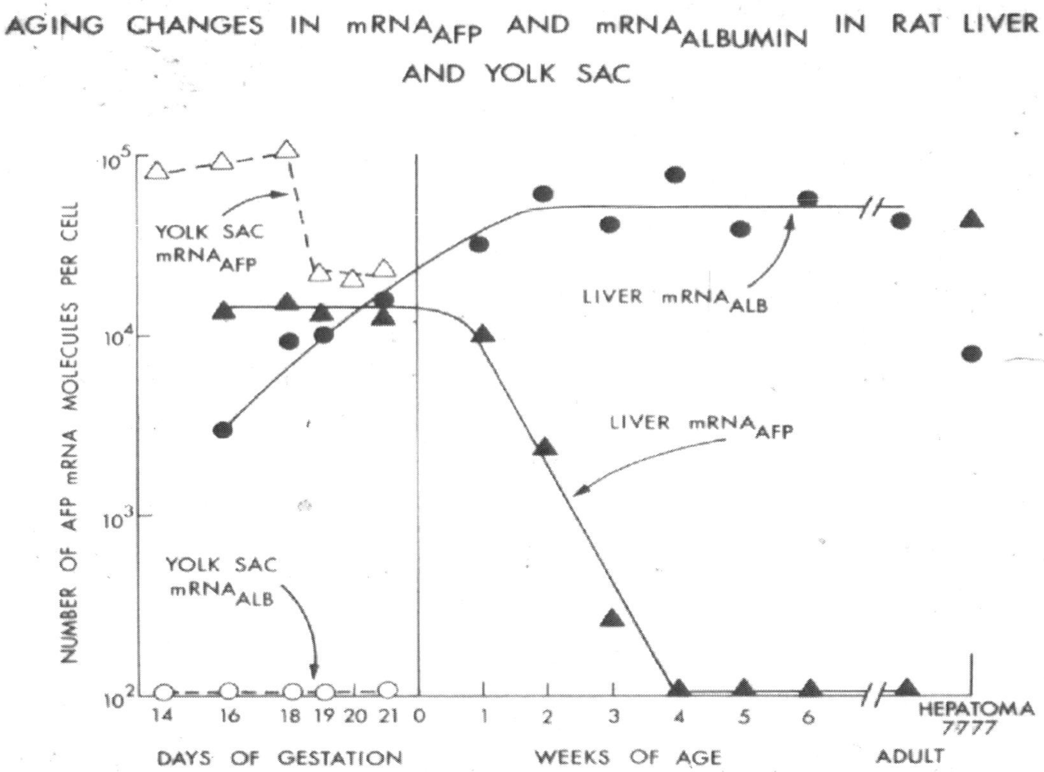

Figure 6. AFP and albumin mRNA levels in yolk sac and liver during gestation and after birth.

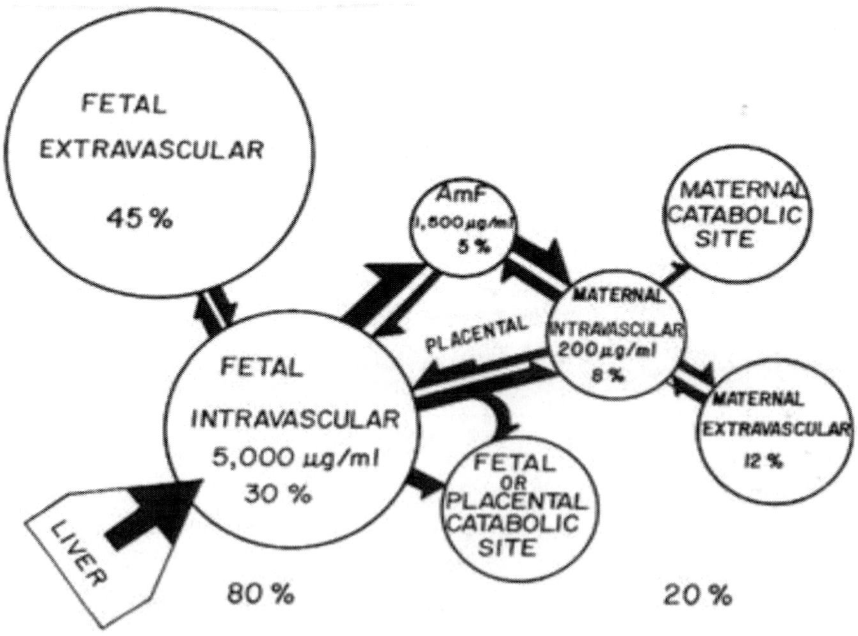

Figure 7. Synthesis and distribution of AFP in the fetal-maternal unit.

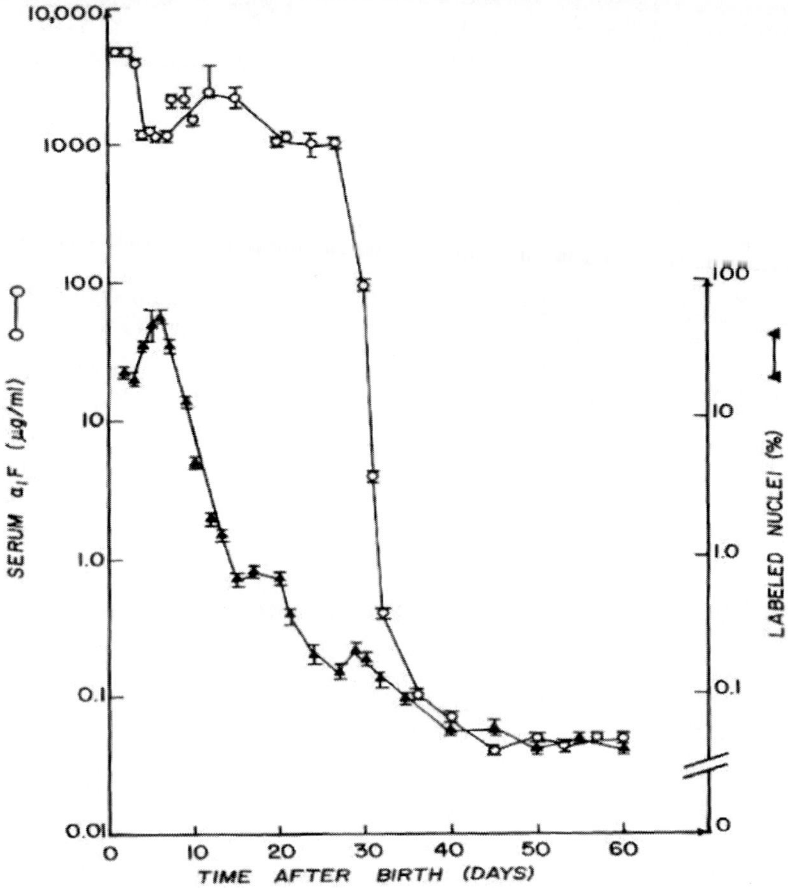

Figure 8. Liver cell proliferation (top curve) and serum AFP (bottom curve) in newborn rats.

AFP in Development

AFP is synthesized in high amounts in the fetal yolk sac and lesser amounts in the fetal GI tract resulting in high serum levels in the fetus and lower levels in the pregnant mother [16]. After birth there is production in the newborn rat liver for about 3 weeks while there is still proliferation of liver cells [17]. This finding was supported by measurement of mRNA for AFP (Figure 6) [18]. This is high in the fetal yolk sac and liver and then falls for 3 weeks after birth. In contrast, there is no mRNA for albumin in the yolk sac, but there is a marked increase in the fetal liver during the last days of gestation with maintenance of high levels in the adult. Using radiolabeled AFP to determine distribution of AFP during gestation, assays for AFP concentrations in various fluids of the fetal-maternal unit and volumes of these fluids the synthesis and distribution of AFP in the fetal maternal unit [19] are shown in Figure 7. The AFP in the fetal-maternal unit is produced in the fetal liver and yolk sac. From here it diffuses from the fetal blood and amniotic fluid into the maternal circulation. There may be sites of catabolism of AFP in both the fetus and mother. The concentration in the fetal serum (5,000 µg/ml) is much higher than in the maternal serum (200 µg/ml), but given the much larger volume of the maternal blood and extravascular space we estimate that about 80% of the AFP is in the fetal unit and 20% in the maternal. After birth there is decreasing synthesis by the newborn liver for about 3 weeks associated with continued proliferation of the newborn rat liver cells which ends about 3 weeks of age (Figure 8). In the adult there is production after liver injury associated with restitutive proliferation of hepatocytes. Varying levels are produced by transplanted hepatomas and after exposure to chemical carcinogens (see below).

AFP IN RESTITUTIVE PROLIFERATION

Partial Hepatectomy (PH)

The liver is a remarkable organ from many standpoints, but one in particular is the ability of the liver to reconstitute itself after removal of up to 2/3rds of its mass. This reconstitution is accomplished with a wave of mitosis that begins about 1 day after PH and ends by 2 days. AFP levels begin to rise about 12 hours after mitosis begins, peak at 3 days and fall back to normal by 5 or 6 days (Figure 9). By immunofluorescence AFP can be seen in proliferating hepatocytes [20].

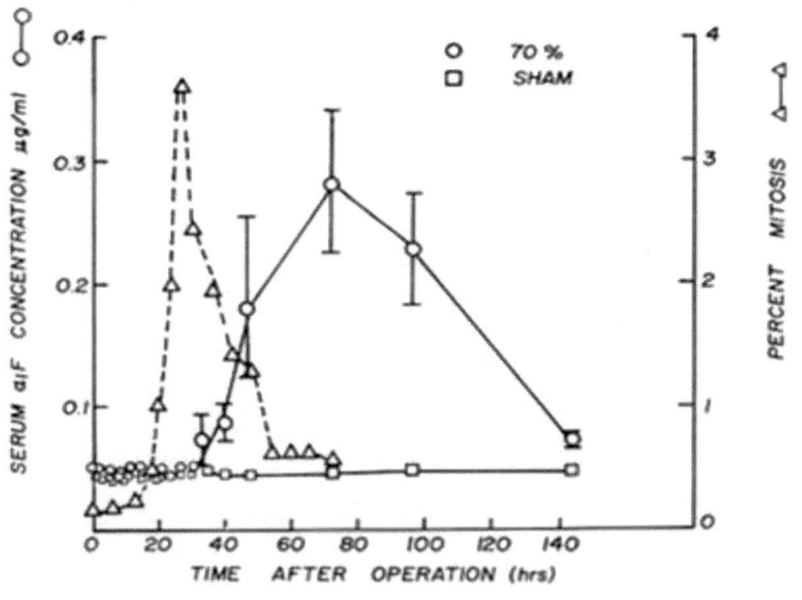

Figure 9. Percent mitotic figures and serum AFP levels after partial hepatectomy.

Galactosamine

Similar AFP serum levels and cellular localization are see during the restitutive proliferation after chemically induced liver necrosis induced by CCl_4 and galactosamine (Figure 10), but usually the AFP levels remain higher for a few more days due to more prolonged injury and restitutive proliferation [21]. By immunofluorescence AFP is seen in the proliferating hepatocytes adjacent to the zones of liver cell necrosis [18].

Allyl Alcohol (AA)

In contrast to most chemical-induced liver cell necrosis which involves central (CCl_4) or midzone (galactosamine, phenobarbital, lead nitrate), AA administered to rats induces necrosis of the liver cells in the immediate periportal zone. Analysis of the cells that proliferated in response to this injury showed null small portal "oval" cells proliferating next to the bile ducts and differentiating into AFP producing cells as they expanded across the necrotic zone [22] (Figure 11). This supports the concept of a lineage of cells in the liver cord with the putative liver stem cells located next to or within the terminal bile ducts with maturing cells extending to the central zone [23]. The "oval cell hypothesis," i.e., that oval cells are either liver stem cells or their immediate progeny is supported by the cellular response to chemical hepatocarcinogens.

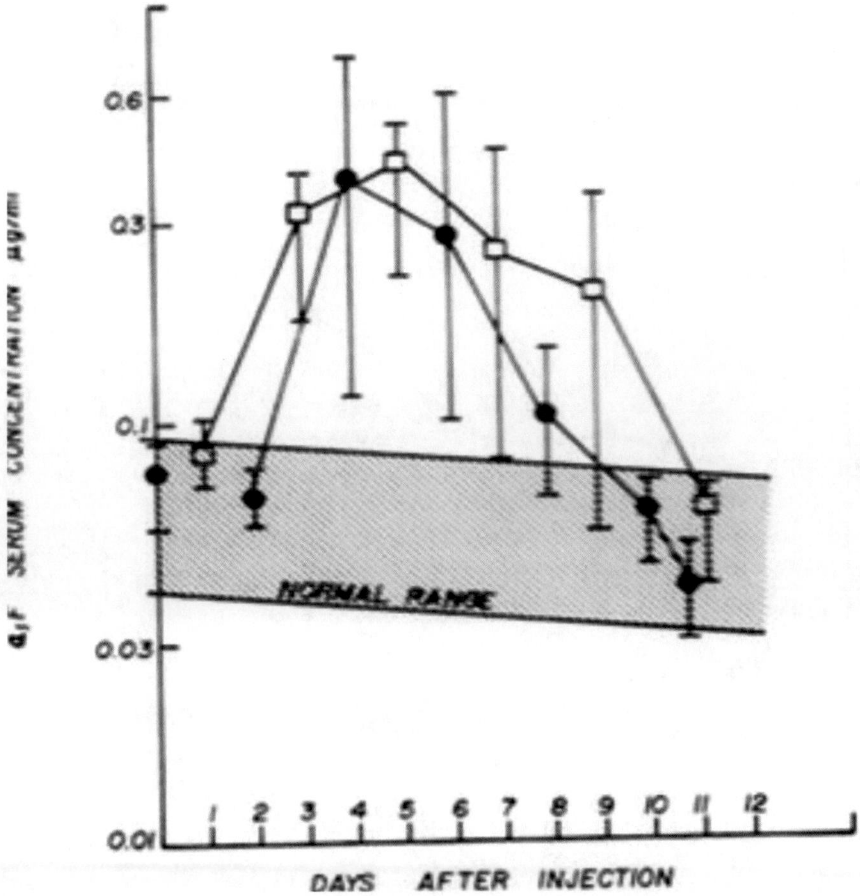

Figure 10. Serum levels of AFP after galactosamine injury.

Chemical Hepatocarcinogenesis

The hepatocellular response to chemical hepatocarcinogenesis shows a similar heterogeneous "cell lineage" response as was seen after injury. The varied kinetics of the serum AFP elevations seen during chemical hepatocarcinogenesis indicated that different cells may be activated to proliferate and form carcinomas [1], (Figures 12 and 13). For this chapter, the cellular responses to diethylnitrosamine (DEN), acetylaminofluorene-CCl_4, and cyclic acetylaminofluorene (AAF) are chosen to illustrate the various levels in the hepatocyte lineage that chemical hepatocarcinogens appear to act.

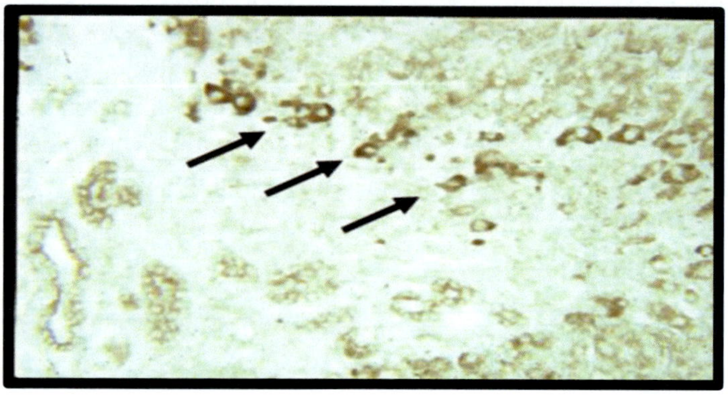

Figure 11. AFP labeled cells between "null" periportal oval cells and transitional hepatocytes after allyl alcohol induced periportal liver injury.

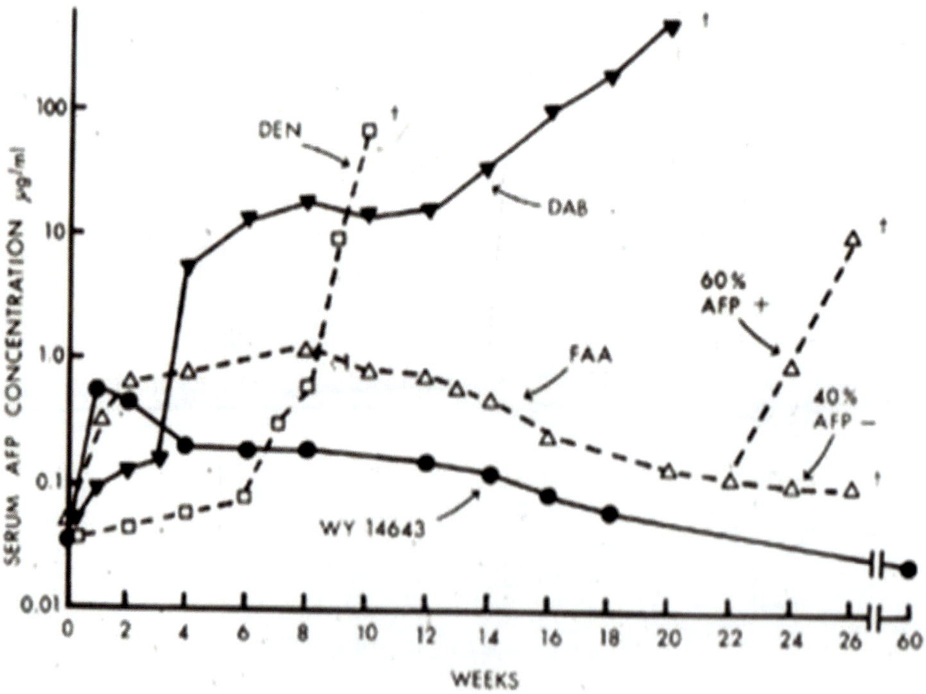

Figure 12. Serum levels of AFP after exposure to various chemical hepatocarcinogens.

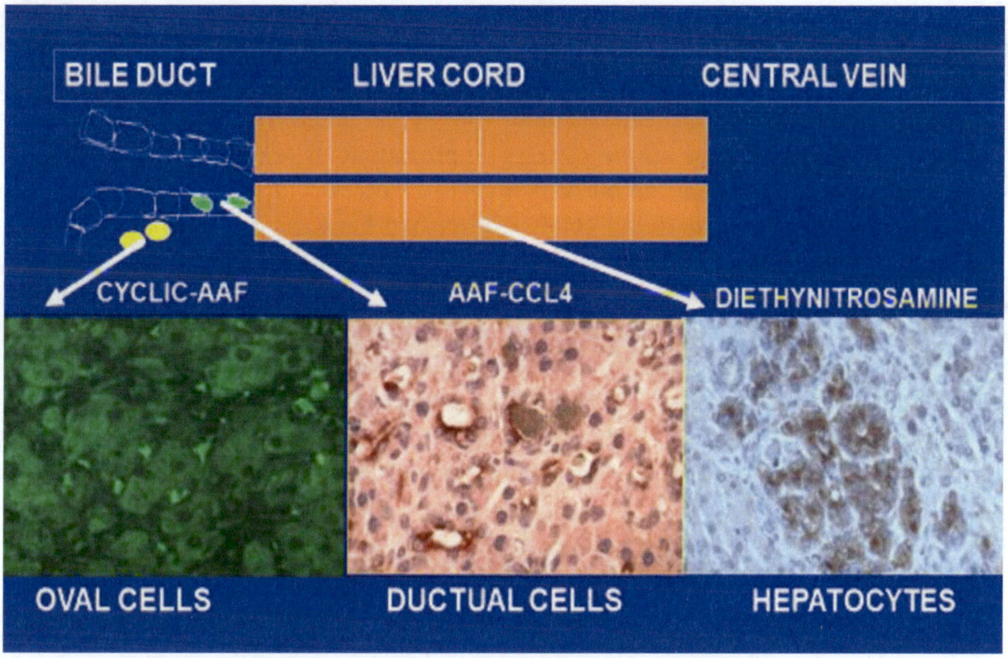

Figure 13. Immunolocalization of AFP in various cell populations after exposure to chemical hepatocarcinogens. DEN: hepatocytes; AAF-CCl4: ductal progenitor cells; cyclic AAF: oval cells.

Diethylnitrosamine (DEN)

Serum AFP levels did not become elevated until after 6 weeks of exposure. Then AFP is seen in individual hepatocytes and in microcarcinomas. Thus, DEN induced hepatoma appear to arise from microcarcinomas formed from nature hepatocytes [24, 25].

Acetylaminofluorene-CCl_4

In this model AAF is given to block proliferation of hepatocytes. Then CCl4 is given to induce injury and restitutive proliferation [25]. Since the hepatocytes cannot proliferate, the liver compensates by the marked proliferation of small oval cells in ductal structures [26]. In this model there is an early elevation of serum AFP followed by a decline for up to 10 months, then a rapid elevation when hepatomas develop.

Cyclic AAF

This model was designed to produce a series of "nodular changes" in the liver prior to hepatoma development [27]. It is based on the concept that hepatomas arise from preneoplastic nodules [8, 9]. AAF is fed for four 3 weeks on; 1 week off cycles. Feeding AAF for more than 3 weeks at a time results in a high fatality rate. After the first cycle foci of basophilic hepatocytes are seen; after 2 cycles these develop into microscopic nodules; after 3 cycles the nodules become visable grossly; after 4 cycles large nodules deform the liver. Hepatomas develop 5 months after the 4th cycle. Serum AFP rises to over 0.1 μg/ml during the first cycle then falls to 0.01 μg/ml for the remaining 4 cycles [25]. Serum AFP then rises rapidly after 6 months to between 10 and 100 μg/ml when the rats develop multiple hepatomas at 9 months. Based on the concept of premalignant lesions, it was expected that the preneoplastic nodules would be positive for AFP. However, the nodules were uniformly negative; AFP is seen in small "oval cells" that expand between the liver cords eventually giving rise to AFP positive hepatomas [28]. Similar staining of oval cells is seen in hepatocarcinogenic models based on choline deficiency [29, 30]. It was from this observation that we concluded that hepatomas arose from cells at different stages in the liver cell lineage [31, 32] (Figure 14). Following the concept of maturation arrest as the biologic mechanism in cancer [33), it is postulated that hepatomas arise from block of differentiation at various levels: infant or fetal liver stem cell (hepatoblastoma), adult liver stem cell (oval cells); adult ductular bipolar cells (oval cell ducts) and mature hepatocytes [32]. This model is consistent with cancers arising from cells at various levels of differentiation in their lineage. Since mature liver cells can divide, we postulate that hepatocellular carcinomas can arise from any stage in the liver cell lineage [32]. Since it had previously been argued that the liver did not have a tissue determined stem cell population, we concluded that all cancers arise from stem cells at the base of the lineage, that all cancers contain stem cells and that forcing these cancer stem cells to differentiate could eliminate the cancer stem cell that is responsible for resistance to therapy [33].

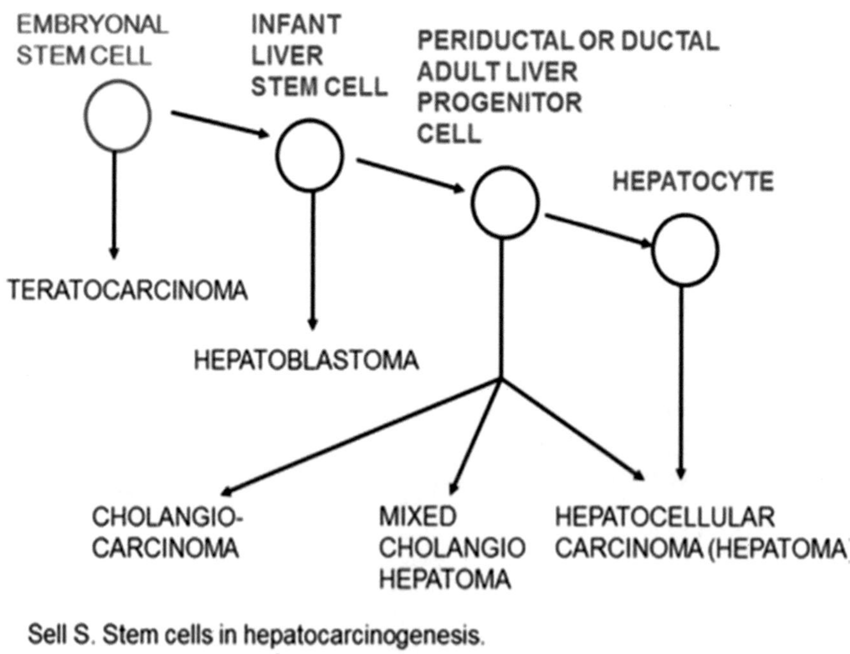

Figure 14. Projected sites of maturation arrest in liver cancer. In the cell lineage from embryonal stem cell to hepatocyte, the intermediate cells are the infant liver stem cells responsible for hepatoblastomas and the periductular or ductular (oval) stem cells.

AFP Production by Hepatomas

In the 1960s Harold Morris at Howard U. in Washington, D.C. derived a number of transplantable hepatomas after chemical treatment of Buffalo rats. The serum levels of AFP varied greatly in rats bearing various hepatomas. In general the serum level was higher in fast growing tumors than slow growing ones, higher in less differentiated tumors than more differentiated tumors, and higher in more aneuploid tumors than with diploid tumors [34] (Figure 15). The serum levels of AFP accurately reflected the growth of the transplanted tumors and predicted the development of lung metastases after removal of the primary tumor [35] and effective prevention of metastases by radiation of the lung at the time of surgical removal of the tumor [36] (Figure 16). This preclinical model is proof of principle that serum levels of AFP may be used to follow growth and the effects of treatment of AFP producing tumors. We are currently using this model to test therapy using liver cancer stem cell directed nanoparticles containing small molecular inhibitors of stemness signaling.

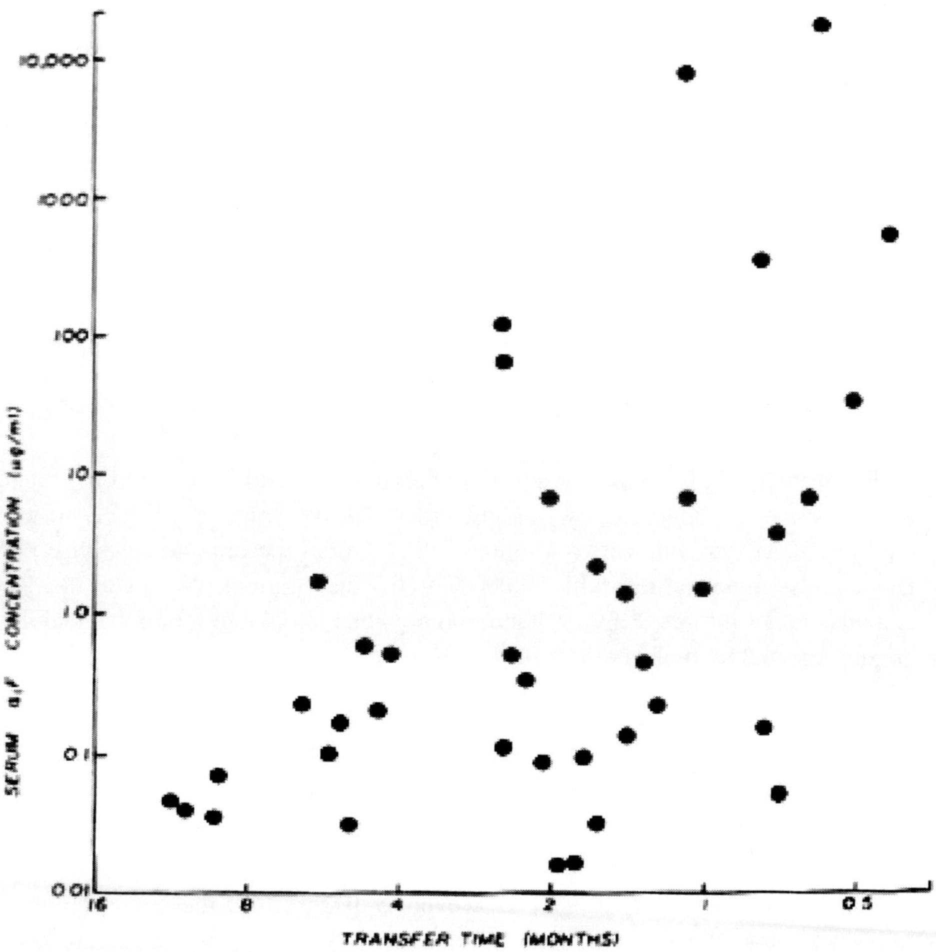

Figure 15. Serum AFP levels correlate with transfer time of transplantable Morris hepatoma. The levels are high with fast growing tumors and low with slow growing tumors.

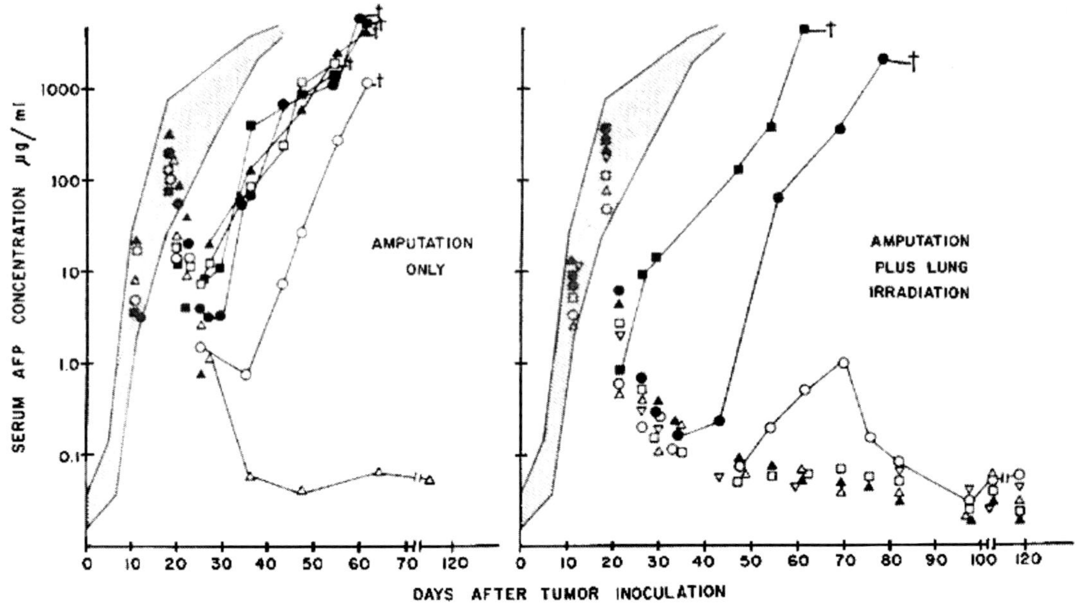

Figure 16. Serum AFO levels during growth, after surgical removal (re-elevation indicates lung metastases) and effect of radiation (right panel) on development of lung metastases (35).

Function of AFP

The finding that AFP is by far the most prominent serum protein during fetal development of many species indicates that the ability to make AFP is evolutionarily preserved and therefore must have an important function. At least 3 functions of AFP have been suggested and are worthy of mention.

I. AFP is immunosuppressive and serves to prevent maternal rejection of the fetus. This hypothesis was supported by findings that passive transfer of AFP to mice or addition of AFP to cultures of immune cells lowered the immune response [37, 38]. However, in mice and rats with 10,000X higher than normal AFP levels, the immune response is not inhibited [39, 40] and homozygous knock out fetal mice lacking AFP are not rejected by their mothers [41].
II. AFP serves to carry selected fatty acids to the brain that are necessary for myelination of nerve fibers. This hypothesis is based on the following findings, AFP does bind and carry long chain fatty acids used for myelination [42]. AFP is identified in developing neurons [43, 44], but AFP is not synthesized [45] by fetal brain tissue and there is no mRNA in fetal brains [46]. At first glance this hypothesis is not supported by the evidence as there is normal development of the fetal brain in knockout mice lacking AFP [41]. However as discussed in the next section, female mice lacking AFP are sterile. Thus, homozygous AFP KO mice must be obtained by breeding homozygous KO males with heterozygous KO females. Thus, the

homozygous KO mice could receive AFP by transfer of AFP from the amniotic fluid made by a heterozygous fetal liver and yolk sac.

III. AFP binds estrogen, and its presence in the brain has a role in determining brain sexual differentiation [47, 48]. It does appear that the principal action of prenatal estrogen exposure is to defeminize and masculinize the brain. AFP binds estradiol circulating in the female mouse fetus with high affinity and capacity and protects the developing mouse brain from this effect and also prevents subsequent sterility [41, 49]. However, although this function is now convincingly documented for mice, human AFP does not bind estrogen [50] so that the protection from masculinization function of AFP in human females remains unclear.

REFERENCES

[1] Sell S. (2008) Alpha-fetoprotein, Stem Cells, and Cancer. The Abbot Award Lecture. *Tumor Biol*, 29:161-180.

[2] Bergstrand CF, Czar B. (1957) Demonstration of a new protein fraction in serum from the human fetus. *Scand J Cllin lab Invest,* 8, 174.

[3] De Muralt G, Roulet DLA. (1961) Etude immunologique des proteines seriques foetales humaines. *Helv. Paediat Acta*, 517-533.

[4] Masopust J. (1962) Development of serum protein spectrum during childhood. *Rev Czech Med,* 8, 214-225.

[5] Abelev GI, Perova SD, Khramkova NI, Postnikova AS, Irlin IS. (1963) Production of embryonal α-globulin by transplantable mouse hematomas. *Transplantation*, 1: 174-180.

[6] Tatarinov YS. (1964) Presence of embryonal α-globulin in the serum of patients with primary hepatocellular carcinomas. *Vopr Med khimii*, 90-91.

[7] Abelev GI. (1983) Alphafetoprotein: The genesis. *Oncodevelopmental Biol. and Med,* 4, 371-381.

[8] Farber E, Cameron R. (1979) The sequential analysis of liver cancer induction. *Adv. Cancer Res.* 31, 125-226.

[9] Farber E. (1984) The multistep nature of cancer development. *Cancer Res,* 44, 4217-4223.

[10] Farr RS. (1958) Quantitative Immunochemical measure of the primary interaction between I*BSA and antibody. *J. Infectious Dis,* 103, 237-262.

[11] Yalow RS, Berson SA. (1960) Immunoassay of endogenous insulin in man. *J Clin Invest,* 39, 1157-1175. (NOBEL PRIZE)

[12] Sell S, Jalowayski I, Bellone C, Wepsic HT. (1972) Isolation and characterization of rat α1-fetoprotein. *Cancer Research*, 32, 1184-1189.

[13] Sell S. (1973) Radioimmunoassay of rat α1-fetoprotein. Cancer Res. 33:1010-1015.

[14] Sell S, Gord D. (1973) Refinement of radioimmunoassay for detection of 1 ng rat alpha-1 fetoprotein. *Immunochemistry,* 10:439-442.

[15] Sell S, Becker FF. (1978) Guest Editorial: Alpha-fetoprotein. *J. National Cancer Inst,* 60,63-69.

[16] Sell S, Skelly H. (1976) Tissue sites of alpha1-fetorpotein synthesis by the rat during pregnancy and hepatoma growth. *J. Natl. Cancer Inst,* 56, 645-648.
[17] Sell S, Nichols M, Becker FF, Leffert JL. (1974) Hepatocyte proliferation and alpha1-fetoprotein in pregnant, neonatal and partially hepatectomized rets. *Cancer Res.* 34, 865-871.
[18] Sell S, Alexander D. (1974) Rat alpha1 fetoprotein. V. Catabolism and fetal-meternal distribution. *J. Nat Cancer Inst.,* 52, 1483.
[19] Sala-Trepat JM, Dever J, Sargent TD, Thomas K, Sell S, Bonner J. (1979) Changes in expression of albumin and α-fetoprotein genes during rat liver development and neoplasia. *Biochemistry,* 18, 2167.
[20] Sell S. (1980) Heterogeneity of alpha-fetoprotein (AFP) and albumin containin cells in normal and ppathological permissive states for AFP production: AFP containing cells induced in adult rats recapitulate the appearance of AFP containing hepatocytes in fetal rats. *Oncodevelopmental Biol and Med,* 1, 93-105.
[21] Sell S, Reynolds RD, Reutter W. J. (1974) Rat alpha1 fetoprotein: Appearance following galactosamine-induced liver injury. *J. Natl. Cancer Inst,* 53, 289-291.
[22] Yavorkovsky L. Lai E, Sell S. (1995) Participation of small intraportal "stem" cells in the restitutive response of the liver to periportal necrosis induced by allyl alcohol. *Hepatology*, 21, 1702-1712.
[23] Sell S. (2001) The role of liver progenitor cells in repair of liver injury and in liver transplantation. *Wound Repair Regener*, 9, 467-482.
[24] Sell, S. and Dunsford, H. (1989) Evidence for the stem cell origin of hepatocellular carcinoma and cholangiocarcinoma. *Am. J. Pathol,* 134, 1347-1363.
[25] Becker, F.F. and Sell, S. (1979) Differences in serum alphafetoprotein concentrations during the carcinogenic sequences resulting from exposure to diethylnitrosamine or acetylaminofluorene. *Cancer Res,* 39, 1437-1442.
[26] Yin, L, Lynch D, Ilic Z, Sell S. (2002) Proliferation and differentiation of ductular liver precursor cells and non-parenchymal cells during the restitutive response of the rat liver to carbon tetrachloride injury after inhibition of hepatocyte proliferation by N-2-acetylaminofluorine. *Histology and Histopathology,* 17, 65-81.
[27] Teebor GW, Becker FF. (1971) Regression and persistence of hyperplastic hepatic nodules induced by N-2-fluorenylacetamide and their relationship to hepatocarcinogenesis. *Cancer Res*, 31, 1-3.
[28] Sell S. (1978) Distribution of alpha-fetoprotein and albumin containing cells in the livers of Fisher rats fed four cycles of N-2-fluorenylacetamide. *Cancer Res,* 38, 3107-3113.
[29] Shinozuka, H., Lombardi, B., Sell, S. and Iammarino, R.M. (1978) Modification of ethionine liver carcinogenesis by choline deficiency: Early histological and functional alterations. *Cancer Res,* 38, 1092-1098.
[30] Shinozuka, H., Sells, M.A., Katyal, S.L., Sell, S. and Lombardi, B. (1979) Effects of a choline-devoid diet on the emergence of alpha-glutamyltranspeptidase positive foci in the liver of carcinogen treated rats. *Cancer Res,* 39, 2515-2521.
[31] Sell S, Sala-Trepat, Sargent T, Thomas K, Nahon J-L, Goodman T, Bonner J. (1980) Molecular mechanisms of control of albumin and alphafetoprotein production: A system to study the early effects of chemical hepatocarcinogens. *Cell Biol Intl Reports*, 4, 235-254.

[32] Sell S. (2006) Stem cells in hepatocarcinogenesis. Cell Science Reviews ISN NO. 1742-8130.

[33] Sell, S. and Pierce, G. B. (1994) Biology of Disease: Maturation arrest of stem cell differentiation is a common pathway for the cellular origin of teratocarcinomas and epithelial cancers. *Lab. Invest*, 70, 6-21. An AACR Scientific Landmark.

[34] Sell S, Morris HP. (1974) Relationship of rat α1-fetoprotein to growth rate and chromosome composition of Morris hepatomas. *Cancer Res,* 34, 1413-1417.

[35] Sell S, Wepsic HT, Nickel R, Nichols M. (1974) Rat alpha1-fetoprotein IV. The effect of growth and surgical removal of Morris hepatoma 7777 upon the serum alpha1-fetoprotein concentrations of Buffalo rats. *J. Natl. Cancer Inst.* 52,133-137.

[36] Sell, S., Stillman, D., Michaelsen, M., Alaimo, J. and VonEssen, C. (1977) Prevention of pulmonary metastases by irradiation of the lung: A model system employing a transplantable hepatoma and alpha1-fetoprotein as an index of tumor growth. *Rad. Res*, 69, 54-64.

[37] Murgita R.A, and Tomasi T.G. (1975) Suppression of the immune response by α-fetoprotein. I. the effect of mouse α-fetoprotein on the primary and secondary antibody response. *J. Exp Med,* 141, 269-271.

[38] Murgita R.A. and Tomasi T.G. (1975) Suppression of the immune response by α-fetoprotein II. The effect of mouse α-fetoprotein on mixed lymphocyte reactivity and mitogen-induced transformation. *J. Exp. Med,* 141, 440-449.

[39] Sheppard HW, Sell S, Threfts P, Bahu R. (1977) Effects of α-fetoprotein on murine immune responses. Studies on mice. *J. Immunol,* 119, 91-97.

[40] Sheppard HW, Sell S, Threfts P, Bahu R. (1977) Effects of α-fetoprotein on murine immune responses. Studies on rats. *J. Immunol,* 119, 98-103.

[41] Gabant P, Forrester L, Nichols J, Reeth T, de Mees C, Pajack B, Watt A, Smitz J, Alexandre H, Szpirer C and Szpirer J. (2002) Alpha-Fetoprotein, the Major Fetal Serum Protein, Is Not Essential for Embryonic Development but Is Required for Female Fertility. *PNAS USA*, 99, 2865-12870.

[42] Parmelee D.C., Evenson, M.A. and Deutsch, H.F. (1978) The presence of fatty acids in human alpha-fetoprotein. *J. Biol. Chem,* 251, 2114-2119.

[43] Benno R.W, and Williams T. H. (1978) Evidence for intracellular localization of Alpha-fetoprotein in the developing brain. *Brain Res*, 142, 182-186.

[44] Uriel J, Trojan J, Moro R, Pineiro A. (1983) Intracelluar uptake of α-fetoprotein: a marker of neural differentiation. *Ann NY Acad Sci*, 321-329.

[45] Schachter BS, and Toran-Allerand, C.D. (1982) Intraneuronal α-fetoprotein and albumin are not are not synthesized locally in the developing brain. *Brain Res*. 281:93-98.

[46] Sell, S., Longley, M.A. and Boulter, J. (1985) Alpha-fetoprotein and albumin gene expression in brain and other tissues of fetal and adult rats. *Dev. Brain Research* 22, 49-53.

[47] Attardi B and Ruoslahti E. (1976) Foetoneonatal oestradiol-binding prodtein in mouse brain cytosol is α-fetoprotein. *Nature London,* 263, 685-687.

[48] McEwen, B.S., Piapinger, L., Chapel, C. Geriach J, and Wallach, G. (1975) The role of fetoneonatal estrogen binding proteins in the association of estrogen with neonatal brain cell antigens. *Brain Res,* 96,400-407.

[49] Bakker J, DeMees C, Douhard Q, Balthazart J, Gabant P, Szpirer J and Szpirer. (2006) Alpha-fetoprotein protects the developing female mouse brain from masculinization and defeminization by estrogens. *Nat. Neurosci,* 9, 220-2018.
[50] Carter CS. (2002) "Neuroendocrinology of sexual behavior in the female." In Becker JB. *Behavioral Endocrinology*. Cambridge, Mass: MIT Press. pp. 88–89. ISBN 0-262-52321-3.

In: Alpha-Fetoprotein: Functions and Clinical Applications ISBN: 978-1-63484-875-6
Editors: Nisha Lakhi and Michael Moretti

Chapter 3

THE ROLE OF ALPHA-FETOPROTEIN IN HUMAN HEPATOCELLULAR CARCINOMA

Wenbo Meng[1,2,3,4], MD, PhD, Bing Bai[1,2,3,4], MD, Yan Li[5], MD, Zhongtian Bai[2,3,4], MD, PhD, Ping Yue[2,3,4], MD, Kexiang Zhu[2,3,4], MD, PhD, Wence Zhou[2,3,4], MD, PhD, and Xun Li[2,3,4,*], MD, PhD

[1]Special Minimally Invasive Surgery, the First Hospital of Lanzhou University, Gansu, China
[2]The Second Department of General Surgery, the First Hospital of Lanzhou University
[3]Hepatopancreatobiliary Surgery Institute of Gansu Province, Clinical Medical College Cancer Center of Lanzhou University, Gansu, China
[4]Biological therapy and regenerative medicine transformation center of Gansu province, Gansu, China
[5]Department of Immunology, Cleveland Clinic, Cleveland, OH, US

ABSTRACT

Human hepatocellular carcinoma (HCC) is the third most common cause of cancer-related death. Various aberrantly regulated signaling pathways, including the PI3K/AKT/mTOR pathway and the RAF/MEK/ERK pathway, have been reported in HCC, and these pathways may serve as a paragon target for HCC therapy and prevention. With this connection, the multikinase inhibitor Sorafenib has demonstrated some therapeutic benefits for patients with advanced HCC. Although, limited outcomes have been proven in clinical usage, it still inspires scientists to develop more efficient methods to conquer the difficulty.

---

* Corresponding author: Xun Li, M.D., Ph.D., E-mail: drlixun@163.com. Professor of Surgery, Department-II of General Surgery, the First Hospital of Lanzhou University, Lanzhou 730000, Gansu Province, China, Hepatopancreatobiliary Surgery Institute of Gansu Province, Lanzhou 730000, Gansu Province, China; Clinical Medical College Anti-Cancer Center of Lanzhou University, Lanzhou 730000, Gansu Province, China. Phone: +86931 8356821; Fax: +86931 8622275.

Alpha-fetoprotein (AFP) is probably the best-defined tumor marker for HCC, and is still widely used as a clinical diagnostic and prognostic tool. Up to 70% of HCC cases exhibit elevated serum levels of AFP; however, its pathophysiological functions in HCC carcinogenesis are poorly defined. It is known that AFP is not just a fetal form of carrier protein and a tumor marker, but it is also preciously involved in the regulation of cellular cell growth, differentiation, apoptosis, and angiogenesis functions. It has been reported that AFP plays a critical role in in regulating the proliferation and apoptosis of liver HCC cells through Fas/FasL, caspase-3, and PI3K/AKT signaling pathways. It has been clarified that PI3K/AKT pathway can be governed in an autocrine manner (cell migration/autocrine stimulation/secretion), by insulin-like growth factor (IGF) and vascular endothelial growth factor (VEGF). Nevertheless, previous studies have demonstrated that AFP may stimulate angiogenesis and induce metastasis of liver cancer. HCC is a kind of highly vascularized tumor in which angiogenesis plays a crucial role in tumor progression and metastasis. Alpha-fetoprotein might regulate the major angiogenic factors and matrix metalloproteinase in human liver cancer cells. Silencing of AFP caused a reduced expression of MMP9 and MMP2 in Huh-7 cells, further supporting the notion that in HCC, AFP not only contributes to angiogenesis but may also play a role in tumor invasion.

AFP not only serves as a valuable tumor marker for patients with HCC but also possesses substantial regulatory effects for several important biological processes which has promising value in molecular target therapy in near future. In this review article, we will systematically discuss the recent development of AFP in hepatocellular carcinoma. The article would greatly interest the fundamental research field as well as the clinic.

Keywords: alpha-fetoprotein, hepatocellular carcinoma, signal pathway, tumor maker

INTRODUCTION

Hepatocellular carcinoma (HCC), is the 5th most common cancers worldwide, and accounts for one of the leading causes of cancer-related death, with a five-year survival rate about 7% [1]. Of the 65,000 HCC cases reported worldwid annually, Asia accounts for nearly 75-80% of cases, and China has the proportion of approximately 55% [2]. The carcinogenesis of HCC is correlated with the integration of several environmental and hereditary factors, such as HBV and HCV virus infection, aflatoxin toxicity, alcohol addiction, smoking, parasites, and intake of nitrosamines [1]. Together these factors atribute to the poor prognosis, high invasion ability, and malignancy of HCC. The outcome of the only curable approach, surgery, is significantly depending on the early diagnosis of HCC. However, occult progression is the reason most patients are diagnosed at a later stage, which leads to an inferior resection rate, often accompanied by massive primary tumor, multiple metastasis sites, and cirrhosis. Hence, it is of great necessity to develop more efficient early diagnosis methods for the purpose of improving resection rate and enhancing post-operation quality of life for HCC patients [3].

Molecular biomarkers, often detected in tissue and blood samples, are specific substances that correlate with tumor progression, and in certain conditions, they can be very sensitive for the diagnosis of early-stage patients. The identified biomarkers are generally classified into several categories, which including hormones, enzymes, receptors, oncogenes, tumor suppressor genes and their products, tumor antigens, and antibodies [4]. Some of them are well known, such as alpha fetoprotein (AFP), glypican-3(GPC3), γ-glutamyltransferase II(γ-GGT II), des-γ-carboxy prothrombin(DCP, lens culinary agglutinin-reactive AFP(AFP-L3), and α-1-antitrypsin [5-7]. Until now AFP is regarded as the most important biomarker used in HCC screening, and AFP combined with imaging tests has proven to be feasible in early HCC detection [8]. High-level of AFP present in serum is not only a reliable signal for risk of cancer, but also a prognostic indicator of disease status post-operation or in patients who received other treatments [9, 10]. In addition to being a biomarker, there are reports that AFP can regulate cell differentiation, proliferation, as well as the process of tumorigenesis by multi-effect activities [11]. By summarizing the results from some preliminary studies, this review will discuss the biological characteristics and functions of AFP, and further highlight its value and prospect as a cancer marker.

1. Biological Characteristics of AFP

AFP is a glycoprotein originally produced in liver and vitelline during the early stage of fetal development. It was first identified in serum of human embryos, and later on detected in the serum of HCC-loaded mice and HCC patients. AFP is composed of a polypeptide chain, which has the molecular weight of about 68 KD and half-life of about five days. The gene coding AFP is located in 4q11-13 with 15 exons and 14 introns [13, 14]. AFP is a member of albumin gene families and highly homologous to serum albumin, alpha albumin, and vitamin D binding proteins [15]. The AFP's molecular structure resembels a 'U' shape. The area that lies in the low-density part of electron cloud is called the "hinge region," which probably is the connector between structural domains, and further promotes domain swinging as well as the interaction with corresponding ligands. AFP has three structural domains named I, II, and III, and there are four, five, and six disulfide bonds, respectively. This structural feature contributes to the high stability of AFP [16].

AFP has various types of carbohydrate chain structures, which could explain the existence of AFP heteroplasmon. These carbohydrates are catalytically synthesized by a complete set of saccharifying enzyme of the host cells. As the quality and quantity of these enzymes are different in human organs, consequently the AFPs originated from various tissues and tumors are different. Plant lectins tests can identify the heterogeneity of AFP. Thus, AFPs are be categorized into three subtypes, which are L1, L2, and L3, according to different affinity of heteroplasmon and lectins, plus the results of electrophoresis [18, 19]. Among those three subtypes, AFP-L1 does not has any affinity with lectins base on previous reports. This is the most common type and generally detected in serum of benign liver disease patients. AFP-L2 has moderate affinity and in most conditions rises in yolk sac tumor patients'serum; AFP-L3 is mostly released from HCC cells that can be bound to lectins. Clinically, the AFP-L3/AFP ratio is widely used to evaluate the level of malignancy; it also has particular value in early diagnosing and judging prognosis of HCC [20, 21].

1.1. Biological Functions of AFP

In vivo and *in vitro* experiments have confirmed that the biological functions of AFP are not necessarily the same in different species. Based on results from previous studies, this review will summarize the research status of AFP in various aspects, such as immunological suppression mechanism and signaling transduction pathways.

2.1. AFP Is Involved in Multiple Signaling Transduction Pathways

Although the mechanism of how AFP affects cancer cells' proliferation is not fully understood, there is evidence showing that AFP receptors are expressed on the membrane surface of cancer cells [22-24]. For example, by binding to certain receptors, AFP can affect cancer cell proliferation through regulating the activity of cAMP-PKA pathway and the expression level of K-ras p21 gene [24]. RAS is paramount in the protein tyrosine kinase related signaling transduction. *In vitro* studies showed that the activation of AFP receptors can promote proliferation in BEL-7402 cell line through the PTK-RAS-MAPK pathway [25]; the inhibition of caspase-3 caused by AFP binding to its receptors would lead to apoptosis signal transduction block in HCC cells [26].

Cytoplasmic AFP is a major regulatory factor in the HCC-related PI3K/AKT signaling transduction pathway. Using fluorescence resonance energy technique to study AFP, PI3K, and Akt mRNA levels in serum and HCC tissues section, current research showed that AFP can promote the capacity of HCC cells' proliferation by activating the PI3K/AKT signaling pathway [27].

Knockdown AFP in Huh-7 cell, the HCC cell line with high AFP expression level, showed that silencing of AFP would lead to PI3K/AKT down-regulation as the phosphorylation level of AKT was remarkably decreased. From this, we assumed that AFP may affect cancer cell proliferation ability by regulating AKT phosphorylation level through PI3K/AKT pathway. *In vitro*, silencing AFP expression will result in up-regulation of tumor suppressor gene PTEN, and a decreased AKT phosphorylation. As a kind of protein phosphatase, the loss of PTEN activity will lead to PI3K accumulation and activation of AKT, and further protein phosphorylation effect will start cell cycle growth and promote cancer cell metastasis [28].

AFP is involved in other signaling pathways related to HCC development as well. The nuclear transcription factor NF-κB is able to induce the expression of anti-apoptosis protein encoding genes, such as Bcl-XL and XIAP. The TNF-α can induce cancer cell death by inhibiting NF-κB. A precious study using HCC cell line HepG2 as test subject showed that AFP acted as a kind of protective factor which can antagonize the cytotoxic effect induced by TNF-α [29]. Also by regulating the expression level of Fas/FasL and TRAIL/TRAILR factors, AFP has the ability to help HCC cells to evade immune system monitoring and activate cancer cell proliferation [30]. On the other hand, high doses of AFP is attributed to a cell apoptosis-inducing effect. This effect may be due to the AFP-induced cytochrome C release, which can substitute for cIAP-2 in the apoptotic complex. Moreover, further activation of caspase3 and caspase9 will initiate cell death processes [31]. It can be realized from above that by regulating the activity of certain signaling pathways, AFP holds potential as a promising drug theraputic agent. As research deepens, more moleculer based medicines which target signaling transducers will be applied to future clinical tumor therapy.

2. AFP Is Involved in Immunosuppression Process

AFP has an immune-suppressing effect, which can be proved by the fact that during the development of the embryo, a high level of serum AFP can protect the embryo from maternal immune attack. During HCC's development, the body system retrieves AFP expression ability, and the subsequent abundantly expressed AFP will compromise body immune response function. Recent studies have shown that AFP has an effect upon macrophages, leading to decreased phagocytic and antigen presenting abilities [32]. AFP also exhibits inhibitory effect on NK cells, and T lymphocytes proliferation [33]. AFP-induced apoptosis of antigen presenting cells could be a reason HCC cells evade immune monitoring [34]. There are research reports indicating that AFP may participate in the process of immune cell apoptosis by up-regulating the expression of pro-apoptotic genes in peripheral lymphocytes, such as Bax, Bid, Bad, and ApaF. Meanwhile during the process of immune cell infiltration, AFP enhances the effect of some apoptins, such as caspase-3, Fas, FasL and TRAIL [35]. The induction of these apoptotic markers is often followed by Foxp3 expression within lymph nodes and CD4+ T cells accumulation.

Dendritic cells are a kind of antigen presenting cell which plays an essential role in tumor immunologic response. The loss of dendritic cell function will result in immunosuppression and tumor progressing. AFP can cause dendritic cell dysfunction by inducing dendritic cell apoptosis. The ability of dendritic cells to produce anti-tumor immune response factors such as IL-12 and TNF-α decreases remarkably after dendritic cells are treated with AFP. HCC patients with high level of AFP have significantly lower TNF-α level than healthy group [36]. Summary of these research achievements together could explain why a high level of AFP may be an inducer for severe dendritic cell dysfunction and subsequent tumor progression. AFP-induced antigen presenting cell apoptosis and cell function loss may be the key mechanism underlying the appearance that HCC cells can evade immune monitoring [34].

As research around AFP immunosuppression mechanisms is pushed further, it has been discovered that based on specific peptide and DNA immune stimulation, AFP-specific cytotoxic T cells can specifically recognize and kill AFP-producing HCC cell lines. This result showed that AFP may act as a kind of tumor specific antigen which holds the potential as a promising T-cell based immunotherapy target. Nowadays research indicates that four types of immunodominant HLA-restrictive T cell epitope are derived from AFP, which can stimulate a specific T cells response, and these epitopes can be further be recognized by CD8+T cells. In autologous cultured T cells originated from HCC patients and transgenic mice HLA-A2.1, AFP epitope sequences transfected dendritic cells are found to have AFP-specific immune response [37]. When the transfecting epitope sequence is located in the third structure domain of AFP, which contains a nine-peptide fragment GVALQTMKQ, T cells present cytotoxic effect and produce cytokines [38]. When the transfecting epitope sequence is located in the second structure domain of AFP, which contains a nine-peptide fragment LLNQHACAV, T cells present increased IL-12 and TNF-α production ability, thus producing a killing effect on HLA-A2 positive HCC cell lines HepG2 and T2 [39].

In another experiment performed, co-culturing adenovirus vector transfected AFP (Ad-hAFP) with CIK cells and dendritic cells which are obtained from HCC patients, about 70% AFP-expressing HCC cells lysed. While lymphocytes were co-cultured with Ad-hAFP transfected DCs, the cytotoxicity was significantly higher than Ad-mock transfected dendritic cells. These results suggested the possibility of inducing AFP-specific anti-HCC immune response [34]. Together these findings indicated that immunodominant epitopes-induced HLA-restricted immunodominant polypeptide could promote a CD8+ T cell response. Tumor immunotherapy aiming at the AFP-positive HCC has vast potential for future development [40, 41].

2.3. AFP Has Dual-Directional Regulation Effect in Cell Proliferation Process

It was elucidated in previous studies that AFP has the ability to stimulate cell proliferation and differentiation in a variety of cells [42, 43]. In recent studies, it was discovered that the regulation pattern of AFP could be both up-regulated and down-regulated in the cell proliferation and differentiation, and the outcome was widely dependent on AFP dose. Low concentration of AFP has a positive stimulative effect on the estrogen-resistant tumor or regular tissues. As in estrogen-resistant HCC patients, AFP can significantly decrease the expression of cyclin-dependent kinase suppressor gene P27 at mRNA and protein level; moreover, the expression level of proliferating cell nuclear antigen(PCNA) is upregulated [44]. This cell proliferation stimulating effect may be coupled to the inhibition of cell apoptosis [45]. On the other hand, AFP presents inhibitory effect on estrogen-sensitive HCC cell lines, such as recombinant AFP can inhibit the proliferation of MCF-7 tumor cells [46]. It was reported natural AFP would expose certain peptide fragments which showed inhibitory effect on estrogen-dependent cancer cells when treated with high concentration of estradiol [47]. This could be regarded similar to the role growth inhibitory factor played through signaling pathway mediated by G protein-coupled receptors [48]. The anti-tumour effect of antiestrogenic drugs such as Tamoxifen could be promoted by combined application of AFP or its derivative; this finding will dedicate new thoughts for the development of novell anti-tumor drugs [49].

3. Clinical Use of AFP

Although AFP has been identified as a ‘golden’ tumor-specific marker which holds great value in HCC diagnosis, the specificity and sensitivity of AFP was somehow not so satisfactory. This was the reason other alternative markers had to be used simultaneously for HCC early diagnosis. Indeed, AFP is significant for the diagnosis and prognostic prediction of highly malignant HCC patients. It has been reported that nearly 70%-80% of HCC patients with lower differentiated tumor have relatively higher serum AFP level than those other types of differentiated tumor. Also in postoperative HCC patients, the serum AFP level can predict early recurrence and metastasis.

With the development of modern genetic engineering techniques, AFP can be manipulated through techniques such as microRNA and RNA interference to meet the purpose of suppressing cell proliferation and tumor growth [50, 51]. Meanwhile, based on comparative proteomics technology, researchers have discovered six protein markers in HCC patients' tumor and non-tumor tissues. These markers may be involved in hepatocellular damage, and the over-expression of these markers correlates with the serum AFP level [52].

The prior study, by processing 21 AFP-producible tumor cell lines, and screening 213 genes, demonstrates that a high level of serum AFP correlates with the over-expression of several liver-specific genes. These genes included growth factors, angiogenesis factors, and factors related to lipids and iron metabolism mediated by Wnt and MAPK signaling pathways [53]. AFP may have a functional relationship with factors regulating cell proliferation, differentiation, adhesion, and migration. Studies have shown that in HCC patients, the over-expression of AFP is often accompanied by the over-expression of phosphatidylinositol chitosan-3 and heparan sulfate proteoglycan, among which AFP and phosphatidyl inositol chitosan-3 are closely associated with p53 gene mutation [54]. These results illustrate that AFP has a synergistic effect with other growth regulatory factors in the HCC development.

In recent years, AFP has been considered to be important in cancer therapy. Based on high sensitivity and specificity of AFP present in tumor cells, cytokines and exogenous genes for therapeutic use can be transfected into certain positions on AFP promoter or enhancer sequences. Trans-acting factors that activate subsequence transcription procedures, are on the way to meet the treatment purpose by selective target gene expression.

AFP coding proteins and antigen epitope peptides can induce specific cytotoxic T lymphocytes to kill AFP-positive tumor cells. Studies have shown that AFP can cause an immune effect to eliminate tumor cells. In addition, AFP can also serve as an anti-cancer drug carrier molocule, through AFP receptors expressed on the surface of tumor cells. The combination of AFP and anti-cancer drugs specifically directed toward cancer cells may improve the effect of anti-cancer drug therapy [55, 56].

REFERENCES

[1] Bosetti C, Turati F, La Vecchia C. (2014) Hepatocellular carcinoma epidemiology[J] *Best practice and Research Clinical Gastroenterology*, 28, 753-70.

[2] Farrell G. (2010) Prevention of hepatocellular carcinoma in the Asia-Pacific region: consensus statements. *Journal of Gastroenterology and Hepatology*, 25, 657-63.

[3] Avila MA, Berasain C, Sangro B, Prieto J. (2006) New therapies for hepatocellular carcinoma. *Oncogene,* 25, 3866-84.

[4] Rich N, Singal AG. (2014) Hepatocellular carcinoma tumour markers: Current role and expectations. *Best practice and Research Clinical Gastroenterology*, 28, 843-53.

[5] Johnson PJ.(2001) The role of serum alpha-fetoprotein estimation in the diagnosis and management of hepatocellular carcinoma. Clinics in liver disease. 5, 145-59.

[6] Li D, Mallory T, Satomura S. (2001) AFP-L3: a new generation of tumor marker for hepatocellular carcinoma. *Clinica chimica acta, International Journal of Clinical Chemistry,* 313,15-9.

[7] Yuen MF, Lai CL. (2005) Serological markers of liver cancer[J] *Best Practice and Research Clinical Gastroenterology*, 19, 91-9.
[8] Debruyne EN, Delanghe JR. (2008) Diagnosing and monitoring hepatocellular carcinoma with alpha-fetoprotein: new aspects and applications. *Clinica chimica acta; International Journal of Clinical Chemistry*, 395, 19-26.
[9] Yeh CN, Lee WC, Chen MF. (2003) Hepatic resection and prognosis for patients with hepatocellular carcinoma larger than 10 cm: two decades of experience at Chang Gung memorial hospital. *Annals of Surgical Oncology*, 10, 1070-6.
[10] Sen G, Rai R, Manas DM. (2003) Factors predictive of 5-year survival after transarterial chemoembolization for inoperable hepatocellular carcinoma. *Br J Surg*, 90, 325-331.
[11] Gillespie JR, Uversky VN. (2000) Structure and function of alpha-fetoprotein: a biophysical overview. *Biochimica et Biophysica Acta,* 1480, 41-56.
[12] Bergstrand CG, Czar B. (1956). Demonstration of a new protein fraction in serum from the human fetus. *Scandinavian Journal of Clinical and Laboratory Investigation*, 8, 174.
[13] Dudich I, Tokhtamysheva N, Semenkova L, Dudich E, Hellman J, Korpela T. (1999) Isolation and structural and functional characterization of two stable peptic fragments of human alpha-fetoprotein. *Biochemistry*, 38, 10406-14.
[14] Pucci P, Siciliano R, Malorni A, Marino G, Tecce MF, Ceccarini C, et al. (1991) Human alpha-fetoprotein primary structure: a mass spectrometric study. *Biochemistry*, 30, 5061-6.
[15] Cooke NE, David EV. (1985) Serum vitamin D-binding protein is a third member of the albumin and alpha fetoprotein gene family. *The Journal of Clinical Investigation*, 76, 2420-4.
[16] He XM, Carter DC. (1992) Atomic structure and chemistry of human serum albumin. *Nature*, 358, 209-15.
[17] Terentiev AA, Moldogazieva NT. (2013) Alpha-fetoprotein: a renaissance Tumour biology. *The journal of the International Society for Oncodevelopmental Biology and Medicine.* 34, 2075-91.
[18] Keel BA, Harms RL, Leal JA, Cho S. (1990) Characterization of human alpha fetoprotein charge microheterogeneity during fetal development. *Molecular Reproduction and Development*, 27, 281-7.
[19] Shimizu K, Katoh H, Yamashita F, Tanaka M, Tanikawa K, Taketa K, et al. (1996) Comparison of carbohydrate structures of serum alpha-fetoprotein by sequential glycosidase digestion and lectin affinity electrophoresis. *Clinica Chimica Acta; International Journal of Clinical Chemistry.* 254, 23-40.
[20] Li YL, Ma HB, Guo JX, Shi JB, Xu J, Liu AX, et al. (2008) [The studies on early warning of Lens culinaris-reactive alpha-Fetoprotein (AFP-L3) detecting in primary hepatocellular carcinoma] *Zhonghua shi yan he lin chuang bing du xue za zhi = Zhonghua shiyan he linchuang bingduxue zazhi = Chinese Journal of Experimental and Clinical Virology*, 22, 336-8.
[21] Tanwandee T, Setthasin S, Charatcharoenwitthaya P, Chainuvati S, Leelakusolvong S, Pausawasdi N, et al. (2009) Clinical utility of lens culinaris agglutinin-reactive alpha-fetoprotein in the diagnosis of hepatocellular carcinoma: evaluation in a Thai referral

population. J*ournal of the Medical Association of Thailand = Chotmaihet thangphaet*, 92 Suppl 2, S49-56.

[22] Villacampa MJ, Moro R, Naval J, Failly-Crepin C, Lampreave F, Uriel J. (1984). Alpha-fetoprotein receptors in a human breast cancer cell line. *Biochemical and Biophysical Research Communications*, 122, 1322-7.

[23] Torres JM, Laborda J, Naval J, Darracq N, Calvo M, Mishal Z, et al. (1989) Expression of alpha-fetoprotein receptors by human T-lymphocytes during blastic transformation. *Molecular Immunology*, 26, 851-7.

[24] Li MS, Li PF, Yang FY, He SP, Du GG, Li G. (2002) The intracellular mechanism of alpha-fetoprotein promoting the proliferation of NIH 3T3 cells. *Cell research*, 12, 151-6.

[25] Li MS, Li PF, Chen Q, Du GG, Li G. (2004) Alpha-fetoprotein stimulated the expression of some oncogenes in human hepatocellular carcinoma Bel 7402 cells. *World Journal of Gastroenterology*, 10, 819-24.

[26] Li M, Li H, Li C, Zhou S, Guo L, Liu H, et al. (2009) Alpha fetoprotein is a novel protein-binding partner for caspase-3 and blocks the apoptotic signaling pathway in human hepatoma cells. *International Journal of Cancer Journal International du Cancer,* 124, 2845-54.

[27] Zheng L, Gong W, Liang P, Huang X, You N, Han KQ, et al. (2014) Effects of AFP-activated PI3K/Akt signaling pathway on cell proliferation of liver cancer. *Tumour biology: The Journal of the International Society for Oncodevelopmental Biology and Medicine*, 35, 4095-9.

[28] Li M, Li H, Li C, Wang S, Jiang W, Liu Z, et al. (2011) Alpha-fetoprotein: a new member of intracellular signal molecules in regulation of the PI3K/AKT signaling in human hepatoma cell lines. *International Journal of Cancer/ Journal International du Cancer*, 128, 524-32.

[29] Semenkova LN, Dudich EI, Dudich IV, Shingarova LN, Korobko VG. (1997) Alpha-fetoprotein as a TNF resistance factor for the human hepatocarcinoma cell line HepG2. *Tumour biology: The Journal of the International Society for Oncodevelopmental Biology and Medici*ne, 18, 30-40.

[30] Li MS, Ma QL, Chen Q, Liu XH, Li PF, Du GG, et al. (2005) Alpha-fetoprotein triggers hepatoma cells escaping from immune surveillance through altering the expression of Fas/FasL and tumor necrosis factor related apoptosis-inducing ligand and its receptor of lymphocytes and liver cancer cells. *World Journal of Gastroenterology*, 11, 2564-9.

[31] Semenkova L, Dudich E, Dudich I, Tokhtamisheva N, Tatulov E, Okruzhnov Y, et al. (2003) Alpha-fetoprotein positively regulates cytochrome c-mediated caspase activation and apoptosome complex formation. *European Journal of Biochemistry / FEBS*, 270, 4388-99.

[32] Atemezem A, Mbemba E, Marfaing R, Vaysse J, Pontet M, Saffar L, et al.(2002) Human alpha-fetoprotein binds to primary macrophages. *Biochemical and Biophysical Research Communications*, 296, 507-14.

[33] Matsuura E, Kang Y, Kitakawa H, Ogata A, Kotani T, Ohtaki S, et al. (1993) Modulation of T cell function by alpha-fetoprotein: An in vivo study on porcine thyroid peroxidase-induced experimental autoimmune thyroiditis in transgenic mice producing

human alpha-fetoprotein. *Tumour biology: The Journal of the International Society for Oncodevelopmental Biology and Medicine*, 20, 162-71.

[34] Gonzalez-Carmona MA, Marten A, Hoffmann P, Schneider C, Sievers E, Schmidt-Wolf IG, et al. (2006) Patient-derived dendritic cells transduced with an a-fetoprotein-encoding adenovirus and co-cultured with autologous cytokine-induced lymphocytes induce a specific and strong immune response against hepatocellular carcinoma cells. *Liver International: Official Journal of the International Association for the Study of the Liver,* 26, 369-79.

[35] Irony-Tur-Sinai M, Grigoriadis N, Tsiantoulas D, Touloumi O, Abramsky O, Brenner T. (2009) Immunomodulation of EAE by alpha-fetoprotein involves elevation of immune cell apoptosis markers and the transcription factor FoxP3. *Journal of the Neurological Sciences*, 279, 80-7.

[36] Um SH, Mulhall C, Alisa A, Ives AR, Karani J, Williams R, et al.(2004). Alpha-fetoprotein impairs APC function and induces their apoptosis. *Journal of Immunology,* 173, 1772-8.

[37] Butterfield LH, Koh A, Meng W, Vollmer CM, Ribas A, Dissette V, et al. (1999) Generation of human T-cell responses to an HLA-A2.1-restricted peptide epitope derived from alpha-fetoprotein. *Cancer Research*. 59, 3134-42.

[38] Butterfield LH, Meng WS, Koh A, Vollmer CM, Ribas A, Dissette VB, et al. (2001) T cell responses to HLA-A*0201-restricted peptides derived from human alpha fetoprotein. *Journal of Immunology*, 166, 5300-8.

[39] Butterfield LH, Ribas A, Meng WS, Dissette VB, Amarnani S, Vu HT, et al.(2003). T-cell responses to HLA-A*0201 immunodominant peptides derived from alpha-fetoprotein in patients with hepatocellular cancer. *Clinical Cancer Research: An Official Journal of the American Association for Cancer Research,* 9, 5902-8.

[40] Butterfield LH. (2007) Recent advances in immunotherapy for hepatocellular cancer. *Swiss Medical Weekly*, 137, 83-90.

[41] Pardee AD, Butterfield LH. (2012) Immunotherapy of hepatocellular carcinoma: Unique challenges and clinical opportunities. *Oncoimmunology,* 1, 48-55.

[42] Toder V, Blank M, Gold-Gefter L, Nebel L. (1983) The effect of alpha-fetoprotein on the growth of placental cells in vitro. *Placenta*, 4, 79-86.

[43] Leffert HL, Sell S. (1974) Alpha1-fetoprotein biosynthesis during the growth cycle of differentiated fetal rat hepatocytes in primary monolayer culture. *The Journal of Cell Biology*, 61, 823-9.

[44] Shehata MA, Nosseir HR, Nagy HM, Farouk G. (2006) Cyclin dependent kinase inhibitor p27(kip1) expression and subcellular localization in relation to cell proliferation in hepatocelluelr carcinoma. *The Egyptian Journal of Immunology/ Egyptian Association of Immunologists*, 13, 115-30.

[45] Laderoute MP, Pilarski LM. (1994) The inhibition of apoptosis by alpha-fetoprotein (AFP) and the role of AFP receptors in anti-cellular senescence. *Anticancer research*, 14, 2429-38.

[46] Boismenu R, Semeniuk D, Murgita RA. (1997) Purification and characterization of human and mouse recombinant alpha-fetoproteins expressed in Escherichia coli. *Protein Expression and Purification*, 10, 10-26.

[47] Mizejewski GJ, Dias JA, Hauer CR, Henrikson KP, Gierthy J. (1996) Alpha-fetoprotein derived synthetic peptides: assay of an estrogen-modifying regulatory segment. *Molecular and Cellular Endocrinology*, 118, 15-23.

[48] Hamza A, Sarma MH, Sarma RH. (2003) Plausible interaction of an alpha-fetoprotein cyclopeptide with the G-protein-coupled receptor model GPR30: docking study by molecular dynamics simulated annealing. *Journal of Biomolecular Structure and Dynamics*, 20, 751-8.

[49] Mizejewski GJ. (2009) Alpha-fetoprotein (AFP)-derived peptides as epitopes for hepatoma immunotherapy: A commentary. *Cancer Immunology, Immunotherapy: CII*, 58, 159-70.

[50] Won YS, Lee SW. (2007) Targeted retardation of hepatocarcinoma cells by specific replacement of alpha-fetoprotein RNA. *Journal of Biotechnology*, 129, 614-9.

[51] Zhang H, Bai Z, Chen J, Wang Z, Li J. (2007) Alpha-fetoprotein-specific transfer factors downregulate alpha-fetoprotein expression and specifically induce apoptosis in Bel7402 alpha-fetoprotein-positive hepatocarcinoma cells. *Hepatology Research: The Official Journal of the Japan Society of Hepatology*, 37, 557-67.

[52] Lee NP, Chen L, Lin MC, Tsang FH, Yeung C, Poon RT, et al. (2009) Proteomic expression signature distinguishes cancerous and nonmalignant tissues in hepatocellular carcinoma. *Journal of Proteome Research*, 8, 1293-303.

[53] Saito S, Ojima H, Ichikawa H, Hirohashi S, Kondo T. (2008) Molecular background of alpha-fetoprotein in liver cancer cells as revealed by global RNA expression analysis. *Cancer science*, 99, 2402-9.

[54] Wang XY, Degos F, Dubois S, Tessiore S, Allegretta M, Guttmann RD, et al. (2006) Glypican-3 expression in hepatocellular tumors: diagnostic value for preneoplastic lesions and hepatocellular carcinomas. *Human Pathology*, 37, 1435-41.

[55] Lutsenko SV, Feldman NB, Finakova GV, Gukasova NV, Petukhov SP, Posypanova GA, et al. (2000) Antitumor activity of alpha fetoprotein and epidermal growth factor conjugates in vitro and in vivo. *Tumour Biology: The Journal of the International Society for Oncodevelop-mental Biology and Medicine*, 21, 367-74.

[56] Mizejewski GJ, Mirowski M, Garnuszek P, Maurin M, Cohen BD, Poiesz BJ, et al. (2010) Targeted delivery of anti-cancer growth inhibitory peptides derived from human alpha-fetoprotein: review of an International Multi-Center Collaborative Study. *Journal of Drug Targeting*, 18, 575-88.

In: Alpha-Fetoprotein: Functions and Clinical Applications ISBN: 978-1-63484-875-6
Editors: Nisha Lakhi and Michael Moretti

Chapter 4

USE OF ALPHA-FETOPROTEIN AND OTHER BIOMARKERS FOR ASSESSING HEPATOCELLULAR CARCINOMA RISK IN VIRAL HEPATITIS

Syed Abdul Basit*[1], *MD, Mahendran Jayaraj*[1], *MD, Christian D. Stone*[2], *MD, MPH, and Robert G. Gish*[1,3,4,*], *MD

[1]Department of Internal Medicine, Section of Gastroenterology and Hepatology, University of Nevada School of Medicine, Las Vegas, Nevada
[2]Comprehensive Digestive Institute of Nevada, Las Vegas, Nevada
[3]Division of Gastroenterology and Hepatology, Department of Medicine, Stanford University Medical Center, Stanford, California, US
[4]Hepatitis B Foundation, 3805 Old Easton Road, Doylestown, PA, US

ABSTRACT

Hepatocellular carcinoma (HCC) diagnosis after symptom onset carries a grim prognosis with five year survival of < 20%. Thus, in a population with risk factors for HCC, our goal should be to diagnose it in an early stage and manage the patient with curative intent. The currently available screening and surveillance modalities, including ultrasound (US), are underutilized in the clinical setting. Current AASLD guidelines only recommend US every 6 months for surveillance of high-risk patients. The biomarkers that have been studied to date are only risk biomarkers and of those discussed here, only alpha-fetoprotein (AFP), AFP-L3%, and des-gamma carboxyprothrombin (DCP) are FDA-qualified as biomarkers for HCC risk. Due to low sensitivity and specificity as screening tests, they should not be used as sole screening or surveillance tests and are not used to diagnose HCC. HCC is diagnosed by imaging with CT scan or MRI with contrast using Liver Imaging Reporting and Data System (LI-RADS) criteria and, rarely, by liver biopsy. One or more of the biomarkers may, however, complement US for screening and surveillance, potentially triggering a CT or MRI with contrast. Elevated or increasing biomarkers should trigger a CT or MRI. AFP has been shown to be a strong prognostic indicator and to complement US for surveillance. A higher AFP value has clearly been

* Corresponding Author: rgish@robertgish.com.

associated with poor clinical outcome while AFP reduction > 50% after treatment predicts an improved response. The pre-treatment AFP value may identify patients who will benefit more from loco-regional therapy than transplant. A higher pre-transplant AFP level has been associated with tumor recurrence and thus may help identify patients who will need close observation for early HCC recurrence. The AFP level (with a ceiling of < 500 ng/ml) has been utilized in some liver allocation models to grant MELD exception points. AFP-L3% has been found to complement AFP in HCC surveillance, especially when the AFP value is indeterminate. As with AFP, a higher AFP-L3% level may be reflective of either advanced disease or aggressive HCC. Elevated levels of both AFP-L3% and DCP have been associated with vascular invasion by the tumor. A high post-ablation AFP-L3% value may be reflective of disease recurrence. Similar to AFP, AFP-L3% may help determine treatment modality. DCP can help differentiate HCC from non-malignant chronic liver disease. Its re-emergence after treatment can be reflective of HCC recurrence. DCP can complement AFP in HCC risk assessment. Data for other biomarkers are limited, and no firm recommendation can currently be made for their use. Several risk assessment models have been proposed to help with some of the limitations of these biomarkers. Multiple large studies are underway on the clinical utility of HCC biomarkers and on the proposed models that will ultimately help answer remaining questions on their usefulness. In the meantime, guidelines from the Japan Society of Hepatology recommend AFP, AFP-L3% and DCP for HCC surveillance. Experts in the field agree that until better screening and surveillance tools are available for HCC, we should utilize the available biomarkers to complement US and help identify patients who would benefit from further radiological workup.

INTRODUCTION

Hepatocellular carcinoma (HCC) is the most common primary cancer of the liver, the fifth most common cancer in men and seventh in women worldwide, and the second leading cause of cancer-related mortality worldwide with more than 650,000 new cases every year [1]. The majority of HCC cases are diagnosed in developing countries where endemic hepatitis B virus (HBV) infection is the most prevalent risk factor [2]. Developing countries account for 85% of the worldwide HCC burden, with more than 80% of cases found in sub-Saharan Africa and Eastern Asia where the incidence rate is more than 20 per 100,000 people, and in Mongolia where the rate is over 60 per 100,000 [3]. The incidence of HCC is on the rise in most parts of the world [4]. Infection with hepatitis C virus (HCV) is the leading cause of HCC in Japan although there has been a recent decrease in HCC incidence attributed to HCV treatment responders [5]. In a recent retrospective study of U.S. veterans, the incidence of HCC and HCC-related mortality has risen since 2001 [6]. There is an alarming 2.5 fold increase in HCC incidence, most of which is attributed to HCV infection. The American Cancer Society predicted 35,660 new liver and intrahepatic bile duct cancers in 2015 in the United States, approximately three-fourths of which would be HCC [7].

Because of its prevalence and high mortality, it is imperative to identify the predisposing factors for the development of HCC. This risk stratification will help identify patients who would benefit from HCC screening and surveillance. The American Association for the Study of Liver Diseases (AASLD) defines screening as "application of diagnostic tests in patients at risk for HCC, but in whom there is no a priori reason to suspect that HCC is present" while surveillance is "the repeated application of screening tests'' in patients at risk for a disease

[4]. Cirrhosis is the major risk factor for the development of HCC with the five-year cumulative risk in the cirrhotic patient varying from 5 to 30%, depending upon the etiology of the liver disease [8]. In the U.S. the incidence of HCC is higher in patients with advanced cirrhosis and high Child-Pugh class, highest in patients with chronic HCV infection, and second-highest in those with chronic HBV (CHB) [8]. The 5-year cumulative incidence for common risk factors is summarized in Table 1. In subjects with compensated liver cirrhosis, HCC is the most common cause of death, accounting for 54-70% of mortality [8]. At autopsy, 80-90% of patients with HCC are found to have liver cirrhosis. It is important for surveillance guidelines to recognize that in approximately 10-20% of patients HCC develops in the absence of cirrhosis [9]. In patients who develop HCC in the absence of cirrhosis there is almost always chronic inflammation; thus, the necroinflammatory process appears to be the most important predisposing factor [8]. The common causes for viral and non-viral chronic hepatitis and HCC are detailed in Table 2.

Table 1. Common risk factors and incidence of HCC [8]

Risk factors	5-year cumulative incidence*
HCV	17% in Europe and United states
	30% in japan
Hemochromatosis	21%
Chronic HBV infection	10% in Europe and United States
	15% in sub-Saharan Africa and Eastern Asia
Alcohol related cirrhosis	8% **
Biliary Cirrhosis	4%

* Cumulative incidence refers to the proportion of population at risk that will develop an outcome in a given period of time.

** In the absence of HBV and HCV infection.

Table 2. Risk factors for cirrhosis and HCC [2, 8, 10, 66, 180-182]

HBV
HCV
Coinfection or tri-infection with HCV/HBV/HIV
Aflatoxin in patients with HBV and HCV infection
Non-viral hepatitis-related risk factors
Alcohol-related liver disease
Nonalcoholic fatty liver disease
Hemochromatosis
Primary biliary cirrhosis
Alpha-1 antitrypsin deficiency
Primary sclerosing cholangitis
Autoimmune hepatitis
Wilson disease
Cryptogenic cirrhosis
Hereditary tyrosinemia
Porphyria
Aflatoxin

HCC diagnosed after symptom onset carries a very grim prognosis, with five-year survival less than 10% [4]. Thus, the approach to the population at risk should be two-pronged:

A. Identification and treatment of the predisposing liver conditions.
B. Initiation of surveillance programs in those patients at risk for HCC.

The main emphasis on preventing HCC should be on identifying modifiable risk factors and treating them promptly. Obesity, diabetes, smoking, and infection with HBV, HCV and human immunodeficiency virus (HIV) infection have been recognized as major risk factors for HCC [10]. Universal vaccination against HBV infection has been shown to decrease the incidence of HCC [11-14]. The differential risks for HCC with CHB are summarized in Table 3. There is conflicting evidence for chemoprevention of HCC with nucleos(t)ide analogues [15]. The Greece cohort study reported that antiviral therapy in cirrhotic patients with HBeAg-negative CHB does not decrease the risk of HCC [16]. Although the nucleoside analogue entecavir is a potent antiviral that can fully suppress HBV, studies assessing its effectiveness as a prophylactic against the development of HCC have yielded conflicting results. Studies from Japan [17] and Hong Kong [18] reported that the use of entecavir in chronic HBV was associated with reduced risk of HCC. Trinchet et al. reported similar findings in France where treatment of viral hepatitis was associated with low HCC incidence [19]. This positive response was not observed in a study by Lampertico et al. [20]. Although controversy has existed about whether antiviral therapy should be recommended for CHB patients with compensated cirrhosis who have detectable, but low, serum HBV DNA levels, a recent analysis found that the 5-year cumulative HCC incidence rate was 2.2%, 8.0%, and 14.0% for patients with undetectable HBV DNA (<12 IU/mL), low HBV DNA levels (<2000 IU/mL) plus normal alanine aminotransferase (ALT) levels, and low HBV DNA levels plus elevated ALT levels at baseline, lending support for initiating antiviral therapy in this patient population [21].

Achieving a sustained viral response (SVR) with HCV treatment has been shown to decrease HCC incidence [22] and all-cause mortality [23] in patients with chronic HCV infection. The substantially higher SVR rates now being achieved with the new direct acting antiviral agents [24] may result in significant reduction of HCC incidence and mortality in this population in future. However, it has been suggested that even with DAA-associated SVRs, HCC incidence could continue to increase until 2030 [25] and that the risk of HCC may persist for 8-10 years after SVR achievement [26].

Table 3. Incidence of HCC in the HBV-infected population [183]

Hepatitis B infection status-Untreated	Incidence in Western population at risk	Incidence in Asian population at risk
Inactive carriers	0.2%	0.2%
Chronic hepatitis without cirrhosis	0.3%	0.6%
Compensated cirrhosis	2.2%	3.7%

Prorok has proposed that there are two components for an effective surveillance program [27]: (1) the surveillance test should detect cancer early; and (2) therapy initiated as a result of early detection by screening must be more effective than therapy undertaken in later stages of disease. Using these criteria, screening and surveillance are justified for HCC [28]. The first randomized controlled trial of HCC screening, a study in which liver ultrasound (US) and AFP testing were offered to the surveillance group every six months, found that, after five years of follow up, biannual surveillance reduced HCC mortality by 37% [29].

Multiple studies have shown that HCC surveillance is underutilized in populations at risk, with 20% or fewer of patients receiving surveillance that follows guidelines [30, 31]. Current AASLD guidelines do not recommend any biomarkers for HCC surveillance although the omission of AFP from surveillance has been questioned [32]. Some authors have concluded that US or AFP are suboptimal surveillance tools if used individually; for optimal sensitivity and specificity imaging and biomarkers should be used in combination [33, 34]. The main goal of surveillance is to find HCC in an earlier stage with small tumor burden to enhance the chance of cure or prolonging life. Increased tumor size alone worsens prognosis. Five-year survival has been shown to be significantly lower in small HCC with tumor size 3.5-5cm compared to smaller tumor mass (< 3.5 cm) [35]. The current recommendations for HCC screening and surveillance from several societies are summarized in Table 4.

Table 4. Surveillance recommendations from liver societies [10, 24, 66, 184-186]

Liver societies	Current surveillance recommendations
AASLD*(2010)	US every 6 months.
	Biomarkers are currently not recommended.
EASL ** (2012)	US every 6 months.
	Biomarkers are currently not recommended.
APASL***	US every 6 months
	AFP every 6 months
WGO****	Ultrasound every 4-12 months based on severity of liver disease. Shorter interval is recommended for advanced cirrhotic patients.
JSH %	US every 3-4 months in extremely high risk groups or every 6 months in high risk groups.
	AFP/AFP-L3%/DCP every 3-4 months in extremely high risk groups and every 6 months in high risk groups.

* American Association for the Study of Liver Diseases. **European Association for the Study of the Liver. *** Asian Pacific Association for the Study of the Liver. **** World Gastroenterology Organization. % Japanese Society of Hepatology. DCP, AFP and AFP-L3 are defined elsewhere in the text.

HCC Biomarkers

Cancer biomarkers either reflect the presence of tumor in the body or predict the risk of developing cancer [36]. The Early Detection Research Network (EDRN) established by the

National Cancer Institute proposed that a biomarker needs to pass through five phases to produce a useful population-screening tool [37]:

Phase 1: Promising directions identified.

Phase 2: Clinical assay detects established disease.

Phase 3: Biomarker detects disease early before it becomes clinical and a "screen positive" rule is defined.

Phase 4: Extent and characteristics of disease detected by the test and the false referral rate are identified.

Phase 5: Impact of screening on reducing the burden of disease on the population is quantified.

The first disease biomarker discovered was the Bence-Jones protein. It was discovered in 1840 [38] and reported as a cancer biomarker in 1848 [39]. With advancements in technology, a growing number of biomarkers have been identified for HCC. Most of these biomarkers are in an early phase of development. Currently identified HCC biomarkers along with the EDRN phase for each are shown in Table 5.

Table 5. EDRN phase and current status of HCC biomarkers [36, 37]

Phase	Category	HCC biomarkers	
		Serum/plasma/blood	**Tissue**
1	Preclinical exploratory studies	Circulating miR-16 EpCAM positive circulating tumor cells IL-6 and IL-10 [187]	Five-gene score, FGF3/FGF4 amplification High Met expression, miR-185 and miR-201, SOUX, AKR1B10, CD34 expression, Human carbonyl reductase2,
2	Clinical assay development and validation, case control studies	DCP (PIVKAII), Glypican 3, GP73, SCCA, Osteopontin, AFP-L3, Canavaninosuccinate	
3	Retrospective longitudinal repository studies	AFP	
4	Prospective screening studies	AFP	
5	Cancer control randomized studies	AFP	

Genome-wide association studies have identified single nucleotide polymorphisms (SNPs), which are associated with HCC susceptibility [36]. Although potentially informative, studies on these SNPs are in preliminary stages. To date, the SNPs known to be associated with increased HCC susceptibility are rs7574865, rs9272015, rs2596542, rs9275319, rs101206, and rs2551677; among those found to be associated with decreased susceptibility are rs2880301 and rs455804 [36]. Three loci, rs17401966, rs7574865 and rs9275319, have been shown to be associated with HBV-related HCC; two loci, rs2596542 and rs9275572, have been shown to be associated with HCV-related HCC [40-44]. At present, SNPs do not provide good prediction at either the individual or population level [45]. These sequencing tests are not yet available clinically.

Role of Currently Available Biomarkers and Their Utilization in HCC Management

The clinical utility and evidence based utilization of currently available HCC biomarkers are summarized in Table 9. Here we discuss the available biomarkers with a special emphasis on those which are currently recommended in society guidelines. Several models of the probability of HCC development have been developed. The GALAD model uses gender, age, AFP-L3% (*AFP-lens culinaris agglutinin),* AFP and DCP (Des-gamma-carboxythrombin). The initial case control and validation study showed promising results [46], and recent validation of this model in a large multinational cohort showed its high sensitivity (82.2-91.6%) and specificity (87.6-89.7) [47].

It is important to note that the biomarkers discussed here are only risk biomarkers. Due to low sensitivity and specificity as screening tests, they should not be used as sole screening or surveillance tests and are not used to diagnose HCC. HCC is diagnosed by imaging with CT scan or MRI with contrast using Liver Imaging Reporting and Data System (LI-RADS) criteria and, rarely, by liver biopsy. One or more of the biomarkers may, however, complement ultrasound (US) for screening and surveillance, potentially triggering a CT or MRI with contrast when either US imaging indicates the need or HCC biomarkers are elevated. Elevated or increasing biomarkers should trigger a CT or MRI. Of the biomarkers discussed here, only AFP, AFP-L3%, and DCP are FDA-cleared as biomarkers for HCC risk.

Alpha-Fetoprotein

The modern era of cancer biomarkers dawned with the discovery of alpha-fetoprotein (AFP), an oncofetal protein with a molecular weight of 68,000 [48] that is a member of the albuminoid gene superfamily [49]. Abelev et al. reported a protein in hepatoma bearing mice that migrated to the alpha 1 region in protein electrophoresis [50]. This protein was later found in the serum of a patient with primary liver cancer [51]. In 1970, the first international multicenter study on AFP reported that it was detected in the serum of approximately 75% of patients with primary liver cancer [52]. AFP is known to be present in small quantities in normal adults [49], some fetal organs, proliferating hepatocytes, and several cancers, including HCC, yolk sac tumors, and germ cell tumors [48]. One of the main arguments against the use of AFP as a surveillance biomarker is its presence in multiple conditions, summarized in Table 6.

Table 6. Causes of Elevated AFP in non-pregnant adults [188]

Liver related conditions	Non-liver related conditions
Hepatocellular carcinoma	Gastric cancer
Acute hepatitis	Esophageal cancer [189]
Chronic hepatitis	Pancreatic cancer [190]
Fatty liver disease [191]	Colon and rectal cancer [189, 192]
Acute liver failure [193]	Germ cell tumors
Cirrhosis [194, 195]	Colitis
Intrahepatic cholangiocarcinoma [196]	Ataxia telangiectasia

AFP is the only HCC biomarker that has been studied in all 5 EDRN phases. AFP is not FDA-cleared as a biomarker for HCC risk when used alone; it is only cleared if used in combination with AFPL3%. Its usefulness may range from identifying the population at risk to predicting response to treatment [36]. AFP's association with the long term risk of HCC was evaluated in a retrospective longitudinal cohort study in which 617 patients with CHB were followed for 22 years (median 6.2 years) [53]. The cumulative incidence of HCC after 18 years of follow up was 40.8% in the higher AFP group vs 17.9% in the lower AFP group. The authors concluded that elevated serum AFP was associated with increased risk of HCC and that this increased risk can persist for many years after the index AFP measurement. The risk was independent of other demographic factors.

Because of sensitivity and specificity issues, trials assessing AFP's association with HCC have used a variety of cutoffs for the AFP level. In a large phase II trial of HCC biomarkers in which chronic viral hepatitis was the main cause of liver disease, an AFP cutoff of 10.9 ng/ml had 66% sensitivity and 81% specificity [54]. A systematic review of AFP studies showed that a cutoff value of 20 mcg/L resulted in sensitivity of 41-65% and specificity of 80-94% while with a cutoff value of 200 mcg/L the sensitivity decreased to 20-45% and the specificity increased to 99-100% [55]. Researchers have reported that combining AFP with US for HCC surveillance improves sensitivity and specificity [33, 34].

To overcome the low sensitivity issue, El-Serag et al. proposed an AFP-based algorithm to predict HCC risk [56]. They found that a predictive model that included data on levels of AFP, ALT, and platelets, along with age at time of AFP test, and interaction terms between AFP and ALT and between AFP and platelets best discriminated between HCV-infected cirrhotic patients who would and would not develop HCC within six months. The authors envisioned that application of this modified AFP model would decrease both false positives and false negatives. In a study of AFP in cirrhotic patients Oka et al. found that the cumulative incidence of HCC was significantly higher in patients who had AFP levels of 20 ng/ml or more but below 200 ng/ml at baseline compared to those with levels below 20 ng/ml, as well as in patients who had repeated transient increases in AFP to above 100 ng/ml compared to those who had levels consistently below 20 ng/ml [57]. The researchers concluded that such patients should be treated as being in a super-high-risk group for HCC and should be provided frequent US examination.

AFP has been shown to increase hepatoma cell proliferation in vitro [58] and cytoplasmic AFP has been shown to function on retinoic acid receptor signaling and thus promote the human hepatoma [59]. The suppression of AFP mRNA expression may inhibit the tumor cell growth [48]. In a preliminary study, epitope-optimized AFP activated CD8 T cells and generated strong antitumor effects in the carcinogen-induced autochthonous HCC mouse model [60], leading researchers to hypothesize that the same type of epitope optimization might enable the development of effective human vaccines to prevent HCC recurrence after liver resection [61]. In a phase I human AFP vaccine study in which 6 patients with advanced HCC were given vaccine twice weekly for three weeks all patients generated T cell responses [62]. However, the T cell responses did not result in any anti-tumor activity in a later study [63].

AFP for Surveillance of CHB Patients

The annual HCC risk for CHB patients with cirrhosis is estimated to be 3-5% [64]. The annual HCC risk in patients without cirrhosis has been estimated to be 0.4–0.6% in Asian males > 40 years of age and 0.3–0.6% in Asian females > 50 years of age [64]. In the United States, the annual HCC incidence in CHB patients has been reported to be 387 per 100,000 for men and 63 per 100,000 for women in an Alaska native cohort [65] and 470 per 100,000 overall and 657 per 100,000 in men only in a heterogeneous urban North American population [65]. The AASLD IDSA HCV Guidance Panel has concluded that surveillance with dual modality AFP and US becomes cost effective when the incidence of HCC exceeds 0.2% [66]. Carr et al. [67] found that compared to HCV-associated HCC, HBV-associated HCC has larger tumor size and higher AFP level [67]. HCC surveillance using AFP has been assessed in several CHB populations. The first surveillance study in a CHB population was conducted in China using combined AFP and US every six months [29]. In the surveillance arm, there was a 37% reduction in HCC-associated mortality. In a prospective 16-year study of 1,487 Alaska natives with CHB, surveillance was done with AFP every 6 months; men and non-pregnant women with an elevated AFP level were evaluated with US. In a comparison of the long-term survival rate for patients whose HCC was detected by the surveillance program to a historical control group of Alaska native HCC patients, screening was shown to be effective in detecting most HCC tumors at a resectable stage and to significantly prolong survival rates [68]. In a recent review of 132 patients diagnosed with HCC during regular surveillance, the primary mode of tumor detection was US only in 51.5%, US and AFP in 22.0%, AFP only in 19.7%, and incidental in 6.8% of patients. It was found that AFP increased sensitivity by 19.7% for all patients and 28.0% for HBV-related early stage HCC patients, suggesting that AFP may play a significant role in increasing sensitivity in HCC surveillance, especially for detecting early stage HBV-related HCC [69]. Using a computerized decision-analytic model to compare various surveillance strategies Thompson et al. concluded that in a mixed-etiology cohort, the most effective surveillance strategy is to survey with AFP and US every 6 months, but noted that cost effectiveness varied considerably depending on the cirrhosis etiology, with surveillance much more likely to be cost-effective in those with HBV-related cirrhosis [70]. El Serag et al. recommend HCC surveillance every six months with the combination of AFP and US in all patients with cirrhosis or advanced hepatic fibrosis irrespective of etiology, and in adult CHB patients irrespective of cirrhosis [64].

AFP for Surveillance of Patients with Chronic Hepatitis C

HCV infection is associated with a 15-20 fold higher risk of HCC compared to non-HCV patients, with patients with advanced fibrosis or cirrhosis at the highest risk [3]. The rate of HCC among patients with chronic hepatitis C (CHC) ranges from 1% to 3% over 30 years [3]. Any level of HCV viremia is a strong risk factor for HCC while treatment that results in a sustained viral response (SVR) substantially decreases the risk; studies have reported a 57% to 75% reduction in risk in patients who achieved an SVR with interferon-based therapy [3]. However, because some level of risk persists for years after treatment, patients should have continued surveillance after achieving an SVR. Post-interferon treatment, a high AFP level has been associated with a higher incidence of HCC [71].

In a comparison of biomarker levels in patients in the HALT-C trial in whom HCC did or did not develop, it was found that mild to moderate elevations in total AFP occur frequently in patients with CHC and advanced fibrosis, appear to relate to factors other than HCC, and are poor predictors of HCC [72]. In patients without HCC, 24.5% had at least one AFP between 20 and 199, and 2.3% had at least one AFP value ≥ 200 ng/ml. The authors reported that if the higher cutoff of 200 ng/ml is used then the specificity of AFP increases to 99% but the sensitivity falls to less than 20%. In a retrospective case-control study that was designed to assess whether AFP levels might accurately detect HCC in subgroups of patients, it was found that AFP most accurately detects HCC in patients without HCV infection [73]. AFP levels of 59 ng/mL or greater most accurately detected HCC in patients with HCV-associated cirrhosis; levels of AFP of 11 ng/mL or greater accurately identified HCC in HCV-negative patients. Chang et al. reported that the addition of AFP (cutoff value of 20 ng/ml) to US for surveillance increased sensitivity from 92 (US alone) to 99.2 (US combined with AFP) without significantly affecting the specificity of surveillance (68.3% vs 71.5%) [74]. Other studies have also shown that AFP is complementary to US in surveillance [75, 76]. In a retrospective study of patients with biopsy-proven cirrhosis who were evaluated for high hepatocyte proliferation (S phase fraction), 74% of subjects had hepatitis C; 39% of patients who had high AFP (> 20ng/ml) and high S phase fraction (≥ 1.8) developed HCC compared to 1% of those with normal AFP and low S phase fraction [77], emphasizing the usefulness of AFP as a risk marker for future HCC.

AFP as Prognostic Indicator for Outcomes after the Diagnosis of HCC

Pre-operative serum AFP level can be predictive of the malignant features and prognosis of HCC. In a retrospective study, patients who had liver resection for HCC were classified based on AFP level into 3 categories: AFP level of ≤ 20, 20-400 and > 400 ng/ml [78]. It was found that pre-operative AFP and tumor size were closely related to HCC post-operative survival. Furthermore, hepatectomy was less effective in patients with AFP value of > 400. The authors concluded that HCC patients with serum AFP higher than 20 ng/mL need comprehensive therapy besides surgical resection and require close follow up. Similar results for HBV-associated HCC have been reported by other investigators [79-82].

A higher AFP level has been reported to be predictive of survival. In a retrospective analysis of HCC patients with baseline AFP > 200 ng/ml that compared patients with AFP value of > 10,000 and < 10,000, it was found that AFP > 10,000 was associated with significantly worse one-month and six-month overall survival and a higher chance of distant metastatic disease [83]. In a study of Egyptian patients with HCV-associated HCC who underwent hepatic resection, elevated AFP was one of the significant variables predicting tumor recurrence [84]. Serum AFP level at the time of diagnosis with HCV-related HCC has also been shown to be an independent predictor of mortality [85]. In a systematic review of 72 studies, Tandon et al. reported that in HCC patients with cirrhosis the predictors of mortality were AFP, portal vein thrombosis, tumor size, and Child-Pugh class [86].

In patients with HCV-related HCC, it has been reported that the two factors associated with post-resection survival are vascular invasion and AFP level [87]. An AFP value of ≥ 1000 ng/ml and the presence of vascular invasion were independent unfavorable prognostic factors affecting overall survival; AFP of ≥ 1000 ng/ml was an independently significant predictor of poor disease-free survival. It has also been reported that pre-operative AFP of > 100 ng/ml, as well as multifocal lesions and history of tumor rupture, were independent risk factors for recurrence of HCC after hepatectomy [88]. Liu et al. reported that AFP > 400 ng/ml was associated with low overall survival post-hepatectomy in a cohort of Chinese HCC patients, most of whom had CHB (89.7% CHB, 2.4% CHC, and 4.2% both CHB and CHC) [89]. In their review, Singhal et al. noted that AFP > 400 ng/ml has been associated with high tumor burden, bilobar disease, portal vein thrombosis and lower median survival [90]. The aggressive nature of AFP producing tumor may be partially attributed to ephrin-A1, an angiogenic factor [91]. Ephrin A-1 induces the expression of AFP by HCC cells and also over expresses matrix metalloproteinase-2 (MMP-2). MMP-2 overexpression by ephrin-A1 may be involved in promoting HCC invasion and metastasis in AFP-producing tumors.

Studies have also evaluated pre-transplant AFP level as a predictor of post-transplant survival. In one study, an AFP value of > 15 was independently associated with low post-transplant survival: the higher the pre-transplant AFP value, the lower the post-transplant survival independent of Milan criteria [92]. In addition, changes in serum AFP level while on the waiting list corresponded closely to changes in post-transplant mortality. In an assessment of 6478 adult transplant recipients registered in the Scientific Registry of Transplant Recipients (SRTR) it was found that only total tumor volume (TTV) and AFP > 400 predicted post-transplant patient survival, leading the authors to recommend that eligibility criteria be extended beyond the Milan criteria, with inclusion of HCC with TTV of 115 cm3 if AFP is < 400 ng/ml [93]. An analysis of 2253 post-transplant patients from the United Network for Organ Sharing (UNOS) registry divided them into low (53.7%), medium (35.7%), and high (10.6%) pre-transplant AFP groups. The low AFP group demonstrated the best 4-year survival (76%) compared with the medium (65%) and high (57%) AFP groups [94].

The utility of AFP for assessing response to treatment has also been assessed. In a study of 51 single tumor HCC patients (61% related to chronic viral hepatitis) with baseline AFP > 200 ng/ml who had received transarterial locoregional therapy the response to treatment was defined as reduction in AFP of > 50% [95]. AFP responders to locoregional therapy were found to have a high likelihood of showing EASL response (> 50 reduction in viable tumor size). Survival correlated better with AFP response than EASL response. In a prospective study, Wang et al. reported that AFP status one month after radiofrequency ablation was a strong predictor of short term HCC recurrence [96].

The AFP response to sorafenib treatment and its correlation with survival is controversial. Personeni et al. suggested that AFP response (> 20% reduction in AFP value during 8 weeks treatment) can help identify a subgroup of advanced HCC patients who can achieve longer survival while on sorafenib therapy [97]. Llovet et al. reported that although baseline AFP predicted survival in the placebo cohort of the Sorafenib HCC Assessment Randomized Protocol (SHARP) trial, change in AFP value neither predicted survival nor time to progression in the sorafenib treatment cohort [98].

AFP has been incorporated in multiple HCC staging systems, summarized in Table 7. A comparison study of the current staging systems for advanced HCC that used survival analysis and relative operating characteristic (ROC) to access the prognostic value of each scoring system found that the Advanced Liver Cancer Prognostic System (ALCPS) performed best, with the largest area under the ROC curve in predicting 3-month overall survival (sensitivity 76.32%, specificity 78.72%) [99]. Cancer of the Liver Italian Program (CLIP) and Chinese University Prognostic Index (CUPI) were similar to ALCPS in prognostic discrimination but had relatively lower power. In 2006, Japanese investigators proposed the BALAD (bilirubin, albumin, Lens culinaris agglutinin-reactive alpha-fetoprotein/AFP-L3%, AFP and DCP) score [100]. In a recent study of patients with HBV-related HCC, it was found that the BALAD score could stratify the cohort into different patient groups with distinct median overall survival, and further stratify outcomes in each Barcelona Clinic Liver Cancer (BCLC) subgroup [101].

Table 7. Different staging systems utilizing AFP

Cancer of the Liver Italian Program (CLIP) variables [197]	**Chinese University Prognostic Index (CUPI) variables** [198]	**French score variables** [198]
Portal vein thrombosis	TNM ***stage	Portal vein thrombosis
Child-Turcotte Pugh	Asymptomatic disease on presentation	
	Ascites	
Tumor morphology	Total bilirubin	Total bilirubin
	Alkaline phosphatase	Alkaline phosphatase
AFP (ng/ml)	AFP	AFP

AFP for Allocation of Liver Organ for Liver Transplantation

When the Model for End-Stage Liver Disease (MELD) scoring system was first introduced in 2002 to prioritize candidates for liver transplantation, MELD exception points were given to patients with HCC by Organ Procurement and Transplantation Network (OPTN) imaging criteria and patients with high AFP (> 500) without any radiological evidence of HCC. A study in 2006 using the UNOS registry revealed that the majority of patients who received extra MELD points based on an elevated AFP did not have HCC, leading the researchers to conclude that AFP correlates poorly with the presence of HCC in patients awaiting liver transplantation [102]. Since then AFP has not been used for determining organ allocation.

UNOS recently approved a proposal to delay granting MELD exception points to HCC patients for 6 months irrespective of their MELD score or AFP level, based on multiple studies showing the advantage gained by HCC patients when compared to non-HCC patients;

specifically, candidates will now be registered at their calculated MELD/PELD scores for the first three months (initial application) as well as for the first three-month extension, as long as the candidate continues to meet the policy criteria; at six months (the second extension), candidates will receive a score of 28 [103]. Currently, in the OPTN allocation system, patients with AFP over 500 ng/mL are not to receive organs until the AFP is under 500. HCC exception points allocation models used in selected organ allocation systems are summarized in Table 8.

Table 8. HCC exception points models used in selected organ allocation systems [104]

Organization	MELD exception points at listing	Points allocation based on AFP level
OPTN*/UNOS	22	No
Eurotransplant	22	No
Human Organ Procurement and Exchange Program	22	Yes
Organització Catalana de Trasplantaments	19	Yes

* Organ Procurement and Transplantation Network.

AFP values or the log of AFP values have been used in calculating scores in newly proposed allocation models such as HCC-MELD, deMELD, MELD-HCC, and new deMELD [104]. A recently published study used the log of AFP values along with the MELD score and the number and size of tumors to develop the MELD equivalent score (MELDEQ) and applied it to UNOS data on adult patients who were added to the wait list between January 22, 2005 and September 30, 2009 [105]. The authors concluded that HCC patients with a combination of a low biochemical MELD score and a low AFP level (MELDEQ $\leq$ 15) would receive a substantial advantage in contrast to patients with chemical MELD scores in a similar range and that a delay of 6 months for listing might be appropriate; however, the survival of patients with MELDEQ scores > 15 would probably be adversely affected by a universal 6-month delay in listing.

Table 9. Clinical utility and evidence based utilization of currently available HCC biomarkers

Risk prediction	Screening and diagnosis	Prognostic indicator	Risk of PVT	HCC recurrence	Response to treatment
AFP	AFP	AFP	AFP-L3	AFP	AFP
AFP-L3	AFP-L3	AFP-L3	DCP	AFP-L3	AFP-L3
DCP	DCP	DCP	High Met expression	DCP	FGF
SNP	OPN	OPN		Five gene score	Five gene score
	GPC3	GPC3		OPN	High Met expression
	FGF	FGF			
	GP73	High Met expression			
	CSA				

AFP-L3%

AFP shows differential affinity for the lectin *Lens culinaris* agglutinin (LCA) [106]. Based on this affinity, AFP electrophoresis results in three distinct glycoforms: AFP-L1, AFP-L2, and AFP-L3. AFP-L1 does not have affinity for LCA and is present in chronic hepatitis and liver cirrhosis; AFP-L2 has moderate affinity for LCA and is predominantly produced by yolk sac tumors; AFP-L3 is more prevalent in HCC patients. Results for AFP-L3 are shown as the ratio of LCA-reactive AFP to total AFP (AFP-L3%). In a cohort of chronic hepatitis patients, AFP-L3% was found to be more sensitive than AFP for HCC surveillance [107]. In the subgroup with AFP < 20, AFP-L3% levels were significantly higher in HCC patients than non-HCC patients; the sensitivity for HCC detection was more than 70%. The sensitivity increased further when the combination of AFP-L3% and des-gamma-carboxy prothrombin (DCP) was used. A systemic review of 12 studies suggested that AFP-L3% can be complementary to AFP in diagnosing HCC [108]. In particular, AFP-L3% may add to the clinical utility of AFP if the AFP level is indeterminate. In a retrospective study that assessed blood samples from subjects with benign and malignant liver disease, using a 35% cutoff for AFP-L3% in a subgroup of patients (predominantly CHC patients) whose AFP level was considered indeterminate (10-200 ng/ml) increased the diagnostic utility and helped diagnose an additional 10% of patients not diagnosed using AFP [109]. AFP-L3% increases early in HCC and has been reported to diagnose HCC earlier than imaging based diagnosis [110]. Sato et al. [111] and Kumada et al. [112] reported similar findings in their prospective studies of cirrhotic patients (predominantly caused by HCV and HBV infection).

AFP-L3% is FDA-cleared as a biomarker for HCC risk when used in combination with AFP. AFP-L3% elevation can be reflective of either advanced HCC or more aggressive tumor. In a multicenter study of newly diagnosed HCC patients, 90% of whom had chronic viral hepatitis, AFP-L3% of 15% or more was found to be indicative of advanced HCC even with small tumor size or low AFP concentration [113]. In this study, elevated AFP-L3% was associated with malignant portal vein thrombosis and overall poor prognosis. Elevated AFP-L3% can be reflective of poor HCC differentiation along with other prognostic indicators of poor prognosis, including malignant portal vein thrombosis [114-116].

With solid tumors there is always a concern that there may be tumor recurrence at the primary target organ or a distant metastatic lesion. AFP-L3% may identify patients with a higher risk of early recurrence. In a single center study, 416 newly diagnosed HCC patients were identified, more than 90% of whom had underlying HCV or HBV infection [117]. All the patients received ablative therapy with curative intent. A positive AFP-L3% value post-ablation was the strongest predictor of recurrence, with all patients with AFP-L3% $\geq$ 15% experiencing recurrence within 18 months.

Changes in AFP-L3% may reflect treatment response. In a study of HCC patients, AFP-L3% level was more reflective of clinical response than AFP [118]. Tamura et al. showed that AFP-L3% status 1 month after treatment with curative intent was an independent predictor of HCC recurrence [119]. The status of HCC biomarkers may help decide the preferred treatment option. In a retrospective study of patients with single nodular < 5cc HCC without any vascular invasion the positive status of HCC biomarkers significantly predicted HCC recurrence in the radiofrequency ablation group [120]. The authors suggested that positive status of three markers may be reflective of micro-invasiveness and, hence, surgical resection might be a preferred treatment modality as compared to radiofrequency ablation.

DCP

Des-gamma carboxyprothrombin (DCP), also known as protein induced by vitamin K absence/antagonist-II (PIVKA-II) [121], is a prothrombin precursor produced in HCC [122]. The exact mechanism of DCP production in HCC has not been fully clarified [123].

First reported as a tumor biomarker in 1984, DCP was detected in the sera of 69 out of 76 patients with biopsy-proven HCC [124]. In three patients there was a decrease in DCP level with treatment and re-elevation in the level was associated with disease recurrence. Marrero et al. [125] and Durazo et al. [126] have reported that DCP was better for differentiating HCC from nonmalignant chronic liver disease than AFP. However, whether DCP is superior to AFP for screening and surveillance of HCC is controversial [64].

Because the mechanisms of DCP and AFP production by HCC are mutually independent [122] the combination of these biomarkers can be complementary [127] and will help increase the sensitivity of screening for early phase HCC. In a study that evaluated the sensitivity and specificity of DCP, alone or in combination with AFP, for the detection of HCC in CHB patients, the combination had optimal sensitivity and specificity for detecting early stage HCC [128]. Similarly, DCP can increase HCC detection in combination with AFP-L3% [129]. In an HCC surveillance study of patients, 70% of whom had chronic viral hepatitis, the combination of DCP and AFP-L3% had optimal sensitivity and specificity for detecting early HCC independent of AFP level [130]. High DCP levels are predictive of tumor aggressiveness [131] and microvascular [129] or portal vein invasion [132]. A retrospective cohort study of HCC patients found that pre-transplant AFP, AFP-L3%, and DCP all predicted HCC recurrence after transplantation and showed that combining biomarkers with Milan criteria may be better than using Milan criteria alone for optimizing liver transplantation eligibility [133]. DCP is FDA-cleared as a biomarker for HCC risk.

Osteopontin

Osteopontin (OPN) is an acidic glycoprotein which has varied physiological functions. Its link to carcinogenesis was described by Senger et al. [134]. Increase in the serum OPN level has been reported in multiple myeloma and colon, pancreatic, breast and ovarian cancer [135-137]. OPN over-production has been reported in HCC [138] and has been associated with poor prognosis [139]. In a study that included HCC patients, patients with chronic liver diseases, and healthy controls Kim et al. reported higher OPN levels in HCC patients compared to patients in the other two groups [140]. In the HCC patients, the OPN level increased significantly with advancing degree of tumor stage and Child-Pugh class. Diagnostic sensitivity and specificity of OPN for HCC were 87% and 82%, respectively. OPN had a greater area under curve value (0.898) than either AFP (0.745) or DPC (0.578), which the researchers concluded suggested the superior diagnostic accuracy of OPN. In a study that measured OPN and AFP levels in patients with HCV-related liver cirrhosis (with and without HCC) and in healthy controls, OPN levels were significantly higher in cirrhotic patients with HCC [141]. In these patients, OPN levels increased significantly with late tumor stage, advanced Child-Pugh class, larger tumor size (≥ 5 cm) and high tumor grade. The sensitivity and specificity of OPN for HCC were 88.3% and 85.6%, respectively, and its diagnostic accuracy was superior to AFP. In a study that assessed postoperative changes in OPN level it was found that OPN levels significantly decreased after curative resection of HCC and that postoperative OPN might serve as a surrogate biomarker for monitoring treatment response and tumor recurrence after resection [142]. OPN has been shown to play an important role in HCC metastasis and tumor growth of HCC [143] and in disease progression [144] and may be a potential therapeutic target for combating these. Osteopontin is not FDA-qualified as a biomarker for HCC risk and is not currently available.

Glypican

Glypican-3 (GPC3) is a member of the glypican family of heparin sulfate proteoglycans that are anchored to the cell surface by glycosyl-phosphatidylinositol [145]. GPC3 is expressed in up to 72% of HCC but is not detectable in sera of healthy donors and patients with hepatitis [146, 147]. High GPC3 expression has been reported in HCV-related HCC [148, 149] and in a study of HBV- and HCV-related HCC patients, GPC 3 was found to be a sensitive and specific marker for diagnosis of early stage HCC [150]. In a similar study of HCC (predominantly due to CHB and CHC), the GPC3 level was significantly elevated in both early and advanced HCC [151]. GPC3 appears to be complementary to AFP for diagnosis of HCC [151, 152]. Positive GPC3 status in HCC is associated with an overall poor prognosis [149, 153] and with early HCC recurrence after resection [153, 154]. Glypican is not FDA-qualified as a biomarker for HCC risk and is not currently available.

Golgi Protein 73

Golgi protein 73 (GP73) is a Golgi protein of unknown function which has been found to be upregulated in viral hepatitis [155]. In normal livers, GP73 is constitutively expressed at low levels by biliary epithelial cells but not by hepatocytes. In advanced liver disease due to both viral causes (HBV, HCV) and non-viral causes (alcohol-induced liver disease, autoimmune hepatitis) hepatocyte expression of GP73 is dramatically up-regulated [156]. In advanced liver disease, the GP73 level is significantly higher in HCC patients compared to non-HCC patients and has significantly higher sensitivity than AFP for diagnosing early HCC [157-159]. It also has better sensitivity and specificity than AFP for diagnosing HBV-related HCC [160, 161]. The combination of AFP and GP73 has been reported to be more sensitive than AFP alone for HCC diagnosis [152]. The conclusion of a systematic review of studies assessing the accuracy of GP73 and AFP in the diagnosis of HCC was that GP73 has accuracy comparable to AFP for HCC diagnosis, but that determining the value of the combination of GP73 and AFP for HCC detection will require further investigation [162]. To date, data is limited and does not satisfy the criteria for a biomarker put forward by EDRN. GP73 is not FDA-qualified as a biomarker for HCC risk and is not currently available.

Fibroblast Growth Factor (FGF) Amplification

FGF3 and FGF4 are receptors for fibroblast growth factor and promote HCC [36]. FGF2 has also been reported to be highly expressed in HCC [163]. In a study of HCC patients undergoing resection a high pre-operative FGF level was found to be predictive of invasive tumor and early postoperative recurrence [164]. FGFR4 contributes significantly to HCC progression [165]. An *in vitro* study that investigated the expression of human FGF19 and its receptor, FGFR4, in HCC specimens found that FGF19 was significantly overexpressed in HCC and that FGF19 recombinant protein could increase the proliferation and invasion capabilities of HCC cell lines and inhibit their apoptosis while decreasing FGF19 and FGFR4 expression by siRNA significantly inhibited proliferation and increased apoptosis [166]. The researchers concluded that FGF19 inhibition might be a therapeutic strategy for HCC. High FGF3/4 amplification has been shown to be an indicator of poor prognosis and there is limited evidence that its expression may predict response to sorafenib in some patients; however, the use of high levels of FGF3/4 expression as a general purpose marker is problematic because the FGF3/4 mutation has only been observed in a very small percentage of HCC patients [36, 167, 168]. Blockade of FGFR4-mediated signaling might be a potential therapeutic target [169] but because FGFR4 plays an important physiological role in the regulation of hepatic bile acid synthesis any possible use of therapeutic FGF19-FGFR4 inhibition must be considered in conjunction with the potential safety implications of FGF19-FGFR4 blockade [170]. Due to limited data, it is difficult to ascertain whether there are any differential changes in FGF3/4 amplification in viral hepatitis-related HCC compared to other etiologies. FGFs are not FDA-qualified as biomarkers for HCC risk and are not currently available.

Canavaninosuccinate

Metabolomic profiling using gas chromatography has been shown to differentiate individuals with HCC from non-HCC subjects using a panel of metabolite biomarkers which could be particularly helpful in subjects with AFP < 20 ng/ml [171]. Canavaninosuccinate (CSA) is an organic acid metabolite produced in liver [36] that has been identified as a metabolite biomarker which could differentiate HCC from cirrhotic and healthy individuals [172]. The combination of AFP and CSA has been shown to have high sensitivity and specificity (92% and 100%, respectively) for discriminating HCC from cirrhosis [36]. CSA is not FDA-qualified as a biomarker for HCC risk and is not currently available.

Five Gene Score

A score using 5 genes (TAF9, RAMP3, HN1, KRT19 and RAN) was developed to predict disease-free survival after curative resection [36]. These genes are involved in dysregulated pathways of HCC development. The score was developed and then validated in HCV-related HCC in a Western population and HBV-related HCC in an Asian population [173]. The score has been shown to accurately predict disease free survival and early recurrence and with its highly predictive accuracy it might help guide therapy [174]. It could stratify the type of therapy based on the score. For example, HCC patients with good genetic prognostic parameters might benefit from resection while in patients with bad prognostic parameters resection would not improve the outcome [174]. Patients with a score of ≥ zero have worse survival compared to HCC patients with a score < zero [36]. The 5-gene score is not FDA-qualified as a biomarker for HCC risk and is not currently available.

High Met Expression

It has been shown that high expression of c-Met, a high affinity receptor for hepatocyte growth factor, is associated with poor prognosis and HCC recurrence [175, 176] and with portal vein invasion [177]. Inhibition of c-Met may have potential as a therapy in HCC patients with strong c-Met expression [176]. In a phase 2 trial, treatment with tivantinib, a selective oral inhibitor of MET, resulted in a median time to progression for patients with MET-high tumors of 2.7 months compared to 1.4 months for patients on placebo [178]. If future trials confirm the results, it might be useful for this high c-MET subgroup of patients. It might also complement primary target therapy based on c-Met expression [179]. However, c-MET is not FDA-qualified as a biomarker for HCC risk and is not currently available.

Conclusion

HCC diagnosis after symptom onset carries a grim prognosis with five-year survival of < 20%. Thus, in a population with risk factors for HCC, our goal should be to diagnose it in an early stage and manage the patient with curative intent. The currently available screening and surveillance modalities, including US, are underutilized in the clinical setting. Current AASLD guidelines only recommend US every 6 months for surveillance of high-risk patients. The biomarkers that have been studied to date are only risk biomarkers and of those discussed here, only AFP, AFP-L3%, and DCP are FDA-qualified as biomarkers for HCC risk. Due to low sensitivity and specificity as screening tests, they should not be used as sole screening or surveillance tests and are not used to diagnose HCC. HCC is diagnosed by imaging with CT scan or MRI with contrast using Liver Imaging Reporting and Data System (LI-RADS) criteria and, rarely, by liver biopsy. One or more of the biomarkers may, however, complement ultrasound (US) for screening and surveillance, potentially triggering a CT or MRI with contrast. Elevated or increasing biomarkers should trigger a CT or MRI.

AFP has been shown to be a strong prognostic indicator and to complement US for surveillance. A higher AFP value has clearly been associated with poor clinical outcome while AFP reduction > 50% after treatment predicts an improved response. The pre-treatment AFP value may identify patients who will benefit more from loco-regional therapy than transplant. A higher pre-transplant AFP level has been associated with tumor recurrence and thus may help identify patients who will need close observation for early HCC recurrence. AFP has been used in multiple HCC staging systems, including ALCPS, CLIP, CUPI and BALAD. The AFP level (with a ceiling of < 500 ng/ml) has been utilized in some liver allocation models to grant MELD exception points. AFP-L3% has been found to complement

AFP in HCC surveillance, especially when the AFP value is indeterminate. As with AFP, a higher AFP-L3% level may be reflective of either advanced disease or aggressive HCC. Elevated levels of both AFP-L3% and DCP have been associated with vascular invasion by the tumor. A high post-ablation AFP-L3% value may be reflective of disease recurrence. Similar to AFP, AFP-L3% may help determine treatment modality. DCP can help differentiate HCC from non-malignant chronic liver disease. Its re-emergence after treatment can be reflective of HCC recurrence. DCP can complement AFP in HCC risk assessment. Data for other biomarkers are limited, and no firm recommendation can currently be made for their use.

Several risk assessment models have been proposed to help with some of the limitations of these biomarkers. Multiple large studies are underway on the clinical utility of HCC biomarkers and on the proposed models that will ultimately help answer remaining questions on their usefulness. In the meantime, guidelines from the Japanese Society of Hepatology recommend AFP, AFP-L3% and DCP for HCC surveillance. Experts in the field agree that until better screening and surveillance tools are available for HCC, we should utilize the available biomarkers to complement US and help identify patients who would benefit from further radiological workup.

REFERENCES

[1] Mittal S, El-Serag HB. (2013) Epidemiology of hepatocellular carcinoma: consider the population. *J Clin Gastroenterol*, 47 Suppl, S2-6.

[2] El-Serag HB. (2011) Hepatocellular carcinoma. *N Engl J Med*, 365, 1118-1127.

[3] El-Serag HB. (2012) Epidemiology of viral hepatitis and hepatocellular carcinoma. *Gastroenterology*, 142, 1264-1273 e1261.

[4] Bruix J, Sherman M. (2010) Management of Hepatocellular Carcinoma: An Update. *Hepatology*, 000, 1-35.

[5] Umemura T, Ichijo T, Yoshizawa K, Tanaka E, Kiyosawa K. (2009) Epidemiology of hepatocellular carcinoma in Japan. *J Gastroenterol*, 44 Suppl 19, 102-107.

[6] Beste LA, Leipertz SL, Green PK, Dominitz JA, Ross D, Ioannou GN. (2015) Trends in burden of cirrhosis and hepatocellular carcinoma by underlying liver disease in US veterans, 2001-2013. *Gastroenterology*, 149, 1471-1482 e1475.

[7] American Cancer Society. Cancer Facts and Figures 2015. Atlanta, Georgia: American Cancer Society; 2015.

[8] Fattovich G, Stroffolini T, Zagni I, Donato F. (2004) Hepatocellular carcinoma in cirrhosis: incidence and risk factors. *Gastroenterology*, 127, S35-50.

[9] Simonetti RG, Camma C, Fiorello F, Politi F, D'Amico G, Pagliaro L. (1991) Hepatocellular carcinoma. A worldwide problem and the major risk factors. *Dig Dis Sci*, 36, 962-972.

[10] European Association for the Study of the Liver, European Organisation for Research and Treatment of Cancer. (2012) EASL-EORTC clinical practice guidelines: management of hepatocellular carcinoma. *J Hepatol*, 56, 908-943.

[11] Chang MH, Chen CJ, Lai MS, Hsu HM, Wu TC, Kong MS, et al. (1997) Universal hepatitis B vaccination in Taiwan and the incidence of hepatocellular carcinoma in children. Taiwan Childhood Hepatoma Study Group. *N Engl J Med*, 336, 1855-1859.

[12] Chang MH, Chen TH, Hsu HM, Wu TC, Kong MS, Liang DC, et al. (2005) Prevention of hepatocellular carcinoma by universal vaccination against hepatitis B virus: the effect and problems. *Clin Cancer Res*, 11, 7953-7957.

[13] Chang MH, You SL, Chen CJ, Liu CJ, Lee CM, Lin SM, et al. (2009) Decreased incidence of hepatocellular carcinoma in hepatitis B vaccinees: a 20-year follow-up study. *J Natl Cancer Inst*, 101, 1348-1355.

[14] Kao JH. (2015) Hepatitis B vaccination and prevention of hepatocellular carcinoma. *Best Pract Res Clin Gastroenterol*, Published online ahead of print.

[15] Colombo M, Iavarone M. (2014) Role of antiviral treatment for HCC prevention. *Best Pract Res Clin Gastroenterol*, 28, 771-781.

[16] Papatheodoridis GV, Manolakopoulos S, Touloumi G, Vourli G, Raptopoulou-Gigi M, Vafiadis-Zoumbouli I, et al. (2011) Virological suppression does not prevent the development of hepatocellular carcinoma in HBeAg-negative chronic hepatitis B patients with cirrhosis receiving oral antiviral(s) starting with lamivudine monotherapy: results of the nationwide HEPNET. Greece cohort study. *Gut*, 60, 1109-1116.

[17] Hosaka T, Suzuki F, Kobayashi M, Seko Y, Kawamura Y, Sezaki H, et al. (2013) Long-term entecavir treatment reduces hepatocellular carcinoma incidence in patients with hepatitis B virus infection. *Hepatology*, 58, 98-107.

[18] Wong GL, Chan HL, Mak CW, Lee SK, Ip ZM, Lam AT, et al. (2013) Entecavir treatment reduces hepatic events and deaths in chronic hepatitis B patients with liver cirrhosis. *Hepatology*, 58, 1537-1547.

[19] Trinchet JC, Bourcier V, Chaffaut C, Ait Ahmed M, Allam S, Marcellin P, et al. (2015) Complications and competing risks of death in compensated viral cirrhosis (ANRS CO12 CirVir prospective cohort). *Hepatology*, 62, 737-750.

[20] Lampertico P, Soffredini R, Vigano M, Minola E, Cologni G, Rizzi M, et al. (2013) 5-year entecavir treatment in NUC-naive, field-practice patients with chronic hepatitis B showed excellent viral suppression and safety profile but no prevention of HCC in cirrhotics. *J Hepatol*, 58, s306.

[21] Sinn DH, Lee J, Goo J, Kim K, Gwak GY, Paik YH, et al. (2015) Hepatocellular carcinoma risk in chronic hepatitis B virus-infected compensated cirrhosis patients with low viral load. *Hepatology*, 62, 694-701.

[22] Singal AG, Volk ML, Jensen D, Di Bisceglie AM, Schoenfeld PS. (2010) A sustained viral response is associated with reduced liver-related morbidity and mortality in patients with hepatitis C virus. *Clin Gastroenterol Hepatol*, 8, 280-288, 288 e281.

[23] Morgan RL, Baack B, Smith BD, Yartel A, Pitasi M, Falck-Ytter Y. (2013) Eradication of hepatitis C virus infection and the development of hepatocellular carcinoma: a meta-analysis of observational studies. *Ann Intern Med*, 158, 329-337.

[24] AASLD IDSA HCV Guidance Panel. (2015) Hepatitis C guidance: AASLD-IDSA recommendations for testing, managing, and treating adults infected with hepatitis C virus. *Hepatology*, 62, 932-954.

[25] Harris RJ, Thomas B, Griffiths J, Costella A, Chapman R, Ramsay M, et al. (2014) Increased uptake and new therapies are needed to avert rising hepatitis C-related end

stage liver disease in England: modelling the predicted impact of treatment under different scenarios. *J Hepatol*, 61, 530-537.

[26] Aleman S, Rahbin N, Weiland O, Davidsdottir L, Hedenstierna M, Rose N, et al. (2013) A risk for hepatocellular carcinoma persists long-term after sustained virologic response in patients with hepatitis C-associated liver cirrhosis. *Clin Infect Dis*, 57, 230-236.

[27] Prorok PC. (1992) Epidemiologic approach for cancer screening. Problems in design and analysis of trials. *Am J Pediatr Hematol Oncol*, 14, 117-128.

[28] Bruix J, Llovet JM. (2003) HCC surveillance: who is the target population? *Hepatology*, 37, 507-509.

[29] Zhang BH, Yang BH, Tang ZY. (2004) Randomized controlled trial of screening for hepatocellular carcinoma. *J Cancer Res Clin Oncol*, 130, 417-422.

[30] Singal AG, Yopp A, C SS, Packer M, Lee WM, Tiro JA. (2012) Utilization of hepatocellular carcinoma surveillance among American patients: a systematic review. *J Gen Intern Med*, 27, 861-867.

[31] Singal AG, Yopp AC, Gupta S, Skinner CS, Halm EA, Okolo E, et al. (2012) Failure rates in the hepatocellular carcinoma surveillance process. *Cancer Prev Res (Phila)*, 5, 1124-1130.

[32] Marrero JA, El-Serag HB. (2011) Alpha-fetoprotein should be included in the hepatocellular carcinoma surveillance guidelines of the American Association for the Study of Liver Diseases. *Hepatology*, 53, 1060-1061; author reply 1061-1062.

[33] Singal AG, Conjeevaram HS, Volk ML, Fu S, Fontana RJ, Askari F, et al. (2012) Effectiveness of hepatocellular carcinoma surveillance in patients with cirrhosis. *Cancer Epidemiol Biomarkers Prev*, 21, 793-799.

[34] Rahimi RS, Yopp AC, Singal AG. (2012) Current issues and future trends in surveillance for hepatocellular carcinoma. *Clinical Liver Disease*, 1, 186-189.

[35] Zhang W, Wang X, Jiang R, Hou J, Mu X, Li G, et al. (2015) Effect of tumor size on cancer-specific survival in small hepatocellular carcinoma. *Mayo Clin Proc*, 90, 1187-1195.

[36] Chaiteerakij R, Addissie BD, Roberts LR. (2015) Update on biomarkers of hepatocellular carcinoma. *Clin Gastroenterol Hepatol*, 13, 237-245.

[37] Pepe MS, Etzioni R, Feng Z, Potter JD, Thompson ML, Thornquist M, et al. (2001) Phases of biomarker development for early detection of cancer. *J Natl Cancer Inst*, 93, 1054-1061.

[38] Hajdu SI. (2006) A note from history: the first biochemical test for detection of cancer. *Ann Clin Lab Sci*, 36, 222-223.

[39] Jones HB. (1848) On a new substance occurring in the urine of a patient with mollities ossium. *Philos Trans R Soc Lond*, 138, 55-62.

[40] Zhang H, Zhai Y, Hu Z, Wu C, Qian J, Jia W, et al. (2010) Genome-wide association study identifies 1p36.22 as a new susceptibility locus for hepatocellular carcinoma in chronic hepatitis B virus carriers. *Nat Genet*, 42, 755-758.

[41] Kumar V, Kato N, Urabe Y, Takahashi A, Muroyama R, Hosono N, et al. (2011) Genome-wide association study identifies a susceptibility locus for HCV-induced hepatocellular carcinoma. *Nat Genet*, 43, 455-458.

[42] Jiang DK, Sun J, Cao G, Liu Y, Lin D, Gao YZ, et al. (2013) Genetic variants in STAT4 and HLA-DQ genes confer risk of hepatitis B virus-related hepatocellular carcinoma. *Nat Genet*, 45, 72-75.
[43] Chen K, Shi W, Xin Z, Wang H, Zhu X, Wu X, et al. (2013) Replication of genome wide association studies on hepatocellular carcinoma susceptibility loci in a Chinese population. *PLoS One*, 8, e77315.
[44] Hoshida Y, Fuchs BC, Tanabe KK. (2012) Genomic risk of hepatitis C-related hepatocellular carcinoma. *J Hepatol*, 56, 729-730.
[45] Nahon P, Zucman-Rossi J. (2012) Single nucleotide polymorphisms and risk of hepatocellular carcinoma in cirrhosis. *J Hepatol*, 57, 663-674.
[46] Johnson PJ, Pirrie SJ, Cox TF, Berhane S, Teng M, Palmer D, et al. (2014) The detection of hepatocellular carcinoma using a prospectively developed and validated model based on serological biomarkers. *Cancer Epidemiol Biomarkers Prev*, 23, 144-153.
[47] Berhane S, Toyoda H, Tada T, Kumada T, Kagebayashi C, Satomura S, et al. (2015) The 'GALAD score' for serological detection of hepatocellular carcinoma: International validation and assessment of the influence of tumour size and aetiology on model utility. *J Hepatol*, 62, S434.
[48] Li M, Li H, Li C, Zhou S, Guo L, Liu H, et al. (2009) Alpha fetoprotein is a novel protein-binding partner for caspase-3 and blocks the apoptotic signaling pathway in human hepatoma cells. *Int J Cancer*, 124, 2845-2854.
[49] Terentiev AA, Moldogazieva NT. (2006) Structural and functional mapping of alpha-fetoprotein. *Biochemistry (Mosc)*, 71, 120-132.
[50] Abelev GI, Perova SD, Khramkova NI, Postnikova ZA, Irlin IS. (1963) Production of embryonal alpha-globulin by transplantable mouse hepatomas. *Transplantation*, 1, 174-180.
[51] Tatarinov IS. (1964) [Detection of Embryo-Specific Alpha-Globulin in the Blood Serum of a Patient with Primary Liver Cancer]. *Vopr Med Khim*, 10, 90-91.
[52] O'Conor GT, Tatarinov YS, Abelev GI, Uriel J. (1970) A collaborative study for the evaluation of a serologic test for primary liver cancer. *Cancer*, 25, 1091-1098.
[53] Hann HW, Fu X, Myers RE, Hann RS, Wan S, Kim SH, et al. (2012) Predictive value of alpha-fetoprotein in the long-term risk of developing hepatocellular carcinoma in patients with hepatitis B virus infection--results from a clinic-based longitudinal cohort. *Eur J Cancer*, 48, 2319-2327.
[54] Marrero JA, Feng Z, Wang Y, Nguyen MH, Befeler AS, Roberts LR, et al. (2009) Alpha-fetoprotein, des-gamma carboxyprothrombin, and lectin-bound alpha-fetoprotein in early hepatocellular carcinoma. *Gastroenterology*, 137, 110-118.
[55] Gupta S, Bent S, Kohlwes J. (2003) Test characteristics of alpha-fetoprotein for detecting hepatocellular carcinoma in patients with hepatitis C. A systematic review and critical analysis. *Ann Intern Med*, 139, 46-50.
[56] El-Serag HB, Kanwal F, Davila JA, Kramer J, Richardson P. (2014) A new laboratory-based algorithm to predict development of hepatocellular carcinoma in patients with hepatitis C and cirrhosis. *Gastroenterology*, 146, 1249-1255 e1241.

[57] Oka H, Tamori A, Kuroki T, Kobayashi K, Yamamoto S. (1994) Prospective study of alpha-fetoprotein in cirrhotic patients monitored for development of hepatocellular carcinoma. *Hepatology*, 19, 61-66.

[58] Wang XW, Xie H. (1999) Alpha-fetoprotein enhances the proliferation of human hepatoma cells in vitro. *Life Sci*, 64, 17-23.

[59] Li M, Li H, Li C, Guo L, Liu H, Zhou S, et al. (2009) Cytoplasmic alpha-fetoprotein functions as a co-repressor in RA-RAR signaling to promote the growth of human hepatoma Bel 7402 cells. *Cancer Lett*, 285, 190-199.

[60] Hong Y, Peng Y, Guo ZS, Guevara-Patino J, Pang J, Butterfield LH, et al. (2014) Epitope-optimized alpha-fetoprotein genetic vaccines prevent carcinogen-induced murine autochthonous hepatocellular carcinoma. *Hepatology*, 59, 1448-1458.

[61] He Y, Hong Y, Mizejewski GJ. (2014) Engineering alpha-fetoprotein-based gene vaccines to prevent and treat hepatocellular carcinoma: review and future prospects. *Immunotherapy*, 6, 725-736.

[62] Butterfield LH, Ribas A, Meng WS, Dissette VB, Amarnani S, Vu HT, et al. (2003) T-cell responses to HLA-A*0201 immunodominant peptides derived from alpha-fetoprotein in patients with hepatocellular cancer. *Clin Cancer Res*, 9, 5902-5908.

[63] Butterfield LH, Ribas A, Dissette VB, Lee Y, Yang JQ, De la Rocha P, et al. (2006) A phase I/II trial testing immunization of hepatocellular carcinoma patients with dendritic cells pulsed with four alpha-fetoprotein peptides. *Clin Cancer Res*, 12, 2817-2825.

[64] El-Serag HB, Davila JA. (2011) Surveillance for hepatocellular carcinoma: in whom and how? *Therap Adv Gastroenterol*, 4, 5-10.

[65] Sherman M, Peltekian KM, Lee C. (1995) Screening for hepatocellular carcinoma in chronic carriers of hepatitis B virus: incidence and prevalence of hepatocellular carcinoma in a North American urban population. *Hepatology*, 22, 432-438.

[66] Bruix J, Sherman M, American Association for the Study of Liver D. (2011) Management of hepatocellular carcinoma: an update. *Hepatology*, 53, 1020-1022.

[67] Carr BI, Guerra V, Steel JL, Lu SN. (2015) A comparison of patients with hepatitis B- or hepatitis C-based advanced-stage hepatocellular carcinoma. *Semin Oncol*, 42, 309-315.

[68] McMahon BJ, Bulkow L, Harpster A, Snowball M, Lanier A, Sacco F, et al. (2000) Screening for hepatocellular carcinoma in Alaska natives infected with chronic hepatitis B: a 16-year population-based study. *Hepatology*, 32, 842-846.

[69] Sinn DH, Yi J, Choi MS, Kim YJ, Gwak GY, Lee JH, et al. (2015) Serum alpha-fetoprotein may have a significant role in the surveillance of hepatocellular carcinoma in hepatitis B endemic areas. *Hepatogastroenterology*, 62, 327-332.

[70] Thompson Coon J, Rogers G, Hewson P, Wright D, Anderson R, Cramp M, et al. (2007) Surveillance of cirrhosis for hepatocellular carcinoma: systematic review and economic analysis. *Health Technol Assess*, 11, 1-206.

[71] Asahina Y, Tsuchiya K, Nishimura T, Muraoka M, Suzuki Y, Tamaki N, et al. (2013) alpha-fetoprotein levels after interferon therapy and risk of hepatocarcinogenesis in chronic hepatitis C. *Hepatology*, 58, 1253-1262.

[72] Sterling RK, Wright EC, Morgan TR, Seeff LB, Hoefs JC, Di Bisceglie AM, et al. (2012) Frequency of elevated hepatocellular carcinoma (HCC) biomarkers in patients with advanced hepatitis C. *Am J Gastroenterol*, 107, 64-74.

[73] Gopal P, Yopp AC, Waljee AK, Chiang J, Nehra M, Kandunoori P, et al. (2014) Factors that affect accuracy of alpha-fetoprotein test in detection of hepatocellular carcinoma in patients with cirrhosis. *Clin Gastroenterol Hepatol*, 12, 870-877.

[74] Chang TS, Wu YC, Tung SY, Wei KL, Hsieh YY, Huang HC, et al. (2015) Alpha-fetoprotein measurement benefits hepatocellular carcinoma surveillance in patients with cirrhosis. *Am J Gastroenterol*, 110, 836-844; quiz 845.

[75] Singal AG, Nehra M, Adams-Huet B, Yopp AC, Tiro JA, Marrero JA, et al. (2013) Detection of hepatocellular carcinoma at advanced stages among patients in the HALT-C trial: where did surveillance fail? *Am J Gastroenterol*, 108, 425-432.

[76] Gebo KA, Chander G, Jenckes MW, Ghanem KG, Herlong HF, Torbenson MS, et al. (2002) Screening tests for hepatocellular carcinoma in patients with chronic hepatitis C: a systematic review. *Hepatology*, 36, S84-92.

[77] Sangiovanni A, Colombo E, Radaelli F, Bortoli A, Bovo G, Casiraghi MA, et al. (2001) Hepatocyte proliferation and risk of hepatocellular carcinoma in cirrhotic patients. *Am J Gastroenterol*, 96, 1575-1580.

[78] Ma WJ, Wang HY, Teng LS. (2013) Correlation analysis of preoperative serum alpha-fetoprotein (AFP) level and prognosis of hepatocellular carcinoma (HCC) after hepatectomy. *World J Surg Oncol*, 11, 212.

[79] Tangkijvanich P, Anukulkarnkusol N, Suwangool P, Lertmaharit S, Hanvivatvong O, Kullavanijaya P, et al. (2000) Clinical characteristics and prognosis of hepatocellular carcinoma: analysis based on serum alpha-fetoprotein levels. *J Clin Gastroenterol*, 31, 302-308.

[80] An SL, Xiao T, Wang LM, Rong WQ, Wu F, Feng L, et al. (2015) Prognostic significance of preoperative serum alpha-fetoprotein in hepatocellular carcinoma and correlation with clinicopathological factors: a single-center experience from China. *Asian Pac J Cancer Prev*, 16, 4421-4427.

[81] Blank S, Wang Q, Fiel MI, Luan W, Kim KW, Kadri H, et al. (2014) Assessing prognostic significance of preoperative alpha-fetoprotein in hepatitis B-associated hepatocellular carcinoma: normal is not the new normal. *Ann Surg Oncol*, 21, 986-994.

[82] Hsu CY, Liu PH, Lee YH, Hsia CY, Huang YH, Lin HC, et al. (2015) Using serum alpha-fetoprotein for prognostic prediction in patients with hepatocellular carcinoma: what is the most optimal cutoff? *PLoS One*, 10, e0118825.

[83] Gupta S, Kutty G, Jagar P, Jain P, Lad T. (2013) Very high alpha-fetoprotein (AFP): A poor prognostic indicator in hepatocellular carcinoma in the modern era. *Ann Oncol*, 24, iv67.

[84] Wahab MA, Shehta A, Hamed H, El Nakeeb A, Salah T. (2014) Predictors of recurrence in hepatitis C virus related hepatocellular carcinoma after hepatic resection: a retrospective cohort study. *Eurasian J Med*, 46, 36-41.

[85] Tyson GL, Duan Z, Kramer JR, Davila JA, Richardson PA, El-Serag HB. (2011) Level of alpha-fetoprotein predicts mortality among patients with hepatitis C-related hepatocellular carcinoma. *Clin Gastroenterol Hepatol*, 9, 989-994.

[86] Tandon P, Garcia-Tsao G. (2009) Prognostic indicators in hepatocellular carcinoma: a systematic review of 72 studies. *Liver Int*, 29, 502-510.

[87] Hanazaki K, Kajikawa S, Koide N, Adachi W, Amano J. (2001) Prognostic factors after hepatic resection for hepatocellular carcinoma with hepatitis C viral infection: univariate and multivariate analysis. *Am J Gastroenterol*, 96, 1243-1250.

[88] Chong CC, Lee KF, Ip PC, Wong JS, Cheung SY, Wong J, et al. (2012) Pre-operative predictors of post-hepatectomy recurrence of hepatocellular carcinoma: can we predict earlier? *Surgeon*, 10, 260-266.
[89] Liu L, Miao R, Yang H, Lu X, Zhao Y, Mao Y, et al. (2012) Prognostic factors after liver resection for hepatocellular carcinoma: a single-center experience from China. *Am J Surg*, 203, 741-750.
[90] Singhal A, Jayaraman M, Dhanasekaran DN, Kohli V. (2012) Molecular and serum markers in hepatocellular carcinoma: predictive tools for prognosis and recurrence. *Crit Rev Oncol Hematol*, 82, 116-140.
[91] Iida H, Honda M, Kawai HF, Yamashita T, Shirota Y, Wang BC, et al. (2005) Ephrin-A1 expression contributes to the malignant characteristics of {alpha}-fetoprotein producing hepatocellular carcinoma. *Gut*, 54, 843-851.
[92] Berry K, Ioannou GN. (2013) Serum alpha-fetoprotein level independently predicts posttransplant survival in patients with hepatocellular carcinoma. *Liver Transpl*, 19, 634-645.
[93] Toso C, Asthana S, Bigam DL, Shapiro AM, Kneteman NM. (2009) Reassessing selection criteria prior to liver transplantation for hepatocellular carcinoma utilizing the Scientific Registry of Transplant Recipients database. *Hepatology*, 49, 832-838.
[94] Mailey B, Artinyan A, Khalili J, Denitz J, Sanchez-Luege N, Sun CL, et al. (2011) Evaluation of absolute serum alpha-fetoprotein levels in liver transplant for hepatocellular cancer. *Arch Surg*, 146, 26-33.
[95] Memon K, Kulik L, Lewandowski RJ, Wang E, Ryu RK, Riaz A, et al. (2012) Alpha-fetoprotein response correlates with EASL response and survival in solitary hepatocellular carcinoma treated with transarterial therapies: a subgroup analysis. *J Hepatol*, 56, 1112-1120.
[96] Wang NY, Wang C, Li W, Wang GJ, Cui GZ, He H, et al. (2014) Prognostic value of serum AFP, AFP-L3, and GP73 in monitoring short-term treatment response and recurrence of hepatocellular carcinoma after radiofrequency ablation. *Asian Pac J Cancer Prev*, 15, 1539-1544.
[97] Personeni N, Bozzarelli S, Pressiani T, Rimassa L, Tronconi MC, Sclafani F, et al. (2012) Usefulness of alpha-fetoprotein response in patients treated with sorafenib for advanced hepatocellular carcinoma. *J Hepatol*, 57, 101-107.
[98] Llovet JM, Pena CE, Lathia CD, Shan M, Meinhardt G, Bruix J, et al. (2012) Plasma biomarkers as predictors of outcome in patients with advanced hepatocellular carcinoma. *Clin Cancer Res*, 18, 2290-2300.
[99] Li X, Dong M, Lin Q, Chen ZH, Ma XK, Xing YF, et al. (2013) Comparison of current staging systems for advanced hepatocellular carcinoma not amendable to locoregional therapy as inclusion criteria for clinical trials. *Asia Pac J Clin Oncol*, 9, 86-92.
[100] Toyoda H, Kumada T, Osaki Y, Oka H, Urano F, Kudo M, et al. (2006) Staging hepatocellular carcinoma by a novel scoring system (BALAD score) based on serum markers. *Clin Gastroenterol Hepatol*, 4, 1528-1536.
[101] Chan SL, Mo F, Johnson P, Li L, Tang N, Loong H, et al. (2015) Applicability of BALAD score in prognostication of hepatitis B-related hepatocellular carcinoma. *J Gastroenterol Hepatol*, 30, 1529-1535.
[102] Kemmer N, Neff G, Kaiser T, Zacharias V, Thomas M, Tevar A, et al. (2006) An analysis of the UNOS liver transplant registry: high serum alpha-fetoprotein does not

justify an increase in MELD points for suspected hepatocellular carcinoma. *Liver Transpl*, 12, 1519-1522.

[103] United Network for Organ Sharing. Policy notice 12-2014. Richmond, Virginia: United Network for Organ Sharing; 2014.

[104] Toso C, Mazzaferro V, Bruix J, Freeman R, Mentha G, Majno P. (2014) Toward a better liver graft allocation that accounts for candidates with and without hepatocellular carcinoma. *Am J Transplant*, 14, 2221-2227.

[105] Marvin MR, Ferguson N, Cannon RM, Jones CM, Brock GN. (2015) MELDEQ: An alternative Model for End-Stage Liver Disease score for patients with hepatocellular carcinoma. *Liver Transpl*, 21, 612-622.

[106] Li D, Mallory T, Satomura S. (2001) AFP-L3: a new generation of tumor marker for hepatocellular carcinoma. *Clin Chim Acta*, 313, 15-19.

[107] Hann HW, Li D, Yamada H, Satomura S, Coben R, DiMarino AJ. (2014) Usefulness of highly sensitive AFP-L3 and DCP in surveillance for hepatocellular carcinoma in patients with a normal alpha-fetoprotein. *J Med Microb Diagn*, 3, 1-6.

[108] Yi X, Yu S, Bao Y. (2013) Alpha-fetoprotein-L3 in hepatocellular carcinoma: a meta-analysis. *Clin Chim Acta*, 425, 212-220.

[109] Leerapun A, Suravarapu SV, Bida JP, Clark RJ, Sanders EL, Mettler TA, et al. (2007) The utility of Lens culinaris agglutinin-reactive alpha-fetoprotein in the diagnosis of hepatocellular carcinoma: evaluation in a United States referral population. *Clin Gastroenterol Hepatol*, 5, 394-402; quiz 267.

[110] Taketa K, Endo Y, Sekiya C, Tanikawa K, Koji T, Taga H, et al. (1993) A collaborative study for the evaluation of lectin-reactive alpha-fetoproteins in early detection of hepatocellular carcinoma. *Cancer Res*, 53, 5419-5423.

[111] Sato Y, Nakata K, Kato Y, Shima M, Ishii N, Koji T, et al. (1993) Early recognition of hepatocellular carcinoma based on altered profiles of alpha-fetoprotein. *N Engl J Med*, 328, 1802-1806.

[112] Kumada T, Toyoda H, Tada T, Kiriyama S, Tanikawa M, Hisanaga Y, et al. (2014) High-sensitivity Lens culinaris agglutinin-reactive alpha-fetoprotein assay predicts early detection of hepatocellular carcinoma. *J Gastroenterol*, 49, 555-563.

[113] Oka H, Saito A, Ito K, Kumada T, Satomura S, Kasugai H, et al. (2001) Multicenter prospective analysis of newly diagnosed hepatocellular carcinoma with respect to the percentage of Lens culinaris agglutinin-reactive alpha-fetoprotein. *J Gastroenterol Hepatol*, 16, 1378-1383.

[114] Yamashita F, Tanaka M, Satomura S, Tanikawa K. (1996) Prognostic significance of Lens culinaris agglutinin A-reactive alpha-fetoprotein in small hepatocellular carcinomas. *Gastroenterology*, 111, 996-1001.

[115] Malaguarnera G, Giordano M, Paladina I, Berretta M, Cappellani A, Malaguarnera M. (2010) Serum markers of hepatocellular carcinoma. *Dig Dis Sci*, 55, 2744-2755.

[116] Khien VV, Mao HV, Chinh TT, Ha PT, Bang MH, Lac BV, et al. (2001) Clinical evaluation of lentil lectin-reactive alpha-fetoprotein-L3 in histology-proven hepatocellular carcinoma. *Int J Biol Markers*, 16, 105-111.

[117] Tateishi R, Shiina S, Yoshida H, Teratani T, Obi S, Yamashiki N, et al. (2006) Prediction of recurrence of hepatocellular carcinoma after curative ablation using three tumor markers. *Hepatology*, 44, 1518-1527.

[118] Yamashita F, Tanaka M, Satomura S, Tanikawa K. (1995) Monitoring of lectin-reactive alpha-fetoproteins in patients with hepatocellular carcinoma treated using transcatheter arterial embolization. *Eur J Gastroenterol Hepatol*, 7, 627-633.

[119] Tamura Y, Suda T, Arii S, Sata M, Moriyasu F, Imamura H, et al. (2013) Value of highly sensitive fucosylated fraction of alpha-fetoprotein for prediction of hepatocellular carcinoma recurrence after curative treatment. *Dig Dis Sci*, 58, 2406-2412.

[120] Ueno M, Hayami S, Shigekawa Y, Kawai M, Hirono S, Okada KI, et al. (2015) Prognostic impact of surgery and radiofrequency ablation on single nodular HCC 5 cm: Cohort study based on serum HCC markers. *J Hepatol.*

[121] Kang KH, Kim JH, Kang SH, Lee BJ, Seo YS, Yim HJ, et al. (2015) The influence of alcoholic liver disease on serum PIVKA-II levels in patients without hepatocellular carcinoma. *Gut Liver*, 9, 224-230.

[122] Zhang YS, Chu JH, Cui SX, Song ZY, Qu XJ. (2014) Des-gamma-carboxy prothrombin (DCP) as a potential autologous growth factor for the development of hepatocellular carcinoma. *Cell Physiol Biochem*, 34, 903-915.

[123] Inagaki Y, Tang W, Makuuchi M, Hasegawa K, Sugawara Y, Kokudo N. (2011) Clinical and molecular insights into the hepatocellular carcinoma tumour marker des-gamma-carboxyprothrombin. *Liver Int*, 31, 22-35.

[124] Liebman HA, Furie BC, Tong MJ, Blanchard RA, Lo KJ, Lee SD, et al. (1984) Des-gamma-carboxy (abnormal) prothrombin as a serum marker of primary hepatocellular carcinoma. *N Engl J Med*, 310, 1427-1431.

[125] Marrero JA, Su GL, Wei W, Emick D, Conjeevaram HS, Fontana RJ, et al. (2003) Des-gamma carboxyprothrombin can differentiate hepatocellular carcinoma from nonmalignant chronic liver disease in american patients. *Hepatology*, 37, 1114-1121.

[126] Durazo FA, Blatt LM, Corey WG, Lin JH, Han S, Saab S, et al. (2008) Des-gamma-carboxyprothrombin, alpha-fetoprotein and AFP-L3 in patients with chronic hepatitis, cirrhosis and hepatocellular carcinoma. *J Gastroenterol Hepatol*, 23, 1541-1548.

[127] Kim do Y, Han KH. (2012) Epidemiology and surveillance of hepatocellular carcinoma. *Liver Cancer*, 1, 2-14.

[128] Yoon YJ, Han KH, Kim do Y. (2009) Role of serum prothrombin induced by vitamin K absence or antagonist-II in the early detection of hepatocellular carcinoma in patients with chronic hepatitis B virus infection. *Scand J Gastroenterol*, 44, 861-866.

[129] Shirabe K, Itoh S, Yoshizumi T, Soejima Y, Taketomi A, Aishima S, et al. (2007) The predictors of microvascular invasion in candidates for liver transplantation with hepatocellular carcinoma-with special reference to the serum levels of des-gamma-carboxy prothrombin. *J Surg Oncol*, 95, 235-240.

[130] Choi JY, Jung SW, Kim HY, Kim M, Kim Y, Kim DG, et al. (2013) Diagnostic value of AFP-L3 and PIVKA-II in hepatocellular carcinoma according to total-AFP. *World J Gastroenterol*, 19, 339-346.

[131] Kobayashi M, Ikeda K, Kawamura Y, Yatsuji H, Hosaka T, Sezaki H, et al. (2009) High serum des-gamma-carboxy prothrombin level predicts poor prognosis after radiofrequency ablation of hepatocellular carcinoma. *Cancer*, 115, 571-580.

[132] Koike Y, Shiratori Y, Sato S, Obi S, Teratani T, Imamura M, et al. (2001) Des-gamma-carboxy prothrombin as a useful predisposing factor for the development of portal

venous invasion in patients with hepatocellular carcinoma: a prospective analysis of 227 patients. *Cancer*, 91, 561-569.

[133] Chaiteerakij R, Zhang X, Addissie BD, Mohamed EA, Harmsen WS, Theobald PJ, et al. (2015) Combinations of biomarkers and Milan criteria for predicting hepatocellular carcinoma recurrence after liver transplantation. *Liver Transpl*, 21, 599-606.

[134] Senger DR, Wirth DF, Hynes RO. (1979) Transformed mammalian cells secrete specific proteins and phosphoproteins. *Cell*, 16, 885-893.

[135] Saeki Y, Mima T, Ishii T, Ogata A, Kobayashi H, Ohshima S, et al. (2003) Enhanced production of osteopontin in multiple myeloma: clinical and pathogenic implications. *Br J Haematol*, 123, 263-270.

[136] Koopmann J, Fedarko NS, Jain A, Maitra A, Iacobuzio-Donahue C, Rahman A, et al. (2004) Evaluation of osteopontin as biomarker for pancreatic adenocarcinoma. *Cancer Epidemiol Biomarkers Prev*, 13, 487-491.

[137] Fedarko NS, Jain A, Karadag A, Van Eman MR, Fisher LW. (2001) Elevated serum bone sialoprotein and osteopontin in colon, breast, prostate, and lung cancer. *Clin Cancer Res*, 7, 4060-4066.

[138] Gotoh M, Sakamoto M, Kanetaka K, Chuuma M, Hirohashi S. (2002) Overexpression of osteopontin in hepatocellular carcinoma. *Pathol Int*, 52, 19-24.

[139] Pan HW, Ou YH, Peng SY, Liu SH, Lai PL, Lee PH, et al. (2003) Overexpression of osteopontin is associated with intrahepatic metastasis, early recurrence, and poorer prognosis of surgically resected hepatocellular carcinoma. *Cancer*, 98, 119-127.

[140] Kim J, Ki SS, Lee SD, Han CJ, Kim YC, Park SH, et al. (2006) Elevated plasma osteopontin levels in patients with hepatocellular carcinoma. *Am J Gastroenterol*, 101, 2051-2059.

[141] El-Din Bessa SS, Elwan NM, Suliman GA, El-Shourbagy SH. (2010) Clinical significance of plasma osteopontin level in Egyptian patients with hepatitis C virus-related hepatocellular carcinoma. *Arch Med Res*, 41, 541-547.

[142] Zhou C, Zhou HJ, Zhang XF, Lou LL, Ye QH, Zheng Y, et al. (2013) Postoperative serum osteopontin level is a novel monitor for treatment response and tumor recurrence after resection of hepatitis B-related hepatocellular carcinoma. *Ann Surg Oncol*, 20, 929-937.

[143] Qin L. (2014) Osteopontin is a promoter for hepatocellular carcinoma metastasis: a summary of 10 years of studies. *Front Med*, 8, 24-32.

[144] Dong QZ, Zhang XF, Zhao Y, Jia HL, Zhou HJ, Dai C, et al. (2013) Osteopontin promoter polymorphisms at locus -443 significantly affect the metastasis and prognosis of human hepatocellular carcinoma. *Hepatology*, 57, 1024-1034.

[145] Filmus J, Selleck SB. (2001) Glypicans: proteoglycans with a surprise. *J Clin Invest*, 108, 497-501.

[146] Capurro M, Wanless IR, Sherman M, Deboer G, Shi W, Miyoshi E, et al. (2003) Glypican-3: a novel serum and histochemical marker for hepatocellular carcinoma. *Gastroenterology*, 125, 89-97.

[147] Wang HL, Anatelli F, Zhai QJ, Adley B, Chuang ST, Yang XJ. (2008) Glypican-3 as a useful diagnostic marker that distinguishes hepatocellular carcinoma from benign hepatocellular mass lesions. *Arch Pathol Lab Med*, 132, 1723-1728.

[148] Yorita K, Takahashi N, Takai H, Kato A, Suzuki M, Ishiguro T, et al. (2011) Prognostic significance of circumferential cell surface immunoreactivity of glypican-3 in hepatocellular carcinoma. *Liver Int*, 31, 120-131.

[149] Chen IP, Ariizumi S, Nakano M, Yamamoto M. (2014) Positive glypican-3 expression in early hepatocellular carcinoma predicts recurrence after hepatectomy. *J Gastroenterol*, 49, 117-125.

[150] Liu H, Li P, Zhai Y, Qu CF, Zhang LJ, Tan YF, et al. (2010) Diagnostic value of glypican-3 in serum and liver for primary hepatocellular carcinoma. *World J Gastroenterol*, 16, 4410-4415.

[151] Tangkijvanich P, Chanmee T, Komtong S, Mahachai V, Wisedopas N, Pothacharoen P, et al. (2010) Diagnostic role of serum glypican-3 in differentiating hepatocellular carcinoma from non-malignant chronic liver disease and other liver cancers. *J Gastroenterol Hepatol*, 25, 129-137.

[152] Wang Y, Yang H, Xu H, Lu X, Sang X, Zhong S, et al. (2014) Golgi protein 73, not Glypican-3, may be a tumor marker complementary to alpha-Fetoprotein for hepatocellular carcinoma diagnosis. *J Gastroenterol Hepatol*, 29, 597-602.

[153] Shirakawa H, Suzuki H, Shimomura M, Kojima M, Gotohda N, Takahashi S, et al. (2009) Glypican-3 expression is correlated with poor prognosis in hepatocellular carcinoma. *Cancer Sci*, 100, 1403-1407.

[154] Ning S, Bin C, Na H, Peng S, Yi D, Xiang-hua Y, et al. (2012) Glypican-3, a novel prognostic marker of hepatocellular cancer, is related with postoperative metastasis and recurrence in hepatocellular cancer patients. *Mol Biol Rep*, 39, 351-357.

[155] Kladney RD, Bulla GA, Guo L, Mason AL, Tollefson AE, Simon DJ, et al. (2000) GP73, a novel Golgi-localized protein upregulated by viral infection. *Gene*, 249, 53-65.

[156] Kladney RD, Cui X, Bulla GA, Brunt EM, Fimmel CJ. (2002) Expression of GP73, a resident Golgi membrane protein, in viral and nonviral liver disease. *Hepatology*, 35, 1431-1440.

[157] Marrero JA, Romano PR, Nikolaeva O, Steel L, Mehta A, Fimmel CJ, et al. (2005) GP73, a resident Golgi glycoprotein, is a novel serum marker for hepatocellular carcinoma. *J Hepatol*, 43, 1007-1012.

[158] Ba MC, Long H, Tang YQ, Cui SZ. (2012) GP73 expression and its significance in the diagnosis of hepatocellular carcinoma: a review. *Int J Clin Exp Pathol*, 5, 874-881.

[159] Zhao XY, Li N, Ding HG, Jiang FF. (2010) [Detection and evaluation of serum GP73, a resident Golgi glycoprotein, as a marker in diagnosis of hepatocellular carcinoma]. *Zhonghua Zhong Liu Za Zhi*, 32, 943-945.

[160] Mao YL, Yang HY, Xu HF, Sang XT, Lu X, Yang ZY, et al. (2008) [Significance of Golgi glycoprotein 73, a new tumor marker in diagnosis of hepatocellular carcinoma: a primary study]. *Zhonghua Yi Xue Za Zhi*, 88, 948-951.

[161] Hu JS, Wu DW, Liang S, Miao XY. (2010) GP73, a resident Golgi glycoprotein, is sensibility and specificity for hepatocellular carcinoma of diagnosis in a hepatitis B-endemic Asian population. *Med Oncol*, 27, 339-345.

[162] Zhou Y, Yin X, Ying J, Zhang B. (2012) Golgi protein 73 versus alpha-fetoprotein as a biomarker for hepatocellular carcinoma: a diagnostic meta-analysis. *BMC Cancer*, 12, 17.

[163] Lee HJ, Kang HJ, Kim KM, Yu ES, Kim KH, Kim SM, et al. (2015) Fibroblast growth factor receptor isotype expression and its association with overall survival in patients with hepatocellular carcinoma. *Clin Mol Hepatol*, 21, 60-70.
[164] Poon RT, Ng IO, Lau C, Yu WC, Fan ST, Wong J. (2001) Correlation of serum basic fibroblast growth factor levels with clinicopathologic features and postoperative recurrence in hepatocellular carcinoma. *Am J Surg*, 182, 298-304.
[165] Ho HK, Pok S, Streit S, Ruhe JE, Hart S, Lim KS, et al. (2009) Fibroblast growth factor receptor 4 regulates proliferation, anti-apoptosis and alpha-fetoprotein secretion during hepatocellular carcinoma progression and represents a potential target for therapeutic intervention. *J Hepatol*, 50, 118-127.
[166] Miura S, Mitsuhashi N, Shimizu H, Kimura F, Yoshidome H, Otsuka M, et al. (2012) Fibroblast growth factor 19 expression correlates with tumor progression and poorer prognosis of hepatocellular carcinoma. *BMC Cancer*, 12, 56.
[167] Arao T, Ueshima K, Matsumoto K, Nagai T, Kimura H, Hagiwara S, et al. (2013) FGF3/FGF4 amplification and multiple lung metastases in responders to sorafenib in hepatocellular carcinoma. *Hepatology*, 57, 1407-1415.
[168] Kudo M. (2014) Biomarkers and personalized sorafenib therapy. *Liver Cancer*, 3, 399-404.
[169] Gauglhofer C, Paur J, Schrottmaier WC, Wingelhofer B, Huber D, Naegelen I, et al. (2014) Fibroblast growth factor receptor 4: a putative key driver for the aggressive phenotype of hepatocellular carcinoma. *Carcinogenesis*, 35, 2331-2338.
[170] Mellor HR. (2014) Targeted inhibition of the FGF19-FGFR4 pathway in hepatocellular carcinoma; translational safety considerations. *Liver Int*, 34, e1-9.
[171] Chen T, Xie G, Wang X, Fan J, Qiu Y, Zheng X, et al. (2011) Serum and urine metabolite profiling reveals potential biomarkers of human hepatocellular carcinoma. *Mol Cell Proteomics*, 10, M110 004945.
[172] Wang B, Chen D, Chen Y, Hu Z, Cao M, Xie Q, et al. (2012) Metabonomic profiles discriminate hepatocellular carcinoma from liver cirrhosis by ultraperformance liquid chromatography-mass spectrometry. *J Proteome Res*, 11, 1217-1227.
[173] Nault JC, De Reynies A, Villanueva A, Calderaro J, Rebouissou S, Couchy G, et al. (2013) A hepatocellular carcinoma 5-gene score associated with survival of patients after liver resection. *Gastroenterology*, 145, 176-187.
[174] Schirmacher P, Calvisi DF. (2013) Molecular diagnostic algorithms in hepatocellular carcinoma: dead-end street or light at the end of the tunnel? *Gastroenterology*, 145, 49-53.
[175] Dang H, Steinway SN, Ding W, Rountree CB. (2015) Induction of tumor initiation is dependent on CD44s in c-Met(+) hepatocellular carcinoma. *BMC Cancer*, 15, 161.
[176] You H, Ding W, Dang H, Jiang Y, Rountree CB. (2011) c-Met represents a potential therapeutic target for personalized treatment in hepatocellular carcinoma. *Hepatology*, 54, 879-889.
[177] Kondo S, Ojima H, Tsuda H, Hashimoto J, Morizane C, Ikeda M, et al. (2013) Clinical impact of c-Met expression and its gene amplification in hepatocellular carcinoma. *Int J Clin Oncol*, 18, 207-213.
[178] Santoro A, Rimassa L, Borbath I, Daniele B, Salvagni S, Van Laethem JL, et al. (2013) Tivantinib for second-line treatment of advanced hepatocellular carcinoma: a randomised, placebo-controlled phase 2 study. *Lancet Oncol*, 14, 55-63.

[179] Abou-Alfa GK. (2013) Approaching the era of personalised therapy for liver cancer? *Lancet Oncol*, 14, 7-8.

[180] Liu Y, Wu F. (2010) Global burden of aflatoxin-induced hepatocellular carcinoma: a risk assessment. *Environ Health Perspect*, 118, 818-824.

[181] Magnussen A, Parsi MA. (2013) Aflatoxins, hepatocellular carcinoma and public health. *World J Gastroenterol*, 19, 1508-1512.

[182] Nissen NN, Martin P. (2002) Hepatocellular carcinoma: the high-risk patient. *J Clin Gastroenterol*, 35, S79-85.

[183] Baran B. (2015) Nucleos(t)ide analogs in the prevention of hepatitis B virus related hepatocellular carcinoma. *World J Hepatol*, 7, 1742-1754.

[184] Omata M, Lesmana LA, Tateishi R, Chen PJ, Lin SM, Yoshida H, et al. (2010) Asian Pacific Association for the Study of the Liver consensus recommendations on hepatocellular carcinoma. *Hepatol Int*, 4, 439-474.

[185] Ferenci P, Fried M, Labrecque D, Bruix J, Sherman M, Omata M, et al. (2010) World Gastroenterology Organisation Guideline. Hepatocellular carcinoma (HCC): a global perspective. *J Gastrointestin Liver Dis*, 19, 311-317.

[186] Kokudo N, Hasegawa K, Akahane M, Igaki H, Izumi N, Ichida T, et al. (2015) Evidence-based Clinical Practice Guidelines for Hepatocellular Carcinoma: The Japan Society of Hepatology 2013 update (3rd JSH-HCC Guidelines). *Hepatol Res*, 45.

[187] Hsia CY, Huo TI, Chiang SY, Lu MF, Sun CL, Wu JC, et al. (2007) Evaluation of interleukin-6, interleukin-10 and human hepatocyte growth factor as tumor markers for hepatocellular carcinoma. *Eur J Surg Oncol*, 33, 208-212.

[188] Wong RJ, Ahmed A, Gish RG. (2015) Elevated alpha-fetoprotein: differential diagnosis - hepatocellular carcinoma and other disorders. *Clin Liver Dis*, 19, 309-323.

[189] Kurihara K, Konishi F, Kanazawa K, Fujii T, Saito K. (1997) Alpha-fetoprotein-producing carcinoma of the colon: report of a case. *Surg Today*, 27, 453-456.

[190] Kawamoto S, Hiraoka T, Kanemitsu K, Kimura M, Miyauchi Y, Takeya M. (1992) Alpha-fetoprotein-producing pancreatic cancer--a case report and review of 28 cases. *Hepatogastroenterology*, 39, 282-286.

[191] Xu P, Xu CF, Wan XY, Yu CH, Shen C, Chen P, et al. (2014) Association between serum alpha-fetoprotein levels and fatty liver disease: a cross-sectional study. *World J Gastroenterol*, 20, 11865-11870.

[192] Nakajima T, Okazaki N, Morinaga S, Tsumuraya M, Shimosato Y, Saiki S. (1985) A case of alpha-fetoprotein-producing rectal carcinoma. *Jpn J Clin Oncol*, 15, 679-685.

[193] Hadem J, Strassburg CP, Manns MP. (2012) Prediction of outcome and selection of the liver transplantat candidate in acute liver failure. *Front Physiol*, 3, 340.

[194] Hu KQ, Kyulo NL, Lim N, Elhazin B, Hillebrand DJ, Bock T. (2004) Clinical significance of elevated alpha-fetoprotein (AFP) in patients with chronic hepatitis C, but not hepatocellular carcinoma. *Am J Gastroenterol*, 99, 860-865.

[195] Bayati N, Silverman AL, Gordon SC. (1998) Serum alpha-fetoprotein levels and liver histology in patients with chronic hepatitis C. *Am J Gastroenterol*, 93, 2452-2456.

[196] Shen WF, Zhong W, Xu F, Kan T, Geng L, Xie F, et al. (2009) Clinicopathological and prognostic analysis of 429 patients with intrahepatic cholangiocarcinoma. *World J Gastroenterol*, 15, 5976-5982.

[197] Noda T, Sasaki Y, Yamada T, Eguchi H, Yano M, Ohigashi H, et al. (2009) Usefulness of the CLIP scoring system for prediction of postoperative prognosis of patients with large hepatocellular carcinoma. *J Hepatobiliary Pancreat Surg*, 16, 538-545.
[198] Subramaniam S, Kelley RK, Venook AP. (2013) A review of hepatocellular carcinoma (HCC) staging systems. *Chin Clin Oncol*, 2, 33.

In: Alpha-Fetoprotein: Functions and Clinical Applications
Editors: Nisha Lakhi and Michael Moretti
ISBN: 978-1-63484-875-6

Chapter 5

Alpha-Fetoprotein in a Protein Interaction Network: Systems Biology Approach in Discovery Biomarkers for Hepatocellular Carcinoma

N. T. Moldogazieva*, PhD, Dr Sci,
and A. A. Terentiev, PhD, Dr Sci
N. I. Pirogov Russian National Research Medical University,
Moscow, Russia

Abstract

This chapter highlights the latest advances, challenges and opportunities in genomic, epigenomic and proteomic approaches in discovering highly specific and sensitive biomarkers for hepatocellular carcinoma (HCC) diagnosis and prognosis. For the last years, growing amount of investigations have been carried out with the use of systems biology approach to study molecular mechanisms of cancer progression and to discover novel biomarker candidates and targets for anti-cancer therapy. Systems biology is focused on studying complex systems such as cancer, preliminary, by network analysis at different spatio/temporal scales. The network biology approaches provide the basis for identifying genomic and proteomic signatures of HCC through the identification of key molecules involved in cancer progression, metastasis and invasion. Alpha-fetoprotein (AFP) has been shown to be one of these molecules expressed in coordinated manner with a number of other cancer-associated proteins that may be cooperatively responsible for uncontrolled cell proliferation, lack of differentiation, increased angiogenesis and metastasis observed in HCC with poor prognosis. Current data obtained in this field allow concluding that AFP remains the golden standard among cancer biomarkers, since other HCC molecular markers identified to date are complementary to AFP.

* To whom correspondence should be addressed: nmoldogazieva@mail.ru.

INTRODUCTION

Cancer is a disease known to be a multi-factorial and complex disorder and a process with non-linear dynamics [1]. Systems biology and computational biology approaches along with various “omics” and structural biology technologies are presently being exploited to understand cancer complexity [2, 3]. Cancer systems biology has arised as a logical extension and application of systems biology to the field of medical sciences [4, 5].

Systems biology is itself an interdisciplinary scientific field to study complex systems. Complexity of biological systems arises from numerous non-linear interactions between cellular components. Complex systems possess so named emergent properties (and functions) when a system of two components has behavior that each individual component does not have. A system of ten components has properties (and functions) that the system of two components does not possess. This indicates that behavior of the entire system is not a result of simple summarizing of all the properties of the system, but it represents the integration of all these properties [6].

The goal of systems biology is to understand structure and functioning of living systems at different temporal and spatial scales. It is primarily focused on construction and analysis of networks at the level of completely organism-specific or tissue-specific genomes, proteomes and metabolomes or at the level of a network sub-circuits [5, 7]. Structural and dynamic properties of such networks are responsible for cell functioning and for controlling cell fate. Systems biology approaches are presently becoming a powerful tool to understand principles, rules and mechanisms that underlie cell proliferation, differentiation, migration and apoptosis, both in normal and in disease.

As the EU-USA Workshop concluded in 2009, systems biology can advance cancer research and are likely to be superior to many other research strategies [8]. Moreover, understanding dynamic processes that occur in living cells is no more possible with the use of a single method both *in vitro* and *in vivo,* even in combination with *in silico* approaches. The development of novel state-of-the-art technologies provides integration of systems biology, structural biology and computational biology approaches in investigation various cellular processes. Through the integrative usage of high-resolution structure determination technologies such as NMR and X-ray crystallography and structural biology approaches (electron microscopy, atomic force microscopy, and mass-spectrometry and single-molecule techniques), researchers of different scientific fields are presently becoming able to study details of protein structures and their complexes as well as dynamic changes in their behavior. Moreover, this allows constructing molecular networks including genomic and protein interaction networks and analyzing interactomes of different tissues, both normal and pathological. Genome-wide association studies, DNA microarray and mRNA global expression analysis, along with computational biology methods and bioinformatics tools are extensively employed in coherence with these approaches.

Analysis of molecular networks has been shown to be a useful tool not only in understanding dynamic interactions between components of a cell, between individual cells and between cells and their microenvironment, but also in disease diagnosis and prognosis, and for assessment of therapeutic strategies as well as biomarker identifying [9]. This chapter is focused on analysis of state-of-the-art and description of novel technologies in hepatocellular carcinoma (HCC) biomarker discovering with emphasis on network biology approaches. The special role of alpha-fetoprotein (AFP) as a primary HCC biomarker is underlined.

BIOMARKER PROFILING TECHNOLOGIES

The identification, standardization and validation of effective tumor biomarkers dramatically influence the cancer diagnosis, prognosis and anti-cancer drug development. In the past, the focus of cancer biomarker research has been on the usage of a single experimental technique such as immunoprecipitation analysis or ELISA. Currently, biomarkers are being identified with the use of integrated technologies including genetics, epigenetic, omics, systems biology and molecular imaging approaches [10, 11].

The process of candidate biomarker identification typically involves analysis of tissue samples and blood serum or plasma to reveal a profile of gene expression, protein and metabolite distribution [8]. This is based on comparative study of tissue or blood samples to identify genes and proteins whose expression levels are changed in patients as compared to those in healthy donors. In this manner, a panel of biomarkers can be constructed with sensitivity and specificity necessary for the detection and monitoring of a disease. The goal of this strategy is to determine unique gene expression patterns specific to a certain disease and to found correlations between gene expression/protein concentration and disease progression levels.

Proteomic profiling has been shown to be useful not only in evaluation of changes in gene expression patterns resulting in shifts in protein amount in tissues and body fluids in normal and in cancer, but also gives additional information to assess a tumor stage. This is especially true for such heterogeneous tumor as hepatocellular carcinoma (HCC), whose early diagnosis, prognosis and reliable tools in assessment of therapeutic strategies are very important to increase survival rate for patients [12]. Comprehensive serum proteomic profiling enables differentiating early and advanced HCCs as well as HCC from chronic liver diseases, even without knowledge of the clinical symptoms.

Correct sample preparation is a critical step in biomarker profiling. This is especially important if a biomarker candidate presents in a sample at a very low concentration (at ng/ml level). Sample fractionation and protein enrichment by immuno-affinity depletion are usually employed to decrease the samle complexity and to increase the assay sensitivity, specificity, high-throughput capability and multiplicity [13].

Currently available proteomic tools typically include integrated usage of 2-dimensional polyacrylamide gel electrophoresis (2-DE), multidimensional liquid chromatography, liquid chromatography-selected reaction monitoring mass spectrometry (LC-SRM MS), MALDI (matrix-assisted laser-desorption ionization), SELDI (surface-enhanced laser desorption/ionization) or tandem mass spectrometry (MS/MS), and protein chip and microarray technologies [14-16]. Capability of proteomic methods has been greatly enhanced by recent advances in quantitative proteomic analysis such as stable isotope labeling by amino acids in cell culture (SILAC), isotope coded affinity tag (ICAT) and isobaric tags for relative and absolute quantitation (iTRAQ) technologies. These approaches have allowed studying quantitative changes in a whole proteome and in the expression level and functions of a certain protein biomarker [17-19].

Additionally, new technologies for discovery novel cancer biomarkers are currently extensively developed and include epigenetic approaches that involve microRNA profiling along with DNA hyper- or hypomethylation and post-translational histone modification analysis [20, 21]. Genome-wide association studies used for DNA methylation and microRNA profiling have advanced discovery of epigenetic biomarkers for the cancer diagnosis and prognosis along with revealing new targets for anti-cancer therapeutics. Presently, it is becoming recognized that epigenome is globally reshaped in HCC and interplay between both genetic and epigenetic factors is important at all stages of the carcinogenesis [22, 23]. It has been also shown that over-expression of microRNAs leads to inhibition of cell proliferation and tumor growth and this may contribute to understanding molecular mechanisms regulating the cancer progression. For instance, the role of microRNAs in regulation of AFP-positive HCC growth has been demonstrated [24].

Alpha-Fetoprotein in Combined Testing for HCC

Hepatocellular carcinoma, in 70–90% of all cases, develops as a consequence of the liver cirrhosis that, in turn, predominantly is caused by inflammation associated with hepatitis B and C viral infections, or by alcohol abuse and metabolic syndrome [25, 26]. HCC is the fifth most commonly diagnosed cancer in men (554000 cases per year, 7.5% of the total amount) and the seventh in women (228000 cases per year, 3.4% of the total amount) worldwide [27, 28]. It is the second most frequent cause of cancer-related death among men and the sixth leading cause of death from cancers among women.

A considerable improvement in 5-year survival rate from 40% up to 70% and in prognosis for HCC patients has been observed when they are diagnosed at the early stage of the disease and are potentially treated with the use of liver transplantation, surgical resection, or tumor ablation. However, because of the early stages of HCC are accompanied by inflammation and cirrhosis, accurate staging of HCC requires distinguishing of HCC from inflammation and cirrhosis to increase survival rate of the cancer patients. This dictates importance of discovery of novel biomarkers for early HCC diagnosis.

Alpha-fetoprotein is the primary molecular biomarker for diagnosis and prognosis of hepatocellular carcinoma and has been recognized as a "golden standard" among the serum tumor markers [29]. Despite wide usage in the clinics as an independent prognostic factor for HCC [30, 31], diagnostic value of the test for AFP is not high because of poor sensitivity and specificity, which have been reported to vary from 39% to 64% and from 76% to 91%, respectively [32]. Besides, approximately 40% of early HCC patients and 15-20% of advanced HCC patients have been shown to be AFP-negative (serum AFP level is less than 20 ng/ml) [33].

Advances in genomics and proteomics approaches and biomarker assay techniques over the last decade have resulted in the identification of numerous novel biomarkers and have improved the diagnosis of HCC, both in cirrhotic and non-cirrhotic patients. However, all newly discovered biomarkers are still complementary to AFP, because diagnostic accuracy increases when these biomarkers are used in combination with AFP and/or with each other [34-36].

The currently recommended combined testing for HCC includes AFP, des-gamma-carboxyl prothrombin (DCP), and Lens culinaris agglutinin-reactive fraction of AFP (AFP-L3); all these three tumor markers are considered to be useful for monitoring treatment responsiveness and tumor recurrence. In addition, alpha-L-fucosidase, glypican-3 (GPC3), squamous cell carcinoma antigen-1 (SCCA-1), osteopontin (OPN), Golgi protein-73 (GOLPH2), carcinoembryonic antigen (CEA), vascular endothelial growth factor (VEGF) and matrix metalloproteinases-2 and -9 (MMP-2 and MMP-9) along with micro-RNAs are presently being extensively studied as novel molecular biomarkers for HCC [37-42].

Serum levels of fucosylated isoform of AFP (AFP-L3) and DCP, also known as prothrombin induced by vitamin K absence or antagonist-II (PIVKA-II), have been shown to correlate with tumor aggressiveness and prognosis [38]. Increase in the percentage of AFP-L3 over the total AFP level (>10%) has been shown to be very specific for small-sized HCC. PIVKA-II has been also reported to be more specific than total AFP level in detecting HCC and has been shown to be more reliable than AFP as prognostic tool for HCC recurrence and patients survival after hepatectomy, liver transplantation, radio-frequency ablation and trans-arterial chemoembolization treatment [43]. However, since the serum levels of these proteins are independent of each other, combined measurement of the two or three biomarkers has been demonstrated to increase their sensitivity and accuracy for HCC diagnosis.

Based on data on aberrant fucosylation during the development of HCC, analysis of relationship between fucosylation index and tumor progression in HBV-associated HCC was performed with the use of MALDI-TOF mass spectrometry for serum proteomic profiling to diagnose HCC at the early stages [44]. Aberrant serum fucosylated glycoproteins allowed distinguishing HCC from control groups: no liver disease, chronic hepatitis B without cirrhosis, and HBV-infected cirrhosis. When combined with serum AFP detection (AFP>20 ng/mL), the sensitivity/specificity is improved to 78/88%, 85/88%, and 89/91%, respectively.

Another tumor biomarker, GPC3 is a heparan sulfate proteoglycan expressed in more than 70% of HCC [45]. Expression of GPC3 was elevated in patients with both HCC and liver cirrhosis in comparison with healthy donors, but there was no difference between HCC and the cirrhosis in GPC level [46]. Additionally, no significant correlation between serum levels of GPC3 and AFP was observed. It is known that AFP is over-expressed in advanced HCC, while both mRNA for glypican-3 and the protein itself are highly expressed at the early stages of HCC.

GPC has not been demonstrated as a promising biomarker for HCC diagnosis and prognosis when used alone. However, a number of studies showed that GPC3 could be complementary to AFP in increasing diagnostic accuracy of the test for HCC. Combination of AFP with GPC3 improved sensitivity of the test for HCC up to 82% or 94%, depending on HCC type [45-47]. Down-regulation of GPC3 with the use of specific siRNA or anti-GPC3 antibodies was resulted in decrease in cell migration, metastasis and invasion [47]. The silencing of GPC3 gene transcription caused inhibition of tumor growth in nude mice xenograft. These data indicated that GPC3 might be a target for anticancer therapy.

Relatively recently, comparative proteomic profiling of highly fractionated plasma from patients with cirrhosis and HCC with the use of mass spectrometry identified osteopontin (OPN) as a new biomarker for the early diagnosis of HCC [48]. Gene expression profiling identified OPN as one of the leading proteins associated with HCC growth and metastasis [49]. OPN-positive HCC cells were shown to have much more chromosomal amplifications, and the regulation of OPN expression was provided by single nucleotide polymorphisms in the *OPN* gene promoter region. OPN sensitivity was better than that of AFP in differentiating all HCC groups tested – hepatitis B virus-related HCC, hepatitis C virus-related HCC and early-stage HCC, from cirrhosis.

The combination of both proteins, AFP and OPN, enhanced sensitivity and specificity of the test for detecting HCC, indicating that these molecular markers are complementary to each other. Mass spectrometry profiling showed increase in diagnostic performance of the test for HCC when OPN was used in combination with latent-transforming growth factor-β-binding protein-2 (LTBP2). Both OPN and LTBP2 were significantly elevated in HCC patients compared to those with other chronic liver diseases and healthy donors [50].

The proteomic approach was exploited to analyze the secretory releasing proteome of HCC tissues and to identify other plasma biomarkers. Overall, in this study 1365 proteins were identified in serum-free conditioned media [51]. There were enzymes and extracellular or membrane-bound proteins, and proteins participating in transforming growth factor-β (TGF-β)-signaling pathway among the identified proteins. Levels of AFP, metalloproteinase-1 (MMP-1), OPN, and pregnancy-specific beta-1-glycoprotein-9 (PSG-9) were shown to be significantly increased in HCC patients. Here, sensitivity and specificity of MMP-1 and OPN in the diagnosis of HCC were greater than that of AFP.

Expression of MMP-2, MMP-9 and VEGF in HCC tissues has been shown to closely correlate to the recurrence of HCC after hepatectomy [52]. The high levels of these proteins in the serum were due to high expression of their mRNAs in the tumor tissue of HCC patients [53]. Additionally, study of expression and clinical significance of other apoptosis- and angiogenesis-related factors showed that neither mutant p53, nor survivin levels correlate to HCC recurrence.

Wide-range global data mining of genomic and proteomic databases has been also used to screen novel biomarker candidates for detecting HCC [54]. cDNA microarray, copy number variation, somatic mutation, epigenetic, and quantitative proteomics data were combined with multiple reaction monitoring (MRM) based on triple quadrupole mass spectrometry technology. MRM data were used to verify HCC candidate biomarkers in individual serum samples from three groups: (i) healthy donors, (ii) HCC patients with no treatment, and (iii) HCC patients who underwent locoregional therapy. Four potential biomarkers: the actin-binding protein anillin (ANLN), filamin-B (FLNB), complementary C4-A (C4A), and AFP were identified as HCC biomarkers in this study. The combination of the biomarkers was shown to improve distinguishing HCC patients with no treatment from healthy donors.

Known Molecular Functions of HCC Biomarkers

AFP. For the last decade, attempts of different groups of investigators have led to revealing roles of different proteins in pathogenesis of HCC at the molecular level. Among these proteins, AFP is the most prominent tumor biomarker [55, 56] whose biological activity has been long studied and extensively reviewed [57-61]. Shortly, AFP has been shown to possess the ability of dual regulation of cell proliferation and survival depending on cell type and AFP concentration, along with immunosuppressive activity and capability of binding and transportation of different hydrophobic ligands [61]. Despite the mechanisms of AFP action is not elucidated, suggestion has been made that structure/function relationships exist between growth factors, cell adhesion proteins and AFP and that the effects of AFP may be provided through the mitogen-activated protein kinase (MAPK)-signaling pathway (Table 1) [59, 60].

Table 1. Signaling pathways experimentally shown to be dysregulated in HCC

№	Pathway	Canonical signaling ligands	Ligands involved in HCC progression	Known inhibitors	Cellular functions	References
1.	Ras/MAPK/ ERK/Myc	EGF, FGF, PDGF	AFP, DCP, VEGF, SCCA-1 and SCCA-2	Raf265, MEK162, sorafenib, vemurafenib	Cell proliferation, differentiation, survival and invasion, angiogenesis, protection from oxidative stress	Fujikawa et al., 2009; Moldogazieva, Terentiev, 2006; Matsubara et al., 2012; Turato et al., 2014; Pontisso, 2014
2.	PI3K/Akt/ mTOR	EGF, IGF-1, insulin, CAM	AFP	PTEN, GSK3B	Cell proliferation and migration, glucose metabolism	Gao et al.,, 2015; Li et al., 2011; Zhu et al., 2015
3.	Met/JAK1/ STAT3	IL-6, IL-10, LIF, INF-γ	DCP, BMP-2	CYT387, JSI-124, SOCS3, jakinibs, ruxolitinib, tofacitinib	Cell proliferation, inflammation, immune response	Inagaki et al., 2011
4.	Wnt/ β-catenin	BMPs, Nodal, FGFs	GPC3	APC, axin, GSK-3	Cell proliferation and migration, cell fate specification, epithelial-to-mesenchymal transition	Capurro et al., 2005; Villanueva et al., 2007; Moeini et al., 2012
5.	Integrins (CD44)/ (FAK)	Collagen, fibronectin, laminin, selectins, vitronectin	OPN, CEA	Y-37632, integrilin, roxifiban	Cell-to-ECM attachment, cell adhesion, migration and survival, angiogenesis	El-Tanani, 2008; Anborgh et al., 2010
6.	SMAD4		TGF-βs, BMPs, FGF, MMP-2 and MMP-9	I-SMADs (SMAD6 and SMAD7), sclerostin, TIMPs, marimastat	Cell cycle, proliferation, differentiation, adhesion and survival, angiogenesis	Lachenmayer, 2012; Hernanda, 2015
7.	TRAIL-I/ TRAIL-II (DR4, DR5 death receptors of TNF receptor family)	TRAIL/ caspases-8, -10 and -3; or Bcl-2/ caspases-9 and -3	TRAIL/ caspase-3	FLIPs, IAP, Bcl-XL, XIAP	Cell apoptosis	Li et al., 2009

Additionally, cytoplasmic AFP has been shown to be able to co-localize and interact with caspase-3, and to block apoptosis induced by TNF-alpha-related apoptosis-inducing ligand (TRAIL) and all-trans retinoic acid (ATRA) [62]. Cytoplasmic AFP has been also shown to function as a regulator of phosphatidyl-inositol-3-kinase (PI3K)/Akt signaling in human hepatoma cell line [63]. Inhibition of the PI3K/Akt pathway and down-regulation of AFP expression was obtained by Mir-1236 followed by inhibition of AFP-stimulated cell proliferation, migration, invasion and vasculogenic mimicry [64]. This was caused by phosphatase and tensin homolog (PTEN) accumulation and led to the inhibition of malignant phenotype of the hepatoma cells. AFP co-localized and interacted with PTEN, thus inducing CXCR4 chemokine receptor expression by activating AKT/mTOR signaling pathway and stimulating the hepatoma cell migration [65].

DCP. It is the abnormal prothrombin without carboxylation of γ-carbon atom in several glutamic acid residues at its N-terminus and, as a result, it does not have coagulation activity [66]. Instead, DCP has been shown to exhibit growth factor activity directly stimulating DNA synthesis and HCC cell proliferation both in autocrine and in paracrine manner, through Jenus kinase-1 (JAK1)-signal transducer and activator of transcription-3 (STAT3) signaling pathway by binding to the Met cell surface receptor. Additionally, DCP can promote vascular endothelial cell proliferation and migration through MAPK-mediated signaling pathway [67]. This was confirmed by immunohistochemical analysis revealing existence of correlation of expression of DCP with tumor size and hyperactive vasculature in HCC patients [68].

Existence of several variants of DCP differed by number of glutamic acid residues and their positions in the γ-carboxyglutamic (Gla) domain has been shown. Serum DCP level increases in patients with vitamin K deficiency or in patients who have obstructive jaundice. DCP elevated in these conditions was found to contain more Gla residues and was designated as NX-DCP that is also produced in HCC tissues, but is differed from DCP in expression level and by biological properties. Levels of DCP or NX-DCP expression and DCP/NX-DCP ratio were demonstrated to be closely related to malignant properties of HCC [69].

GPC3. Another HCC biomarker, GPC3, similarly to AFP, has been recognized as an oncofetal protein whose expression may be regulated in the manner similar to that of AFP [70]. GPC3 biosynthesis has been shown to increase in the regenerating liver and to be suppressed by zink fingers and homeoboxes protein-2 (Zhx2). It has been shown that GPC3 stimulates HCC growth through canonical Wnt-mediated signaling [71]. Remarkably, the Wnt-mediated pathways are involved in several processes including embryonic development, cell fate determination, proliferation, polarity, migration, and stem cell maintenance and are well-characterized oncogenic pathways in HCC [72-74].

Osteopontin. OPN is an additional HCC biomarker candidate that has been shown to be very acidic secreted glycophosphoprotein found in extracellular matrix (ECM). It has diverse functions and is normally expressed by variety of cells and tissues, including fibroblasts, osteoblasts, osteocytes, dendritic cells, macrophages, myoblasts, endothelial cells, brain, kidney and placental cells [75]. Its normal physiological roles include bone mineralization, regulation of immune response, vascular remodeling, wound repair and controlling the developmental processes. OPN has been shown to enhance adhesion, migration, invasion, and survival of cells and their attachment to ECM [76]. Hence, OPN has an important role in regulating cell signaling during tumor progression and metastasis [77, 78]. These functions are provided by binding of OPN to integrin and CD44 receptors to initiate signaling cascades.

Squamous cell carcinoma antigen. Newly proposed biomarker for HCC, squamous cell carcinoma antigen, was initially discovered as a 390 amino acid-containing member of the ovalbumin serine protease inhibitor (serpin) family and a tumor marker of squamous cell carcinoma [79]. Afterwards, existence of its two isoforms, SCCA-1 and SCCA-2, encoded by almost identical tandemly arranged genes were revealed [80, 81]. Expression of both SCCA-1 and SCCA-2 has been shown not to be restricted to the squamous epithelium; they also were found in other tissues.

These serpins may coordinately regulate cysteine and serine proteinase activity in both normal and transformed cells. SCCA-1, also known as SERPINB3, protects cells from oxidative stress, but in chronic liver damage this serpin may lead to hepatocellular carcinoma through inhibition of apoptosis, induction of epithelial-to-mesenchymal transition and through decrease in desmosomal junctions, increase in cell proliferation and invasiveness. Over-expression of SERPINB3 was found in the most aggressive HCCs with poor prognosis [82]. Mechanisms of tumor growth promotion induced by SERPINB3 include the inhibition of intra-tumor infiltration of natural killer cells, up-regulation of Myc oncogene and participation in Ras-mediated signaling [83].

A significant correlation of SCCA-1 with transforming growth factor-β (TGF-β) concentration at both mRNA and protein levels was observed. TGF-*β-initiated* signaling associated with Wnt target gene expression was also identified as one of the important features of the most aggressive HCCs [84]. Microarray studies have shown transcriptional over-expression of Smad4, the intracellular effector of TGF-β and bone morphogenesis protein (BMP) signaling pathway, in a subset of HCCs [85].

Evaluability of Network Analysis in HCC

Network theory allows describing relationships such as direct interactions or correlations between every pair of cellular components – genes, their protein products, or intermediates of metabolic processes. Modeling a system through the network construction provides understanding detailed mechanisms of the system functioning. Protein-protein interaction networks along with gene regulatory, signal transduction (receptor and non-receptor protein kinases and their intracellular effectors) and metabolic networks belong to molecular networks that may have relevance for cancer development [3-5].

Molecular networks are objects for study by systems biology. Systems biology approaches reveal the up- and down-regulated gene signatures, associated with biological processes and interactions between genes and their protein products, and provided information for understanding the mechanisms underlying development of a disease. Network construction based on analysis of co-expression and co-localization of different cancer-associated genes and their protein products with revealing their interaction partners allows: (i) studying molecular mechanisms underlying cancer pathogenesis, (ii) discovery of novel biomarkers, and (iii) finding targets for effective anticancer-therapy [9]. Formation of a common interaction network by different cancer-associated proteins may evidence existence of common functions and ability to exhibit their effects through the same signaling pathways and/or existence of a crosstalk between the pathways.

HCC is the heterogeneous disease, pathogenesis of which has not been fully elucidated and which demonstrates high resistance to conventional chemotherapy [73]. Several groups of investigators have shown participation of multiple signaling pathways in HCC progression (Table 1). Activation of Ras/MAPK, Akt/PI3K/mTOR and Wnt-mediated signaling pathways has been shown to be risk factors for early recurrence and poor prognosis for the disease [86-88]. These pathways initiated by growth factors such as epidermal growth factor (EGF), insulin-like growth factor (IGF), platelet-derived growth factor (PDGF), hepatocyte growth factor (HGF), and vascular endothelial growth factor (VEGF) are involved in regulation of cell proliferation, migration and differentiation, angiogenesis, tumor invasion and metastasis during HCC.

Evaluability of network analysis in HCC research may be demonstrated by the following example. Current usage of MAPK-mediated signaling inhibitors such as sorafenib, sunitinib, or gefitinib for the cancer treatment is based on the strategy to target a gene or several genes, and to affect the entire pathway. However, this approach fails to give promising results because the MAPK-mediated signaling belongs to redundant pathways, i.e., it has crosstalks that involve other pathways. Redundancy has been shown to be one of the mechanisms underlying the anti-cancer drug resistance [89]. Redundancy is an ability of two or more elements of a system to perform the same function so that inactivation of one of them could have no effect on dynamic processes described by the entire network. Computational framework developed to analyze gene expression data from HCC subsets with different drug responsiveness showed that network connectivity may be dramatically changed by a drug. Thus, describing network structure and features such as redundancy may be used as a basis for development of a new strategy in anti-cancer therapy.

With the use of DNA microarray in combination with quantitative PCR analysis, text mining and immunohistochemistry, Saito et al. [90] reveled genes whose mRNA expression was highly correlated with that of AFP in different liver cancer cell lines. Among these genes were those encode proteins of albumin family and transcription factors along with genes involved in lipid and iron metabolism, angiogenesis and blood coagulation. Intracellular effectors of MAPK- and Wnt-mediated pathways were also identified. Interestingly, the fifth highest correlation coefficient with AFP was shown for glypican-3 that is a component of Wnt-mediated signaling and the recognized biomarker used in combination with AFP for HCC diagnosis and prognosis. Protein interaction network constructed in this work showed existence of functional interlinkage between AFP and GPC3 through their common participation in tumor suppressor p53 transcription factor network (Figure 1).

Based on data demonstrating that almost all currently used biomarker proteins, including AFP, cancer antigen 125 (CA125) and prostate specific antigen (PSA), belong to secreted glycoproteins, Li et al. [91] analyzed the secretome of human metastatic HCC and assessed the up-regulation of glycosylation sites. This approach was also provided by observation that protein glycosylation is linked to the cancer development. Approximately half of 300 glycoproteins revealed in this study were shown to be related to either cancer as a whole, or liver cancer, or neoplastic metastasis. This was demonstrated by construction of the network of the up-regulated glycoproteins.

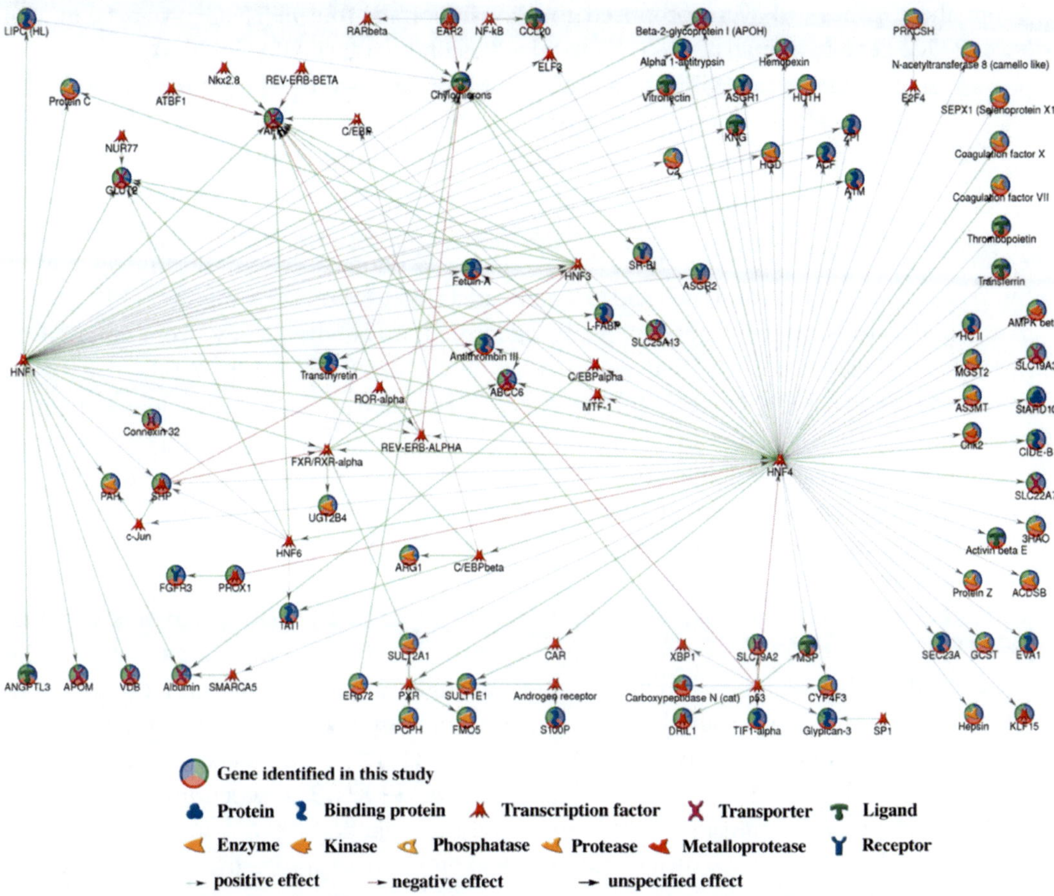

Figure 1. Network of genes expressed in a manner of coordination with AFP and implicated in AFP-positive HCC pathogenesis revealed by global mRNA expression analysis. This figure shows 213 genes whose expression level was significantly correlated with that of AFP as identified by Saito et al. [90]. Permission to use the figure is provided by Tadashi Kondo and by the publisher, Wiley Global.

Albumin family proteins including AFP were found to belong to liver cancer related glycoproteins. Gene ontology annotations allowed categorizing these proteins based on intracellular localization. Those proteins localized in a cell membrane composed 36% of total amount of proteins and included growth factor receptors such as ERBB3, FGFR4, ADAM17, ADAM10, CD152, and NRP1, while cytoplasmic proteins composed 27% of total amount of proteins and comprised intracellular effectors of growth factor-initiated signaling pathways – ERK1/2, Akt, p38/MAPK, and PI3K. Growth factors themselves, including EGF, FGF, and TGF-beta along with AFP, BMP and LTBP2 exhibited extracellular localization.

Another group of researchers developed molecular classifier for HCC diagnosis based on the identification of up-regulated and down-regulated differentially expressed genes located on network hubs, i.e., nodes characterized by the largest amount of connections [92]. For the network of up-regulated genes, the following 10 hubs were identified: MAPK1, SP1, HDAC1, YY1, ABL1, PTK2, SMAD2, NCOA3, CDC25A, and NCOA2. Seven hubs were revealed in the network of down-regulated genes: FOS, ESR1, JUNB, EGFR, SOCS3, FOLH1, and IGF1 among 27 hub genes.

From these genes, 14 were proposed to play important role in HCC progression. For instance, PTK2 gene in humans encodes a focal adhesion kinase (FAK) subfamily protein that participates in cell attachment to ECM and its inhibition leads to blocking tumor growth and recovery in HCC patients [93]. Suppressor of cytokine signaling-3 (SOCS3) inhibits JAK/STAT-mediated pathway and contributes to the tumor growth suppression. The epidermal growth factor receptor (EGFR) was shown to be expressed frequently in HCC, contributing to the tumor aggressiveness. Blockage of EGFR by cetuximab resulted in cell cycle arrest in G_1/G_0-stage due to increase in expression of cyclin-dependent kinase inhibitors, p21(Waf1/CIP1) and p27(Kip1), and decrease in cyclin D1 expression [94].

Among the 100 top up-regulated genes identified in another study from microarray datasets, the hallmark cancer-related genes were GPC3, CDK1, CD109, AFP, SERPINI1, CCNB1 and CCNB2 [95]. Gene ontology analysis revealed 43 hub genes that may play important role in HCC progression. The network hubs were remarkably enriched in genes involved in the following pathway: p53 signaling, cell cycle, oocyte meiosis, and progesterone-mediated oocyte maturation. Genes that are involved in drug metabolism, linoleic acid metabolism, retinol metabolism and metabolism of xenobiotics by cytochrome-P450 were, mainly, down-regulated and were not located in the network hubs.

Dynamic co-expression analysis of differentially expressed genes in hepatitis C virus-induced Huh7 HCC cells was performed by Song et al. [96]. Specific functional modules were analyzed at different time points to study changes in expression level of infection-related and hepatocarcinogenesis-related genes. Gene ontology analysis revealed the network functional modules simultaneously enriched in aminoacyl-tRNA synthetases, heat shock 70 kDa protein 2 (HSPA2), thioredoxin-interacting protein (TXNIP), cytochrome P-450 1A1 (CYP1A1), and glycerol-3-phosphate acyltransferase1 mitochondrial (GPAM).

In additional to traditional differential gene expression analysis, the identification of differentially co-expressed gene pairs to investigate the changes in expression correlations between genes has been proposed as a promising alternative to uncover gene modules involved in HCC pathogenesis [97]. Subnetworks enriched in differentially co-expressed gene pairs at each HCC stage were revealed and the network modules enriched in the cancer-related genes with corresponding functional annotations were found. In particular, APC, YWHAZ and IKBKG genes were found to be the most remarkable genes in the network modules, and their dynamic interaction partnership was shown to play an important role in HCC development.

At the pre-cancerous stages of HCC, the network modules were enriched in proliferation-inhibiting genes, simultaneously dismissing some apoptosis-inhibiting genes. Transition stage from early HCC to advanced HCC was shown to be enriched with the greatest number of functional terms. Seven genes in the advanced stage of HCC were involved in "negative regulation of apoptotic process," "wound healing" and "MAPK cascade" terms, functions frequently implicated in carcinogenesis or cancer metastasis. Altogether, 16 differentially co-expressed protein interaction pairs were revealed in multiple subnetworks; seven of them were connected to tumor suppressor APC (adenomatous polyposis coli)-centered module, in which four proteins: APC itself, cytohesin-2 (CYTH2), arrestin beta-2 (ARRB2), and catenin alpha-1 (CTNNA1), were shown to interact with each other and to be involved in the tumor metastasis [98].

Multidimensional gene expression analysis in combination with protein interaction network and clinical data analysis were recently used to demonstrate involvement of ECM and programmed cell death signaling in HCC progression [99]. Study of detailed mechanisms of ECM involvement in signal transduction revealed the role of chemokine CCL20-CCR6 axis in HCC progression.

Contribution of ECM into heterogeneous character of HCC has been additionally demonstrated by another study through the investigation of functional and topological features of HCC transcriptome. Gene clustering analysis identified involvement of the following functional terms in HCC pathogenesis: "extracellular matrix organization," "cell-cycle," and "translational elongation" [100]. Co-expression of the following well-connected genes were related to the cell cycle regulation: ASP homolog (ASPM), cyclin-dependent kinase inhibitor 3 (CDKN3), never in mitosis gene a (NIMA), Rac GTPase activating protein 1 (RACGAP1), and TOP2A topoisomerase. The authors proposed these results to be a resource for further identification and validation of potential target genes implicated in HCC pathogenesis.

Conclusion

Alpha-fetoprotein has been proposed to function in coordinated and cooperative manner with other embryo-specific and cancer-associated proteins and, hence, interactions of all the proteins may be described with the use of network theory. To date, there is low overlap in genomic and proteomic signatures involved in pathogenesis of HCC across multiple studies. Some studies have revealed *AFP* gene to be dysregulated in HCC, while others have not. This may be explained by heterogeneous character of HCC and by the involvement of wide range of genes and variety of redundant signaling pathways in HCC development. Changes in the expression level of some genes may be specific for HCC, while other gene signatures may by common for the most tumors or for the process of tumorigenesis, as a whole. Additionally, it has been commonly observed that some portion of HCC subsets are AFP-negative (AFP serum concentration is less than 20 ng/ml).

Nevertheless, several years after Debruyne and Delange [29], we would conclude that AFP is still remaining the "golden standard" among molecular biomarkers used in HCC diagnosis and prognosis. Capability of all other biomarkers identified to-date to increase the specificity and sensitivity of the test for HCC is observed only in combined usage with AFP or with each other. Further investigations are necessary to elucidate molecular mechanisms underlying HCC pathogenesis and to discover novel highly specific and sensitive biomarker candidates in testing for HCC. This requires usage of integrated approaches including systems biology, structural biology and computational biology technologies. Network analysis provides very promising approach to reach this goal.

References

[1] Wang, E. (2010) *Cancer systems biology.* Boca Raton, FL: CRC Press, Taylor and Francis Group, 456 p.

[2] Wolkenhauer, O., Auffray, C., Baltrusch, S., Blüthgen, N., Byrne, H., Cascante, M., Ciliberto, A., Dale, T., Drasdo, D., Fell, D., Ferrell, J.E. Jr., Gallahan, D., Gatenby, R.,

Günther, U., Harms, B.D., Herzel, H., Junghanss, C., Kunz, M., van Leeuwen, I., Lenormand, F., Levi, F., Linnebacher, M., Lowengrub, J., Maini, P.K., Malik, A., Rateitschak, K., Sansom, O., Schäfer, R., Schürrle, K., Sers, K., Schnell, S., Shibata, D., Tyson, J., Vera, J., White, M., Zhivotovsky, B., and Jaster R. (2010) Systems biologists seek fuller integration of systems biology approaches in new cancer research programmes. *Cancer Res., 70(1),* 12-13.

[3] Wolkenhauer, O., Auffray, C., Jaster, R., Steinhoff, G., Damman, O. (2013) The road from systems biology to systems medicine. *Pediatr. Res., 73(4 Pt.2),* 502-507.

[4] Friboulet, A., Thomas, D. (2005) Systems biology – an interdisciplinary approach. *Biosens. Bioelectron., 20(12),* 2404-2407.

[5] Ferrel, J.E. Jr. (2009) Q and A: Systems biology. *BMC J. Biol., 8,* 2.

[6] Coveney, P.V., Fowler, P.W. (2005) Modelling biological complexity: a physical scientist's perspective. *J. R. Soc. Interface, 2(4),* 267-280.

[7] Terentiev, A.A., Moldogazieva, N.T., Shaitan, K.V. (2009) Dynamic proteomics in modeling of the living cell. Protein-protein interactions. *Biochemistry (Mosc.), 74(13),* 1586-1607.

[8] Aebersold, R., Auffray, C., Baney, E., Barillot, E., Brazma, A., Brett, C., Brunak, S., Butte, A., Califano, A., Celis, J., Čufer, T., Ferrell, J., Galas, D., Gallahan, D., Gatenby, R., Goldbeter, A., Hace, N., Henney, A., Hood, L., Iyengar, R., Jackson, V., Kallioniemi, O., Klingmüller, U., Kolar, P., Kolch, W., Kyriakopoulou, C., Laplace, F., Lehrach, H., Marcus, F., Matrisian, L., Nolan, G., Pelkmans, L., Potti, A., Sander, C., Seljak, M., Singer, D., Sorger, P., Stunnenberg, H., Superti-Furga, G., Uhlen, M., Vidal, M., Weinstein, J., Wigle, D., Williams, M., Wolkenhauer, O., Zhivotovsky, B., Zinovyev, A., Zupan, B. (2009) Report on EU–USA Workshop: How Systems Biology Can Advance Cancer Research. *Mol. Oncol. 3(1),* 9-17.

[9] Somvanshi, P.R., Venkatesan, K.V. (2014) A conceptual review on systems biology in health and diseases: from biological networks to modern therapeutics. *Syst. Synth. Biol., 8(1),* 99-116.

[10] Aghoram, R., Cai, P., Dickinson, J.A. (2012) Alpha-foetoprotein and/or liver ultrasonography for screening of hepatocellular carcinoma in patients with chronic hepatitis B. *Cochrane Database Syst. Rev., 9,* CD002799.

[11] Shao, Y.Y., Hsu, C.H., Cheng, A.L. (2015) Predictive biomarkers of sorafenib efficacy in advanced hepatocellular carcinoma: Are we getting there? *World J. Gastroenterol., 21(36),* 10336-10347.

[12] Poon, T.C., Yip, T.T., Chan, A.T., Yap, C., Yip, V., Mok, T.S., Lee, C.C., Leung, T.W., Ho, S.K., Johnson, P.J. (2003) Comprehensive proteomic profiling identifies serum proteomic signatures for detection of hepatocellular carcinoma and its subtypes. *Clin. Chem., 49(5),* 752-760.

[13] Mustafa, G.M., Larry, D., Petersen, J.R., Elferink, C.J. (2012) Targeted proteomics for biomarker discovery and validation of hepatocellular carcinoma in hepatitis C infected patients. *World J. Hepatol., 8(10),* 1312-1324.

[14] Anderson, N.L. (2005) The roles of multiple proteomic platforms in a pipeline for new diagnostics. *Mol. Cell Proteomics, 4(10),* 1441-1444.

[15] El-Aneed, A., Banoub, J. (2006) Proteomics in the diagnosis of hepatocellular carcinoma: focus on high risk hepatitis B and C patients. *Anticancer Res. 26(5A),* 3293-3300.

[16] Rafalko, A., Dai, S., Hancock, W.S., Karger, B.L., Hincapie, M. (2012) Development of a Chip/Chip/SRM platform using digital chip isoelectric focusing and LC-Chip mass spectrometry for enrichment and quantitation of low abundance protein biomarkers in human plasma. *J. Proteome Res., 11(2),* 808-817.

[17] He, X., Wang, Y., Zhang W., Li, H., Luo, R., Zhou, Y., Liao, C.L., Huang, H., Li, X., Xie, Z., He, M. (2014) Screening differential expression of serum proteins in AFP-negative HBV-related hepatocellular carcinoma using iTRAQ-MALDI-MS/MS. *Neoplasma, 61(1),* 17-26.

[18] Huang, X., Zeng, Y., Xing, X., Zeng, J., Cao, Y., Cai, Z., Xu, B., Liu, X., Huang, A., Liu, J. (2014) Quantitative proteomics analysis of early recurrence/metastasis of huge hepatocellular carcinoma following radical resection. *Proteome Sci., 12,* 22.

[19] Liu, T., Xue, R., Huang, X., Zhang, D., Dong, L., Wu, H., Shen, X. (2011) Proteomic profiling of hepatitis B virus-related hepatocellular carcinoma with magnetic bead-based matrix-assisted laser desorption/ionization time-of-flight mass spectrometry. *Acta Biochim. Biophys. Sin. (Shanghai), 43(7),* 542-550.

[20] Banaudha, K.K., Verma, M. (2015) Epigenetic biomarkers in liver cancer. *Methods Mol. Biol., 1238,* 65-76.

[21] Zhang, Y. (2015) Detection of epigenetic aberrations in the development of hepatocellular carcinoma. *Methods Mol. Biol., 1238,* 709-731.

[22] Pogribny, I.P., Rusyn, I. (2014) Role of epigenetic aberrations in the development and progression of human hepatocellular carcinoma. *Cancer Letts., 342(2),* 223-230.

[23] Anwar, S.L., Lehmann, U. (2014) DNA methylation, microRNAs, and their crosstalk as potential biomarkers in hepatocellular carcinoma. *World J. Gastroenterol., 20(24),* 7894-7913.

[24] Ma, L., Chua, M.S., Andrisani, O., So, S. (2014) Epigenetics in hepatocellular carcinoma: an update and future therapy perspectives. *World J. Gastroenterol., 20(2),* 333-345.

[25] El-Serag, H.B., Rudolf, K.L. (2007) Hepatocellular carcinoma: epidemiology and molecular carcinogenesis. *Gastroenterology, 132(7),* 2557-2576.

[26] Zhang, D.Y., Friedman, S.L. (2012) Fibrosis-dependent mechanism of hepatocarcinogenesis. *Hepatology, 56(2),* 769-775.

[27] Caldwell, S., Park, S.H. (2009) The epidemiology of hepatocellular cancer: from the perspectives of public health problem to tumor biology. *J. Gastroenterol., 44(Suppl. 19),* 96-101.

[28] Kim do, Y., Han, K.H. (2012) Epidemiology and surveillance of hepatocellular carcinoma. *Liver Cancer, 1(1),* 2-14.

[29] Debruyne, E.N., Delange, J.R. (2008) Diagnosing and monitoring hepatocellular carcinoma with alpha-fetoprotein: new aspects and applications. *Clin. Chim. Acta, 395(1-2),* 19-26.

[30] Tyson, G.L., Duan, Z., Kramer, J.R., Davila, J.A., Richardson, P.A., El-Serag, H.B. (2011) Level of α-fetoprotein predicts mortality among patients with hepatitis C-related hepatocellular carcinoma. *Clin. Gastroenterol. Hepatol., 9(11),* 989-994.

[31] Peng, S.Y., Chen, W.J., Lai, P.L., Jeng, Y.M., Sheu, J.C., Hsu, H.C. (2004) High alpha-fetoprotein level correlates with high stage, early recurrence and poor prognosis of hepatocellular carcinoma: significance of hepatitis virus infection, age, p53 and beta-catenin mutations. *Int. J. Cancer, 112(1),* 44-50.

[32] Gupta, S., Bent, S., Kohlwes, J. (2003) Test characteristics of alpha-fetoprotein for detecting hepatocellular carcinoma in patients with hepatitis C. A systematic review and critical analysis. *Ann. Intern. Med., 139,* 46-50.

[33] Xu, J., Liu, C., Zhou, L., Tian, F., Tai, M.H., Wei, J.C., Qu, K., Meng, F.D., Zhang, L.Q., Wang, Z.X., Zhang, J.Y., Chang, H.L., Liu, S.N., Xu, X.S., Song, Y.Z., Liu, J., Zhang, P. (2012) Distinctions between clinicopathological factors and prognosis of alpha-fetoprotein negative and positive hepatocelluar carcinoma patients. *Asian Pac. J. Cancer Prev., 13,* 559-562.

[34] Donati, M., Brancato G., Donati, A. (2004) Clinical biomarkers in hepatocellular carcinoma (HCC). *Front. Biosci. (Schol. Ed.), 2,* 571-577.

[35] Lim, T.S., Kim, do Y., Han, K.H., Kim, H.S., Shin, S.H., Jung, K.S., Kim, B.K., Kim, S.U., Park, J.Y., Ahn, S.H. (2016) Combined use of AFP, PIVKA-II, and AFP-L3 as tumor markers enhances diagnostic accuracy for hepatocellular carcinoma in cirrhotic patients. *Scand. J. Gastroenterol., 51(3),* 344-353.

[36] Tsuchiya, N., Sawada, Yu., Endo, I., Saito, K., Uemura, Y., Nakatsura, T. (2015) Biomarkers for the early diagnosis of hepatocellular carcinoma. *World J. Gastroenterol., 21(37),* 10573-10583.

[37] Nakatsura, T., Nishimura, Y. (2005) Usefulness of the novel oncofetal antigen for diagnosis of hepatocellular carcinoma and melanoma. *BioDrug, 19(2),* 71-77.

[38] Beale, G., Chattopadhyay, D., Gray, J., Stewart, S., Hudson, M., Day, C., Trerotoli, P., Giannelli, G., Manas, D., Reeves, H. (2008) AFP, PIVKAII, GP3, SCCA-1 and follisatin as surveillance biomarkers for hepatocellular cancer in non-alcoholic and alcoholic fatty liver disease. *BMC Cancer, 8,* 200.

[39] Gan, Y., Liang, Q., Song, X. (2014) Diagnostic value of alpha-L-fucosidase for hepatocellular carcinoma: a meta-analysis. *Tumor Biol., 35(5),* 3953-3960.

[40] Chen, I.P., Ariizumi, S., Nakano, M., Yamamoto, M. (2014) Positive glypican-3 expression in early hepatocellular carcinoma predicts recurrence after hepatectomy. *J. Gastroenterol., 49(1),* 117-125.

[41] Ho, M., Kim, H. (2011) Glypican-3: a new target for cancer immunotherapy. *Eur. J. Cancer, 47(3),* 333-338.

[42] Mao, Y., Yang, H., Xu, H., Lu, X., Sang, X., Du, S., Zhao, H., Chen, W., Xu, Y., Chi, T., Yang, Z., Chen, J., Zhong, S., Mohanti, S.R., Lopez-Soler, R., Millis, J.M., Huang, J., Zhang, H. (2010) Golgi protein 73 (GOLPH2) is a valuable serum marker for hepatocellular carcinoma. *Gut, 59(12),* 1687-1693.

[43] Seo, S.I., Kim, H.S., Kim, W.J., Shin, W.G., Kim, D.J., Kim, K.H., Jang, M.K., Lee, J.H., Kim, J.S., Kim, H.Y., Kim, D.J., Lee, M.S., Park, C.K. (2015) Diagnostic value of PIVKA-II and alpha-fetoprotein in hepatitis B virus-associated hepatocellular carcinoma. *World J. Gastroenterol., 21(13),* 3928-3935.

[44] Liao, J., Zhang, R., Qian, H., Cao, L., Zhang, Y., Xu, W., Li, J., Wu, M., Yin, Z. (2012) Serum profiling based on fucosylated glycoproteins for differentiating between chronic hepatitis B and hepatocellular carcinoma. *Biochem. Biophys. Res. Commun., 240(2),* 308-314.

[45] Capurro, M., Wanless, I.R., Sherman, M., Deboer, G., Shi, W., Miyoshi, E., Filmus, J. (2003) Glypican-3: a novel serum and histochemical marker for hepatocellular carcinoma. *Gastroenterology, 125(1),* 89-97.

[46] Jia, X., Gao, Y., Zhai, D., Liu, J., Cai, J., Wang, Y., Jing, L., Du, Z. (2015) Assessment of the clinical utility of glypican-3 as a serum marker for the diagnosis of hepatocellular carcinoma. *Technol. Cancer Res. Treat., pii: 1533034615605248.*

[47] Wang, L., Yao, M., Pan, L.H., Qian, Q., Yao, D.F. (2015) Glypican-3 is a biomarker and a therapeutic target of hepatocellular carcinoma. *Hepatobiliary Pancreat. Dis. Int., 14(4),* 361-366.

[48] Shang, S., Plymoth, A., Feng, Z., Rosen, H.R., Sangrajrang, S., Hainaut, P., Marrero, J.A., Beretta, L. (2012) Identification of osteopontin as a novel marker for early hepatocellular carcinoma. *Hepatology, 55(2),* 483-490.

[49] Qin, L. (2014) Osteopontin is a promoter for hepatocellular carcinoma metastasis: a summary of 10 years of studies. *Front. Med., 8(1),* 24-32.

[50] Da Costa, A.N., Plymoth, A., Santos-Silva, D., Ortiz-Cuaran, S., Camey, S., Guilloreau, P., Sangrajrang, S., Khuhaprema, T., Mendy, M., Lesi, O.A., Chang, H.K., Oh, J.K., Lee, D.H., Shin, H.R., Kirk, G.D., Merle, P., Beretta, L., Hainaut, P. (2015) Osteopontin and latent-TGF β binding-protein 2 as potential diagnostic markers for HBV-related hepatocellular carcinoma. *Int. J. Cancer, 136(1),* 172-181.

[51] Yang, L., Rong, W., Xiao, T., Zhang, Y., Xu, B., Liu, Y., Wang, L., Wu, F., Qi, J., Zhao, X., Wang, H., Han, N., Guo, S., Wu, J., Gao, Y., Cheng, S. (2013) Secretory/releasing proteome-based identification of plasma biomarkers in HBV-associated hepatocellular carcinoma. *Sci. China Life Sci., 56(7),* 638-646.

[52] Guo, R.P., Zhong, C., Shi, M., Zhang, C.Q., Wei, W., Zhang, Y.Q., Li, J.Q. (2006) [Expression and clinical significance of certain apoptosis and angiogenesis factors in hepatocellular carcinoma]. *Zhonghua Wai Ke Za Zhi, 44(23),* 1626-1630.

[53] Niu, Q., Tang, Z., Ma, Z., Qin, L., Bao, W., Zhang, L. (2001) [Relationship between serum matrix metalloproteinase-2 and metastasis and recurrence following radical hepatic resection in hepatocellular carcinoma]. *Zhonghua Gan Zang Bing Za Zhi, Suppl.1,* 58-60.

[54] Kim, H., Kim, K., Yu, S.J., Jang, E.S., Yu, J., Cho, G., Yoon, J.H., Kim, Y. (2013) Development of biomarkers for screening hepatocellular carcinoma using global data mining and multiple reaction monitoring. *PLoS One 8(5),* e63468.

[55] Mizejewski, G.J. (2001) Alpha-fetoprotein structure and function: relevance to isoforms, epitopes and conformational variants. *Exp. Biol. Med., 226,* 377-408.

[56] Mizejewski, G.J. (2002) Biological role of alpha-fetoprotein in cancer: prospects for anticancer therapy. *Expert Rev. Anticancer Ther., 2(6),* 709-735.

[57] Mizejewski, G.J. (2015) Alpha-fetoprotein and inflammation: is AFP an acute and/or chronic phase reactant? *J. Hematol. Thrombo. Dis., 3(1),* 1000191.

[58] Terentiev, A.A., Moldogazieva, N.T. (2006) Structural and functional mapping of alpha-fetoprotein. *Biochemistry (Mosc.), 71(2),* 120-132.

[59] Moldogazieva, N.T., Terentiev, A.A. (2006) Alpha-fetoprotein and growth factors. Structure/function relationships and analogies. *Adv. Biol. Chem., 46,* 99-148 (Rus.).

[60] Terentiev, A.A., Moldogazieva, N.T. (2007) Cell adhesion proteins and alpha-fetoprotein. Similar structural motifs as prerequisites for common functions. *Biochemistry (Mosc.), 72(9),* 920-935.

[61] Terentiev, A.A., Moldogazieva, N.T. (2013) Alpha-fetoprotein: a renaissance. *Tumor Biol., 34(4),* 2075-2091.

[62] Li, M., Li, H., Li, C., Zhou, S., Guo, L., Liu, H., Jiang, W., Liu, X., Li, P., McNutt, M.A., Li, G. (2009) Alpha fetoprotein is a novel protein-binding partner for caspase-3 and blocks the apoptotic signaling pathway in human hepatoma cells. *Int. J. Cancer, 124(12),* 2845-2854.

[63] Li, M., Li, H., Li, C., Wang, S., Jiang, W., Liu, Z., Zhou, S., Liu, X., McNutt, M.A., Li, G. (2011) Alpha-fetoprotein: a new member of intracellular signal molecules in regulation of the PI3K/AKT signaling in human hepatoma cell lines. *Int. J. Cancer, 128,* 524-532.

[64] Gao, R., Cai, C., Gan, J., Yang, X., Shuang, Z., Liu, M., Li, S., Tang, H. (2015) Mir-1236 down-regulates alpha-fetoprotein, thus causing PTEN accumulation, which inhibits the PI3K/AKT pathway and malignant phenotype in hepatoma cells. *Oncotarget, 6,* 6014-6028.

[65] Zhu, M., Guo, J., Xia, H., Li, W., Lu, Y., Dong, X., Chen, Y., Xie, X., Fu, S., Li, M. (2015) Alpha-fetoprotein activates AKT/mTOR signaling to promote CXCR4 expression and migration of hepatoma cells. *Oncoscience, 2,* 59-70.

[66] Inagaki, Y., Tang, W., Makuuchi, M., Hasegawa, K., Sugawara, Y., Kokudo, N. (2011) Clinical and molecular insights into the hepatocellular carcinoma tumour marker des-γ-carboxyprothrombin. *Liver Int., 31(1),* 22-35.

[67] Fujikawa, T., Shiraha, H., Yanamoto, K. (2009) Significance of des-gamma-carboxy prothrombin production in hepatocellular carcinoma. *Acta Med. Okayama, 63(6),* 299-304.

[68] Matsubara, M., Shiraha, M., Kataoka, J., Iwamuro, M., Horiguchi, S., Nishina, S., Takaoka, N., Uemura, M., Takaki, A., Nakamura, S., Kobayashi, Y., Nouso, K., Yamamoto, K. (2012) Des-γ-carboxyl prothrombin is associated with tumor angiogenesis in hepatocellular carcinoma. *J. Gastroenterol. Hepatol., 27(10),* 1602-1608.

[69] Sumi, A., Akiba, J., Ogasawara, S., Nakayama, M., Nomura, Y., Yasumoto, M., Sanada, S., Nakashima, O., Abe, T., Yano, H. (2015) Des-γ-carboxyprothrombin (DCP) and NX-DCP expressions and their relationship with clinicopathological features in hepatocellular carcinoma. *PLoS One, 10(3),* e0118452.

[70] Morford, L.A., Davis, C., Jin, L., Dobierzewska, A., Peterson, M.L., Spear, B.T. (2007) The oncofetal gene glypican-3 is regulated in the postnatal liver by zinc fingers and homeoboxes 2 and in the regenerating liver by alpha-fetoprotein regulator 2. *Hepatology, 46,* 1541-1547.

[71] Capurro, M.I., Xiang, Y.Y., Lobe, C., Filmus, J. (2005) Glypican-3 promotes the growth of hepatocellular carcinoma by stimulating canonical Wnt signaling. *Cancer Res., 65(14),* 6245-6254.

[72] Villanueva, A., Newell, P., Chiang, D.Y., Freiedman, S.L., Llovet, J.M. (2007) Genomics and signaling pathways in hepatocellular carcinoma. *Semin. Liver Dis., 27(1),* 55-76.

[73] Moeini, A., Cornella, H., Villanueva, A. (2012) Emerging signaling pathways in hepatocellular carcinoma. *Liver Cancer, 1(2),* 83-93.

[74] Pez, F., Lopez, A., Kim, M., Wands, J.R., Caron de Fromentel, C., Merle, P. (2013) Wnt signaling and hepatocarcinogenesis: molecular targets for the development of innovative anticancer drugs. *J. Hepatol., 59(5),* 1107-1117.

[75] Denhardt, D.T., Guo, X. (1993) Osteopontin: a protein with diverse functions. *FASEB J., 7,* 1475-1482.

[76] El-Tanani, M.K. (2008) Role of osteopontin in cellular signaling and metastatic phenotype. *Front. Biosci., 13,* 4276-4284.

[77] Anborgh, P.H., Mutrie, J.C., Tuck, A.B., Chambers, A.F. (2010) Role of the metastasis-promoting protein osteopontin in the tumour microenvironment. *J. Cell. Mol. Med., 14,* 2037-2044.

[78] Rangaswami, H., Bulbule, A., Kundu, G.C. (2006) Osteopontin: role in cell signaling and cancer progression. *Trends Cell. Biol. 16(2),* 79-87.

[79] Suminami, Y., Kishi, F., Sekiguchi, K., Kato, H. (1991) Squamous cell carcinoma antigen is a new member of the serine protease inhibitors. *Biochem. Biophys. Res. Commun., 181(1),* 51-58.

[80] Suminami, Y., Kishi, F., Murakami, A., Sakaguchi, Y., Nawata, S., Numa, F., Kato, H. (2001) Novel forms of squamous cell carcinoma antigen transcripts produced by alternative splicing. *Biochim. Biophys. Acta, 1519 (102),* 122-126.

[81] Cataltepe, S., Gornstein, E.R., Schick, C., Kamachi, Y., Chatson, K., Fries, J., Silverman, G.A., Upton, M.P. (2000) Co-expression of the squamous cell carcinoma antigens 1 and 2 in normal adult human tissues and squamous cell carcinomas. *J. Histochem. Cytochem. 48(1),* 113-122.

[82] Turato, C., Vitale, A., Fasolato, S., Ruvoletto, M., Terrin, L., Quarta, S., Ramirez Morales, R., Biasiolo, A., Zanus, G., Tan, P.S., Hoshida, Y., Gatta, A., Cillo, U., Pontisso, P. (2014) SERPINB3 is associated with TGF-β1 and cytoplasmic β-catenin expression in hepatocellular carcinomas with poor prognosis. *Br. J. Cancer, 110(11),* 2708-2715.

[83] Pontisso, P. (2014) Role of SERPINB3 in hepatocellular carcinoma. *Ann. Hepatol. 13(6),* 722-727.

[84] Lachenmayer, A., Alsinet, C., Savic, R., Cabellos, L., Toffanin, S., Hoshida, Y., Vilanueva, A., Minquez, B., Newell, P., Tsai, H.W., Barretina, J., Thung, S., Ward, S.C., Bruix, J., Mazzaffero, V., Schwartz, M. (2012) Wnt-pathway activation in two molecular classes of hepatocellular carcinoma and experimental modulation by sorafenib. *Clin. Cancer Res., 18(18),* 4997-5007.

[85] Hernanda, P.Y., Chen, K., Das, A.M., Sideras, K., Wang, W., Li, J., Cao, W., Bots, S.J., Kodach, L.L., de Man, R.A., Ijzermans J.N., Janssen, H.L., Stubbs, A.P., Sprengers, D., Bruno, M.J., Metselaar, H.J., Ten Hagen, T.L., Kwekkeboom, J., Peppelenbosch, M.P., Pan, Q. (2015) SMAD4 exerts a tumor-promoting role in hepatocellular carcinoma. *Oncogene, 34(39),* 5055-5068.

[86] Nakanishi, K., Sakamoto, M., Yamasaki, S., Todo, S., Hirohashi, S. (2005) Akt phosphorylation is a risk factor for early disease recurrence and poor prognosis in hepatocellular carcinoma. *Cancer, 103,* 307-312.

[87] Lachenmayer, A., Alsinet, C., Savic, R., Cabellos, R., Toffanin, S., Hoshida, Y., Villanueva, A., Minquez, B., Newell, P., Tsai, H.W., Barreta, J., Thung, S., Ward, S.C., Bruix, J., Mazzaferro, V., Scwartz, M., Freidman, S.L., Llovet, J.M. (2012) Wnt-pathway activation in two molecular classes of hepatocellular carcinoma and experimental modulation by sorafenib. *Clin. Cancer Res., 18(18),* 4997-5007.

[88] Zhao, L.J., Wang, L., Ren, H., Cao, J., Li, L., Ke, J.S., Qi, Z.T. (2005) Hepatitis C virus E2 protein promotes human hepatoma cell proliferation through the MAPK/ERK signaling pathway via cellular receptors. *Exp. Cell Res., 305(1),* 23-32.

[89] Lavi, O., Skinner, J., Gottesman, M.M. (2014) Network features suggest new hepatocellular carcinoma treatment strategies. *BMC Syst. Biol., 8,* 88.

[90] Saito, S., Ojima, H., Ichikawa, H., Hirohashi, S., Kondo, T. (2008) Molecular background of alpha-fetoprotein in liver cancer cells as revealed by global RNA expression analysis. *Cancer Sci., 99,* 2402–2409.

[91] Li, X., Jiang, J., Zhao, X., Wang, J., Han, H., Zhao, Y., Peng, B., Zhong, R., Ying, W., Qian, X. (2013) N-glycoproteome analysis of the secretome of human metastatic hepatocellular carcinoma cell lines combining hydrazide chemistry, HILIC enrichment and mass spectrometry. *PLoS One, 8(12),* e81921.

[92] Zhang, Y., Wang, S., Li, D., Zhnag, J., Gu, D., Zhu, Y., He, F. (2011) A systems biology-based classifier for hepatocellular carcinoma diagnosis. *PLoS One, 6(7),* e22426.

[93] Bagi, C., Christensen, J., Cohen, D., Roberts, W., Wilkie, D., Swanson, T., Tuthill, T., Andresen, C.J. (2009) Sunitinib and PF-562,271 (FAK/Pyk2 inhibitor) effectively block growth and recovery of human hepatocellular carcinoma in a rat xenograft model. *Cancer Biol. Ther., 8(9),* 856-865.

[94] Huether, A., Hopfner, M., Baradari, V., Schuppan, D., Scherubl, H. (2005) EGFR blockade by cetuximab alone or as combination therapy for growth control of hepatocellular cancer. *Biochem. Pharmacol., 70,* 1568-1578.

[95] Jin, B., Wang, W., Du, G., Huang, G.Z., Han, L.T., Tang, Z.Y., Fan, D.G., Li, J., Zhang, S.Z. (2015) Identifying hub genes and dysregulated pathways in hepatocellular carcinoma. *Eur. Rev. Med. Pharmacol. Sci., 19(4),* 592-601.

[96] Song, Q., Zhao, C., Ou, S., Meng, Z., Kang, P., Fan, L., Qi, F., Ma, Y. (2015) Co-expression analysis of differentially expressed genes in hepatitis C virus-induced hepatocellular carcinoma. *Mol. Med. Rep., 11(1),* 21-28.

[97] Yu, H., Lin, C-C., Li, Y-Y., Zhao, Z. (2013) Dynamic protein interaction modules in human hepatocellular carcinoma progression. *BMC Syst. Biol., 7(S5),* S2.

[98] Su, L.K., Vogelstein, B., Kinzler, K.W. (1993) *Association of the APC tumor suppressor protein with catenins. Science, 262(5140),* 1734-1737.

[99] Zhuang, L., Wu, Y., Han, J., Ling, X., Wang, L., Zhu, C., Fu, Y. (2014) A network biology approach to discover the molecular biomarker associated with hepatocellular carcinoma. *Biomed. Res. Int., 2014,* 278956.

[100] Drozdov, I., Bornschein, J., Wex, T., Valeyev, N.V., Tsoka, S., Maifertheiner, P. (2012) Functional and topological properties in hepatocellular carcinoma transcriptome. *PLoS One, 7(4),* e35510.

In: Alpha-Fetoprotein: Functions and Clinical Applications ISBN: 978-1-63484-875-6
Editors: Nisha Lakhi and Michael Moretti

Chapter 6

ALPHA-FETOPROTEIN IN MALIGNANT PEDIATRIC TUMORS

Lisa VanHouwelingen*[1], *MD, and John A. Sandoval*[2,*], *MD
[1]St. Jude Children's Research Hospital, Memphis, TN, US
[2]Baptist Children's Hospital, Miami, FL, US

ABSTRACT

This concise chapter summarizes the role of serum α-fetoprotein (AFP) as a useful biomarker in malignant pediatric tumors. The fetal yolk sac and liver generate high levels of AFP during gestation, decline over the next 12 months of infancy, and only trace amounts are detected in childhood. As a result, persistent elevation of AFP correlates with a number of select pediatric malignant conditions. Serum AFP is over expressed in a considerable fraction of germ cell tumors (GCTs), hepatoblastoma (HB) and hepatocellular carcinoma (HCC). We provide the reader with a review of AFP as a useful specific marker for the diagnosis, management and follow-up in pediatric cancers.

Keywords: α-fetoprotein, oncofetal protein, germ cell tumors, hepatoblastoma, and hepatocellular carcinoma

INTRODUCTION

The accurate diagnosis of pediatric solid tumors and the timely recognition of disease recurrence involve the combination of clinical suspicion, imaging techniques, tissue biopsy, and serum tumor markers. Traditional cancer biomarkers are biologically measured substances that are expressed by malignant tissues, or generated by the host in response to the tumor and aid the clinician in diagnosis, staging, assessing response to treatment and detecting disease recurrence [1]. Tumor markers can be measured in the blood, cerebrospinal fluid, or serous effusions. In general, elevated levels of these biomarkers have been

* Corresponding Author: John.Sandoval@BMG.MD.

implicated in playing an important role in different types of cancer, and some play a significant role as a prognostic marker [2].

Alpha-fetoprotein (AFP) was identified in human fetuses in 1956 by Bergstrand and Czar [3]. Normally produced during gestation by the yolk sac and the liver during fetal life, AFP is a glycoprotein that is highly elevated in the circulation of newborns and its concentration decreases (half-life, 5 days) to 10-20 ug/L during the first 12 months of infant life [4]. A major mammalian oncofetal protein, AFP is a member of the albunioid gene superfamily, which consists of AFP, serum albumin, vitamin D-binding protein, and alpha-albumin (afamin) on chromosome 4. AFP is found in monomeric as well as dimeric and trimeric forms, and binds copper, nickel, fatty acids, and bilirubin [5]. In normal fetuses, AFP binds the hormone estradiol. Altered serum AFP levels have been observed concurrent with aberrant growth manifestation in some birth defects (increased in open neural tube defects, and ompaholocele and decreased in Down syndrome) and a subset of endodermal derived malignancies, most frequently hepatoblastoma (HB), hepatocellular carcinoma (HCC) and germ cell tumors (GCT), some esophageal and pancreatic carcinomas, and also some benign conditions, particularly those associated with liver damage and regeneration [6]. The importance of considering infants' ages while interpreting AFP levels should be underscored, as age-dependent changes in serum AFP levels occur. AFP is initially synthesized in the yolk sac and is subsequently made by hepatocytes and gastrointestinal mucosa during embryogenesis, it reaches its peak concentration at approximately 12–14 weeks of gestation, and then declines thereafter, reaching normal concentration at 8–12 months of age [7]. As a result, AFP serum levels are highly elevated in neonates, and premature infants can have very high levels of AFP, which is accepted as normal (term neonates: 41,687 *l* g/L, preterm infants: 158,125 *l* g/L, median levels) [8].

Due to the age-related changes in AFP values, and the fact that this biomarker is a tumor-associated, not tumor-specific protein, elevation of AFP levels alone do not allow for the diagnosis of malignancy. Additional examination, including pathologic diagnosis, imaging, and exclusion of other conditions associated with elevations of AFP must compliment tumor marker evaluation. An AFP serum level above 500ug/L is rarely associated with a benign diagnosis [2].

Germ Cell Tumors

The exact incidence of pediatric GCTs is not precisely known, approximately 50% of GCTs are benign (pure mature teratoma), which do not secrete AFP and thus go unreported. Malignant GCTs account for 3% of all childhood cancers, with 350 new pediatric germ cell tumors diagnosed each year. These tumors develop from variations in normal differentiation (gonadal GCT) and/or aberrant migration (extragonadal GCT) of primordial germ cells. Extragonadal GCTs occur in a variety of midline locations (retroperitoneal, genital, or cranial) but most commonly affect the mediastinum and the sacrococcygeal region. Morphologically, they are divided into seminomas and non-seminomatous germ cell tumors (NSGCTs). Most NSGCTs are a mixture of histologies such as embryonal, choriocarniomas, teratomas, and yolk sac tumors (YST).

AFP secretion is characteristic for malignant GCT with yolk sac tumor differentiation [2]. While predicting behavior in GCT can be confusing secondary to varying factors such as patient age, histologic subtype, anatomic site, and clinical stage, serum levels of AFP is well known in its utility as non-invasive diagnostic indicator of GCT with yolk sac component [9]. This marker is well validated in its utility as a diagnostic indicator, staging, monitoring of therapeutic response to therapy and follow up [10]. In infants and adolescents who are without underlying hepatic disease, a significant elevation of AFP indicates significant component of YST, and rules out pure mature teratoma or seminoma [1].

The diagnosis of malignant GCT involves the evaluation of serum AFP in conjunction with review of histopathology and diagnostic imaging. At times, and often due to sampling error, the pathology may fail to identify a yolk sac component even when the serum AFP is elevated. This is particularly true for intracranial GCT where only limited tissue biopsies are possible [2, 11, 12].

The clinical use of AFP extends beyond initial diagnosis. Tumor marker follow up examination is also performed during neoadjuvant chemotherapy to assess for tumor response to therapy. A failure of AFP to decline during chemotherapy may cause clinicians to re-evaluate need for earlier surgical intervention or adjustment of chemotherapy regimen. The ability to evaluate tumor response to chemotherapy may help define prognostic subgroups. The biological importance of this tumor marker has been reinforced by the International Germ Cell Classification Consensus (IGCCC) which divides adults into three prognostic groups: good, intermediate, and poor based on the presence of primary mediastinal tumor, non-pulmonary metastases and level of tumor markers [10]. Identification of risk groups allows for tailoring of treatment as well as comparison of results across populations of similarly defined patients and ultimately the collaboration of international clinical trials [13]. This has not yet translated to pediatric oncology. In the United States, pediatric GCTs have typically been divided into risk groups according to stage, histologic grade, and, more recently, site (gonadal vs. extragonadal). Serum markers have not contributed to risk stratification. The importance of age and level of AFP at diagnosis have varied among studies. Several groups, including the Children's Cancer and Leukemia Group, the French Society of Pediatric Oncology and COG have previously identified serum AFP (>10, 000 ng/ml) as a prognostic factor [11, 10, 13]. An analysis by the Children's Oncology Group (COG), and the Children's Cancer and Leukemia Group (CCLG) compared the application of the adult IGCC System to pediatric malignant NSGCT by pooling 25 years worth of clinical trial data. The authors sought to determine if the tumor marker criteria developed in adults would be prognostic among pediatric patients and whether tumor marker data may be relevant in pediatric risk stratification. This study found that AFP was not found to be prognostic of outcome in pediatric patients, and therefore its use in risk stratification in pediatric malignant GCT remains unclear [13].

Another rare finding while on neo-adjuvant chemotherapy, is the growing teratoma syndrome in which a patient with a mixed GCT demonstrates radiographic progression of disease with a reduction of tumor marker. This occurs when mature or immature teratoma components of the tumor continue to proliferate while the YST component responds to chemotherapy. In this case, immediate surgical excision of tumor is recommended [2].

A final use of AFP as a tumor marker is found following completion of therapy (surgery, chemotherapy, or combination) to assess for recurrence of disease. The goal after therapy is to have normalization of serum AFP. Tumor recurrence is demonstrated by an abrupt elevation of the AFP level. YST and mixed GCTs require close follow-up with serial AFP level monitoring as part of the surveillance strategy for this patient population.

HEPATOBLASTOMA AND HEPATOCELLULAR CARCINOMA

Liver malignancies account for slightly more than 1% of all pediatric tumors, with approximately 100–150 new cases of liver tumors being diagnosed annually in the US [14]. Pediatric malignant liver tumors include hepatoblastoma (HB), hepatocellular carcinoma (HCC), sarcomas, GCTs, and rhabdoid tumors. The pediatric embryonal tumor HB accounts for 66% of malignant hepatic neoplasms.

Serum AFP levels are an important tumor marker for HB and HCC. In addition, serum AFP levels may also be elevated in other nonmalignant conditions such as mesenchymal hamartoma, infantile hepatic hemangioendothelioma, viral and chronic active hepatitis, cirrhosis, ulcerative colitis, tyrosinemia, and in some immune-deficiency conditions [19, 20].

Most patients with HB have extremely high serum levels of AFP [up to 100,000 ± 1,000,000 times the normal value (<10 ng/mL)], but approximately 5%–10% of patients have unexpectedly low or even normal AFP values [21]. For HB, the measurement of serum AFP levels is a valuable diagnostic test, owing to the correlation between AFP level and disease activity. Serial determinations of AFP determine the effectiveness of therapy. The half-life of the protein in the circulation is between 4 and 9 days, and levels usually return to the normal range (<10 ng/mL) by 4 ± 6 weeks of complete resection [22, 23]. The failure of serum AFP levels to return to normal after surgery, or a secondary increase in AFP levels, is an indication of either incomplete tumor resection or metastases.

Serum AFP is also used a screening tool to detect the development of HB in children with Beckwith-Wiedemann syndrome (BWS) or isolated hemihyperplasia (HH). The development of HB in these cancer predisposition syndromes has been observed in 1-3% of patients with BWS and HH. Improved survival is dependent on early detection. This can be inferred by the improved prognosis of early stage (stage I and II) HB as compared to late stage (stage III, IV) tumors: 91% five-year event-free survival vs. 25%-64% five year event-free survival, respectively. This is accomplished with routine screening with hepatic ultrasound and serum AFP at 3-month intervals up to the age of four years [24].

A final use of serum AFP in children with HB is its use in prognostication. Although the best predictor of survival in HB is complete surgical resection, more than half of children present with initially unresectable tumors (Stage III or IV) (North American Staging System). In an effort to achieve cytoreduction and improved resectability, children with up-front unresectable HB receive neo-adjuvant cisplatin-based chemotherapy [25]. Patients with HB responding poorly to induction therapy (as demonstrated by persistently elevated AFP levels) have been shown to have a significantly worse outcome [22]. In fact, the strongest predictor of survival in unresectable or metastatic HB is the timing and magnitude of the change in AFP level during the administration neo-adjuvant chemotherapy. A large early decrease in AFP level correlates with the best outcome. It has been further suggested that monitoring the changes in AFP during neo-adjuvant chemotherapy may help identify poor responders and, in those patients, we should consider a change in therapeutic approach [26].

Another poor predictor of survival is low serum AFP at diagnosis. Although most patients present with elevated levels, serum AFP is low or normal in approximately 5-10% of cases. Children with HB and low AFP level at diagnosis have been identified as a high-risk group. This group of patient will generally present with more extensive disease at diagnosis and have a high risk of treatment failure [27].

HB can also be associated with the inappropriate secretion of peptide hormones and other associated proteins. Marked thrombocytosis secondary to release of tumor-derived factors with thrombopoietin-like activity, growth factors, platelet-derived microparticles, factors released from bone marrow endothelial cells, and growth factors secreted by megakaryocytes can influence this process [28, 29, 30]. Likewise, ectopic bHCG production can cause precocious puberty in boys (virilizing HB) or a *forme fruste* in females [31]. A case of hormonally active HB causing both ectopic ACTH syndrome and PTH-related peptide-induced hypercalcemia has also been reported [32]. Although these paraneoplastic syndromes are not frequently linked to HB, they underscore the importance of considering these oncofetal proteins as tumor-associated rather than tumor-specific, and the diverse presentations that can pose diagnostic and therapeutic challenges in HB.

HCC is the second most common hepatic malignancy in childhood accounting for less than 0.5% of all pediatric cancer [33]. Its relative frequency is 0.5 to 1.0 cases per million children [34]. Unlike HB, it is more commonly found in older children and teenagers with underlying liver disease (viral hepatitis, inborn errors of metabolism (tyrosinemia, type I glycogen storage disease)). It is also associated with a very poor prognosis (54% 5-year disease-free survival). The use of AFP levels in HCC has been primarily for diagnostic purposes as most HCC are associated with elevated levels of AFP. Although not found to be an independent predictor of survival, it has been observed that HCC with serum AFP <100ng/mL at diagnosis tend to have a worse outcome and are frequently observed in the fibrolamellar HCC variant [35]. Serum AFP may also be used for surveillance purposes in the "at risk" patient. Patients treated for biliary atresia (obstructive cholangiopathy of the newborn) in infancy should be monitored for the development of cirrhosis and subsequently HCC as well as cholangiocarcinoma. This can be accomplished with abdominal ultrasound and serum AFP levels at 6-12 month intervals [36]. Children who are diagnosed with inborn errors of metabolism, congenital malformations, or who have cirrhosis induced by chronic liver disease require screening for the development of HCC.

AFP and Associated Rare Pediatric Solid Cancers

While serum AFP is primarily used in the diagnosis and management of GCT (with yolk sac component), HB, and HCC, elevated levels have also been found in select cases of nephroblastoma (Wilms tumor, WT) and pancreatoblastoma (PB). For example, Roth and colleagues first described WT presenting with elevated serum AFP [34]. Since this initial description, only four other WT cases have been documented with histologically classic type WT and elevated AFP [37-40]. More notably, elevated AFP is seen in the teratoid WT, in which more than 50% of the tumor is comprised of teratoid elements [41]. Nevertheless, due to the extremely rare occurrence, serum AFP is not a typical tumor marker for pediatric renal tumors.

PB is a rare childhood malignancy with only 62 cases reported in the literature [42] and has been described as an embryonal hamartomatous neoplasm arising from multipotential stem cells [43]. Typically presenting as an abdominal mass with progressive jaundice, PB has been associated with an elevated serum AFP at diagnosis in 25% of the described cases. While clinically affiliated with a prolonged, indolent course with favorable outcomes, serum AFP may contribute to the diagnosis and monitoring of therapy for PB [44].

CONCLUSION

Whereas AFP is a well-known classical serum biomarker that can assist with diagnosis of certain malignant solid tumors in children, it remains limited in sensitivity and specificity [45]. Other promising biomarkers for tumor diagnosis, especially GCT, include microRNAs (miRNAs), which are short nucleic acid molecules synthesized in the cellular nucleus. miRNAs modify post-transcriptional gene activity by targeting mRNA molecules. Once released from the nucleus, miRNAs can enter almost all biological fluids (*e.g.*, serum, saliva, urine, and milk), where they are thought to mediate intercellular communication. After departure from the cell of origin, miRNAs show remarkable stability because of their inclusion in membrane vesicles (exosomes). As a result, blood-based miRNAs show considerable promise for cancer diagnosis and monitoring [46]. miRNAs encoded by the small miR-371-3 cluster that maps to the terminal region of the long arm of chromosome 19 are among the most promising candidates for diagnosing and monitoring malignant GCTs, as these tumors overexpress the miR-371-373 and miR-302/367 clusters, regardless of the age of the patient, the histological subtype of the tumor, or its anatomical site [46]. Murray et al. have reviewed the current understanding of miRNAs and their role in diagnosing and monitoring childhood GCTs [47]. As the range of applications for miRNAs includes screening and diagnosis to clinical decision-making aids, surveillance biomarkers and as a source of real-time molecular characterization, it is tempting to speculate these biomolecules may emerge as the next generation "smart" biomarkers for selected pediatric solid malignancies.

REFERENCES

[1] Sandoval, J.A., Malkas, L.H. Hickey, R.J. (2012) Clinical Significance of Serum Biomarkers in Pediatric Solid Mediastinal and Abdominal Tumors. *Int J Mol Sci*, 13, 1126-1153.

[2] Schnieder, D.T., Calaminus, G., Gobel, U. (2001) Diagnostic Value of Alpha-Fetoprotein and Beta-Human Chorionic Gonadotropin in infancy and Childhood. *Pediatric Hematology and Oncology*, 18, 11-26.

[3] Bergstrand C.G., Czar B. (1956) Demonstration of a new protein fraction in serum from the human fetus. *Scand J Clin Lab Invest,* 8,174.

[4] Blohm, M.E., Vesterling-Horner, D, Calaminus,G, Gobel, U. (1998) Alpha 1-Fetoprotein (AFP) reference values in infants up to 2 years of age. *Pediatric Hematology and Oncology*, 15(2), 135-142.

[5] Aoyagi Y, Ikenaka T, Ichida F. (1979). Alpha-Fetoprotein as a carrier protein in plasma and its bilirubin-binding ability. *Cancer Res*, 39(9), 3571-3574.

[6] Marshall W, Bangert S. (2008) Clinical Biochemistry: Metabolic and Clinical Aspects. 2 ed ed. Philadelphia: Churchill Livingstone, Elsevier.

[7] Mueller, B.U., Lopez-Terrada, D., Finegold, M.J. (2006) Tumors of the Liver. In *Principles and Practices of Pediatric Oncology*, 5th ed, Pizzo, P., Poplack, D., Eds, Lippincott Williams and Wilkins: Philadelphia, PA, USA, pp. 887-904.

[8] Wu, J.T., Book, L., Sudar,K.(1981) Serum alpha fetoprotein (AFP) levels in normal infants. *Pediatr. Res,* 15, 50-52.

[9] Billmire D, Vinocur C, Rescorla F et al. (2004) Outcome and staging evaluation in malignant germ cell tumors of the ovary in children and adolescents: an intergroup study. *J Pediatr Surg,* 39(3), 424-429.

[10] International Germ Cell Consensus Classification: a prognostic factor-based staging system for metastatic germ cell cancers. (1997) International Germ Cell Cancer Collaborative Group. *J Clin Oncol*, 15, 594-603.

[11] Gobel, U., Calaminus, G., Engert, J.(1988) Teratomas in infancy and childhood. *Med Pediatr Oncol*, 31, 8-15.

[12] Harms D, Janig U. (1986) Germ cell tumors of childhood: report of 170 cases including 59 pure and partial yolk sac tumors. *Virchows Arch A*, 409, 223-239.

[13] Frazier, A.L., Rumcheva, P., Olson,T., Giller, R., Cushing, B., Cullen, J., Marina, N., London, W.B. (2008) Application of the Adult International Germ Cell Classification System to Pediatric Malignant Non-Seminomatous Germ Cell Tumors: A report from the children's oncology group. *Pediatr Blood Cancer*, 50, 746-751.

[14] Mann, J.R., Raafat, F., Robinson, K., Imerson, J., Gornall, P., Sokal, M., Gray, E., McKeever, P., Hale, J., Bailey, S., Oakhill, A. (2000) The United Kingdom Children's Cancer Study Group's second germ cell tumor study: Carboplatin, Etoposide, and Bleomycin are effective treatment for children with malignant extracranial germ cell tumors, with acceptable toxicity. *J Clin Oncol*, 18, 3809-3818.

[15] Baranzelli, M.C., Kramar, A., Bouffet, E., Quintana, E., Rubie, H., Edan, C., Patte, C. (1999) Prognostic factors in children with localized malignant nonseminomatous germ cell tumors. *J Clin Oncol*, 17, 1212.

[16] Marina, N., London, W.B., Frazier, A.L., Lauer, S., Rescorla, R., Cushing, B., Malogolowkin, M.H., Castleberry, R.P., Womer, R.B., Olson, T. (2006) Prognostic factors in children with extragonadal malignant germ cell tumors: A pediatric intergroup study. *J Clin Oncol*, 24, 2544-2548.

[17] Frazier, A.L., Hale, J.P., Rodriguez-Galindo, C., Dang, H., Olson, T., Murray, M.J., Amatruda, J.F., Thornton, C., Arul, G.S., Billmire, D., Shaikh,F., Farzana,P., Stoneham, S., Krailo, M., Nicholso, J.C. (2015) Revised Risk Classification for Pediatric Extracranial Germ Cell Tumors Based on 25 Years of Clinical Trial Data From the United Kingdom and United States. *J Clin Oncol*, 33,195-201.

[18] Bulterys, M., Goodman, M.T., Smith, M., Buckley, J.D. Hepatic Tumors. In Cancer Incidence, Survival Among Children, Adolescents, United States SEER Program 1975-1995. Available online: http://seer.cancer.gov/publications/childhood/ (access on 12 November 2015)

[19] Urbach, A.H., Zitelli, B.J., Blatt, J., Gartner, J.C., Malatack, J.J. (1987) Elevated alpha-fetoprotein in a neonate with benign hemangioendothelioma of the liver. *Pediatrics*, 80, 596-597.

[20] Sari, N., Yalcin, B., Akyuz, C., Halioglu, M., Buyukpamukcu M. (2006) Infantile hepatic hemangioendothelioma with elevated serum alpha-fetoprotein. *Pediatric Hematology and Oncology*, 23, 639-647.

[21] Kelli, S. (2004) Serum markers in tumor diagnosis and treatment. In *Paediatric Oncology- Clinical Practice and Controversies*, 3rd ed,, Pinkerton, R., Plowan, P.N., Pieters, R., Eds,, Edward Arnold Publishers: London, UK, pp. 169-188.

[22] Van Tornout, J.M., Buckley, J.D., Quinn, J.J, Freusner, J.H., Krailo, M.D., King, D.R., Hammond, G.D., Ortega, J.A. (1997) Timing and magnitude of decline in alpha-fetoprotein levels in treated children with unresectable or metastatic hepatoblastoma are predictors of outcome: a report from the Children's Cancer Group. *J Clin Oncol*, 15, 1190-1197.

[23] Lovvorn, H.N. III, Ayers, D., Zhao, Z., Hilmes, M., Prasad, P., Shinall, M.C. Jr., Berch, B., Neblett, W.W., III, O'Neill, J.A., Jr. (2010) Defining hepatoblastoma responsiveness to induction therapy as measured by tumor volume and serum alpha-fetoprotein kinetics. *J Pediatr Surg*, 45,121-128.

[24] Clericuzio, C.L., Chen, E., McNeil, D.E., O'Connor, T., Zackai, E., Medne, L., Tonlinson, G., DeBaun, M. (2003) Serum alpha fetoprotein screening for hepatoblastoma in children with beckwith-wiedemann syndrome or isolated hemihyperplasia. *J Pedaitr*, 143(2), 270-272.

[25] Ortega, J.A., Douglass, E.C., Feusner, J.H., Reynolds, M., Quinn, J.J., Finegold, M.J., Haas, J.E., King, D.R., Liu-Mares, W., Sensel, M.G., Krailo, M.D. (2000) Randomized Comparison of Cisplatin/Vincristine/Fluoruracil and Cisplatin/continuous infusion Doxorubicin for treatment of pediatric hepatoblastoma: A report from the children's cancer group and the pediatric oncology group. *J Clinl Oncol*, 18(14), 2665-2675.

[26] Sayar, D., Yaniv, I., Goshen, Y., Cohen, I.J. (2001). Treatment of alpha-fetoprotein secreting hepatoblastoma by response of serum alpha-fetoprotein levels: a new concept. *Pediatric Hematology and Oncology*, 18, 509-518.

[27] De Ioris, M., Brugieres, L., Zimmermann, A., Keeling,J., Brock, P., Maibach, R., Pritchard, J., Shafford, L., Zsiros, J., Czaudzerna, P., Perilongo, G. (2008) Hepatoblastoma with a low serum alpha-fetoprotein level at diagnosis: The SIOPEL group experience. *European Journal of Cancer*, 44, 545-550.

[28] Nickerson, H.J., Silberman, T.L., McDonald, T.P. (1980) Hepatoblastoma, thrombocytosis, and increased thrombopoietin. *Cancer*, 45, 315-317.

[29] Shafford, E.A. (1993) Pritchard, J. Extreme thrombocytosis as a diagnostic clue to hepatoblastoma. *Arch Di. Child*, 69, 171.

[30] Komura, E., Matsumura, T., Kato, T., Tahara, T., Tsunoda, Y., Sawada, T. (1988) Thrombopoietin in patients with hepatoblastoma. *Stem Cells*, 16, 329-333.

[31] Perilongo, G., Rigon, F., Murgia, A. (1989) Oncologic causes of precocious puberty. *Pediatr Hematol Oncol*. 6, 331-340.

[32] Grunewald, T.G., von Luettichau, I., Welsch, U., Dorr, H.G., Hopner, F., Kovacs, K., Burdach, S., Rabl, W. (2010) First report of ectopic ACTH syndrome and PTHrP-induced hypercalcemia due to a hepatoblastoma in a child. *Eur J Endocrinol*, 162, 813-818.

[33] Czauderna, P., Mackinlay, G., Perilongo, G., Brown, J., Shafford, E., Aronson, D., Pritchard, J., Chapchap, P., Keeling, J., Plaschkes J., Otte, J.B. (2002) Hepatocellular Carcinoma in Children: Results of the First Prospective Study of the International Society of Pediatric Oncology Group. *J Clin Oncol*, 12, 2798-2804.

[34] Bellani, F.F., Massimino, M. (1993) Liver tumors in childhood: Epidemiology and Clinics. *J Surg Onc*, Suppl 3, 119-121.

[35] Walther, A., Tiao,G. (2003) Approach to Pediatric Hepatocellular Carcinoma. *Clinical Liver Disease,* 2, 219-222.

[36] Fukuda, A., Sakamoto, S., Kanazawa, H., Shigeta, T., Karaki, C., Hamano, I., Uchida, H., Kitagawa, U., Okuse, C., Miyazaki, O., Nosaka, S., Nakazawa, A., Kasahara, A. (2013) Incidentally detected cholangiocarcinoma in an explanted liver with biliary atresia after Kasai operation. *Pediatr Transplantation*, E, 62-66.

[37] Roth, A. (1984) An unusual renal tumor, associated nephroblastoma, nephroblastomatosis, teratoma, and cystic dysplasia with arterial hypertension and secretion of alpha-fetoprotein. *J Urol (Paris)*, 90, 7-10.

[38] Erol, K., Vedat, K., Erkanm D., Mustafa, G., Duygu, H., Avni, A.A. (2005) Wilms tumor associated with elevated alpha-fetoprotein level. *Pediatr Blood Cancer*, 44, 423-424.

[39] Patriarca, C., Orazi, A., Massimino, M., Luksch, R. (1992) A cystic partially differentiated nephroblastoma producing alpha-fetoprotein. *Am J Pediatr Hematol Oncol*, 12, 352-355.

[40] Crocoli, A., Madafferi, S., Jenkner, A., Zaccara, A., Inserra, A. (2008) Elevated serum alpha-fetoprotein in Wilms tumor may follow the same pattern of other fetal neoplasms after treatment: evidence from three cases. *Pediatr Surg Int*, 24, 499-502.

[41] Dhungel, S., Cheng, L.J., Hai, Z.Z. (2014) Elevated serum alpha-fetoprotein in Wilms' tumor: A case report with review of literature. *Journal of Pediatric Surgery Case Reports,* 2, 153-155.

[42] Chan, M.H.M., Shing, M.M.K., Poon, T.C.W., Johnson, P.J., Lam, C.W.K. (2000) Alpha-fetoprotein variants in a case of pancreatoblastoma. *Ann Clin Biochem*, 37, 681-685.

[43] Iseki, M., Suzuki, T., Koizumi, Y., Hirose, M., Laskin, W.B., Nakazawa, S., Ohaki, Y. (1986) Alpha-Fetoprotein-Producing Pancreatoblastoma. *Cancer*, 57, 1833-1835.

[44] Morohoshi, T., Sagawa, F., Mitsuya, T. (1990) Pancreatoblastoma with marked elevation of serum alpha-fetoprotein. *Virchows Archiv A Pathol Anat,* 416, 265-270.

[45] Murray, M.J., Nicholson, J.C. (2011) α-Fetoprotein. *Arch Dis Child Educ Pract Ed*, 96, 141-147.

[46] Murray, M.J., Halsall, D.J., Hook, C.E., Williams, D.M., Nicholson, J.C., Coleman, N. (2011) Identification of microRNAs rom the miR-371-373 and miR-302 clusters as potential serum biomarkers of malignant germ cell tumors. *Am J Clin Pathol*, 135,119-125.

[47] Murray, M.J., Nicholson, J.C., Coleman, N. (2015) Biology of childhood germ cell tumour, focusing on the significance of microRNAs. *Andrology*, 3(1), 129-3.

SECTION 2: THE ROLE OF ALPHA-FETOPROTEIN IN MATERNAL SERUM SCREENING

In: Alpha-Fetoprotein: Functions and Clinical Applications ISBN: 978-1-63484-875-6
Editors: Nisha Lakhi and Michael Moretti

Chapter 7

ROLE OF MATERNAL SERUM ALPHA-FETOPROTEIN AS A SCREENING MARKER AFTER THE INTRODUCTION OF CELL FREE-DNA IN ANEUPLOIDY SCREENING

Howard Cuckle[1,*], D Phil, MSc, BA, and Eugene Pergament[2], MD, PhD, FACMG
[1]Columbia University Medical Center, New York, NY, US
[2]Northwestern Reproductive Genetics, Chicago, IL, US

ABSTRACT

Maternal serum Alpha-fetoprotein (AFP) levels can be used to screen for fetal neural tube defects (NTDs) in the second trimester of pregnancy and is part of multi-marker screening protocols for Down's syndrome and other aneuploidies in the second trimester. It is possible to use the analyte in combination with ultrasound measurement of bipariental diameter to screen for spina bifida in the first trimester. Cell free (cf)DNA is the most powerful determinant of aneuploidy risk but it is currently too expensive to implement routine testing in a public health setting. Instead a Contingent cfDNA protocol is recommended whereby up to 20% of women are selected for cfDNA on the basis of aneuploidy risk calculated from first trimester markers. The addition of first trimester AFP to these markers enhances the accuracy of risk estimation and improves the performance of the Contingent cfDNA test. Even if the price of cfDNA testing falls there would be strong arguements to retain the first trimester aneuploidy risk assessment using serum and ultrasound markers. Specifically, two of the markers together with an additional ultrasound measurement could be used to screen for pre-eclamsia, a common and preventable adverse outcome of pregnancy.

Keywords: AFP, screening, risk, spina bifida, Down's syndrome, adverse outcome

---

* Corresponding Author address: Department of OBGYN, Columbia University Medical Center, 622 W 168th Street, PH1666, New York, NY 10032-3725, US, E-mail: hsc2121@columbia.edu.

INTRODUCTION

The discovery in the mid-70s that second trimester maternal serum Alpha-fetoprotein (AFP) levels are increased on average in pregnancies affected by fetal neural tube defects (NTDs) ushered in the modern era of antenatal screening. Many of the principles and techniques that are now an integral part of multi-marker screening for fetal aneuploidy were first developed for use in screening for NTDs. And the same principles are beginning to be used in antenatal screening for adverse outcomes of pregnancy such as pre-eclampsia (PE).

But in recent decades there were a number of developments which have changed the classic paradign of AFP screening for NTDs: primary prevention has led to a fall in NTD prevalence; detailed ultrasound could replace invasive prenantal diagnosis; and screening for ultrasound markers is an alternative option.

Maternal serum AFP was also the first marker of aneuploidy to be found and the discovery stimulated the search for additional markers leading to multi-marker screening programs. However, there has been a gradual shift in the focus of aneuploidy screening from the second to the first trimester with reduced reliance on AFP in the multi-marker profile.

More recently, an even bigger change has been initiated with the discovery that maternal plasma cell-free (cf)DNA is by itself the most predictive marker of common trisomies. The exact role of this new marker in screening protocols is yet to be established but it is already clear that all current screening protocols have to be reconsidered in light of this discovery, including the future role of AFP.

NEURAL TUBE DEFECTS

NTD prevalence varied worldwide with rates in the past ranging from about 1 per 1000 to 4-8 per 1000 in areas such as Ireland, with population largely of Celtic origin. In recent decades, dietary changes and chemo-prevention by folate supplementation or food fortification have reduced the prevalence in most developed countries. Anencephaly and spina bifida each account for about half of NTDs. Spina bifida lesions can be classidied as 'open' (OSB) when that there is some exposure of neural tissue or covered by a thin transparent membrane, and 'closed' when covered by skin or a thick opaque membrane. OSB is more readily detected by maternal serum AFP and has a poorer prognosis.

Classic AFP Screening Pradigm

In the second trimester AFP is produced by the fetal liver, voided by the fetus into the amniotic fluid and subsequently broken down by the fetal kidneys. A small proportion of the amniotic fluid AFP enters the maternal circulation across the membranes. The concentration of AFP in fetal cerebrospinal fluid is more than 100 times greater than in amniotic fluid so that in anencephaly and OSB leakage through the lesion into the amniotic fluid leads to a large increase in levels. Consequenctly, there is an increase, albeit smaller, in maternal serum AFP.

In unaffected pregnancies the AFP concentration increases with gestation, in the second trimester doubling about every four week. To allow for this, levels are expressed as multiples of the gestation specific median (MoMs) for unaffected pregnancies. The UK Collaborative AFP Study, a large international project published in 1977, found the greatest increase at 16-18 weeks [1]. Using a 2.5 MoM cut-off yielded an observed detection rate (proportion with AFP exceeding the cut-off) for anecephaly was 88% and for OSB 79%; the false-positive rate (proportion of unaffected pregnancies exceeding the cut-off) was 3.3%. The aim of screening at that time was to offer those with positive tests an invasive prenatal diagnosis using amniotic fluid analytes, AFP and subsequently acetylcholinesterase (AChE).

The best estimate of AFP screening performance today is derived by statistical modelling with revised UK Collaborative AFP Study parameters taking account of improved AFP assay precision, ultrasound gestational dating, and maternal weight adjustment [2]. The Collaborative study used menstrual dates, but second trimester dating with a biparietal diameter (BPD) scan is capable of detecting all cases of anecephaly, providing more precise gestational assessment and hence improved MoMs. Also BPD is reduced on average in spina bifida leading to a systematic underestimation of gestation and a beneficial increase in AFP MoMs; 43% in one study [3]. AFP concentrations decline with increasing maternal weight [4] presumably due to a fixed mass of the analyte in a variable volume of blood. Adjustment for weight further improves the precision of AFP MoMs.

In the improved model AFP levels in OSB peak at 18 weeks with a median of 5.8 MoM. Using a 2.5 MoM cut-off the model predicted detection rate is 85% and the false-positive rate 1.4%. Some centres have used the improved precision to increase detection by lowering the cut-off level. Using 2.0 MoM the detection and false-positive rates are 94% and 4.8%.

Ultrasound 'Anomaly' Scan at 18-20 Weeks

A detailed ultrasound scan of fetus structures is now routinely performed in most centres. In a meta-analysis of 10 studies including more than 100,000 pregnancies, the overall OSB detection rate was 95% [5], but this result needs to be interpreted with caution. There is a possibility that the ultrasound observations were influenced by the degree of prior risk, including AFP results. Also, it is remarkable that none of the studies reported false-positives. This could reflect a low prior suspicion of OSB, on average, but it may also indicate bias.

Ultrasound Dating Scan at 14-15 Weeks

In spina bifida, as a result of the associated Arnold–Chiari malformation, there is a tendency for the frontal bones to be scalloped (the 'lemon' sign) or there may be absence or curvature of the cerebellum (the 'banana' sign) in the second trimester. In contrast these markers appear to be relatively rare in unaffected foetuses, making routine screening effective.

A meta-analysis of six retrospective and six preospective studies found that the overall detection rate of the lemon sign was 81% and the banana sign 94% with a 0.8% false-positive rate [5]. The reported studies will not be without bias because it is difficult to have the ultrasonographer report the cranial signs without looking at the spine. A large open spinal defect may be observed before the BPD had been located or the discovery of the spinal lesion may subsequently lead to the cranial result being reviewed before reporting.

Ultrasound Scan at 10-13 Weeks

At this time scanning in the saggital plane can detect several markers of OSB in the anterior brain, which also relate to the Arnold–Chiari malformation. The reported markers mainly concern to the fourth ventricle, which is compressed, and the brainstem (BS) which is thickened in affected pregnancies.

The intracranial translucency (IT) of the fourth ventricle can be obliterated, reduced or displaced as fluid in the posterior brain. The fourth ventricle structure that will later develop into the chroroid plexus can also be reduced. In OSB the brainstem thickness can be measured as a single marker, but superior results have been reported by calculating the ratio of the BS thickness to the distance between the posterior edge of the brainstem and the occipital bone (BSOB). Also in the mid-sagittal plane, absence of the cisterna magna has been reported to be a good marker and three dimensional ultrasound imaging of the entire ventricular system has revealed other markers. Another first trimester consequence of the Arnold-Chiari malformation is caudal displacement of the forehead so that the frontal-maxiliary facial (FMF) angle is reduced and this too can be used as a marker.

Recently, there has been interest in using first trimester BPD as a marker and this can be determined without the considerable skill and experience needed to determine IT, BS, BSOB and FMF angle. In one first trimester study of 80 cases, BPD was expressed in MoMs and half had levels below the 5th centile based on 34,951 unaffected pregnancies [6]. In another study of 26 OSB cases, results were not adjusted directly for gestation but expressed as the ratio between the BPD and trans-abdominal diameter and three-quarters had low BPD [7].

AFP Screening at 10-13 Weeks

First trimester maternal serum AFP may also be a marker of OSB that could be combined with ultrasound markers, but there are conflicting results. In the Collaborative study, at 10-12 weeks gestation 21% of all spina bifida cases have AFP levels above the 97th percentile [1]. In three first trimester series combined, the average AFP level in a total of 64 OSB cases was 1.28 MoM [6, 8, 9]; but in a fourth series of 39 cases, the median level was 0.92 MoM [10]. First trimester maternal serum levels of free β-hCG, a marker of aneuploidy (see below), have been reported to be reduced in two OSB series and one statistical model predicted that adding both AFP and free β-hCG to BPD would increase detection from 50% to 58% for a 5% false-positive rate [6].

Incidental Diagnoses

Second trimester maternal serum AFP levels are increased on average in pregnancies with open abdominal wall defects. One study included 13 omphalocele cases and 20 with gastroschisis [11]: the median AFP levels were 4.1 MoM and 7.0 MoM, respectively. Another study found similar AFP results with medians of 4.2 MoM (17 cases) of omphalocele and 9.4 MoM (23 cases) of gastroschisis [12]. Maternal serum AFP levels are essentially undetectable in cases of complete hydatidiform molar pregnancies because there is no fetus. Fetal death is associated with elevated or low AFP.

Aneuploidy

Aneuploidy is a common event in pregnancy with a wide spectrum of medical consequences from the lethal to the relatively benign. Most affected embryos abort spontaneously early in the first trimester; many of them even before there are clinical signs of pregnancy. Those that survive into the second trimester also experience high late intrauterine mortality and increased risk of infant death. By far the most frequent of these is Down's syndrome (DS), with a birth prevalence of 1-2 per 1,000 in developed countries. Consequently it is the main focus of antental aneuploidy screening.

AFP Screening for Down's Syndrome

Maternal serum AFP was the first analyte to be used in DS screening after the observation that levels were reduced, on average, in pregnancies with chromosome abnormalities in general and DS in particular, in both the second and first trimester. On the basis of meta-analysis the median AFP level in DS pregnancies is 0.73 MoM in the second trimester and changes with first trimester gestation from 0.82 MoM at 11 weeks to 0.80 MoM at 13 weeks [13].

The reason for decreased AFP synthesis in DS is unknown, but in the second trimester it may reflect hepatic immaturity. Histological study has revealed under-vascularization of the placentae of fetuses with various chromosomal defects, possibly representing placental immaturity with arrested or delayed angiogenesis.

Other Markers of Down's Syndrome

Six other DS markers are widely used in routine screening today: maternal serum human chorionic gonadotrophin (hCG), the free-β subunit of hCG, unconjugated estriol (uE_3), inhibin A and pregnancy associated plasma protein (PAPP)-A, and ultrasound nuchal translucency (NT).

Maternal serum hCG and free β-hCG levels are increased on average in DS pregnancies; the latter in both the second and first trimester although the extent of increase is greater as pregnancy progresses. From meta-analysis the median values are 2.0 and 2.3 MoM respectively in the second trimester; whilst in the first trimester they increase from 1.2 to 1.8 MoM and 1.9 to 2.1 MoM between 11 and 13 weeks [13]. Abnormality in cytotrophoblast differentiation may be the basis of the elevated hCG levels in affected pregnancies.

Maternal serum uE_3 levels are reduced on average in DS pregnancies, with meta-analysis medians: second trimester 0.73 MoM; 11 weeks 0.89 MoM and 13 weeks 0.78 MoM [13]. In the DS fetus there is adrenal hypoplasia and the adrenal cortex produces dehydroepiandrosterone sulphate (DHEAS), which the fetal liver hydroxylates. The newly formed product, 16-α-hydroxy-DHEAS, is formed in the fetal liver by hydroxylation of DHEAS and transported to the placenta, where it undergoes desulphation and aromatzation into estriol.

Inhibin levels are increased on average in DS pregnancies, using assays that detect all species and those specific for inhibin-A. These increases are not as marked in the first trimester as the second when the meta-analysis median is 1.8 MoM [13]. Inhibin is considered to have a role in the regulation of gonadotropin biosynthesis and secretion, ovarian and placental steroidogenesis, and oocyte maturation.

PAPP-A levels are reduced on average in first trimester DS pregnancies but this reduction diminishes as pregnancy progresses and there is little or no difference by the second trimester. The meta-analysis first trimester median is 0.45 MoM at 11 weeks and 0.65 MoM at 13 weeks [13]. The reason for the low levels in Down syndrome is not known but it is likely to be connected with placental insufficiency and may be the same mechanism that leads to low levels in non-viable pregnancies.

NT is increased on average in DS pregnancies but there is a narrow window at 11-13 weeks when subcutaneous edema can readily be measured in the fetal neck. The meta-analysis median at 11 weeks is 2.3 MoM and at 13 weeks 1.9 MoM [13]. NT is visualised in the saggital section used for crown-rump length and it is recommended that a standardised technique is adopted for measurement. The reasons for the increased edema in DS are not known but the most plausible explanations are: altered composition of the cellular matrix, abnormal or delayed development of the lymphatic system and cardiac insufficiency.

Four additional ultrasound markers of DS can be determined at the same time as NT measurement: absence of the fetal nasal bone (NB); abnormal blood flow in the ductus venosus (DV) and tricuspid regurgitation (TR); both requiring pulse-wave Doppler; and, the FMF angle. Currently, relatively few centres are sufficiently proficient to determine these markers routinely.

There are also early second trimester ultrasound markers of DS: nuchal skin-fold (NF); nasal bone length (NBL) and pre-nasal thickness (PT) as well as the long bone measurements. And there are also so called 'soft' markers which can be seen on the 18-20 week anomaly scan.

Another maternal serum marker of DS is placental growth factor (PlGF). Meta-analysis of first trimester DS yielded a median of 0.70 MoM [13]. This is not currently used but is likely to have an increasing role in future screening protocols (see below). Others serum markers have been considered in the past but are not included in current protocols: A disintegrin and metalloprotease (ADAM)12s; pregnancy specific glycoprotein-1; other hCG species such as urinary β-core fragment of hCG, it's major metabolic product and hyperglycosylated hCG, also known as invasive trophoblast antigen.

Multi-Marker Screening

It can be shown statistically that the optimal way of interpreting the multi-marker profile is to estimate the risk of Down's syndrome from the individual marker levels. This is done by modifying the prior (maternal age-specific) risk, that pertaining to the situation before testing, by a factor known as the 'likelihood ratio' derived from the marker profile. This posterior risk is then compared to a fixed cut-off risk. This approach will yield a higher detection rate for a given false-positive rate than any other method of test interpretation developed to date. It also provides a way of encapsulating the result for purposes of genetic counselling.

All seven widely used markers are expressed in MoMs and the serum markers are adjusted for maternal weight. Some require adjustment for additional co-variables including ethnicity, smoking, assisted reproduction and maternal conditions such as insulin dependant diabetes. The marker profiles follow an approximately log Gaussian distribution over most of their range for both DS and unaffected pregnancies. These distributions are used to estimate individual likelihood ratios (LRs) and to predict the detection and false-positive rates, using either a national population whose maternal age structure has been published or a simply Gaussian distribution of maternal ages [14].

Second Trimester Screening Protocols

Most experience with serum screening for DS is in the second trimester, where it became a natural extension to established AFP screening programs for NTDs, largely at 15-19 weeks gestation. At this time most centres use a two marker combination of either hCG or free β-hCG and AFP ('Double' test), a three marker set that adds uE_3 ('Triple' test), or four markers with the inclusion of inhibin A ('Quad' test). Modelling predicts that in a population with a standardised maternal age distribution using the three tests with free β-hCG, for a 5% false-positive rate the detection rate would be 61%, 65% and 71% respectively [13]. Predicted performance is poorer with intact hCG instead of free β-hCG. Observed detection rates for all second trimester protocols are generally higher than prediction due to 'viability' bias. This arises because detected cases will usually be terminated whilst one-quarter of missed cases will end in miscarriage and hence lost to follow-up. When this bias is allowed for in a meta-analysis, observed performance is close to model predictions.

It is possible to improve detection by the routine combination of serum and ultrasound markers. One approach is to combine the Quad markers with the facial profile markers NF, NBL and PT. Modelling with a standardised maternal age distribution and using free β-hCG, for a 5% false-positive rate the detection rate would be 93% [13]. The short stature associated with children with Down syndrome is reflected *in utero* by smaller than average long bone lengths measured by ultrasound.

For women who do not present until late in the second trimester DS screening can be provided by the anomaly scan. Data collected in the FaSTER trial has been used to evaluate this policy [15]. LRs were derived for the presence and absence of either a major malformation or each of eight soft markers. The use of these LRs to modify the maternal age-specific DS risk yielded a 69% detection rate for a 5% false-positive rate. Improved LRs for soft markers are now available from meta-analysis [16].

First Trimester Screening

Over the last decade there has been a gradual realisation of the benefits of moving prenatal screening from the second to the first trimester. These advantages include earlier diagnosis; less traumatic and safer termination of pregnancy if requested; and, earlier reassurance. Using serum markers alone would be limited to PAPP-A plus either hCG or free β-hCG, which are weaker combinations than any of the second trimester protocols. Some centres initially used ultrasound NT alone but the best results are obtained by the use of the two serum markers and NT ('Combined' test). Modelling predicts that, using free β-hCG, for a 5% false-positive rate the DS detection rate is 87% at 11 weeks and 80% at 13 weeks [13]. The effect of viability bias is higher for first compared with second trimester protocols as about half of DS cases will end in miscarriage, so the observed detection rates are much higher than predicted. Adding ultrasound NB as an additional marker will increase the predicted detection rate to 95% and 93% at 11 and 13 weeks respectively. Similar benefits have been found for adding DV, PT and FMF angle.

Sequential Screening

The combination of first and second trimester serum markers with and without NT will yield even better results than combinations within the first trimester. One approach is to measure all markers when they are most discriminatory, namely, PAPP-A and NT in the first trimester but delay hCG or free β-hCG measurement until the second trimester together with AFP, uE_3 and inhibin. This six-marker combination, known as the 'Integrated' test, requires non-disclosure of any intermediate risk based on the PAPP-A and NT levels. A serum only version, the 'Serum Integrated' test, has also been proposed. The non-disclosure has been considered unethical, or at least impractical due to the difficulty for the clinician not to act on intermediate findings which would of themselves be abnormal - particularly the NT - for any increase in detection is offset by sacrificing early diagnosis and early reassurance. Alternative two-stage seven marker strategies have been suggested to overcome these limitations. One approach is the 'Step-wise Sequential' test where the first stage is the same as the Combined test and women with risks below the cut-off are offered the same second trimester markers as the Quad test with the final risk based on all markers. To avoid a very high false-positive rate this is best done using a much higher than usual Combined test cut-off. The 'Contingent' test is similar except that only women whose risk is boarderline after the first stage are offered the second stage markers. The model predicted detection rate for a 5% false-positive rate, at 11 weeks using free β-hCG, with these protocols is: Serum Intergrated 78%; Integrated 93%; Step-wise Sequential 94%; and Contingent 92% [13].

Other Aneuploidies

Many centres have extended their multiple marker screening for aneuploidy to include a risk for Edwards' syndrome as well as DS. The meta-analysis average levels of maternal serum AFP, uE_3, hCG or free β-hCG, inhibin and PAPP-A, and ultrasound NT are 0.68, 0.44,

0.31, 0.81, 0.14 and 2.8 MoM respectively [13]. Modelling predicts that for protocols using first trimester markers almost as many Edwards' syndrome cases will be detected 'incidentally' as a result of a high DS risk alone compared with using an explicit Edwards' syndrome risk cut-off [13]. But with second trimester protocols a separate cut-off is needed. A further protocol extension could be made for Patau's syndrome where in the second trimester uE_3 levels are reduced and inhibin increased, whilst in the first trimester, NT is markedly increased with very low free β-hCG and AFP. As with Edwards' syndrome first trimester DS protocols will incidentally detect a large proportion of cases.

Other common aneuploidies have abnormal marker profiles and are often detected incidentally from the DS or Edwards' syndrome risk. A meta-analysis of ten large DS screening studies has been published [17]. DS was estimated to comprise only 58% of detected aneuploidies; DS and Edwards' syndrome comprised 72%; and all common trisomies and sex chromosome abnormalities comprised 87%. These proportions were higher for second trimester serum protocols compared with those carried out wholly or partly in the first trimester.

Women with positive DS screening results should be counselled about all of the types of aneuploidy that may be identified following invasive prenatal diagnosis. To facilitate this, risks for each type of aneuploidy can be estimated from the screening marker profile and even summed to indicate a woman's total aneuploidy risk.

Other Abnormalities

X-linked ichthyosis can be identified through second trimester serum screening because STS deficiency causes abnormal estriol biosynthesis resulting in extremely low maternal serum uE_3 levels. Smith-Lemli-Opitz syndrome (SLOS) is an autosomal recessive disorder in which there is a defect in cholesterol biosynthesis. Because cholesterol is a precursor of estriol, affected pregnancies are characterized by low second trimester maternal serum uE_3 [18]. AFP and hCG also appear to be reduced and an algorithm has been proposed for second trimester screening for SLOS [19].

Several studies have reported increased NT in euploid pregnancies with major cardiac abnormalities; the studies have a wide range of observed detection rates, possibly reflecting referral ascertainment and viability biases as well as different NT cut-off levels. A recent meta-analysis of 20 studies estimated the detection rate for major cardiac defects was 44% for a 5.5% false-positive rate [20]. Given these rates, it is standard practice to offer follow-up fetal echocardiography to women with increased NT, regardless of the serum marker results and normal karotype.

Abnormal DV at 11-14 weeks has also been reported in association with fetal cardiac defects. The routine use of DV in screening for cardiac defects regardless of NT results has been advocated. In a recent meta-analysis the DR and FPR were estimated to be 19% and 4% for those with normal NT compared with 83% and 20%, respectively, if the NT is raised [21]. One study has evaluated different algorithms to optimally combine NT and DV in the assessment of risk for serious cardiac defects [22]. It was estimated that half of fetal cardiac defects could be detected with a false-positive rate of 2.7%.

Adverse Outcomes of Pregnancy

Adverse maternal-fetal complications of pregnancy are more common than chromosomal and structural fetal abnormalities combined. Whilst generally presenting late in pregnancy, many such complications can be shown to originate from events in the first trimester and are candidates for screening. Consequently, a paradigm shift has been proposed inverting the current 'pyramid of care' by shifting the focus of maternal-fetal examinations from the third to the first trimester [23].

The most thoroughly studied complication of pregnancy is pre-eclampsia (PE), a common cause of fetal and maternal morbidity and mortality and accounting for at least 2% and up to 7% of births in certain countries. A recent meta-analysis of nine published trials has shown that low dose aspirin prevents PE, provided treatment is started before 16 weeks gestation [24]. Prophylactic aspirin (75 mg/day) started before 12 weeks is now recommended in pregnancies identified at high risk due to various factors that include body mass index, parity and family history. However, these factors alone are insufficient and too broad.

Instead multi-marker risk screening has been proposed using prior factors, together with 11-13 week uterine artery Doppler pulsatility index (PI), mean arterial pressure (MAP), maternal serum PAPP-A and PlGF. Modeling predicts a detection rate of 93% for early PE, requiring delivery before 34 week, and 38% for all PE, with a false-positive rate of 5% [25].

The aspirin meta-analysis also indicated that early treatment reduced the risk of fetal growth restriction (FGR, or intrauterine growth retardation) or small for gestational age (SGA) fetuses. In the first trimester PAPP-A levels are low as are free-β hCG levels. In the second trimester, AFP, hCG and inhibin-A levels are elevated and uE_3 levels are low. First trimester screening with ultrasound, biophysical and serum markers can be used to identify pregnancies at high risk of this disorder, in the absence of pre-eclampsia. In one study a combination of ultrasound NT, maternal uterine artery PI, MAP together with serum PAPP-A, free β-hCG, PLGF and other serum markers, could identify almost three-quarters of early SGA pregnanacies, requiring delivery before 37 weeks and almost half at term [26].

Preterm birth is the most important cause of perinatal morbidity and mortality with an incidence that appears to be increasing in many developing countries. Maternal race, weight, socio-economic status, family health, and genetic factors all appear to affect rates. First trimester markers that are associated with preterm birth include low PAPP-A, increased NT and increased inhibin A. Second trimester low AFP, or high AFP, hCG and inhibin A have all been reported in pregnancies ending in a preterm birth.

Maternal Plasma Cell-Free DNA

In response to the encouraging published results of early studies, professional bodies cautiously recommended only a secondary use of the cfDNA test, in those already regarded as at high risk. This approach restricted testing to women with positive conventional screening tests and, in some localities, those of advanced maternal age, with a family history of aneuploidy or ultrasound markers. Since then further confirmatory studies have been published, particularly in unselected general risk populations, and there can now be no doubt that primary cfDNA screening could achieve vastly superior performance for the detection of DS than any of the conventional protocols or secondary cfDNA screening.

The best estimate of performance is meta-analysis of studies in high risk women. These studies were carried out before invasive testing so that all outcomes are known. The studies in general risk populations are subject to viability bias as well as limited follow-up. In one such meta-analysis the DS detection rate was 99.3% and false-positive rate 0.11% [27]. The corresponding rates for the other common trisomies were: Edwards' syndrome 96.9% and 0.09%; Patau's syndrome 87.3% and 0.23%.

Despite the superior performance, a policy of routine primary cfDNA is currently impractical in most public health settings because of the high list price leading to unacceptably high 'marginal' cost – that of detecting cases missed by existing policies like the Combined test [28]. The fact that in many localities the tests are available only by shipping samples to commercial laboratories in the USA and China is also an impediment.

To overcome the funding problem a compromise approach has been suggested, the 'Contingent cfDNA' test [28]. This approach proposes that all women, including those of advanced maternal age, initially have a conventional screening test and about 20% with high risks are selected for cfDNA screening. In most localities the Combined test would be used for the initial risk assessment, and modelling predicts that the DS detection rate would be 94% with a false-positive rate of 0.02% [27]. Performance could be further enhanced by including additional maternal serum markers such as PlGF and AFP to improve the initial risk assessment [29]. Modelling predicts that the DS detection rate would increase to 96% [27]. For those with sufficient training additional ultrasound markers could also be included in the initial risk assessment. And instead of selecting all 20% for cfDNA, it would be more practical to offer it to less than 0.5% of them with extremely high risk. All these protocols are vastly more cost-effective than primary cfDNA.

Contingent cfDNA screening also overcomes the problem of cfDNA test failure. About one-third of tests that fail also do so on repeat testing with a fresh sample and those with test failures have an increased chance of fetal aneuploidy. Sample turnaround time is extensive – about a week – so that many tests are not only unresolved but considerable anxiety is generated. The contingent protocol will ameliorate this as the Combined test result is available and could by itself be used to inform decisions in these cases.

FUTURE ROLE OF AFP

Public health decisions on screening policy need to be considered within a broad framework of services and not a single clinical outcome alone. This is particularly so for AFP screening since diverse outcomes are involved: NTDs, common aneuploidies and adverse pregnancy outcomes. Moreover, whilst it is clear that cfDNA testing is now clinically established, optimal implementation is limited by the high price of the test. However, the price is expected to fall in the near future. Thus there is likely to be more variability in screening policy between different centres worldwide than has been seen in the past.

Some centres which can afford the cost of cfDNA – an increasing number as the price falls – may decide to replace conventional multi-marker screening for aneuploidy with primary cfDNA. Since cfDNA can be determined both in the first and second trimester they could schedule the test at 14-15 so that a dating scan can be performed at the same time. Screening for NTDs could be achieved at the same time using the lemon and banana signs of spina bifida. There would be no role for AFP.

Centres that adopt a Contingent cfDNA screening policy based on first trimester Combined test markers may decide to additionally use PlGF and AFP to improve Down's syndrome detection. Such a centre could use the maternal serum AFP level together with first trimester BPD in screening for NTDs. Screening for pre-eclampsia could then be readily incorporated by measuring the uterine artery PI at the same time as the NT scan, and simply measuring blood pressure.

Those with insufficient resources to do any cfDNA screening at all could use the Combined test for aneuploidy and ultrasound scanning either simultaneously or in the second trimester for NTDs. But a proportion of women present too late for the Combined test and may be offered a second trimester Double, Triple or Quad test instead and use the AFP for NTD screening. These tests could be enhanced with ultrasound markers such as those measured in the facial profile.

References

[1] Wald NJ, Cuckle H. (1977) UK Collaborative AFP Study. Maternal serum alpha-fetoprotein measurement in antenatal screening for anencephaly and spina bifida in early pregnancy. *Lancet*, 1, 1323-1332.

[2] Wald NJ, Hackshaw AK, Cuckle HS. (2000) Maternal serum alphafetoprotein screening for open neural tube defects: revised statistical parameters. *Br J Obstet Gynaecol,* 107, 296-298.

[3] Wald N, Cuckle H, Boreham J, et al. (1980) Small biparietal diameter of fetuses with spina bifida: implications for antenatal screening. *Br J Obst Gynaecol*, 87, 219-221.

[4] Wald NJ, Cuckle HS, Densem JW, et al. (1992) Maternal Serum Screening for Down's Syndrome: The Effect of Routine Ultrasound Scan Determination of Gestational Age and Adjustment for Maternal Weight. *Br J Obstet Gynaecol*, 99, 144-149.

[5] Cuckle HS, Thornton JG. (1995) Antenatal diagnosis and management of neural tube defects. In: Levene MI, Lilford RJ, eds. Fetal and Neonatal Neurology and Neurosurgery. London: Churchill Livingstone, pp. 295-309.

[6] Bernard JP, Cuckle HS, Bernard M, et al. (2013) Combined screening for open spina bifida at 11-14 weeks using fetal biparietal diameter and maternal serum markers. *Am J Obstet Gynecol*, 209, 223. e1-5.

[7] Simon EG, Arthuis CJ, Haddad G, et al. (2015) A Biparietal/Transverse Abdominal Diameter (BPD/TAD) Ratio ≤1: a potential hint for open spina bifida at 11-13 weeks scan. *Ultrasound Obstet Gynecol*, 45, 267-72.

[8] Aitken DA, Mccaw G, Crossley JA, et al. (1993) First trimester biochemical screening for fetal chromosome abnormalities and neural tube defects. *Prenat Diagn*, 13,681-689.

[9] Bredaki FE, Poon LC, Birdir C, et al. (2012) First-trimester screening for neural tube defects using alpha-fetoprotein. *Fetal Diagn*, 31, 109-14.

[10] Spencer K, Khalil A, Brown L, et al. (2014) First trimester maternal serum alpha-fetoprotein is not raised in pregnancies with open spina bifida. *Prenat Diagn*, 34, 168-71.

[11] Saller DN Jr, Canick JA, Palomaki GE, et al. (1994) Second-trimester maternal serum alpha-fetoprotein, unconjugated estriol, and hCG levels in pregnancies with ventral wall defects. *Obstet Gynecol*, 84, 852-5.

[12] Saller DN Jr, Canick JA, Palomaki GE, et al. (1994) Second-trimester maternal serum alpha-fetoprotein, unconjugated estriol, and hCG levels in pregnancies with ventral wall defects. *Obstet Gynecol*, 84, 852-5.

[13] Cuckle HS, Pergament E, Benn P. Multianalyte Maternal Serum Screening for Chromosomal Abnormalities and Neural Tube Defects. In: Genetic Disorders and the Fetus: Diagnosis, Prevention and Treatment - 7th edition. (Ed A Milunsky and JM Milunsky).

[14] Cuckle H, Aitken D, Goodburn S, et al. UK National Down's Syndrome Screening Programme, Laboratory Advisory Group. (2004) Age-standardisation when target setting and auditing performance of Down syndrome screening programmes. *Prenat Diagn*, 24, 851-856.

[15] Aagaard-Tillery KM, Malone FD, Nyberg DA, et al. FaSTER Research Consortium. (2009) Role of second trimester genetic sonography following Down syndrome screening. *Obstet Gynecol*, 114, 1189-96.

[16] Agathokleous M, Chaveeva P, Poon LCY, et al. (2013) Meta-analysis of second-trimester markers for trisomy 21. *Ultrasound Obstet Gynecol*, 41, 247-261.

[17] Davis C, Cuckle H, Yaron Y. (2014) Screening for Down syndrome - incidental diagnosis of other aneuploidies. *Prenat Diagn*, 34, 1044-8.

[18] Bradley LA, Palomaki GE, Knight GJ, et al. (1999) Levels of unconjugated estriol and other maternal serum markers in pregnancies with Smith-Lemli-Opitz (RSH) syndrome fetuses. *Am J Med Genet*, 82, 355-358.

[19] Palomaki GE, Bradley LA, Knight GJ, et al. (2002) Assigning risk for Smith-Lemli-Opitz syndrome as part of 2nd trimester screening for Down's syndrome. *J Med Screen*, 9, 43-44.

[20] Sotiriadis A, Papatheodorou S, Eleftheriades M, et al. (2013) Nuchal translucency and major congenital heart defects in fetuses with normal karyotype: a meta-analysis. *Ultrasound Obstet Gynecol*, 42, 383-9.

[21] Papatheodorou SI, Evangelou E, Makrydimas G, et al. (2011) First-trimester ductus venosus screening for cardiac defects: a meta-analysis. *Br J Obstet Gynaecol*, 118, 1438-45.

[22] Borrell A, Grande M, Bennasar M, et al. (2013) First trimester detection of cardiac defects with the use of the ductus venosus blood flow. *Ultrasound Obstet Gynecol,* 42, 51-7.

[23] Nicolaides KH. (2011) A model for a new pyramid of prenatal care based on the 11 to 13 weeks' assessment. *Prenat Diagn*, 31, 3-6.

[24] Bujold E, Roberge S, Lacasse Y, et al. (2010) Prevention of preeclampsia and intrauterine growth restriction with aspirin started in early pregnancy: a meta-analysis. *Obstet Gynecol*, 116 (2 Pt 1), 402-14.

[25] Akolekar R, Syngelaki A, Poon L, et al. (2013) Competing risks model in early screening for preeclampsia by biophysical and biochemical markers. *Fetal Diagn Ther*, 33, 8-15. Erratum in: *Fetal Diagn Ther*, 34, 43.

[26] Karagiannis G, Akolekar R, Sarquis R, et al. (2011) Prediction of small-for-gestation neonates from biophysical and biochemical markers at 11-13 weeks. *Fetal Diagn Ther*, 29, 148-54.

[27] Cuckle HS, Benn P, Pergament E. (2015) Cell-free DNA screening for aneuploidy as a clinical service. *Clin Biochem*, 48, 932-41.

[28] Cuckle H, Benn P, Pergament E. (2014) Clinical Utility and Cost of Non-Invasive Prenatal Testing. *J Matern Fetal Neonatal Med*, 27, 320-1.

[29] Nicolaides KH, Wright D, Poon LC, et al. (2013) First-trimester contingent screening for trisomy 21 by biomarkers and maternal blood cell-free DNA testing. *Ultrasound Obstet Gynecol*, 42, 41-50.

In: Alpha-Fetoprotein: Functions and Clinical Applications
Editors: Nisha Lakhi and Michael Moretti
ISBN: 978-1-63484-875-6

Chapter 8

Maternal Serum Screening in Women with Pre-Existing Diseases and Congenital Disorders: Effects of Alpha-Fetoprotein on False Positive Screens and Course of Disease

G. J. Mizejewski, PhD
Division of Translational Medicine, Molecular Diagnostics Laboratory,
Wadsworth Center, New York State Department of Health, Albany, NY, US

Abstract

Prenatal screening laboratories attempt to maintain a planned balance between the detection rate (DR) and the false positive rate (FPR) for their screening tests. Such labs aim at achieving a constant FPR of 5% or less while sustaining a screen positive detection of 70-90% depending on the disorder. In the present report, eight examples of patient screening are presented which produced real-time false positive (FP) results. The FP cases had been screened using either the triple or quad test for Down syndrome and maternal serum (MS)/amniotic fluid alpha-fetoprotein (AFP) analysis for neural tube defects. The eight diseases and disorders encompass four pre-existing maternal autoimmune diseases and four congenital disorders. The autoimmune diseases include: 1) rheumatoid arthritis (RA); 2) multiple sclerosis (MS); 3) myasthenia gravis (MG); and 4) systemic lupus erythromatosus. The four congenital disorders encompass: 1) epidermolysis bullosa (EB); 2) ventral wall defects (VWD); 3) AFP congenital deficiency (AFP-CD); and 4) AFP hereditary persistence (AFP-HP). While RA, MG, EB, VWD and HP-AFP resulted in false positive screens for neural tube defects (NTD); MS, SLE, and AFP-CD produced false positive screens for Down syndrome (DS). The screening results and case histories of the representative patients are discussed, in addition to a description of some of the clinical characteristics and traits of each of the diseases/disorders involved. In addition, the role of AFP as a soluble immunoprotective factor during pregnancy is presented and discussed. Prior knowledge of the clinical case history of the patients can forewarn the screening lab and greatly aid the practitioner in the subsequent diagnosis, treatment, and management of each pregnancy disease/disorder.

Keywords: hepatoma, biomarkers, chorionic gonadotrophin, congenital disorders, dimeric inhibin, maternal serum, down syndrome, pregnancy, alpha-fetoprotein, neural tube defect

I. Introduction

Alpha-fetoprotein (AFP) is a tumor-associated fetal protein (i.e., an oncofetal protein) which functions both as a carrier/transport molecule and a regulator of growth [1]. AFP consists of a 69 kD single chain polypeptide containing a 4% carbohydrate content [2]. It is produced largely in the liver and is secreted into the serum wherein the glycoprotein circulates with a half-life of 3 to 5 days. In the clinic, AFP has served as a biomarker for both fetal defects as well as tumor growth of hepatomas, germ cell cancer, and gastrointestinal tumors. In non-malignant liver conditions, AFP is an indicator of cirrhosis, viral and alcoholic hepatitis, and liver dysfunction [3].

Elevated AFP in pregnancy is associated with various fetal malformations such as anencephaly, spina bifida, omphalocoele, gastroschisis, nephrosis, renal agenesis, esophageal/phyoric atresia, aplasia cutis congenital, and epidermolysis bullosa. Increased serum levels of AFP have been ascribed to five possible mechanisms [4]. First, neural tube defect (NTD) disorders can result from AFP leakage from the cerebral fluid in the amniotic sac. Second, AFP transudates from the blood vessels of the viscera in the event of defects such as omphalocoele and gastroschisis. Third, congenital nephrosis produces a state of fetal proteinuria with subsequent leakage of AFP into amniotic fluid. Fourth, blockages in the gastrointestinal tract (duodenal/esophageal atresia) cause impairment of fetal swallowing/digestion with the accumulation of fetal proteins (AFP) into the amniotic sac. Lastly, elevated maternal serum and amniotic fluid AFP levels may arise due to pore penetration and leakage from the placenta and skin lesions in aplasia cutis congenita and in epidermolysis bullosa [5].

Laboratories that screen for Down syndrome (DS) and neural tube defects (NTD) conventionally report their false positive rate (FPR) together with their detection rate (rate) due to their interdependency. Many prenatal screening groups try to reduce or limit their FPR (2-5%) at the expense of the DR (75-90%) for the well-being of the patient. False positive (FP) screen results can cause stress, depression, and anxiety in the patient; clinically it has been associated with an increased risk for fetal demise, which largely stem from an unnecessary amniocentesis [6]. Moreover, a FP screen result for DS in the second trimester has been considered an increased risk factor for indicating a multitude of adverse pregnancy outcomes such as pre-eclampsia, small-for-gestational age (SGA) newborns, placenta previa, fetal death, low birthweight, gestational diabetes, and pre-term delivery [7-10]. Certain demographic factors are known to influence biomarker median values which can contribute to increases in the FPR; these include maternal age, body weight, and ethnicity; fetal gender; a maternal vegetarian diet, prior pregnancy outcomes, and assisted reproductive procedures [11-14]. Some of these well-known factors have algorithm adjustments for calculation of the multiple of the medians (MoM). However, other maternal health conditions that produce FP screens are less obvious and yet can manifest in the patient. For example, maternal pre-existing chronic disease and spontaneous malformations in the fetal/placental compartments may affect biomarker profiles producing screens that mimic DS and NTD prenatal biomarker profiles. In such conditions, adjunct clinical procedures such as scanning/imaging techniques, Doppler ultrasound, amniotic fluid biomarker and chromosome analysis, and placental angiography may serve to explain complications in pregnancy.

II. OBJECTIVES

In lieu of the above discourse, the objectives of the present report are three-fold in scope. First, this review will enumerate and exemplify eight known pregnancy conditions that result in FP screen results. These states include: a) four pre-existing autoimmune diseases, and b) four fetal congenital disorders. The four autoimmune diseases included rheumatoid arthritis (RA), multiple sclerosis (MS), myasthenia gravis (MG), and systemic lupus erythromatosus (SLE). The four congenital disorders encompassed epidermolysis bullosa (EB), ventral wall defects (VWD), AFP-congenital deficiency (AFP-CD), and AFP-hereditary persistence (AFP-CD). Secondly, the precise analyte profile of each biomarker set (which led to the FP screen result) is presented utilizing the analyte MoM values, laboratory cutoff numbers, and the DS and NTD relative risk ratios. Thirdly, AFP will be discussed as a potential circulating immunoprotective agent to aid in preserving the fetus as a homograft within the mother's body. Finally, following demonstration of the MoM values leading to the FP screen, each individual disease and disorder is discussed regarding the patient's course of disease during pregnancy. All clinical case studies discussed below were derived from published historical cases mimicked in a New York State Health Department Proficiency Testing Program.

III. AUTOIMMUNE DISEASES (AD)

1. Rheumatoid Arthritis (RA)

A case history specimen at 16 weeks provided an interesting AD example involving elevated levels of MSAFP (MoM = 3.1) together with normal MSuE3 (MoM = 0.63), MSHCG (MoM = 1.16), and MSDIA (MoM = 1.15) levels; this profile resulted in a false positive screen for neural tube defects (NTD) (risk = 1 in 90) and a negative screen for Down syndrome (DS; T21) (Table 2). This specimen mimicked one of eight such literature case studies of pregnant women with autoimmune rheumatoid arthritis (RA) that exhibited similar aberrant levels of AFP as an NTD biomarker [15, 16]. Thus, the above-representative RA specimen produced a false positive prenatal screen for NTD due to high MSAFP levels; the elevation was thought to be caused by placental vasculitis leading to increased passage of AFP into the maternal bloodstream.

RA is an autoimmune disease that results in a systemic disorder affecting many tissues and organs, especially flexible synovial joints [17]. It has been known for many years that about 70% of patients with RA remit during pregnancy [18-20]. The mechanism of remission of RA in pregnancy is incompletely understood and has been proposed as an induction of an immunosuppressive state caused by local soluble immunoprotective factors yet to be elucidated [21]. Such local non-specific soluble factors produced by the mother, fetus, and/or placenta are thought to play a major role in retaining the fetus as an allograft within the mother. These soluble factors possess immunomodulatory properties and are produced in excess at the fetoplacental interface in order to suppress fetal rejection [20]. One such factor could be alpha-fetoprotein.

Alpha-fetoprotein (AFP) is a gestational-age-dependent biomarker known to exhibit immunomodulatory properties. Human AFP has been shown to suppress the mitogenic stimulatory response produced by various lectins as well as inhibiting the mixed leukocyte reaction (MLR) by inducing the production of suppressor T-cells [22, 23]. It was further demonstrated the AFP suppressed Class-II major histocompatibility complex (MHC) expression on macrophages, which may allow the fetus to acquire tolerance to self-antigens during development *in utero*. AFP has also been implicated in the blocking of Class-II MHC antigen expression by human myelocyte macrophages [24]. Recent studies have shown that AFP can also inhibit natural killer cell activity [25] and enhance cytokine, chemokine, and growth factor activity [26]. Finally, it was further demonstrated that AFP can shield target (hepatoma) cells from TNF-alpha induced apoptosis [27, 28] and can protect such cells from cytotoxic lymphocyte attack by blocking progression of the caspase apoptotic pathway [3, 29]. Further discussions on the immunomodulatory properties of AFP in autoimmune disease are presented below in other AD case studies.

Table 1. The case studies and demographics of the eight representative pregnant patients with pre-existing diseases/disorders are listed according to their reproductive histories

Patient-Disease/Disorder	Ethnic Group	Body Weight	Maternal Age	Gravida/Parity	Reproductive History
I. All patient specimens resulted in false positive screen results.					
1) Rheumatoid Arthritis	White	150	26	1/0	No pregnancy adverse outcomes
2) Multiple Sclerosis	White	160	30	3/2	No history of pregnancy complications
3) Myasthenia Gravis	White	136	25	1/0	No previous pregnancy losses
4) Systemic Lupus Erythematosus	Hispanic	150	28	3, 0	No miscarriages, renal dysfunction, autoantibodies
5) Epidermolysis Bullosa	White	145	28	3/0	No previous pregnancy losses
6) Ventral Wall Defects	White	150	25	1/0	No miscarriages
7) AFP Congenital Deficiency	White	150	33	2/1	Pregnancy complications and adverse outcomes
8) AFP Hereditary Persistence	White	145	25	1/0	Accelerated overall body growth and advanced bone age

2. Multiple Sclerosis (MS)

A case history specimen at 15 weeks presented another interesting AD example involving low levels of MSAFP (MoM = 0.70), MSuE3 (MoM = 0.63) and slightly elevated MShCG levels (MoM = 1.7). This biomarker profile resulted in a false positive screen for Down syndrome (see Table 2). Using the triple biomarker test, the T21 risk was 1 in 218 which is greater than the risk cutoff of 1 in 270. This sample was meant to mimic one of several similar literature case studies of pregnant women with pre-existing multiple sclerosis (MS) disease whose serum displayed aberrant triple test biomarker levels [30, 31]. In prior pregnancies, most of the MS case study women had delivered normal term infants free of anatomical defects. In the presently discussed prenatal screen, the representative specimen produced a false positive screen for T21 due to a triple test profile of low MSAFP and MSuE3 together with elevated hCG.

Table 2. The autoimmune diseases/disorders of pregnant women screened via the perinatal quad versus the triple biomarker assays producing false positive results

Autoimmune Disease or Disorder	Biomarker MoM Values				Screen Assessment		Risk Estimate (1 in X)	
	MSAFP	MSuE3	MShCG	MSDIA	NTD	T21	Quad	Triple
I. Autoimmune Diseases								
1) Rheumatoid Arthritis	3.10	0.63	1.16	1.60	Pos	Neg	1 in 90	None
2) Multiple Sclerosis	0.70	0.63	1.72	0.64	Neg	Pos	None	1 in 218
3) Myasthenia Gravis	3.00	1.70	1.90	1.00	Pos	Neg	1 in 90	None
4) Systemic Lupus Erythematosus	1.13	0.50	3.17	2.90	Neg	Pos	1 in 16	1 in 15
II. Congenital Disorders								
5) Epidermolysis Bullosa	5.62	0.65	1.04	1.00	Pos	Neg	1 in 13	None
6) Ventral Wall Defects	3.00	1.70	1.00	1.00	Pos	Neg	1 in 10	None
7) AFP Hereditary Persistence	2.55	0.93	0.86	0.89	Pos	Neg	1 in 295	None
8) AFP Congenital Deficiency	0.02	0.74	1.67	0.84	Neg	Pos	1 in 71	1 in 27

The eight diseases/disorders of pregnant women screened via the prenatal quad biomarker assay are listed according to their analyte profile and multiple of median (MoM) values. MoM = Multiple of Median; MS = maternal serum; AFP = alpha-fetoprotein; uE3 = unconjugated Estriol; hCG human chorionic gonadotrophin; DIA = dimeric inhibin-A.

During pregnancy, MS activity produces two major manifestations, a period of profound reduction of MS symptoms (remission) in the third trimester followed by an exacerbation of disease (relapses) in the postpartum period before returning to a pre-pregnancy disease state [32]. In a study of 935 pregnant women with MS, relapses occurred 40-50% in the postpartum period, while only 10% of relapses took place in the third trimester [33]. Other investigators found a notable decrease in relapses in the third trimester of pregnancy, in contrast to a high increase in the first three months following delivery [32, 34]. As discussed above for RA, there appears to be soluble immuno-protective factors produced during pregnancy that contribute to causing the MS disease to be less active.

Alpha-fetoprotein (AFP) has a well-known history as an immunoregulatory glycoprotein but is also associated with normal fetal growth and development. Recombinant human AFP has been reported to reduce experimental allergic encephalomyelitis (EAE)-induced neuroinflammation (animal version of MS), to increase apoptosis of activated immune cells by inhibiting BCL-2-related pathways, and to increase the presence of FAS-related (CD95) ligands [35]. Furthermore, AFP increased both FOXp3 expression in lymph nodes and T-reg cell numbers in the central nervous system (CNS) [36]. AFP has been studied in animal models of MS and was found to both treat and prevent the induction of EAE. AFP exerts significant immunosuppressive effects on T-cells *in vitro* at physiologic concentrations and has the capacity to induce suppressor T-cells [37]. Overall, several investigators have shown a beneficial effect of AFP on the course of EAE in guinea pigs and other animal models [38, 39]. During human pregnancy, maternal serum AFP levels gradually increase until 30-32 weeks of the third trimester; this is a gestational time in which 50% of MS patients experience disease remission. Thereafter, maternal serum AFP levels decrease to low nanogram/ml levels at postpartum when most relapses were reported to occur.

In regard to the patient case history represented by the above specimen and the seven additional case women, all such patients were followed by immunoassays throughout pregnancy [31]. The number of CD8 suppressor T-cells had decreased, and the CD4 helper cell-to-CD8 suppressor T-cell ratios had increased in the pregnant MS patients compared to normal pregnant women. A reduction in the peripheral population of CD4 helper T-cells have been reported during MS pregnancy with a consequent lowering of the helper-to-suppressor T-cell ratios; these may be related to AFPs suppressive effect on T-cells as described above [37, 40].

3. Myasthenia Gravis (MG)

A case history specimen at 15 weeks represents yet another interesting study involving highly elevated levels of MSAFP (MoM = 3.0), high normal MSuE3 (MoM = 1.70), and a high normal MShCG level (MoM = 1.9); this biomarker profile resulted in a false positive screen for NTD (Risk = 1 in 90) and a negative screen for T21 (Table 2). This sample was intended to mimic one of several literature case studies of pregnant women with myasthenia gravis disease (MGD) that exhibited similar aberrant levels of second trimester biomarkers for NTD [41-43]. All women in these studies had pre-existing MG, delivered previous normal term infants, but had experienced cycles of remissions and relapses prior to term pregnancy. In summary, the representative specimen shown above produced a false positive prenatal screen for NTD. AFP as a major biomarker is discussed below concerning its anti-inflammatory and immunoprotective values during MG pregnancy.

Regarding the patient case MG history linked to the NTD false positive specimen, patients similar to this specimen and other affected pregnant women have been followed throughout pregnancy [41]. The underlying pathology involves maternal autoimmune IgG antibodies binding to the α-subunit of the AchR and preventing or obstructing nerve transmission in the mother. Such IgG autoantibodies are capable of crossing the placenta and entering the fetal compartment. As seen above in RA and MS, MG also produces a notable reduction of MG symptoms (remission) in the second and third trimester followed by an exacerbation of disease (relapses) in the mother's postpartum period before returning to pre-pregnancy disease status [44]. In a study of 531 pregnant women with MG, remissions occurred 30% in the second and third trimesters, while 39% overall relapses occurred in late pregnancy and 30% in the postpartum period [45]. Investigators found a decrease in relapses in the latter trimesters of pregnancy and increases in occurrence during the first three months following delivery [46]. As in other ADs, soluble immunoprotective factors have been proposed to explain the disease remissions and this includes AFP [47].

Alpha-fetoprotein (AFP) is an immunoregulatory factor known to either enhance or inhibit immune responses. Recombinant human AFP has been reported to reduce autoimmune-induced visceral- and neuro-inflammation and to increase apoptosis of activated immune cells by reducing access to BCL-2-related pathways; AFP also increases the expression of FAS-related (CD95) ligands [35]. As shown above for MS, AFP increased both FOXp3 (forkhead) expression in lymph nodes and T-reg cell numbers in some autoimmune disorders [36]. AFP has been studied in animal models of MG and found effective in treating and preventing its induction. AFP at physiological concentrations exerts significant immunosuppressive effects on T-cells *in vitro* and has the ability to induce suppressor T-cells [48]. Overall, several investigators have shown a beneficial effect of AFP on the course of MG in both animal models and in human patients [49, 50]. Pregnancy fluids enriched with AFP have been shown to inhibit *in vitro* interactions of antibodies to the AchE receptor binding to the AchE receptor itself [49]. Removal of AFP from such fluids nullified the inhibitory autoantibody effect. Purified AFP has also been reported to suppress the phytohemagglutinin (PHA) mitogen reaction induced in the proliferative response of lymphocytes in experimental models of MG [50]. During human pregnancy, maternal serum AFP levels gradually increase and peak at the 30-32 week period of the third trimester prior to its decline. The third trimester is the gestational age in which 50% of MG patients experience disease remission [51]. Thereafter, maternal serum AFP levels decrease to low nanogram levels at postpartum when most MG patient relapses were found to occur. Transitory MG occurs in some neonates following delivery presumably due to the continued presence of autoantibodies in the mother and newborn [52].

In further studies, lab animals receiving intravenous injections of AFP failed to develop experimental MG following disease induction [53]. Also, animals with established experimental MG showed clinical improvements in response to purified AFP injections in which anti-AchE receptor antibody production was suppressed [54, 55]. Rats immunized with Torpedo eel AchE receptor develop MG and show both early acute and late phase chronic disease similar in humans with MG; in such instances both phases can be prevented by injections of purified AFP [36]. Not only did AFP treatments result in reduced clinical and electromyographic manifestations, AFP also decreased the serum autoantibody titers against AchE receptors [42]. In like manner, the absence of MG symptoms in the fetus and the delayed onset of worsening in clinical transient neonatal MG may be partially due to the presence of fetal and neonatal AFP in infants following birth [56]. For example, in premature and small-for-gestational age (SGA) infants, the levels of AFP remain high as compared to average birthweight newborns. Moreover, premature and SGA newborns are more common in patients with MG than in the normal pregnancy population.

4. Systemic Lupus Erythromatosus (SLE)

A maternal serum specimen at 18 weeks provided yet another interesting case involving normal levels of MSAFP (MoM = 1.13), low MSuE3 (MoM = 0.50), and elevated MSHCG

(MoM = 3.17) and MSDIA (MoM = 2.9) levels; this profile resulted in a negative screen for NTD but yielded a positive screen risk for T21 (1 in 16; risk cutoff is 1 in 270). This specimen mimics one example of several literature case studies of pregnant women with autoimmune systemic lupus erythematosus (SLE) that exhibited aberrant levels of DS biomarkers [57-59]. Similar to other SLE published reports, the maternal serum biomarkers of the case patient demonstrated MSAFP MoM of 1.16, MSuE3 MoM of 0.7, MSHCG MoM of 3.6, and MSDIA MoM of 3.5 [6, 7] resulting in a positive screen for T21. In this case, the MSAFP levels were low or normal.

Previous reports in the literature had demonstrated that the above biomarker profile also predicted pregnancy complications and adverse outcomes [58, 60]. A combination of thrombophilic defects in SLE patients can result from the effects of antinuclear, anti-phospholipid, anti-RNA, and anti-cardiolipin autoantibodies. It was found that pregnancies with SLE autoantibodies affected hCG MoM and uE3 MoM values both of which are associated with low fetal weight and increased thrombotic event. SLE is a chronic, multisystem autoimmune connective tissue disorder of unknown etiology [61, 62]. As in other autoimmune diseases, the immune system attacks the body's cells and tissues resulting in inflammation and damage to multiple organ systems. SLE is characterized by the formation of antigen: antibody immune complexes which can precipitate within blood vessels and interstitial spaces causing organ destruction [63]. The organ/tissue damage encompasses multiple organs including the kidneys, lungs, brain, spinal cord, heart, and blood vascular system.

Clinical manifestations in SLE patients include skin and facial rashes, arthritis, hemolytic anemia, oral/nasal ulcers, seizures, psychotic symptoms, pericardial inflammations, leucopenia/thrombocytopenia, and kidney dysfunction [63]. Gender differences in SLE patients are quite significant in that the disease occurs nine times more often in women than in men. The laboratory diagnosis for SLE involves the use of several different serum autoantibody assays including antinuclear antibody, antiphospholipid antibodies, lupus anticoagulant, anti-RNA antibodies, and the anti-cardiolipin antibodies SS-A/Ro and SS-B/LA [64, 65]. It is germane to this discussion that AFP injected into NZB mice (SLE disease in animals) was found to reduce the titers of IgG1, IgG2, and IgA circulating autoantibodies [66].

IV. Congenital Disorders

1. Epidermolysis Bullosa (EB)

A specimen at 19 weeks offered another interesting case involving extremely high levels of MSAFP (MoM = 5.6) with normal MShCG, MSuE3, and MSDIA. This biomarker profile resulted in a positive screen for NTD (risk = 1 in 13). Sample MS328 was mimicked after several case studies of pregnant women with Epidermolysis Bullosa (EB) which manifested highly elevated levels of MSAFP and AFAFP and an aberrant presence of acetylcholinesterase (AchE) in amniotic fluid [67, 68]. EB was first reported in 1968 and only 144 pregnancy cases have been described thus far in the literature. EB is an inherited malformation of the skin collectively called genodermatosis; however, EB can occur in fetuses without a previous known risk. In some cases, mothers had given birth to a previous child afflicted with EB or a related sporadic skin disorder termed aplasia cutis congenita (ACC) [69]. Most of the women in the case studies had raised levels of MSAFP determined after their first obstetrician visit. Most of the women with EB did not deliver normal term infants, but rather experienced spontaneous second and/or third trimester abortion or pre-term birth with adverse outcomes involving massive fetal skin lesions. In a few instances, the skin lesions can appear later in the newborn and neonatal period. In conclusion, the specimen above produced a false positive screen for NTD due to the extremely elevated levels of MSAFP and AFAFP.

Epidermis bullosa is an inherited skin connective tissue disorder causing blisters, scarring, erosions, and eruptions within and about the skin and epithelial mucous membranes. This skin disorder can be highly lethal in the fetal, perinatal, and neonatal periods; however, milder forms do exist that allow patients to live into adulthood [70]. The major defect in EB lies in a separation of the epidermis from the dermis layer due to disruption of anchoring filaments between the two layers. Gene mutations occur in several connective tissue proteins including collagen, keratin, integrins, laminins, integrin alpha-6 and beta-4, and in collagenase enzymes [71]. These mutated, altered proteins result in the formation of defective anchors and filaments that attach the basement membrane of the epidermis to the upper layer of the dermal layer. There is no cure for this blistering and denudation skin disorder and it renders the surviving patients susceptible to skin cancers later in life [72]. EB occurs worldwide (50 in 1 million people) affecting every racial and ethnic group, especially Lebanese, Turkish, Arab, Bedouin, and American Indian populations largely due to consanguineous marriages [67, 73].

Diagnostic tests for EB displayed highly elevated levels of both MSAFP and AFAFP. Ultrasound imaging revealed the presence of polyhydramnios, dilated stomach, pyloric atresia, and anatomical defects in the brain ventricles of some fetuses [74, 75]. Pregnancies are often spontaneously or therapeutically aborted in the third trimester (25-33 weeks) with the delivery of dead fetuses or fetuses that die within several days following birth. Second trimester prenatal biomarkers for EB patients include extremely elevated MSAFP levels (24 MoMs) that are mirrored by AFAFP MoMs ranging up to 21 MoMs [76]. AchE is usually present and detected in the amniotic fluid, which prompts fetal skin sampling at 20-22 weeks. However, all EB fetuses display normal karyotypes of both sexes.

2. Ventral Wall Defects (VWD)

A specimen at 20 weeks proved to be a highly interesting case involving extremely high levels of MSAFP (MoM = >3.0) with normal MShCG, MSuE3, and MSDIA (MoM = 1.0). This biomarker profile resulted in a positive screen for NTD (risk = 1 in 13). This specimen was mimicked after several case studies of pregnant women with Ventral Wall Defects (VWD) which manifested highly elevated levels of MSAFP and AFAFP and an aberrant pattern of acetylcholinesterase (AchE) in amniotic fluid [77]. All of the women in the case studies had raised levels of MSAFP determined after their first obstetrician visit. Moreover, their ultrasound results showed some gross anatomical malformations displaying gastrointestinal obstructions such as atresias, stenosis, gastroschisis, and emphalocoele [78].

In overview, this specimen produced a false positive screen for NTD due to the extremely elevated levels of MSAFP and AFAFP.

Ventral wall defects consist of two major types, namely, gastroschisis (GTS) and omphalocele (OPC) [79]. GTS is a full-thickness defect (opening) in the abdominal wall lateral to the insertion of the umbilical cord [80]. As a result of the abdominal wall separation, the viscera protrude out through the opening and float uncovered (lack of membrane) in the amniotic fluid. The bowels are shorted as a consequence and acquire an inflammatory layer or coating over the surfaces of the viscera. Intestinal function in such newborns is slow to recover and the lesion requires surgical repair at birth [81].

In contrast, an OPC is a persistent and vast enlargement of the body stalk of the umbilical cord itself protruding from the abdominal wall. A membranous sac covers the herniated viscera which emerge alongside the umbilical cord and opens into the abdominal wall. OPC is associated with a high incidence of fetal malformations compared to GTS and is co-morbid with chromosomal anomalies. Immature infant birth (<36 weeks), low birthweight (<1500 g), and high mortality rates are also associated with OPC [82].

MSAFP screening leads to the detection of 95% of fetal GTS cases at 15-20 weeks gestation. In a study of 73,782 pregnancies (Maine and Rhode Island), 20 cases of GTS and 13 cases of OPC were identified [83]. The median ranges of MSAFP MoMs were 3.6 to 13.5 MoMs for GTS and 0.5 to 29.8 MoMs for OPC cases. The combined rate of VWD was 5.4 per 10,000 livebirths for this particular study. In another study of 35 VWD pregnancies and 200 controls, the median ranges for triple screen markers were determined [84]. The MSAFP median MoM for GTS was 9.42 MoMs, while the median for OPC was 4.14 MoMs. In contrast, the median MShCG for GTS was 1.10 MoM while that for OPC was 1.13; the median MSuE3 MoM for GTS was 1.10 MoM and OPC was 0.9 MoM. The amniotic fluid biomarkers and ultrasound values for VWDs have also been reported [85, 86]. In GTS specimens, 89% of specimens displayed elevated AFAFP levels, while OPC specimens showed only 29% elevations. Regarding the presence of AchE in AF, 73% of GTS specimens showed a positive presence, but only 27% of OPC specimens did so [87, 88].

3. Alpha-Fetoprotein Hereditary Persistence (AFP-HP)

When specimens have an unexplained elevated MSAFP but lack an abnormal AchE and/or fetal hemoglobin result from AF analysis, such results are problematic for screening laboratories (Table 2). If other family members are found to exhibit elevated AFP levels (MoM = > 2.5), the patient is considered a candidate for Hereditary Persistence of AFP, an autosomal genetic disorder. When amniocentesis and ultrasound results in such patients show a normal karyotype and unremarkable level-II diagnostic sonography for NTD or other structural or anatomic anomalies, the course of action for such patients could be directed toward a presumptive positive case for HP-AFP. Due to the advanced body/bone growth and early puberty in some HP-AFP women, some endocrinologists would suggest that other family members undergo a serum test for AFP and possibly other reproductive biomarkers. Serum elevated adult AFP levels have been found in multiple family members spanning over three generations following pedigree blood testing; such patients are called probands.

Patients without NTD having a positive screen result for NTD involve a rare benign autosomal dominant trait with complete penetrance called HP-AFP. The first clinical case of

HP-AFP was described in 1983 from a prenatal screening program for spina bifida [89, 90]. A 38 year old Caucasian woman in her 21st week of pregnancy displayed unexplained highly elevated (>2.55 MoMs) levels of MSAFP, uE3 MoM = 0.93, and hCG MoM = 0.86 (Table 2). Ultrasound revealed a normal singleton fetus at the correct gestational age together with normal amniotic fluid AFP levels. This baby was anatomically unremarkable at time of delivery with no visible malformations or clinical disabilities. Subsequently, 70 family members of this patient were serum tested and 21 members (including nine males) were found to have elevated levels of serum AFP in adulthood. Since that time, nineteen other such families have been identified in the biomedical literature, some cases of which displayed testicular abnormalities in the males. It is also of interest that females frequently displayed various reproductive endocrine problems. The inherited trait can be present in both male and female descendants within a given family line, however, females demonstrate less life-threatening clinical problems. Both males and females have been reported to be reproductively fertile. Elevated AFP levels in adults without HP-AFP may be present in a variety of conditions other than pregnancy which include liver and germ cell disorders, cirrhosis, hepatitis, malignancies, anemias, immunodeficiencies, and ataxia telangiectasias [81, 91, 92]. Since HP-AFP can be a cause of persistently elevated AFP levels into adulthood, the condition can be discerned by the testing of serum AFP in first and second generation family members followed by molecular DNA studies to determine the genetic basis of the trait. However, adult elevated AFP levels may be difficult to distinguish in HP-AFP patients already being serum tested for malignancies, liver dysfunctions, and pregnancy.

Serial levels of serum AFP are monitored following surgical tumor ablation and these usually recede to normal adult AFP levels (less than 10 ng/ml). Non-pregnant patients with non-receding AFP values following tumor ablation are presumed to have recurrent or persistent hepatic disease and are sometimes treated with chemotherapy at least in cases of hepatoma. Such a case was reported in a 20 month old male child (AFP = 2,659 ng/ml) who underwent surgery for a testicular yolk sac tumor and needlessly received multiple rounds of chemotherapy without the presence of a tumor [93, 94]. The authors of these reports proposed that, in retrospect, levels of AFP in family members should be assayed before deciding on chemotherapy based on mildly or moderately elevated AFP in adult patients.

4. Alpha-Fetoprotein Congenital Deficiency (AFP-CD)

The first clinical cases of AFP-CD were reported in two patients in 1992 following prenatal triple screening for NTD and trisomy-21 (T21) [95]. These screened specimens were of special interest in that the samples produced T21 false positive screens for DS with the virtual absence of AFP (MSAFP MoM = 0.02-0.05) (Table 2). Moreover, these specimens showed normal levels of hCG and only marginally lowered uE3. These patients had no prior histories of pregnancy complications and adverse outcomes, and the subsequent Stage-II ultrasound, acetylcholinesterase (AchE), AFAFP, and karyotyping (amniocentesis) were unremarkable. At full term, these patients delivered normal infants with no apparent physical or developmental problems. However, both specimens displayed one unusual biomarker in that the MSAFP was very low or virtually absent. As discussed below, extremely low or absent AFP levels have been associated with a rare and supposedly benign genetic condition termed congenital deficiency of AFP (AFP-CD).

In 1997, Sher and Shohat published a Down syndrome (DS) screening case report which produced a borderline DS screen result involving the total absence of MSAFP [96]. The DS screening assays resulted in a MSAFP MoM = 0.00, uE3 MoM = 0.63, and an hCG MoM = 1.63 producing a borderline DS of 1 in 290 (cutoff = 1 in 270). Chromosome analysis revealed that the genotype was 47 XX + 21 which was confirmatory for Down syndrome. The authors stated that despite publication of previous cases showing that AFP-CD was a benign condition, amniocentesis should be recommended for patients with extremely low AFP screening results. In a later case study, two patients with congenital absence of AFP from first cousin pregnancies were reported by Sharony et al. [97]. The first patient also displayed inflammatory infiltrates located in the placenta (deciduitis) accompanied by Monosomy-16 in trophoblast cells; however, this patient delivered a normal fetus and AFP-CD was confirmed by genetic procedures. Histochemically-localized AFP was also found to be present in fetal, placental and maternal tissues from this patient.

In another patient described by Sharony, the triple screen demonstrated an AFP MoM = 0.02, uE3 MoM = 0.74, hCG MoM = 1.67 (Table 2), and a cord blood AFP level = 0.00; however, a normal karyotype was present. This patient's test results revealed inflammatory pockets in the placenta (chorioamnionitis) again accompanied by a placental cell Monosomy-16. Interestingly, the two latter families were unrelated as confirmed by case history and both pregnancies resulted in normal, healthy, term newborns. A further study emerged from Sharony's group in 2004 in which they addressed the genetic etiology of AFP-CD in a search of an AFP gene mutation [98].

In summary, it can be seen from the preceding paragraphs that the Down syndrome (DS) positive screen from an AFP-CD woman can potentially produce a false positive screen result. By analysis of triple test results, it was evident that the MoM screening results predicted DS as follows: AFP MoM = 0.0, uE3 MoM = 0.63, and hCG MoM = 1.63. Thus, an AFP-CD is one condition that should be first ruled out when a very low or absent AFP screen result is encountered following triple or quad assays for Down syndrome.

CONCLUSION

The present review has enumerated, discussed, and provided eight known pregnancy conditions that can produce false positive results in prenatal biomarker screens for DS and NTD (Table 2). These conditions encompassed: a) pre-existing maternal disease and b) congenital malformations in the fetus. All cases exhibited normal karyotype determinations and variable AFP amniotic fluid levels following amniocentesis. The biomarker analytes (especially AFP) showed altered levels (MoM values) which could be separated into those that produced NTD versus those resulting in trisomy-21 false positive screens. The five disorders producing elevated MSAFP MoM values, and thus NTD false positive screens, were VWD (MoM = 3.5), AFP-HP (MoM = 2.55), RA (MoM = 3.1), MG (MoM = 3.0), and EB (MoM 5.6). The diseases/disorders with reduced AFP and elevated MShCG MoMs resulting in trisomy-21 were FP screens for AFP-CD (MoM = 0.02) and MS (MoM 1.9), respectively. The lone exception was autoimmune SLE which produced a low/normal AFP giving a false positive FP screen for trisomy-21. Maternal serum DIA was less important in the DS screens, but served to contribute to the DS risk in both MS and SLE diseases.

Correspondingly, the calculated risks for NTD ranged from 1 in 13 to 1 in 295 in RA, MG, EP, VWD, and AFP-HP disorders, while the risk for trisomy-21 in MS, SLE and AFP-CD were 1 in 15 to 1 in 218, respectively. The generation of FP results for these diseases/disorders in the screening lab is unavoidable for the most part, without prior knowledge of the medical history of the patient. However, the clinical case history of the patient is often blinded to the screening lab unless a special notation is attached to the incoming specimen (not a usual scenario). A partial solution to the problem may reside with the screening lab report form. A disclosure or warning (alert) statement could be included on the intake and the report form listing which pre-existing diseases and disorders of a patient could produce a FP screen.

Acknowledgments

The author wishes to extend his thanks and gratitude to Ms. Tracy Godfrey for the typing and processing of this manuscript.

Disclosures

1. Financial: None; no U.S. federal grants were used in the preparation of this paper.
2. Interest: The author declares that there are no known conflicts of interest in the preparation of this manuscript.

References

[1] Mizejewski, G. J. (2001) Alpha-fetoprotein structure and function: relevance to isoforms, epitopes, and conformational variants. *Exp Biol Med (Maywood),* 226, 377-408.

[2] Mizejewski, G. J. (2004) Biological roles of alpha-fetoprotein during pregnancy and perinatal development. *Exp Biol Med (Maywood),* 229, 439-63.

[3] Bei, R. and Mizejewski, G. J. (2011) Alpha fetoprotein is more than a hepatocellular cancer biomarker: from spontaneous immune response in cancer patients to the development of an AFP-based cancer vaccine. *Curr Mol Med,* 11, 564-81.

[4] Brock, D. J. (1976) Mechanisms by which amniotic-fluid alpha-fetoprotein may be increased in fetal abnormalities. *Lancet,* 2, 345-6.

[5] Drugan, A., Vadas, A., Sujov, P. and Gershoni-Baruch, R. (1995) Markedly elevated alpha-fetoprotein and positive acetylcholinesterase in amniotic fluid from a pregnancy affected with dystrophic epidermolysis bullosa. *Fetal Diagn Ther,* 10, 37-40.

[6] Srikrishna, G. and Freeze, H. H. (2009) Endogenous damage-associated molecular pattern molecules at the crossroads of inflammation and cancer. *Neoplasia,* 11, 615-28.

[7] Huang, T., Hoffman, B., Meschino, W., Kingdom, J. and Okun, N. (2010) Prediction of adverse pregnancy outcomes by combinations of first and second trimester

biochemistry markers used in the routine prenatal screening of Down syndrome. *Prenat Diagn,* 30, 471-7.

[8] Guven, S., Karahan, S. C., Kandemir, O., Ucar, U., Cora, A. O. and Bozkaya, H. (2010) Occult inflammation and/or ischemia may be responsible for the false positivity of biochemical Down syndrome screening test. *J Perinat Med,* 38, 367-71.

[9] Mueller, V. M., Huang, T., Summers, A. M. and Winsor, S. H. (2005) The effect of fetal gender on the false-positive rate of Down syndrome by maternal serum screening. *Prenat Diagn,* 25, 1258-61.

[10] Hall, S., Bobrow, M. and Marteau, T. M. (2000) Psychological consequences for parents of false negative results on prenatal screening for Down's syndrome: retrospective interview study. *BMJ,* 320, 407-12.

[11] Ford, C., Moore, A., Bartlett, W. A. and Jones, A. F. (1996) Ethnic differences in outcome of serum screening for Down's syndrome. Differences in median values may account for differences in false positive rates. *BMJ,* 312, 1040-1.

[12] Holloway, P. J. and Bulusu, S. (1996) Ethnic differences in outcome of serum screening for Down's syndrome. For appreciable ethnic differences in false positive rates, use different cut off points. *BMJ,* 312, 1041.

[13] Marcotte, M. P. (1995) Adverse pregnancy outcome after a false-positive screen for Down syndrome using multiple markers. *Obstet Gynecol,* 86, 1035-6.

[14] Abdul-Hamid, S., Fox, R. and Martin, I. (2004) Maternal serum screening for trisomy 21 in women with a false positive result in last pregnancy. *J Obstet Gynaecol,* 24, 374-6.

[15] Andrews, L. G. and Souma, J. A. (1989) Elevated serum alpha-fetoprotein in a pregnant woman with rheumatoid arthritis. *N Engl J Med,* 321, 262-3.

[16] Fialova, L., Kohoutova, B., Peliskova, Z., Malbohan, I. and Mikulikova, L. (1991) [Serum levels of trophoblast-specific beta-1-globulin (SP1) and alpha-1-fetoprotein (AFP) in pregnant women with rheumatoid arthritis]. *Cesk Gynekol,* 56, 166-70.

[17] Scott, D. L., Wolfe, F. and Huizinga, T. W. (2010) Rheumatoid arthritis. *Lancet,* 376, 1094-108.

[18] Helmick, C. G., Felson, D. T., Lawrence, R. C., Gabriel, S., Hirsch, R., Kwoh, C. K., Liang, M. H., Kremers, H. M., Mayes, M. D., Merkel, P. A., Pillemer, S. R., Reveille, J. D., Stone, J. H. and National Arthritis Data, W. (2008) Estimates of the prevalence of arthritis and other rheumatic conditions in the United States. Part I. *Arthritis Rheum,* 58, 15-25.

[19] Lozano, R., Naghavi, M., Foreman, K., Lim, S., Shibuya, K., Aboyans, V., Abraham, J., Adair, T., Aggarwal, R., Ahn, S. Y., Alvarado, M., Anderson, H. R., Anderson, L. M., Andrews, K. G., Atkinson, C., Baddour, L. M., Barker-Collo, S., Bartels, D. H., Bell, M. L., Benjamin, E. J., Bennett, D., Bhalla, K., Bikbov, B., Bin Abdulhak, A., Birbeck, G., Blyth, F., Bolliger, I., Boufous, S., Bucello, C., Burch, M., Burney, P., Carapetis, J., Chen, H., Chou, D., Chugh, S. S., Coffeng, L. E., Colan, S. D., Colquhoun, S., Colson, K. E., Condon, J., Connor, M. D., Cooper, L. T., Corriere, M., Cortinovis, M., de Vaccaro, K. C., Couser, W., Cowie, B. C., Criqui, M. H., Cross, M., Dabhadkar, K. C., Dahodwala, N., De Leo, D., Degenhardt, L., Delossantos, A., Denenberg, J., Des Jarlais, D. C., Dharmaratne, S. D., Dorsey, E. R., Driscoll, T., Duber, H., Ebel, B., Erwin, P. J., Espindola, P., Ezzati, M., Feigin, V., Flaxman, A. D., Forouzanfar, M. H., Fowkes, F. G., Franklin, R., Fransen, M., Freeman, M. K., Gabriel,

S. E., Gakidou, E., Gaspari, F., Gillum, R. F., Gonzalez-Medina, D., Halasa, Y. A., Haring, D., Harrison, J. E., Havmoeller, R., Hay, R. J., Hoen, B., Hotez, P. J., Hoy, D., Jacobsen, K. H., James, S. L., Jasrasaria, R., Jayaraman, S., Johns, N., Karthikeyan, G., Kassebaum, N., Keren, A., Khoo, J. P., Knowlton, L. M., Kobusingye, O., Koranteng, A., Krishnamurthi, R., Lipnick, M., Lipshultz, S. E., Ohno, S. L., Mabweijano, J., MacIntyre, M. F., Mallinger, L., March, L., Marks, G. B., Marks, R., Matsumori, A., Matzopoulos, R., Mayosi, B. M., McAnulty, J. H., McDermott, M. M., McGrath, J., Mensah, G. A., Merriman, T. R., Michaud, C., Miller, M., Miller, T. R., Mock, C., Mocumbi, A. O., Mokdad, A. A., Moran, A., Mulholland, K., Nair, M. N., Naldi, L., Narayan, K. M., Nasseri, K., Norman, P., O'Donnell, M., Omer, S. B., Ortblad, K., Osborne, R., Ozgediz, D., Pahari, B., Pandian, J. D., Rivero, A. P., Padilla, R. P., Perez-Ruiz, F., Perico, N., Phillips, D., Pierce, K., Pope, C. A., 3rd, Porrini, E., Pourmalek, F., Raju, M., Ranganathan, D., Rehm, J. T., Rein, D. B., Remuzzi, G., Rivara, F. P., Roberts, T., De Leon, F. R., Rosenfeld, L. C., Rushton, L., Sacco, R. L., Salomon, J. A., Sampson, U., Sanman, E., Schwebel, D. C., Segui-Gomez, M., Shepard, D. S., Singh, D., Singleton, J., Sliwa, K., Smith, E., Steer, A., Taylor, J. A., Thomas, B., Tleyjeh, I. M., Towbin, J. A., Truelsen, T., Undurraga, E. A., Venketasubramanian, N., Vijayakumar, L., Vos, T., Wagner, G. R., Wang, M., Wang, W., Watt, K., Weinstock, M. A., Weintraub, R., Wilkinson, J. D., Woolf, A. D., Wulf, S., Yeh, P. H., Yip, P., Zabetian, A., Zheng, Z. J., Lopez, A. D., Murray, C. J., AlMazroa, M. A. and Memish, Z. A. (2012) Global and regional mortality from 235 causes of death for 20 age groups in 1990 and 2010: a systematic analysis for the Global Burden of Disease Study 2010. *Lancet,* 380, 2095-128.

[20] Nicholas, N. S. and Panayi, G. S. (1988) Rheumatoid arthritis and pregnancy. *Clin Exp Rheumatol,* 6, 179-82.

[21] Pope, R. M. (1990) Immunoregulatory mechanisms present in the maternal circulation during pregnancy. *Baillieres Clin Rheumatol,* 4, 33-52.

[22] Murgita, R. A. and Tomasi, T. B., Jr. (1975) Suppression of the immune response by alpha-fetoprotein on the primary and secondary antibody response. *J Exp Med,* 141, 269-86.

[23] Yachnin, S. and Lester, E. (1976) Inhibition of human lymphocyte transformation by human alpha-foetoprotein (HAFP); comparison of foetal and hepatoma HAFP and kinetic studies in vitro immunosuppression. *Clin Exp Immunol,* 26, 484-90.

[24] Lu, C. Y., Changelian, P. S. and Unanue, E. R. (1984) Alpha-fetoprotein inhibits macrophage expression of Ia antigens. *J Immunol,* 132, 1722-27.

[25] Yamamoto, M., Tatsumi, T., Miyagi, T., Tsunematsu, H., Aketa, H., Hosui, A., Kanto, T., Hiramatsu, N., Hayashi, N. and Takehara, T. (2011) alpha-Fetoprotein impairs activation of natural killer cells by inhibiting the function of dendritic cells. *Clin Exp Immunol,* 165, 211-9.

[26] Li, M. S., Li, P. F., He, S. P., Du, G. G. and Li, G. (2002) The promoting molecular mechanism of alpha-fetoprotein on the growth of human hepatoma Bel7402 cell line. *World J Gastroenterol,* 8, 469-75.

[27] Li, M., Liu, X., Zhou, S., Li, P. and Li, G. (2005) Effects of alpha fetoprotein on escape of Bel 7402 cells from attack of lymphocytes. *BMC Cancer,* 5, 96.

[28] Li, M., Zhou, S., Liu, X., Li, P., McNutt, M. A. and Li, G. (2007) alpha-Fetoprotein shields hepatocellular carcinoma cells from apoptosis induced by tumor necrosis factor-related apoptosis-inducing ligand. *Cancer Lett,* 249, 227-34.

[29] Mizejewski, G. J. (2009) Alpha-fetoprotein (AFP)-derived peptides as epitopes for hepatoma immunotherapy: a commentary. *Cancer Immunol Immunother,* 58, 159-70.

[30] Popova, E. V., Kukel, T. M., Muravin, A. I., Boiko, A. N., Murashko, A. V. and Gusev, E. I. (2013) [Pregnancy and delivery in women with multiple sclerosis: a retrospective analysis]. *Zh Nevrol Psikhiatr Im S S Korsakova,* 113, 52-6.

[31] Birk, K., Ford, C., Smeltzer, S., Ryan, D., Miller, R. and Rudick, R. A. (1990) The clinical course of multiple sclerosis during pregnancy and the puerperium. *Arch Neurol,* 47, 738-42.

[32] Coyle, P. K. (2014) Multiple sclerosis in pregnancy. *Continuum (Minneap Minn),* 20, 42-59.

[33] Birk, K., Smeltzer, S. C. and Rudick, R. (1988) Pregnancy and multiple sclerosis. *Semin Neurol,* 8, 205-13.

[34] Korn-Lubetzki, I., Kahana, E., Cooper, G. and Abramsky, O. (1984) Activity of multiple sclerosis during pregnancy and puerperium. *Ann Neurol,* 16, 229-31.

[35] Dudich, E. (2007) MM-093, a recombinant human alpha-fetoprotein for the potential treatment of rheumatoid arthritis and other autoimmune diseases. *Curr Opin Mol Ther,* 9, 603-10.

[36] Irony-Tur-Sinai, M., Grigoriadis, N., Lourbopoulos, A., Pinto-Maaravi, F., Abramsky, O. and Brenner, T. (2006) Amelioration of autoimmune neuroinflammation by recombinant human alpha-fetoprotein. *Exp Neurol,* 198, 136-44.

[37] Abramsky, O., Brenner, T., Mizrachi, R. and Soffer, D. (1982) Alpha-fetoprotein suppresses experimental allergic encephalomyelitis. *J Neuroimmunol,* 2, 1-7.

[38] Abramsky, O. (1994) Pregnancy and multiple sclerosis. *Ann Neurol,* 36 Suppl, S38-41.

[39] Evron, S., Brenner, T. and Abramsky, O. (1984) Suppressive effect of pregnancy on the development of experimental allergic encephalomyelitis in rabbits. *Am J Reprod Immunol,* 5, 109-13.

[40] Sridama, V., Pacini, F., Yang, S. L., Moawad, A., Reilly, M. and DeGroot, L. J. (1982) Decreased levels of helper T cells: a possible cause of immunodeficiency in pregnancy. *N Engl J Med,* 307, 352-6.

[41] Hatada, Y., Munemura, M., Matsuo, I., Fujisaki, S., Okamura, H. and Yamanaka, N. (1987) Myasthenic crisis in the puerperium: the possible importance of alpha-fetoprotein. Case report. *Br J Obstet Gynaecol,* 94, 480-2.

[42] Brenner, T., Beyth, Y. and Abramsky, O. (1980) Inhibitory effect of alpha-fetoprotein on the binding of myasthenia gravis antibody to acetylcholine receptor. *Proc Natl Acad Sci U S A,* 77, 3635-9.

[43] Molzer, B., Kainz-Korschinsky, M., Sundt-Heller, R. and Bernheimer, H. (1989) Phytanic acid and very long chain fatty acids in genetic peroxisomal disorders. *J Clin Chem Clin Biochem,* 27, 309-14.

[44] Kalidindi, M., Ganpot, S., Tahmesebi, F., Govind, A., Okolo, S. and Yoong, W. (2007) Myasthenia gravis and pregnancy. *J Obstet Gynaecol,* 27, 30-2.

[45] Eymard, B., Stojkovic, T., Sternberg, D., Richard, P., Nicole, S., Fournier, E., Behin, A., Laforet, P., Servais, L., Romero, N., Fardeau, M., Hantai, D. and Membres du reseau national Syndromes Myastheniques, C. (2013) [Congenital myasthenic

syndromes: difficulties in the diagnosis, course and prognosis, and therapy--The French National Congenital Myasthenic Syndrome Network experience]. *Rev Neurol (Paris),* 169 Suppl 1, S45-55.

[46] Djelmis, J., Sostarko, M., Mayer, D. and Ivanisevic, M. (2002) Myasthenia gravis in pregnancy: report on 69 cases. *Eur J Obstet Gynecol Reprod Biol,* 104, 21-5.

[47] Stoll, C., Ehret-Mentre, M. C., Treisser, A. and Tranchant, C. (1991) Prenatal diagnosis of congenital myasthenia with arthrogryposis in a myasthenic mother. *Prenat Diagn,* 11, 17-22.

[48] Abramsky, O. and Brenner, T. (1979) Alpha-fetoprotein inhibits antibody binding to acetylcholine receptor. Significance for myasthenia gravis and other autoimmune diseases. *Isr J Med Sci,* 15, 943-5.

[49] Brenner, T. and Abramsky, O. (1981) Immunosuppression of experimental autoimmune myasthenia gravis by alpha-fetoprotein rich formation. *Immunol Lett,* 3, 163-7.

[50] Brenner, T., Abramsky, O. and Lisak, R. P. (1981) Influence of alpha-fetoprotein on the in vitro and in vivo immune response to acetylcholine receptor. *Ann N Y Acad Sci,* 377, 208-21.

[51] Gliaschera, M., Biver, P., Brunori, E., Romani, L., Tarani, A. and de Luca Brunori, I. (1989) [Myasthenia gravis and pregnancy. Review of the literature and considerations on the immunoprotective effect of alpha-fetoprotein through the observation of a clinical case]. *Ann Ostet Ginecol Med Perinat,* 110, 20-7.

[52] Donaldson, J. O., Penn, A. S., Lisak, R. P., Abramsky, O., Brenner, T. and Schotland, D. L. (1981) Antiacetylcholine receptor antibody in neonatal myasthenia gravis. *Am J Dis Child,* 135, 222-6.

[53] Buschman, E., van Oers, N., Katz, M. and Murgita, R. A. (1987) Experimental myasthenia gravis induced in mice by passive transfer of human myasthenic immunoglobulin. Evidence for an ameliorating effect by alpha-fetoprotein. *J Neuroimmunol,* 13, 315-30.

[54] Brenner, T., Zielinski, A., Argov, Z. and Abramsky, O. (1984) Prevention of experimental autoimmune myasthenia gravis in rats by fetal alpha-fetoprotein-rich fractions. *Tumour Biol,* 5, 263-74.

[55] Donalson, K., Turner, S., Wastell, H. and Cuckle, H. (2008) Second trimester maternal serum ADAM12 levels in Down's syndrome pregnancies. *Prenat Diagn,* 10, 10.

[56] Heininger, K., Edler, C. and Toyka, K. V. (1984) alpha-Fetoprotein and neonatal transient myasthenia gravis. *Neurology,* 34, 403-4.

[57] La Montagna, G., Baruffo, A., Buono, G. and Tirri, R. (2000) False positivity of prenatal Down's syndrome and neural-tube tests in SLE. *Lancet,* 356, 1194-5.

[58] Ferriman, E. L., Sehmi, I. K., Jones, R., Railton, A., Hilton, R. C. and Cuckle, H. S. (2000) False-positive maternal serum screening in systemic lupus erythematosis: a case report. *Prenat Diagn,* 20, 851.

[59] Petri, M., Ho, A. C., Patel, J., Demers, D., Joseph, J. M. and Goldman, D. (1995) Elevation of maternal alpha-fetoprotein in systemic lupus erythematosus: a controlled study. *J Rheumatol,* 22, 1365-8.

[60] Mizejewski, G. J. Secondary Use of maternal serum alpha-fetoprotein in predicting pregnancy complications and adverse outcomes: contribution of supplemental biomarkers., In *Alpha-Fetoprotein, Function, and Health Implications* (Nova Publishers, New York, 2011), pp. 97-124.

[61] Yaron, Y., Cherry, M., Kramer, R. L., O'Brien, J. E., Hallak, M., Johnson, M. P. and Evans, M. I. (1999) Second-trimester maternal serum marker screening: maternal serum alpha-fetoprotein, beta-human chorionic gonadotropin, estriol, and their various combinations as predictors of pregnancy outcome. *Am J Obstet Gynecol,* 181, 968-74.

[62] Classen, S. R., Paulson, P. R. and Zacharias, S. R. (1998) Systemic lupus erythematosus: perinatal and neonatal implications. *J Obstet Gynecol Neonatal Nurs,* 27, 493-500.

[63] Rahman, A. and Isenberg, D. A. (2008) Systemic lupus erythematosus. *N Engl J Med,* 358, 929-39.

[64] Hemminki, K., Li, X., Sundquist, J. and Sundquist, K. (2009) Familial associations of rheumatoid arthritis with autoimmune diseases and related conditions. *Arthritis Rheum,* 60, 661-8.

[65] Goldblatt, F. and O'Neill, S. G. (2013) Clinical aspects of autoimmune rheumatic diseases. *Lancet,* 382, 797-808.

[66] Gershwin, M. E., Gaxtles, J. J. and Makishima, R. (1979) Alpha-fetoprotein-associated alterations of New Zealand mouse immunopathology. *Immunopharmacology,* 1, 331-41.

[67] Leschot, N. J., Treffers, P. E., Becker-Bloemkolk, M. J., van Zanten, S., de Groot, W. P. and Verjaal, M. (1980) Severe congenital skin defects in a newborn. Case report and relevance of several obstetrical parameters. *Eur J Obstet Gynecol Reprod Biol,* 10, 381-8.

[68] Yacoub, T., Campbell, C. A., Gordon, Y. B., Kirby, J. D. and Kitau, M. J. (1979) Maternal serum and amniotic fluid concentrations of alphafetoprotein in epidermolysis bullosa simplex. *Br Med J,* 1, 307.

[69] Bick, D. P., Balkite, E. A., Baumgarten, A., Hobbins, J. C. and Mahoney, M. J. (1987) The association of congenital skin disorders with acetylcholinesterase in amniotic fluid. *Prenat Diagn,* 7, 543-9.

[70] Frohlich, S. and O'Sullivan, E. (2011) Airway management in adult patients with epidermolysis bullosa dystrophica: a case series. *Anaesthesia,* 66, 842-3.

[71] Lepinard, C., Descamps, P., Meneguzzi, G., Blanchet-Bardon, C., Germain, D. P., Larget-Piet, L., Beringue, F., Berchel, C., Muller, F. and Dumez, Y. (2000) Prenatal diagnosis of pyloric atresia-junctional epidermolysis bullosa syndrome in a fetus not known to be at risk. *Prenat Diagn,* 20, 70-5.

[72] Spaggiari, E., Ruas, M., Dreux, S., Valat, A. S., Czerkiewicz, I., Guimiot, F., Schmitz, T., Delezoide, A. L. and Muller, F. (2013) Management strategy in pregnancies with elevated second-trimester maternal serum alpha-fetoprotein based on a second assay. *Am J Obstet Gynecol,* 208, 303 e1-7.

[73] Carmi, R., Sofer, S., Karplus, M., Ben-Yakar, Y., Mahler, D., Zirkin, H. and Bar-Ziv, J. (1982) Aplasia cutis congenita in two sibs discordant for pyloric atresia. *Am J Med Genet,* 11, 319-28.

[74] Achiron, R., Hamiel-Pinchas, O., Engelberg, S., Barkai, G., Reichman, B. and Mashiach, S. (1992) Aplasia cutis congenita associated with epidermolysis bullosa and pyloric atresia: the diagnostic role of prenatal ultrasonography. *Prenat Diagn,* 12, 765-71.

[75] De Jenlis Sicot, B., Deruelle, P., Kacet, N., Vaillant, C. and Subtil, D. (2005) Prenatal findings in epidermolysis bullosa with pyloric atresia in a family not known to be at risk. *Ultrasound Obstet Gynecol,* 25, 607-9.

[76] Bass, H. N., Miranda, C., Oei, R. and Crandall, B. F. (1993) Association of generalized dystrophic epidermolysis bullosa with positive acetylcholinesterase and markedly elevated maternal serum and amniotic fluid alpha-fetoprotein. *Prenat Diagn,* 13, 55-9.

[77] Langer, J. C. (2003) Abdominal wall defects. *World J Surg,* 27, 117-24.

[78] Laurence, K. M., Dew, J. O., Dyer, C. and Downey, K. H. (1983) Amniocentesis carried out for neural tube indications in South Wales 1973-1981: outcome of pregnancies and findings in the malformed abortuses. *Prenat Diagn,* 3, 187-201.

[79] Ionescu, S., Mocanu, M., Andrei, B., Bunea, B., Carstoveanu, C., Gurita, A., Tabacaru, R., Licsandru, E., Stanescu, D. and Selleh, M. (2014) Differential diagnosis of abdominal wall defects - omphalocele versus gastroschisis. *Chirurgia (Bucur),* 109, 7-14.

[80] Barsoom, M. J., Prabulos, A., Rodis, J. F. and Turner, G. W. (2000) Vanishing gastroschisis and short-bowel syndrome. *Obstet Gynecol,* 96, 818-9.

[81] Mann, L., Ferguson-Smith, M. A., Desai, M., Gibson, A. A. and Raine, P. A. (1984) Prenatal assessment of anterior abdominal wall defects and their prognosis. *Prenat Diagn,* 4, 427-35.

[82] Chen, C. P., Liu, F. F., Jan, S. W., Sheu, J. C., Huang, S. H. and Lan, C. C. (1996) Prenatal diagnosis and perinatal aspects of abdominal wall defects. *Am J Perinatol,* 13, 355-61.

[83] Palomaki, G. E., Hill, L. E., Knight, G. J., Haddow, J. E. and Carpenter, M. (1988) Second-trimester maternal serum alpha-fetoprotein levels in pregnancies associated with gastroschisis and omphalocele. *Obstet Gynecol,* 71, 906-9.

[84] Saller, D. N., Jr., Canick, J. A., Palomaki, G. E., Knight, G. J. and Haddow, J. E. (1994) Second-trimester maternal serum alpha-fetoprotein, unconjugated estriol, and hCG levels in pregnancies with ventral wall defects. *Obstet Gynecol,* 84, 852-5.

[85] Holmgren, G. and Sigurd, J. (1984) Prenatal diagnosis of two cases of gastroschisis following alpha-fetoprotein (AFP) screening. *Acta Obstet Gynecol Scand,* 63, 325-8.

[86] Tucker, J. M., Brumfield, C. G., Davis, R. O., Winkler, C. L., Boots, L. R., Krassikoff, N. E. and Hauth, J. C. (1992) Prenatal differentiation of ventral abdominal wall defects. Are amniotic fluid markers useful adjuncts? *J Reprod Med,* 37, 445-8.

[87] Goldfine, C., Haddow, J. E., Knight, G. J. and Palomaki, G. E. (1989) Amniotic fluid alpha-fetoprotein and acetylcholinesterase measurements in pregnancies associated with gastroschisis. *Prenat Diagn,* 9, 697-700.

[88] Nadel, A. S., Green, J. K., Holmes, L. B., Frigoletto, F. D., Jr. and Benacerraf, B. R. (1990) Absence of need for amniocentesis in patients with elevated levels of maternal serum alpha-fetoprotein and normal ultrasonographic examinations. *N Engl J Med,* 323, 557-61.

[89] Ferguson Smith, M. A., May, H. M., O'Hare, E. and Aitken, D. A. (1983) Hereditary persistence of alphafetoprotein: a new autosomal dominant trait identified in an antenatal screening programme for spina bifida. *J Med Genet,* 20, 454-8.

[90] Houwert, A. C., Giltay, J. C., Lentjes, E. G. and Lock, M. T. (2010) Hereditary persistence of alpha-fetoprotein (HPAF P): review of the literature. *Neth J Med,* 68, 354-8.

[91] Gross, S., Castillo, W., Crane, M., Espinosa, B., Carter, S., DeVeaux, R. and Salafia, C. (2003) Maternal serum alpha-fetoprotein and human chorionic gonadotropin levels in women with human immunodeficiency virus. *Am J Obstet Gynecol,* 188, 1052-6.

[92] van Deuren, M., Verhagen, M. and Weemeas, C. (2011) Hereditary persistence of alpha-fetoprotein. *Neth J Med,* 69, 96.

[93] Cochran, P. K., Chauvenet, A. R., Hart, P. S., de Graaf, S. S., Cushing, B., Kroovand, L. and Wofford, M. M. (1999) Hereditary persistence of alpha-fetoprotein in a child with testicular germ cell tumor. *Med Pediatr Oncol,* 32, 436-7.

[94] Talerman, A., Haije, W. G. and Baggerman, L. (1980) Serum alphafetoprotein (AFP) in patients with germ cell tumors of the gonads and extragonadal sites: correlation between endodermal sinus (yolk sac) tumor and raised serum AFP. *Cancer,* 46, 380-5.

[95] Greenberg, F., Faucett, A., Rose, E., Bancalari, L., Kardon, N. B., Mizejewski, G., Haddow, J. E. and Alpert, E. (1992) Congenital deficiency of alpha-fetoprotein. *Am J Obstet Gynecol,* 167, 509-11.

[96] Sher, C. and Shohat, M. (1997) Congenital deficiency of AFP and Down syndrome screening. *Prenat Diagn,* 17, 884-5.

[97] Sharony, R., Amiel, A., Bouaron, N., Kidron, D., Itzhaky, D. and Fejgin, M. (2003) Congenital deficiency of alpha-fetoprotein and associated chromosomal abnormality in the placenta. *Am J Med Genet A,* 121, 113-7.

[98] Sharony, R., Zadik, I. and Parvari, R. (2004) Congenital deficiency of alpha feto-protein. *Eur J Hum Genet,* 12, 871-4.

In: Alpha-Fetoprotein: Functions and Clinical Applications ISBN: 978-1-63484-875-6
Editors: Nisha Lakhi and Michael Moretti

Chapter 9

USE OF ALPHA-FETOPROTEIN FOR DETECTION OF PREECLAMPSIA

Jonathan B. Carmichael*[*], PhD, and David A. Krantz, MA
PerkinElmer Labs/NTD, Melville, NY, US

ABSTRACT

Preeclampsia is a leading cause of morbidity and mortality in pregnancy. Alpha-fetoprotein has been used traditionally in the identification of individuals at risk for open neural tube disorders and aneuploidy. Recent evidence suggests that AFP has proangiogenic properties, potentially at the stage of spiral artery invasion. As such, AFP may play an integral role in identification of pregnancies at risk for early onset preeclampsia.

Keywords: prenatal screening, alpha-fetoprotein, preeclampsia, angiogenesis, placental growth factor (PlGF), pregnancy-associated plasma protein A (PAPP-A)

INTRODUCTION

In clinical laboratory medicine, diagnostic testing is often utilized to diagnose or rule-out disease in individuals who are at high risk or who are suspected of having a particular disease. On the other hand, screening is a form of testing that does not diagnose an actual disease but rather identifies individuals who are at high-risk for the disorder. Those patients who are screened into a high-risk category can then receive diagnostic follow-up testing or treatment where possible. Screening is often performed when it is not feasible to provide diagnostic testing to an entire population due to either high expense or because the diagnostic test carries a degree of risk to the patient. In prenatal care, the initial use of prenatal screening began in

[*] Corresponding Author address: PekinElmer Labs/NTD, 80 Ruland Road, Suite 1, Melville, N.Y., 11747, US, Email: jon.carmichael@perkinelmer.com.

the 1970's with the use of maternal serum alpha-fetoprotein (AFP) testing to identify patients at high risk for open neural tube defects [1, 2].

AFP in Normal Pregnancy

AFP is a 67,500 dalton single-chain sialylated glycoprotein produced by the yolk sac and the liver during fetal development [3]. Human alpha-fetoprotein shares 39% identity with human albumin [4] and is a member of a multigenic family that comprises the related genes encoding albumin and afamin. While in adult serum, albumin is the most abundant protein, AFP is one of the most abundant proteins in fetal plasma [5, 6]. In normal human adults, AFP levels are very low and its function is not well-defined, however in pregnancy, maternal serum AFP levels are significantly increased. Not surprisingly, there appears to be some overlap in function within this gene family, although the exact the role for AFP has not been elucidated in fetal development.

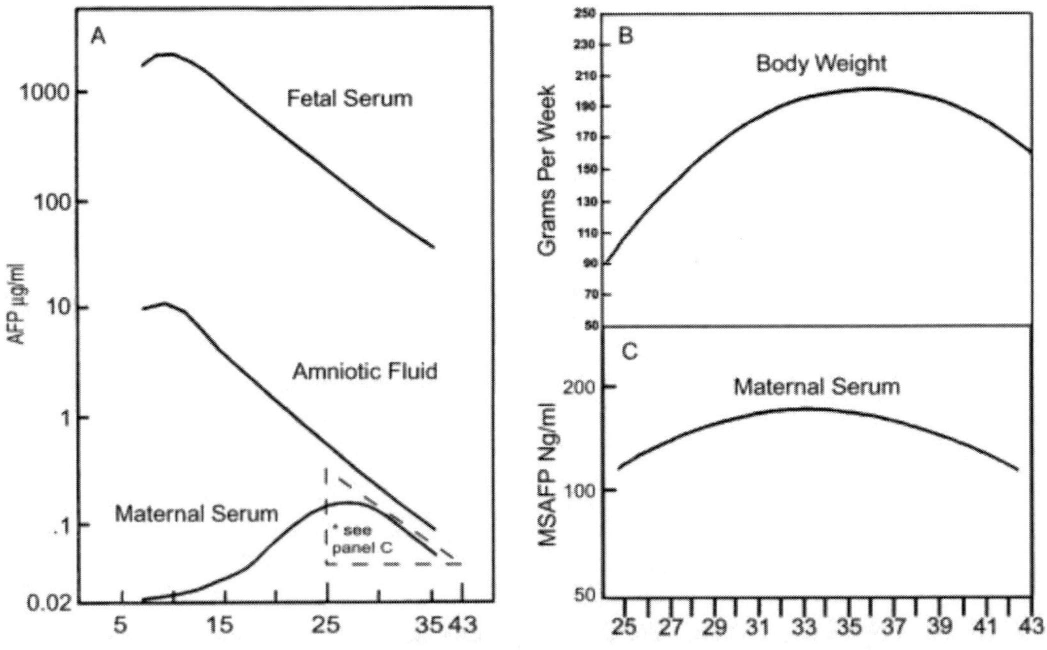

Figure 1. Alpha-fetoprotein (AFP) concentrations in 3 biologic compartments are plotted versus gestational age, whereas maternal serum AFP is compared to fetal weight gain in the last trimester. (A) AFP concentrations are plotted versus gestational age for the 3 biologic compartments (fetal blood, amniotic fluid, maternal serum) in which AFP resides. As shown, both fetal and amniotic fluid AFP levels decline in a parallel fashion throughout pregnancy, whereas maternal serum AFP rises to a peak at 28 to 30 weeks and declines thereafter. Note that the 2 fetal AFP concentrations plunge through 3 levels of measurement units (milligram, microgram, and nanogram) as pregnancy nears termination at 38 to 42 weeks. Data were extracted from references 210 and 212. (B) The weight gain of the fetus in grams per week is plotted versus age from the 26th week of pregnancy to term. (C) The maternal serum AFP concentrations are plotted versus age during the same pregnancy period. The graphs in (B) and (C) show similar geometric similarity in the third trimester of pregnancy. Figure from Mizjewski 2003 [8].

Most of the maternal serum AFP is derived from the flow of fluids across the placenta in a hydrostatic-gradient dependent manner [7]. Although levels of AFP in fetal blood and amniotic fluid peak in the first trimester, maternal serum levels steadily increase during pregnancy until about 32 weeks gestation at which point they begin to decrease (see Figure 1).

Prenatal Screening with AFP and Other Markers

In pregnancies associated with open neural tube defects, elevated levels of amniotic fluid AFP are observed due to the leakage of AFP from the opening in the spine of the fetus. Similarly, maternal serum AFP levels are elevated in cases of open neural tube defects, as well, although the degree of increase is less than that seen with amniotic fluid AFP [2]. Nonetheless, this increase formed the basis of screening programs for open neural tube defects. These screening programs utilizing maternal serum AFP classify 5% of patients as at high risk and within this high risk group 90% of open spina bifida and 98% of anencephaly cases are correctly detected [9].

Since open neural tube defect screening was essentially an effort to determine a risk of delivering a child with an open neural tube defect, more complex calculations involving Bayes' rule were implemented so that the pregnant patient could be provided with a risk that her fetus had a neural tube defect instead of a simple numeric result with a binary interpretation of being above or below a particular cutoff [10]. In order to convert the analytical result into a 'patient-specific' risk not only is the biochemical result used but prior risk factors such as maternal history and ethnicity are factored in as well.

After several years of routine screening for open neural tube defects, it was discovered in the early 1980's that low levels of alpha-fetoprotein, although indicative of low risk for open neural tube defects were actually associated with an increased likelihood of carrying a fetus with a chromosomal abnormality, such as Down syndrome or Trisomy 18 [11]. In the initial studies, it was observed that the median MoM in Down syndrome was about 70% of the median in unaffected pregnancy. Although this represented a significant difference the overall performance of such screening was rather poor with detection rates of 25-30%. As with screening for open neural tube defects, screening for Down syndrome involved not simply using a cut-off value, but rather calculating a risk for Down syndrome. This approach was facilitated since the prior risk of Down syndrome is associated with maternal age and thus the risk associated with the patient's age could be easily factored into the risk calculation.

Eventually, efforts were made to identify additional markers that were associated with increased risk of Down syndrome. These efforts led to the observation that total hCG [12], free Beta hCG [13], unconjugated estriol [14] and inhibin [15] were markers for Down syndrome and eventually multiple marker screening using two, three or four marker protocols were implemented [16]. These multiple marker protocols extended the mathematical approach used with the single marker AFP to use multivariate statistical distributions and led to improved detection rates of 60-70% [16].

In the early 1990's, it was discovered that the first trimester ultrasound marker, nuchal translucency was associated with an increased risk of Down syndrome [17]. At the same time, two biochemical markers free beta hCG [18] and PAPP-A [19] were also seen to be associated with an increased risk of Down syndrome. Eventually, a combined approach was implemented in which nuchal translucency, free Beta hCG and PAPP-A were utilized simultaneously to determine the risk of Down syndrome. Such an approach resulted in the detection of approximately 90% of Down syndrome pregnancies at a 5% false positive rate [20-25]. During this time period, several studies also evaluated the use of AFP as a first trimester marker for Down syndrome but it was not incorporated into the screening protocol [26-30]. More recent studies, performed since 2011 have shown that evaluation of first trimester AFP can improve screening performance [31-35].

Over time, interest arose in whether maternal serum markers were associated with other adverse outcomes of pregnancy aside from neural tube defects and aneuploidies. Morris et al. [36] performed a systematic meta-analysis of cohort studies evaluating second trimester markers and preeclampsia. There was significant variation among studies in the threshold used to identify patients at high-risk as well as significant variation in screening performance. However, for alpha-fetoprotein (AFP), the most effective threshold was 2.0 multiples of the median (MoM) resulting in a positive likelihood ratio (LR) of 2.36 and a negative likelihood ratio of 0.96. More recently, *Bredaki* et al. [37] showed that the level of AFP is influenced by a number of factors across the first, second and third trimesters of pregnancy and that these factors should be accounted for when incorporating AFP into a risk-based algorithm. Overall the authors showed that AFP was significantly elevated in cases of preeclampsia.

Preeclampsia

Preeclampsia is a frequent medical complication during pregnancy, characterized by development of hypertension (systolic blood pressure ≥140 and/or diastolic BP ≥90 mmHg measured at two occasions with at least 4 hours in between) and proteinuria (≥300 mg/day) after 20 weeks gestation [38], and is one of the leading causes worldwide of maternal and perinatal mortality [39]. It is estimated that preeclampsia develops in 5-7.5% of all pregnancies [40, 41], leading to more than 4 million cases worldwide every year [42]. Current estimates suggest more than 72,000 maternal deaths a year are attributable to preeclampsia [42] and up to 40% of all fetal mortality is due to preeclampsia [43].

AFP and Angiogenesis

Much of the pathological conditions observed during pregnancy can be traced to placental dysfunction, but where does AFP impact placental function? Under normal developmental conditions cytotrophoblast invasion of the uterine wall is followed by remodeling of vessels to enable maternal-fetal interaction [44]. In the preeclamptic condition incomplete spiral artery invasion results in development of vessels that are lower in capacity and higher in resistance than required [45]. A number of molecules involved in angiogenesis have been shown to be dysregulated in the preeclamptic state. For instance, attenuation of AKT activity has been shown in systems having increased levels of circulating soluble fms-like tyrosine kinase 1 (sFlt1) [46] and soluble endoglin (sEng) [47]. Further, Akt activity has been shown to be directly associated with increased sEng [47]. sFlt and sEng have been shown to bind angiogenic factors TGF□1, VEGF and PlGF [45] in an antagonistic manner [48, 49]. Significantly, AFP has been shown to activate the AKT/TOR signaling pathway in

HCC cell lines [50] indicating some intersection with the angiogenic process. In addition, Meng et al. [51] reported that silencing of AFP through siRNA activity resulted in a marked loss of expression of VEGF and subsequent increased apoptosis in Huh-7 cells and subsequent reduced invasive activity. These observations lend one to entertain the notion that AFP may be a proangiogenic factor exerting activity through activation of the AKT pathway, potentially through either a direct or indirect VEGF dependent manner [52]. However, to date the exact role of AFP in placental development is not known.

Valensise [53] suggests that preeclampsia can be of two distinct origins; early onset preeclampsia exhibiting defective trophoblast invasion and late onset preeclampsia showing greater association with maternal constitutional factors such as BMI. Although AFP may be credited with activities in both scenarios, it is probable that it's utility in screening is likely to be more effective in the identification of individuals at risk for early onset preeclampsia.

The distinction between early and late onset preeclampsia is a more modern concept and is becoming widely accepted as a better indicator of disease significance than the classic "mild" versus "severe" terminology. Although it is less frequent than the more commonly occurring late onset preeclampsia, occurring in approximately 0.5% of pregnancies [54, 55], there is evidence that more severe disease is associated with the early onset form, variously defined but generally recognized as symptom onset prior to 34 weeks gestation [56] coupled with an increased risk of delivery prior to 34 weeks [57, 58].

EOPE can be associated with IUGR, abnormal uterine and umbilical artery Doppler waveforms as well as other adverse maternal and neonatal outcomes. When compared to late onset preeclampsia, which exhibits milder maternal disease and lower rates of fetal involvement [59], the identification of at risk for EOPE provides an opportunity for intervention and treatment options that would not have been previously available.

Prenatal Screening for Preeclampsia

A large number of factors have been identified as increasing the risk of developing preeclampsia. A recent study by Wright et al. [60] including 58,884 pregnancies of which 1426 were complicated by preeclampsia identified age, weight, height, ethnicity, prior history of preeclampsia, mother with preeclampsia, in vitro fertilization, systemic lupus erythematosus, antiphospholipid syndrome, chronic hypertension and type I diabetes as risk factors for preeclampsia. These risk factors can be included into a risk algorithm which can determine how likely it is that preeclampsia can occur prior to a particular gestational age. Figure 2 shows the degree to which the risk of onset of preeclampsia before a given gestational age increases or decreases based on changes to each of the risk factors. The impact is greatest for early-onset preeclampsia.

A number of randomized control trials and meta-analyses have indicated that women with various risk factors who take low-dose aspirin daily can reduce their risk of preeclampsia. Bujold et al. [61] and Roberge et al. [62] showed that initiation of low-dose

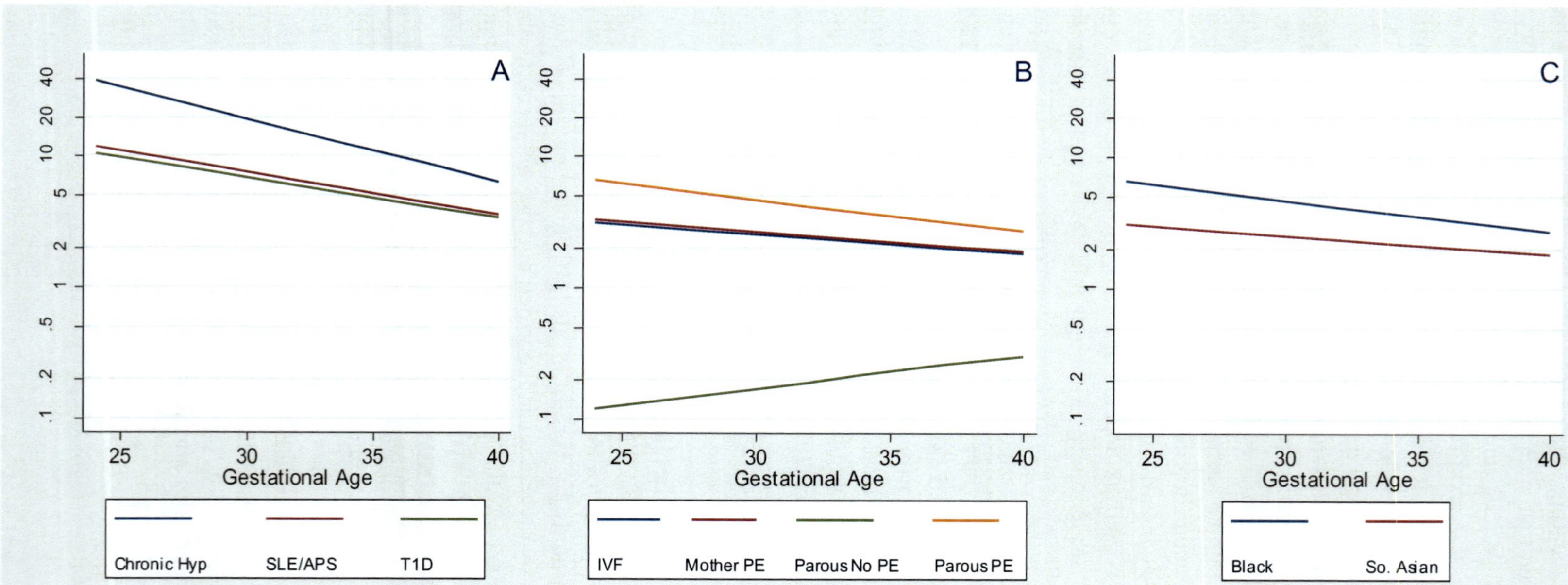
A
B
C
40
20
10
5
2
1
.5
.2
.1
25
30
35
40
Gestational Age
Chronic Hyp
SLE/APS
T1D
IVF
Mother PE
Parous No PE
Parous PE
Black
So. Asian

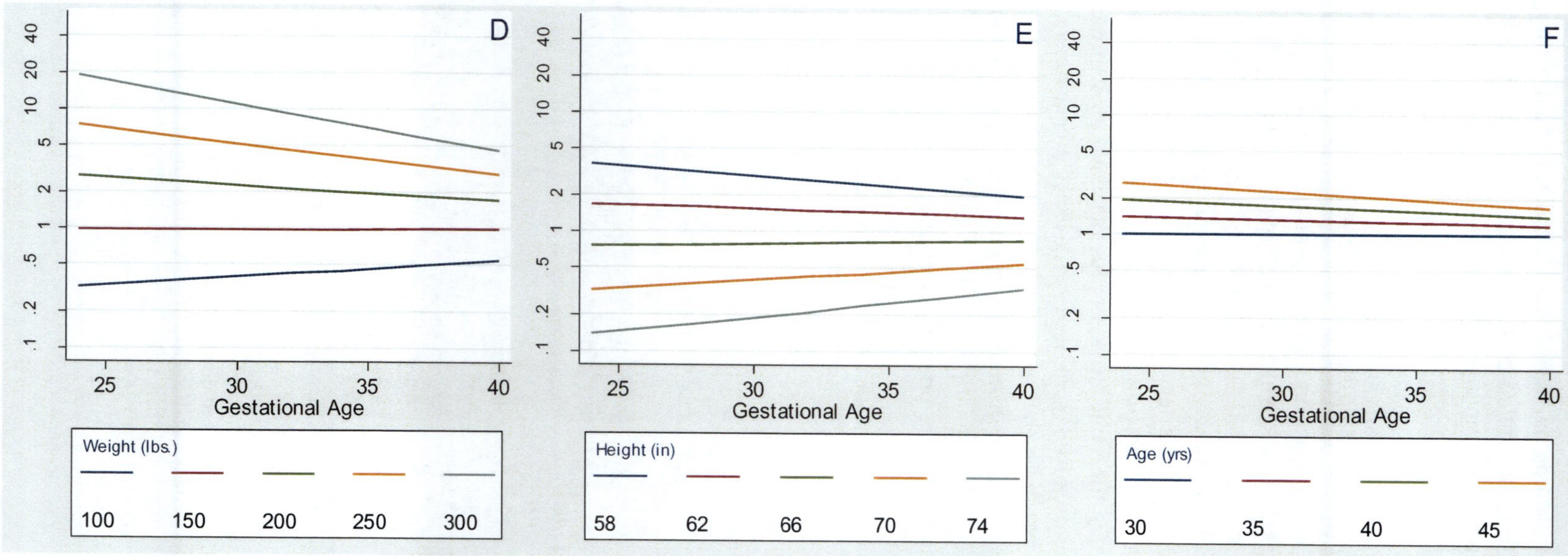

Figure 2. Relative Risk of onset of preeclampsia at or before given gestational age based on A) disease state B) medical history C) ethnicity D) maternal weight E) maternal height and F) maternal age. The baseline risk is associated with no disease state, non-IVF conceived nulliparous pregnancies in which patient's mother did not have history of preeclampsia, Caucasian, weight of 152 lbs. (69 kegs), height of 64.57 inches (1.64 meters) and age of 30. RR = relative risk, Hyp = hypertension, SLE = systematic lupus erythematosus, APS = antiphospholipid syndrome, T1D = type 1 diabetes mellitus, IVF = in vitro fertilization, PE = preeclampsia, Afr. Carib = African Caribbean, So. Asian = South Asian, lbs = pounds, in = inches, yrs = years. Graphs based on data from Wright et al. [60].

aspirin treatment is significantly more effective when such treatment starts prior to 16 weeks of pregnancy. A number of organizations, including the United Kingdom National Institute for Health and Care Excellence [63], the United States Preventative Services Taskforce [64] and the American College of Obstetricians and Gynecologists [65, 66] have recommended the use of low-dose aspirin for women with various risk factors for preeclampsia. These recommendations have tended to be either too narrow, thus limiting the potential to reduce the incidence of preeclampsia or overly broad thus providing aspirin to a large percentage of the population. The potential to combine maternal serum markers, biophysical markers and demographic markers can create an optimize the screening process in which both a limited number of women are identified as high-risk while the vast majority of pregnancies destined to have early-onset preeclampsia are identified. A study by Park et al. [67], identified 92% of early-onset preeclampsia pregnancies (<34 weeks of pregnancy) with a 10% screen positive rate using mean-arterial-pressure, uterine artery Doppler and PAPP-A. Another study by O'Gorman et al. [68] identified 82% of early-onset preeclampsia pregnancies (<32 weeks) with a 5% screen positive rate using the same markers plus placental growth factor (PlGF). With our own data, we were able to identify 91% of early-onset preeclampsia cases (<34 weeks) at a 5% screen positive rate with the same markers as O'Gorman et al. [68] plus alpha-fetoprotein. Figure 3 shows the effect of multiple marker screening in identifying high risk pregnancies amongst the general population and thus allowing for a more focused approach on treatment and prevention of early-onset preeclampsia.

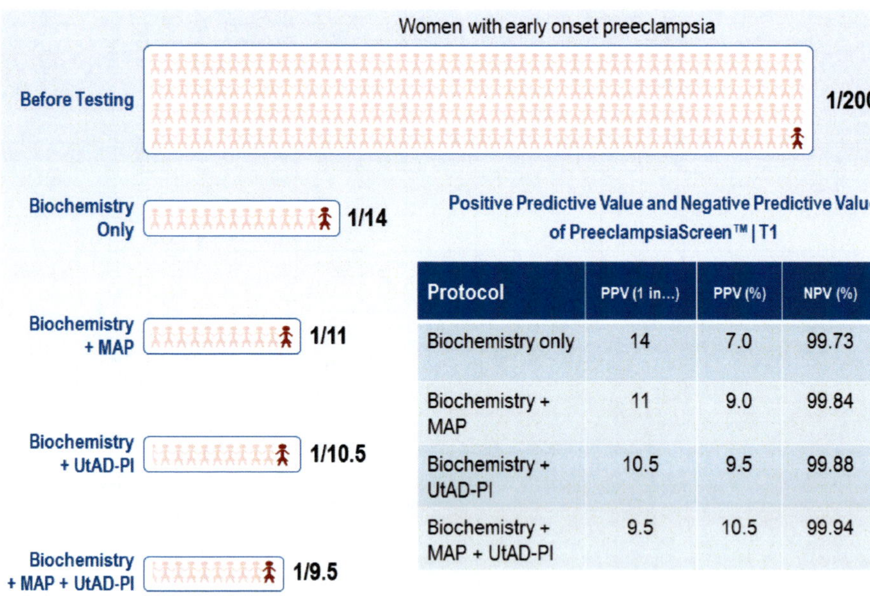

Positive Predictive Value and Negative Predictive Value of PreeclampsiaScreen™ | T1

Protocol	PPV (1 in...)	PPV (%)	NPV (%)
Biochemistry only	14	7.0	99.73
Biochemistry + MAP	11	9.0	99.84
Biochemistry + UtAD-PI	10.5	9.5	99.88
Biochemistry + MAP + UtAD-PI	9.5	10.5	99.94

Figure 3. Based upon a 0.5% incidence of early onset preeclampsia, in the typical obstetrical population, 1 out of every 200 patients will develop early onset preeclampsia, but we don't know which ones. With screening, this pool of high-risk women can be reduced down, anywhere from 14 to 9.5 patients, depending on which components of the assessment are included. Biochemistry includes Pregnancy-associated plasma protein A (PAPP-A), placental growth factor (PlGF) and alpha-fetoprotein (AFP). MAP = mean arterial pressure, UtAD-PI = uterine artery Doppler pulsativity index. PPV = positive predictive value. NPV = negative predictive value. Modeled on unpublished data, PerkinElmer Labs/NTD.

A recent publication by Li et al. [69] reinforces the use of aspirin in the treatment of early onset preeclampsia. They observed that CTBs exposed to a hypoxic environment showed elevated sFlt1 expression. When aspirin was added sFlt1 expression was inhibited [69]. Although the role of AFP has not been elucidated with regards to placental development and angiogenesis it is likely that the protein acts in a proangiogenic manner during placental development. This concept supports the role of AFP as a marker of preeclampsia. Screening protocols incorporating AFP with other biochemical, biophysical and demographic factors will aid in the identification of patients at high-risk for early onset preeclampsia, reduce the number of high risk patients evaluated solely on history and enable a significant benefit brought about by the effects of aspirin treatment prior to 16 weeks of pregnancy. Implementation of such screening protocols can result in a reduction in the incidence of early onset preeclampsia and a significant improvement in pregnancy outcome.

REFERENCES

[1] Brock, D. J., Scrimgeour, J. B., Steven, J., Barron, L. and Watt M. (1978) Maternal plasma alpha-fetoprotein screening for fetal neural tube defects. *Br J Obstet Gynaecol, 85*, 575-81.

[2] Macri, J. N. and Weiss, R. R. (1982) Prenatal serum alpha-fetoprotein screening for neural tube defects. *Obstet Gynecol, 59*, 633-9.

[3] Mizejewski, G. J. (2007) Physiology of alpha-fetoprotein as a biomarker for perinatal distress: relevance to adverse pregnancy outcome. *Exp Biol Med (Maywood), 232*, 993-1004.

[4] Morinaga, T., Sakai, M., Wegmann, T. G. and Tamaoki, T. (1983) Primary structures of human alpha-fetoprotein and its mRNA. *Proc Nat Acad Sci, 80*, 4604-4608.

[5] Tomasi, T. B. (1977) Structure and function of alpha-fetoprotein. *Annu Rev Med, 28*, 453-65.

[6] Yachnin, S. (1978) The clinical significance of human alpha-fetoprotein. *Ann Clin Lab Sci, 8*, 84-90.

[7] Lau, H. L. and Linkins, S. E. (1976) Alpha-fetoprotein. *Am J Obstet Gynecol, 124*, 533–538.

[8] Mizejewski, G. J. (2003) Levels of Alpha-Fetoprotein During Pregnancy and Early Infancy in Normal and Disease States. *Obstet Gynecol Surv, 58*, 804-826.

[9] Milunsky, A., Jick, S. S., Bruell, C. L., MacLaughlin, D. S., Tsung, Y. K., Jick, H., Rothman, K. J. and Willett, W. (1989) Predictive values, relative risks, and overall benefits of high and low maternal serum alpha-fetoprotein screening in singleton pregnancies: new epidemiologic data. *Am J Obstet Gynecol, 161*, 291-7.

[10] Adams, M. J., Jr., Windham, G. C., James, L. M., Greenberg, F., Clayton-Hopkins, J. A., Reimer, C. B. and Oakley, G. P., Jr. (1984) Clinical interpretation of maternal serum alpha-fetoprotein concentrations. *Am J Obstet Gynecol, 148*, 241-54.

[11] Merkatz, I. R., Nitowsky, H. M., Macri, J. N. and Johnson, W. E. (1984) An association between low maternal serum alpha-fetoprotein and fetal chromosomal abnormalities. *Am J Obstet Gynecol, 148*, 886-94.

[12] Bogart, M. H., Pandian, M. R. and Jones, O. W. (1987) Abnormal maternal serum chorionic gonadotropin levels in pregnancies with fetal chromosome abnormalities. *Prenat Diagn, 7*, 623-30.

[13] Macri, J. N., Kasturi, R. V., Krantz, D. A., Cook, E. J., Moore, N. D., Young, J. A., Romero, K. and Larsen, J. W., Jr. (1990) Maternal serum Down syndrome screening: free beta-protein is a more effective marker than human chorionic gonadotropin. *Am J Obstet Gynecol, 163*, 1248-53.

[14] Canick, J. A., Knight, G. J., Palomaki, G. E., Haddow, J. E., Cuckle, H. S. and Wald, N. J. (1988) Low second trimester maternal serum unconjugated oestriol in pregnancies with Down's syndrome. *Br J Obstet Gynaecol, 95*, 330-3.

[15] Wallace, E. M., Swanston, I. A., McNeilly, A. S., Ashby, J. P., Blundell, G., Calder, A. A. and Groome, N. P. (1996) Second trimester screening for Down's syndrome using maternal serum dimeric inhibin A. *Clin Endocrinol (Oxf), 44*, 17-21.

[16] Cuckle H. S. (2000) Biochemical screening for Down syndrome. *Eur J Obstet Gynecol Reprod Biol, 92*, 97-101.

[17] Nicolaides, K. H., Azar, G., Byrne, D., Mansur, C. and Marks, K. (1992) Fetal nuchal translucency: ultrasound screening for chromosomal defects in first trimester of pregnancy. *BMJ, 304*, 867-9.

[18] Spencer, K., Macri, J. N., Aitken, D. A. and Connor, J. M. (1992) Free beta-hCG as first-trimester marker for fetal trisomy. *Lancet, 339*, 1480.

[19] Brambati, B., Macintosh, M. C., Teisner, B., Maguiness, S., Shrimanker, K., Lanzani, A., Bonacchi, I., Tului, L., Chard, T. and Grudzinskas, J. G. (1993) Low maternal serum levels of pregnancy associated plasma protein A (PAPP-A) in the first trimester in association with abnormal fetal karyotype. *Br J Obstet Gynaecol, 100*, 324-6.

[20] Krantz, D. A., Hallahan, T. W., Orlandi, F., Buchanan, P., Larsen, J. W., Jr. and Macri, J. N. (2000) First-trimester Down Syndrome screening using dried blood biochemistry and nuchal translucency. *Obstet Gynecol, 96*, 207–13.

[21] Malone, F. D., Canick, J. A. and Ball, R. H. (2005) First-trimester or second-trimester screening, or both, for Down's syndrome. *N Engl J Med, 353*, 2001–11.

[22] Nicolaides. K. H., Spencer, K., Avgidou, K., Faiola, S. and Falcon, O. (2005) Multicenter study of first-trimester screening for trisomy 21 in 75 821 pregnancies: results and estimation of the potential impact of individual risk-orientated two-stage first-trimester screening. *Ultrasound Obstet Gynecol, 25*, 221–6.

[23] Perni, S. C., Predanic, M., Kalish, R. B., Chervenak, F. A. and Chasen, S. T. (2006) Clinical use of first-trimester aneuploidy screening in a United States population can replicate data from clinical trials. *Am J Obstet Gynecol, 194*, 127–30.

[24] Kirkegaard, I., Petersen, O. B., Uldbjerg, N. and Tørring, N. (2008) Improved performance of first-trimester combined screening for trisomy 21 with the double test taken before a gestational age of 10 weeks. *Prenat Diagn, 28*, 839–44.

[25] Schaelike, M., Kossakiewicz, M., Kossakiewicz, A. and Schild, R. L. (2009) Examination of a first-trimester Down Syndrome screening concept on a mix of 11,107 high- and low-risk patients at a private center for prenatal medicine in Germany. *Eur J Obstet Gynecol Reprod Biol, 144*, 140–45.

[26] Brambati, B., Simoni, G., Bonacchi, I. and Piceni, L. (1986) Fetal chromosomal aneuploidies and maternal serum alpha-fetoprotein levels in first trimester. *Lancet, 328*, 165–6.

[27] Barkai, G., Shaki, R., Pariente, C. and Goldman, B. (1987) First trimester alpha fetoprotein levels in normal and chromosomally abnormal pregnancies. *Lancet, 330*, 389.

[28] Milunsky, A., Wands, J., Brambati, B. and Currie K. (1988) First-trimester maternal serum a-fetoprotein screening for chromosome defects. *Am J Obstet Gynecol, 159*, 1209–13.

[29] Cuckle, H. S., Wald, N. J., Barkai, G., Fuhrmann, W., Altland, K., Brambati, B., Knight, G., Palomaki, G., Haddow, J. E. and Canick, J. (1988) First-trimester biochemical screening for Down syndrome. *Lancet, 2*, 851–2.

[30] Cuckle, H. S. and van Lith, J. M. (1999) Appropriate biochemical parameters in firsttrimester screening for Down syndrome. *Prenat Diagn, 19*, 505–12.

[31] Bredaki, F. E., Wright, D., Matos, P., Syngelaki, A. and Nicolaides, K. H. (2011) First-trimester screening for trisomy 21 using alpha-fetoprotein. *Fetal Diagn Ther, 30*, 215-8.

[32] Johnson, J., Pastuck, M., Metcalfe, A., Connors, G., Krause, R., Wilson, D. and Cuckle, H. (2013). First-trimester Down syndrome screening using additional serum markers with and without nuchal translucency and cell-free DNA. *Prenat Diagn, 33*, 1044-9.

[33] Carmichael, J., Krantz, D., Liu, H. P., Janik, D. and Hallahan, T. (2015). Incorporation of dried blood alpha fetoprotein into traditional first trimester Down syndrome screening service. *Prenat Diagn, 35*, 703-8.

[34] Palomaki, G. E., Eklund, E. E., Neveux, L. M. and Lambert Messerlian, G. M. (2015) Evaluating first trimester maternal serum screening combinations for Down syndrome suitable for use with reflexive secondary screening via sequencing of cell free DNA: high detection with low rates of invasive procedures. *Prenat Diagn, 35*, 789-96.

[35] Wright, D., Syngelaki, A., Bradbury, I., Akolekar, R. and Nicolaides, K. H. (2014) First-trimester screening for trisomies 21, 18 and 13 by ultrasound and biochemical testing. *Fetal Diagn Ther, 35*, 118-26.

[36] Morris, R. K., Cnossen, J. S., Langejans, M., Robson, S. C., Kleijnen, J., Ter Riet, G., Mol, B. W., van der Post, J. A. and Khan, K. S. (2008) Serum screening with Down's syndrome markers to predict pre-eclampsia and small for gestational age: systematic review and meta-analysis. *BMC Pregnancy Childbirth, 8*, 33.

[37] Bredaki, F. E., Mataliotakis, M., Wright, A., Wright, D. and Nicolaides, K. H. (2015) Maternal serum alpha fetoprotein at 12, 22, 32 weeks' gestation in screening for preeclampsia. *Ultrasound Obstet Gynecol, Nov 19*, doi: 10.1002/uog.15818.

[38] Report of the National High Blood Pressure Education Program Working Group on High Blood Pressure in Pregnancy (2000) *Am J Obstet Gynecol, 183*, S1-22.

[39] Sibai, B. and Ghulmiyyah, L. (2012) Maternal mortality from preeclampsia/eclampsia. *Perinatology, 36*, 56-9.

[40] Srinivas, S. K., Edlow, A. G., Neff, P. M., Sammel, M. D., Andrela, C. M. and Elovitz, M. A. (2009) Rethinking IUGR in preeclampsia: dependent or independent of maternal hypertension? *J Perinatol, 29*, 680-4.

[41] Wallis, A. B., Saftlas, A. F., Hsia, J. and Atrash, H. K. (2008) Secular trends in the rates of preeclampsia, eclampsia, and gestational hypertension, United States, 1987-2004. *Am J Hypertens, 21*, 521-6.

[42] AbouZahr, C. (2003) Global burden of maternal death and disability. *Br Med Bull, 67*, 1-11.

[43] Anderson, U. D., Olsson, M. G., Kristensen, K. H., Åkerström, B. and Hansson, S. R. (2012) Review: Biochemical markers to predict preeclampsia. *Placenta, 33 Suppl*, S42-7.
[44] Damsky, C. H., Fitzgerald, M. L. and Fisher, S. J. (1992) Distribution patterns of extracellular matrix components and adhesion receptors are intricately modulated during first trimester cytotrophoblast differentiation along the invasive pathway, in vivo. *J Clin Invest, 89*, 210-22.
[45] Powe, C. E., Levine, R. J. and Karumanchi, S. A. (2011) Preeclampsia, a disease of the maternal endothelium: the role of antiangiogenic factors and implications for later cardiovascular disease. *Circulation, 123*, 2856-69.
[46] Palmer, K. R., Kaitu'u-Lino, T. J., Hastie, R., Hannan, N. J., Ye, L., Binder, N., Cannon, P., Tuohey, L., Johns, T. G., Shub, A. and Tong, S. Placental-Specific sFLT-1 e15a Protein Is Increased in Preeclampsia, Antagonizes Vascular Endothelial Growth Factor Signaling, and Has Antiangiogenic Activity. *Hypertension, 66*, 1251-9.
[47] Cudmore, M. J., Ahmad, S., Sissaoui, S., Ramma, W., Ma, B., Fujisawa, T., Al-Ani, B., Wang, K., Cai, M., Crispi, F., Hewett, P. W., Gratacós, E., Egginton, S. and Ahmed, A. (2012) Loss of Akt activity increases circulating soluble endoglin release in preeclampsia: identification of inter-dependency between Akt-1 and heme oxygenase-1. *Eur Heart J, 33*, 1150-8.
[48] Maynard S. E., Min, J. Y., Merchan, J., Lim, K. H., Li, J., Mondal, S., Libermann, T. A., Morgan, J. P., Sellke, F. W., Stillman, I. E., Epstein, F. H., Sukhatme, V. P. and Karumanchi, S. A. (2003) Excess placental soluble fms-like tyrosine kinase 1 (sFlt1) may contribute to endothelial dysfunction, hypertension, and proteinuria in preeclampsia. *J Clin Invest, 111*, 649-58.
[49] Perucci, L. O., Gomes, K. B., Freitas, L. G., Godoi, L. C., Alpoim, P. N., Pinheiro, M. B., Miranda, A. S., Teixeira, A. L., Dusse, L. M. and Sousa, L. P. (2014) Soluble endoglin, transforming growth factor-Beta 1 and soluble tumor necrosis factor alpha receptors in different clinical manifestations of preeclampsia. *PLoS One, 9*, e97632.
[50] Zhu M., Guo, J., Xia, H., Li, W., Lu, Y., Dong, X., Chen, Y., Xie, X., Fu, S. and Li, M. (2015) Alpha-fetoprotein activates AKT/mTOR signaling to promote CXCR4 expression and migration of hepatoma cells. *Oncoscience, 2*, 59-70.
[51] Meng, W., Li, X., Bai, Z., Li, Y., Yuan, J., Liu, T., Yan, J., Zhou, W., Zhu, K., Zhang, H. and Li, Y. (2014) Silencing alpha-fetoprotein inhibits VEGF and MMP-2/9 production in human hepatocellular carcinoma cell. *PLoS One, 9*, e90660.
[52] Liang, O. D., Korff, T., Eckhardt, J., Rifaat, J., Baal, N., Herr, F., Preissner, K. T. and Zygmunt, M. (2004) Oncodevelopmental alpha-fetoprotein acts as a selective proangiogenic factor on endothelial cell from the fetomaternal unit. *J Clin Endocrinol Metab., 89*, 1415-22.
[53] Valensise, H., Vasapollo, B., Gagliardi, G. and Novelli, G. P. (2008) Early and late preeclampsia: two different maternal hemodynamic states in the latent phase of the disease. *Hypertension, 52*, 873-80.
[54] Espinoza, J., Romero, R., Nien, J. K., Gomez, R., Kusanovic, J. P., Gonçalves, L. F, Medina, L., Edwin, S., Hassan, S., Carstens, M. and Gonzalez, R. (2007) Identification of patients at risk for early onset and/or severe preeclampsia with the use of uterine artery Doppler velocimetry and placental growth factor. *Am J Obstet Gynecol, 196*, 326. e1-13.

[55] Espinoza, J., Kusanovic, J. P., Bahado-Singh, R., Gervasi, M. T., Romero, R., Lee, W., Vaisbuch, E., Mazaki-Tovi, S., Mittal, P., Gotsch, F., Erez, O., Gomez, R., Yeo, L. and Hassan, S. S. (2010) Should bilateral uterine artery notching be used in the risk assessment for preeclampsia, small-forgestational-age, and gestational hypertension? *J Ultrasound Med, 29*, 1103-15.

[56] Gastrich, M. D., Faro, R. and Rosen, T. (2010) Markers of preeclampsia and the relationship to cardiovascular disease: review of the twenty-first century literature. *J Matern Fetal Neonatal Med., 23*, 751-69.

[57] Palmer, K., Saglam, B., Whitehead, C., Stock, O., Lappas, M. and Tong, S. (2011) Severe early-onset preeclampsia is not associated with a change in placental catechol O-methyltransferase (COMT) expression. *Am J Pathol., 178*, 2484-8.

[58] Schaaf, J. M., Bruinse, H. W., van der Leeuw-Harmsen, L., Groeneveld, E., Koopman, C., Franx, A. and van Rijn, B. B. (2011) Reproductive outcome after early-onset pre-eclampsia. *Hum Reprod, 26*, 391-7.

[59] Costa Fda, S., Murthi, P., Keogh, R. and Woodrow, N. (2011) Early screening for preeclampsia. *Rev Bras Ginecol Obstet, 33*, 367-75.

[60] Wright, D., Akolekar, R., Syngelaki, A., Poon, L. C. and Nicolaides, K. H. (2012) A competing risks model in early screening for preeclampsia. *Fetal Diagn Ther, 32*, 171-8.

[61] Bujold, E., Roberge, S., Lacasse, Y., Bureau, M., Audibert, F., Marcoux, S., Forest, J. C. and Giguère, Y. (2010) Prevention of preeclampsia and intrauterine growth restriction with aspirin started in early pregnancy: a meta-analysis. *Obstet Gynecol, 116*, 402-14.

[62] Roberge, S., Nicolaides, K. H., Demers, S., Villa, P. and Bujold, E. (2013) Prevention of perinatal death and adverse perinatal outcome using lowdose aspirin: a meta-analysis. *Ultrasound Obstet Gynecol, 41*, 491-9.

[63] Visintin, C., Mugglestone, M. A., Almerie, M. Q., Nherera, L. M., James, D. and Walkinshaw, S., Guideline Development Group. (2010) Management of hypertensive disorders during pregnancy: summary of NICE guidance. *BMJ, 341*, c2207.

[64] LeFevre, M. (2014) Low-dose aspirin use for the prevention of morbidity and mortality from preeclampsia: U.S. Preventive Services Task Force recommendation statement. *Ann Intern Med., 161*, 819-26.

[65] American College of Obstetricians and Gynecologists, Task Force on Hypertension in Pregnancy (2013). Hypertension in pregnancy. Report of the American College of Obstetricians and Gynecologists' Task Force on Hypertension in Pregnancy. *Obstet Gynecol, 122*, 1122-31.

[66] American College of Obstetricians and Gynecologists (2015). Committee Opinion Summary No. 638: First-Trimester Risk Assessment for Early-Onset Preeclampsia. *Obstet Gynecol, 126*, 689.

[67] Park, F., Russo, K., Williams, P., Pelosi, M., Puddephatt, R., Walter, M., Leung, C., Saaid, R., Rawashdeh, H., Ogle, R. and Hyett, J. (2015) Prediction and prevention of early-onset pre-eclampsia: impact of aspirin after first-trimester screening. *Ultrasound Obstet Gynecol, 46*, 419-23.

[68] O'Gorman, N., Wright, D., Syngelaki, A., Akolekar, R., Wright, A., Poon, L. C. and Nicolaides, K. H. (2015) Competing risks model in screening for preeclampsia by maternal factors and biomarkers at 11-13 weeks gestation. *Am J Obstet Gynecol, pii*, S0002-9378(15)00900-X.

[69] Li, C., Raikwar, N. S., Santillan, M. K., Santillan, D. A. and Thomas, C. P. (2015) Aspirin inhibits expression of sFLT1 from human cytotrophoblasts induced by hypoxi, via cyclo-oxegenase 1. *Placenta, 36*, 446-453.

SECTION 3: ALPHA-FETOPROTEIN IN ADVERSE OUTCOMES AND PREGNANCY COMPLICATIONS

In: Alpha-Fetoprotein: Functions and Clinical Applications ISBN: 978-1-63484-875-6
Editors: Nisha Lakhi and Michael Moretti

Chapter 10

ALPHA-FETOPROTEIN AND FETAL ANEMIA

Jose L. Bartha*, MD, PhD

Professor and Head Department of Obstetrics and Gynecology,
University Hospital La Paz, Universidad Autónoma de Madrid,
Madrid, Spain

ABSTRACT

This chapter reviews current evidence about the relationships between alpha-fetoprotein and both fetal erythropoiesis and anemia. The exact role of alpha-fetoprotein (AFP) in human development remains unclear but there are several findings pointing to a relationship between AFP and fetal hematopoiesis. AFP is synthesized in hematopoietic organs in the fetus (yolk sac and fetal liver) and its synthesis changes parallel to the development and decline of extramedullary hematopoiesis. Maternal serum AFP (MSAFP) levels are increased in cases of fetal anemia due to different reasons such as alloimmunisation, Bart hemoglobin or parvovirus infection. In cases of maternal alloimmunisation it has been suggested that elevations of MSAFP may precede the appearance of other signs of fetal anemia such as elevations in the peak systolic velocity of the fetal middle cerebral artery. In cases of Bart hemoglobin, the combination of MSAFP with sonographic markers has the highest prediction for affected fetus. Future research is needed to determine the exact mechanisms involved in the action of AFP on fetal erythropoiesis and the potential clinical usefulness of this relationship.

INTRODUCTION

Alpha-fetoprotein (AFP) has been wrongly considered for years as simply a form of "fetal albumin" emphasizing only on the transporting function of this complex protein. It is currently known that variations in its tertiary structure are related to multifaceted biological mechanisms and functions including growth, differentiation and regulation of different cells and tissues [1, 2].

* Email: joseluisbartha@me.com.

Unexplained elevations of maternal serum AFP (MSAFP) have been for long time associated with adverse perinatal outcome. In many of these pregnancy complications, elevations of this protein in the maternal serum may be explained by an increased transfer from the fetus to the mother due to injuries in the maternal – fetoplacental interface, especially in those cases in which either the placenta or the placental bed are affected particularly in all the conditions associated to abnormal placentation such as fetal loss, intrauterine growth restriction, pre-eclampsia or placental abruption. But this simple explanation cannot justify many other cases in which the maternal-fetal interface is thought to be normal. A reasonable alternative hypothesis is that the increase in MSAFP may be due to an increase fetal serum levels due to an overproduction of this protein in the fetal compartment. To progress along this way there is a need to know more about the biological role of this protein in fetal development. One of the potential functions of this protein could be related with early stages of fetal erythropoiesis. In fact, MSAFP levels have been described to be elevated in some cases of fetal anemia.

The purpose of this chapter is to review the current evidence about the relationships between alpha-fetoprotein and both fetal erythropoiesis and anemia underlying the potential mechanisms and the clinical utility of these findings.

Fetal Erythropoiesis and Serum AFP Levels

The exact role of alpha-fetoprotein (AFP) in human development remains unclear. It appears that AFP might serve as a modulator/modifier of various cell growths regulatory pathways during embryonic and fetal development in vertebrates [1, 2, 3]. There are several findings pointing to a relationship between AFP and fetal hematopoiesis. AFP is synthesized in hematopoietic organs in the fetus (yolk sac and fetal liver) and its synthesis changes parallel to the development and decline of extramedullary hematopoiesis, and in mice in which hepatic hematopoietic foci are maintained well into adulthood and serum AFP levels are significantly higher than in normal mice [4].

Out of pregnancy there are also evidence of a relationship between the process of erythropoiesis and AFP. Occasionally patients with hepatocellular carcinoma present with erythrocytosis, and high serum levels of both AFP and erythropoietin are associated with this paraneoplastic syndrome, whereas elevations of erythropoietin alone are not enough to cause this syndrome [5-7]. Light microscopic immunohistochemistry has shown that AFP is located in hepatocellular carcinoma cells that are erythropoietin positive in serial sections [6]. Also Yolk sac tumors are related to high concentrations of AFP. If a high AFP concentration is related to erythrocytosis in these pathologic conditions, it can be hypothesized that AFP plays a physiologic role in fetal erythropoiesis.

If so, AFP should be related with fetal hemoglobin levels and its production and levels should considerably vary in cases of fetal anemia, most of all those due to fetal hemolysis when a marked erythropoiesis is expected.

In 1999, we evaluated the relationships between both maternal and fetal serum AFP levels (MSAFP and FSAFP respectively) and fetal hemoglobin, hematocrit, red blood cell count, mean corpuscular volume, mean corpuscular hemoglobin and mean corpuscular hemoglobin concentration in 63 fetuses, 19 of them by prenatal cordocentesis and 44 at elective cesarean section before labor [8]. We also evaluated the relationships between fetal AFP and fetal hemoglobin, hematocrit, red blood cell count, mean corpuscular volume, mean corpuscular hemoglobin, mean corpuscular hemoglobin concentration, erythroblasts, erythropoietin, serum iron, transferrin, ferritin, bilirubin, total proteins and albumin in 51 umbilical cords immediately after singleton childbirth, in healthy, pregnant women at term. After adjustment according to gestational age, significant correlations were found between maternal AFP and fetal hemoglobin (r = -.30, P = .017) (Figure 1) and hematocrit (r = -.25, P = .044). We also found a significant negative correlation between fetal serum alpha-fetoprotein and fetal red blood cells (r = -.55, P = .000026), hemoglobin (r = -.40, P = .0035), hematocrit (r = -.46, P = .00052), erythropoietin (r = -.40, P = .005) and transferrin (r = -.33, P = .016). We then, concluded that it was reasonable to speculate that AFP could play a role in fetal erythropoiesis.

The main limitations of this first study were the absence of any proposed mechanisms for this action and the lack of cases of severe fetal anemia.

In an attempt to elucidate some of the mechanisms that may be involved in the action of AFP on erythropoiesis, we determined the association between alpha-fetoprotein and hematopoietic growth factors in amniotic fluid in 41 women at 15 weeks' gestation when amniocentesis was performed for clinical reasons [9]. All women subsequently delivered anatomically and chromosomally normal infants. The concentrations of stem cell factor, interleukin 3, interleukin 6, erythropoietin, and granulocyte colony-stimulating factor (G-CSF) were measured using commercially available immunoassay kits. We found a significant correlation between AFP and stem cell factor (r =.47, P =.002). No significant correlations between AFP and the rest of the studied cytokines were found (r = -.07, r =.02, r = -.02, and r = -.11 for erythropoietin, G-CSF, interleukin 3, and interleukin 6, respectively). So, we concluded that alpha-fetoprotein could play a role in fetal hematopoiesis in the very early stages of this process.

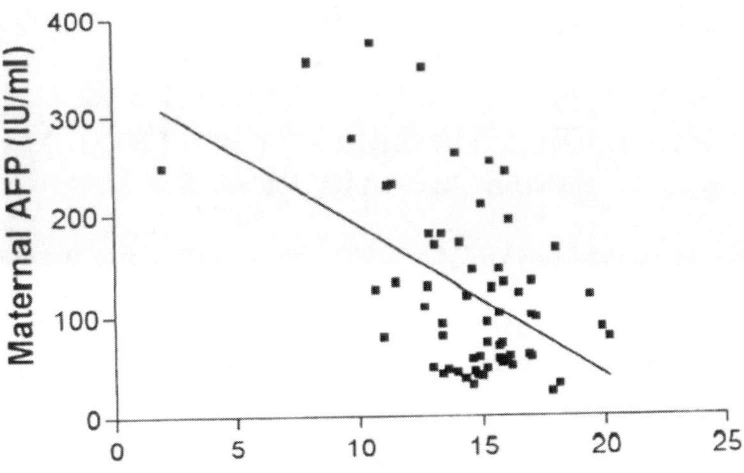

Figure 1. Correlation between fetal hemoglobin and MSAFP in normal fetuses (r = -0.48, p = 0.00004 and p = -0-30, p = 0.01 before and after adjustment by gestational age respectively) (From Bartha JL et al. J Reprod Med 1999 [9], with permission).

In vivo hematopoiesis is a complex system that involves several stages including the activation of stem cells, division of progenitors, proliferation and differentiation. It is regulated by various mechanisms that appear to be under the control of many hematopoietic growth factors. One of them, the stem cell factor (SCF), is expressed on both primitive and mature progenitor cells, preserving their viability, preventing their apoptosis, and influencing their proliferation and differentiation. SCF exerts its effect predominantly on pluripotent stem cells and on CFU-GEMM (colony-forming unit-granulocyte, erythrocyte, macrophage and megakaryocyte) cells, which are previous stages to BFU-E formation [10]. Since AFP has been found to be significantly correlated with SCF, it is thought that it may have a role in fetal hematopoiesis by producing some effect on these primitive stem cells.

Supporting the role of AFP on hematopoiesis, the effect of a preparation containing human α-fetoprotein, on proliferative activity and differentiation of mononuclear leukocytes isolated from donor peripheral blood has been studied in cultures with serum-free media. Proliferative activity of mononuclear leukocytes increased significantly (by 55-58%) after treatment with AFP in comparison with the control. AFP-activated ML formed two cell subpopulations, one consisting predominantly (50-79%) from CD3+ lymphocytes and the other, consisting of large cells, containing macrophage CD14+ and CD34+ precursor cells. The counts of these latter cells in this subpopulation reached 28%. Cultured intact mononuclear leukocytes form a population consisting of CD3+/CD34 cells. The results suggest that human α-fetoprotein in effective doses stimulates proliferation and blast transformation of cultured peripheral blood mononuclear leukocytes with the formation of CD34+/CD45+ hemopoietic precursor cells [11]. In addition, it has been demonstrated that addition of AFP to cultured human bone marrow and splenic cultures increased 2- to 3- fold the efficiency of cloning of stromal precursor cell indicating the possibility of an effect of AFP on the number of stromal precursor cell in hemopoietic and lymphoid organ [12]. All these results suggest that this phenomenon of ex vivo generation of hemopoietic precursor cells using human α-fetoprotein can be important for clinical oncology, specifically, for the development of adaptive cell therapies.

Evidence of a Relationship between Fetal Anemia and Serum AFP Levels

Since long time ago increased AFP levels have been found in two conditions associated with fetal anemia: parvovirus B19 infection and Rh immunization. In the first condition, increased MSAFP levels were attributed to viral hepatic injury or to placental leak secondary to immunologic damage at the maternal-fetal interface, but the increase in AFP was also related to fetal aplastic crisis [13].

MSAFP levels had been known to be increased in red blood cell alloimmunisation. In the past, it was reported that the level of alpha-fetoprotein (AFP), either in maternal serum or in amniotic fluid was increased in Rh-alloimmunised pregnancies and it was proposed that AFP may be used as a marker in the monitoring of this condition [14, 15]. However, the inconsistency in the preliminary studies and the development of new invasive and non-invasive procedures with direct or indirect measurements of fetal hemoglobin for monitoring the disease led to this research line being abandoned. The cause of AFP elevation in anemia remained unclear. Since most of the conditions associated with elevations in maternal serum AFP are explained by an increased fetal–maternal transfer [16], it was thought that the presence of a large and edematous placenta in women with Rhesus disease allowed greater AFP transfer from the fetus to the mother. However, this hypothesis could not explain increased levels of AFP in amniotic fluid. An alternative explanation is that fetal AFP production was increased, which would be reflected in increased levels in fetal serum. Increased production of AFP in fetal liver may reflect immature liver function as a consequence of the activation of hematopoietic cells [17].

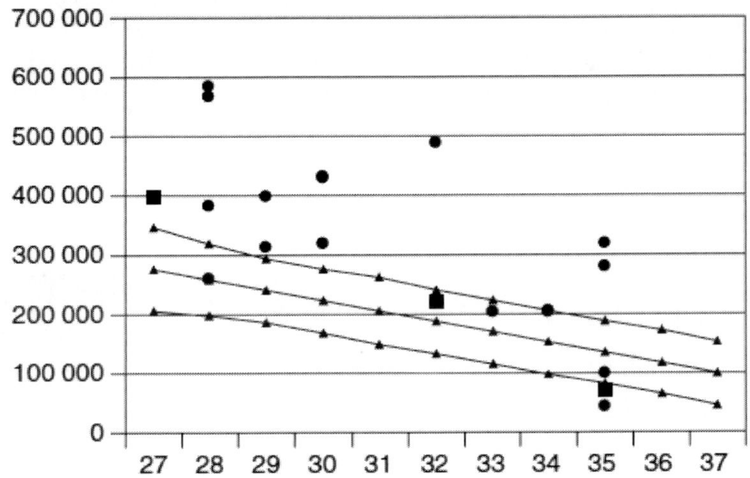

Figure 2. Absolute values of fetal serum AFP plotted against normal range in alloimmune non-hydropic (square) and hydropic (circle) cases after 26 weeks of gestation (from Bartha JL et al. Prenat Diagn 2003 with permission[18]).

To know whether this fact was just the reflect of an increase transfer of AFP from the fetal to the maternal site due to abnormal placentation or this was caused by an increase in fetal AFP production, we evaluated FSAFP levels in cases of alloimmunised pregnancies [18]. For that we studied AFP levels in 32 fetal serum samples from women with red blood cell alloimmunisation at the time of the first fetal blood sampling. This sample included several cases of severe anemia and some cases of fetal hydrops. We expressed the levels of AFP and hemoglobin as absolute numbers and as delta values (number of SDs by which the observed value differed from the normal mean for the same gestation). Main outcome measures were fetal serum AFP levels and fetal hemoglobin concentration. What we found was that FSAFP levels were higher than normal in the anemic cases (delta values 2.4 +/- 5.5 SD), especially in those cases after 26 weeks of gestation (Figure 2) and a high correlation between FSAFP and fetal hemoglobin levels after adjusting for gestational age (delta values) (Figure 3). However, surprisingly mildly affected non-hydropic cases had even higher levels than severely affected fetuses with hydrops. Even fetuses without anemia at the time of their FBS (but with a positive Coombs test) showed the highest levels of AFP. Since AFP levels are thought to be related to fetal hematopoiesis, our results suggest that this response is activated in Rh disease before anaemia is present (presumably indicating successful adaptation to hemolysis by increased red cell production). As the disease worsens, AFP levels fall and this is associated with the worst affected fetuses, which could follow failure of hematopoiesis. There is some evidence supporting a declining fetal hematopoiesis with advancing stages of anemia. It has been reported that both erythroblast and reticulocytes count, indirect measures of fetal hematopoiesis, may be decreased or even absent in the case of severely anemic fetuses reflecting a failure in hematopoiesis [19]. The existence of hyporegenerative anemia in alloimmunised neonates is also well known. The pathogenesis of this anemia remains unclear. These patients have normal or increased erythropoietin levels [20] and it has been suggested that persisting anti-D antibodies may destroy erythrocyte precursors [21]. However, we cannot rule out that our finding of decreased AFP levels in the most severe cases could be explained by other reasons, including fetal liver dysfunction, which has been reported in Rh-immunised fetuses [22].

So we concluded that fetuses affected by red blood cell alloimmunisation have increased levels of serum AFP but these levels fall back to low levels in severe anemia especially with hydrops, which could represent the failing of a compensatory erythropoietic mechanism. Our results suggested that fetal hematopoiesis is activated early in red blood cell alloimmunisation but was subsequently impaired.

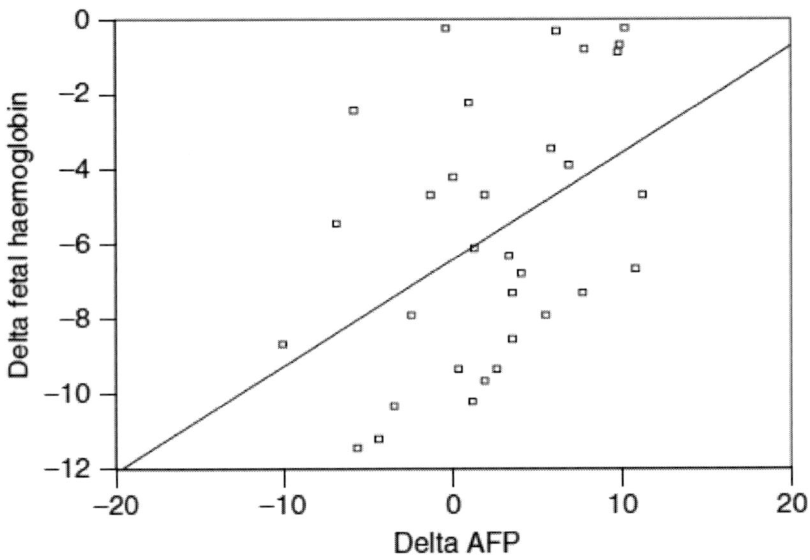

Figure 3. Correlation between delta-AFP and delta-hemoglobin values in alloimmunised fetuses (from Bartha JL et al. Prenat Diagn 2003 with permission[18]).

USING MSAFP LEVELS FOR MANAGING PREGNANT WOMEN AT RISK OF FETAL ANEMIA

Currently the management of women at risk of fetal anemia due to alloimmunisation include determining fetal Rh status by non-invasive testing of fetal DNA in maternal blood and in positive cases serial (normally weekly or biweekly) measurements of peak systolic velocity (PSV) in the fetal middle cerebral artery (MCA). It is known that when PSV in the MCA rises above 1.5 multiples of the median (MoM) for the normal range for gestational age, the risk of moderate to severe fetal anemia is so high that cordocentesis, fetal blood sampling and if needed fetal transfusion should be done. We evaluated the value of adding MSAFP measurement, in the management of these cases. So, we evaluated the relationships between MSAFP levels and both MCA PSV and fetal hemoglobin in women at risk of fetal anemia. For that we studied 41 measurements of MSAFP carried out in 22 women at risk of fetal anemia (16 alloimmunised patients and 6 cases of parvovirus infection) who were monitored by using MCA Doppler measurements [23]. The relationships between MSAFP (MoM) and both MCA peak systolic velocity (z-scores) and fetal hemoglobin (MoM) were studied. What we found were significant correlations between MSAFP and both MCA Doppler measurements ($r = 0.56$, $p = 0.00017$) (Figure 4) and fetal hemoglobin ($n = 13$, $r = -0.71$, $p = 0.006$). MSAFP was higher in cases with fetal anemia ($n = 10$) than in those with normal hemoglobin levels ($n = 3$) (1.7 +/- 0.4 vs 0.8 +/- 0.1 MoM; $p = 0.006$). In cases of alloimmunised pregnancies with fetal anemia, MSAFP elevations preceded the presence of increased MCA Doppler velocity by 2.7 weeks (range 0-9 weeks), but this was not so in cases of parvovirus infection. So we concluded that MSAFP is significantly correlated with both MCA Doppler measurements and fetal hemoglobin. Elevations of MSAFP may appear earlier than MCA Doppler abnormalities in cases of fetal anemia associated with red blood cell alloimmunisation than in those due to parvovirus infection. So potentially measurements of MSAFP could be clinically more useful in cases of immune hemolytic anemia.

Several studies have focused in the utility of measuring MSAFP in cases of parvovirus infection. Komischke et al. [24] evaluated 137 sera from 65 pregnant women with an acute parvovirus B19 infection. They divided the sample between two groups according to the presence or absence of fetal complications. In group I with fetal complications, significant elevations of MSAFP were detected in 13 of the 35 women (37%). But just in one of the nine cases in which sera were obtained before occurrence of fetal complications, MSAFP was elevated. In group 2 without complications, significant elevations of MS-AFP were found only in two of the 30 women (6-7%). So the authors concluded that MS-AFP does not appear to be a regular early marker for poor pregnancy outcome in parvovirus B19-infected pregnancies, although its concentration is frequently elevated at the time of complications.

Simms et al. [25] could not find increased MSAFP concentrations in cases of anemia due to parvovirus infection. There are several potential explanations for that. First, parvovirus infection affects erythroid progenitors and considering that AFP is related to fetal erythropoiesis increases in MSAFP should not be the same as in cases of hemolytic anemia such as in cases of Rhesus disease. Second, as shown by the own authors parvovirus infection use cause only severe anemia when seroconversion is before 21 weeks and we showed in our previous paper, elevations of MSAFP accounts most of them when anemia occurs after 26 weeks. Third, elevations of MSAFP un cases of anemia due to parvovirus infection is not preceding elevations in middle cerebral artery peak systolic velocity and cordocentesis is usually performed when this happens, so probably when fetal blood sampling is performed is not yet the time when MSAFP rise in these cases.

Other groups have evaluated the usefulness of measuring MSAFP in cases at risk of fetal anemia due to other reasons. Fetal homozygous α-thalassemia-1 or hemoglobin (Hb) Bart's disease is the most common cause of hydrops fetalis in Southeast Asia. As a consequence of fetal anemia, hydropic changes, cardiomegaly, hepatomegaly, and placentomegaly can occur as early as in late first trimester. Second-trimester MSAFP may be clinically a useful screening test for fetal Hb Bart's disease among pregnancies at risk. In a first study looking for predictors of hemoglobin Bart at mid-pregnancy, it was found that MSAFP concentrations were significantly higher in women with fetal Hb Bart's disease than those with unaffected fetuses (median 99.53 vs 50.83, P < 0.001), whereas the concentrations of free β-hCG and PAPP-A were not significantly different between the two groups (P = 0.543 and 0.777, respectively) [26]. Further the same authors evaluated 199 pregnancies at risk of fetal Hb Bart's disease at 18 to 22 weeks. Before performing cordocentesis for the Hb typing analysis, the MSAFP levels were analyzed, and sonographic markers (middle cerebral artery peak systolic velocity, cardiothoracic ratio and placental thickness) were measured. The detection rates of MSAFP and the sonographic markers in predicting fetuses affected by Hb Bart's disease were evaluated. What they found was that the MSAFP levels were significantly higher in pregnant women carrying affected fetuses (47 cases) than in those with unaffected fetuses (2.09 MoM vs 1.18 MoM, P < 0.001). MSAFP as a single marker effectively predicted fetal Hb Bart's disease (area under curve receiver operating characteristic 0.832, 95% CI 0.752-0.911), with 87.2% sensitivity and 74.5% specificity using a cutoff value greater than 1.50 MoM. A combination of MSAFP with sonographic markers gave detection rates in the range of 91.5% to 100% [27]. So they concluded that second trimester MSAFP level as single marker is relatively effective in Hb Bart's prenatal screening and in combination with sonographic markers reached a very high detection rate in these cases.

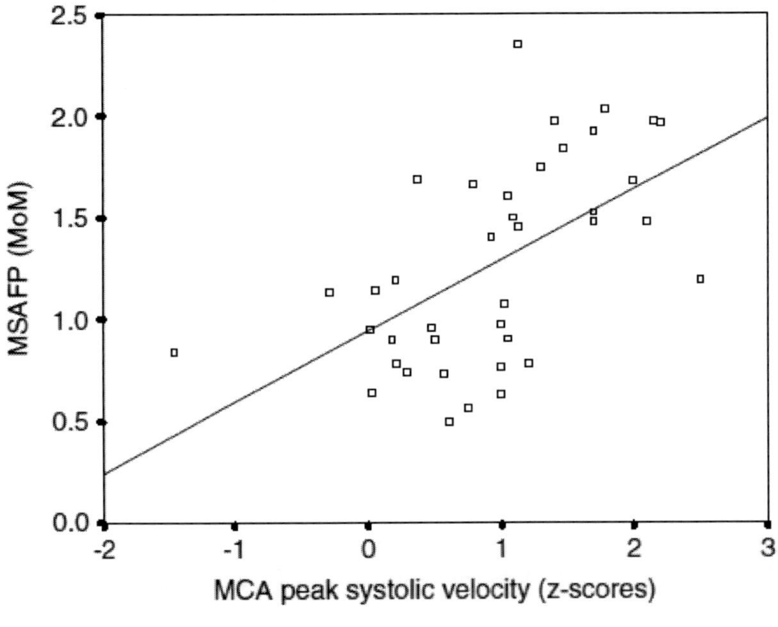

Figure 4. Correlation between MSAFP levels after adjustment for gestational age (MoM) and fetal middle cerebral artery peak systolic velocity according to the normal range (Z-scores) (r = 0.56, p = 0.0001). (From Bartha JL et al. Prenat Diagn 2006[23]).

CONCLUSION

AFP seems to have a role in fetal hematopoiesis. It has been suggested that this action may involved early stages of the erythropoiesis process but further studies are needed to establish the exact mechanisms by which AFP may influence this.

It seems to be clear that serum levels of AFP are increased in many cases of fetal anemia such as immune hemolytic anemia, Bart hemoglobin and probably at some stages of parvovirus infection. There is a need to evaluate AFP in other conditions causes fetal anemia and further research is also needed to evaluate the potential clinical applications of these findings.

REFERENCES

[1] Mizejewski GJ. (1994) Alpha-fetoprotein binding proteins: Implications for transmembrane passage and subcellular localization. *Life Sci,* 56, 1-9.

[2] Mizejewski GJ. (2004) Biological roles of alpha-fetoprotein during pregnancy and perinatal development. *Exp Biol Med (Maywood),* 229, 439-63.

[3] Geuskens M, Torres JM, Esteban C, Urtiel J et al. (1994) Endocytosis of three serum proteins of a multigene family and of arachidonic acid in human lectin-stimulated T lymphocytes. *Microsc Res Tech*, 28, 297-307.

[4] Serova IA, Yunker VM, Kaledin VI. (1982) High level of alphafetoprotein (AFP) and persistence of hemopoiesis in the liver of nude mice. *Oncodev Biol Med*, 3, 351–363.

[5] Hwang SJ, Lee SD, Wu JC, Chang CF, Lu CL, Tsay SH, Lo KJ. (1994) Clinical evaluation of erythrocytosis in patients with hepatocellular carcinoma. *Chuang Jua Hsueh Tsa Chih Tapei*, 53, 262-9.

[6] Sakisaka S, Watanabe M, Tateishi H, Harada M, Shakado S, Mimura Y, Gondo K, Yoshitake M, Noguchi K, Hino T et al. (1993) Erythropoietin production in hepatocellular carcinoma cells associated with polycythemia: immunohistochemical evidence. *Hepatology*, 18, 1357-62.

[7] Kew MC, Fishet JW. (1986) Serum erythropoietin concentrations in patients with hepatocellular carcinoma. *Cancer,* 58, 2485-8.

[8] Bartha JL, Comino-Delgado R, Arce F, Alba P, Broullon JR, Barahona M. (1999) Relationship between alpha-fetoprotein and fetal erythropoiesis. *J Reprod Med,* 44, 689-97.

[9] Bartha JL, Romero-Carmona R, Comino-Delgado R, Arce F, Arrabal J. (2000) Alpha-fetoprotein and hematopoietic growth factors in amniotic fluid. *Obstet Gynecol.* 96, 588-92.

[10] Ohls RK. (1998) Developmental erythropoiesis. In *Fetal and Neonatal Physiology* (2nd edn), Polin RA, Fox WW (eds). WB Saunders company: 1762–1786.

[11] Lebedinskaya OV, Velizheva NP, Donenko FV, Chereshnev VA, Rodionov SY, Akhmatova NK, Shubina IZh, Lebedinskaya EA, Kiselevskii MV. (2006) Effect of "Profetal" on differentiation and functional activity of human mononuclear leukocytes. *Bull Exp Biol Med*, 141, 536-43.

[12] Gorskaya YF, Lebedinskaya OV, Nesterenko VG, Chailakhyan RKh, Gerasimov YV, Latsinik NV, Kuralesova AI, Genkina EN. (2007) Effect of alpha-fetoprotein on the count of bone marrow and splenic stromal precursor cells and proliferation of their cultural descendants. *Bull Exp Biol Med*, 143, 140-2.

[13] Carrington D, Gilmore DH, Whittle MJ, et al. (1982) Maternal serum alpha-fetoprotein: A marker of fetal aplastic crisis during intrauterine human parvovirus infection. *Lancet,* 1, 433-4.
[14] Seppälä M, Ruoslahti E. Alpha-fetoprotein in Rh-immunized pregnancies. Obstet Gynecol 1973; 42: 701–705.
[15] Whyley GA, Hardy NR. (1974) Alpha-fetoprotein levels in amniotic fluids in pregnancies complicated by Rhesus isoimmunizations. *J Obstet Gynaecol Brit Commonwealth,* 81, 459–465.
[16] Katz VL, Chescheir NC, Cefalo RC. (1990) Unexplained elevations of maternal serum alpha-fetoprotein. *Obstet Gynecol Surv*, 45, 719–726.
[17] Seppälä M. (1975) Fetal pathophysiology of human alpha-fetoprotein. *Ann N Y Acad Sci*, 259, 59–73.
[18] Bartha JL, Harrison G, Kyle PM, Soothill PW. (2003) Fetal serum alpha-fetoprotein in alloimmunised pregnancies. *Prenat Diagn*, 23, 17-20.
[19] Nicolaides KH, Thilaganathan B, Rodeck CH, Mibashan RS. (1988) Erythroblastosis and reticulocytosis in anemic fetuses. *Am J Obstet Gynecol* 159: 1063–1065.
[20] Dallacasa P, Ancora G, Miniero R, Gavella B, Brondelli L, Conte R, Salvioli GP. (1996) Erythropoietin course in newborns with Rh hemolytic disease transfused and not transfused in utero. *Pediatr Res*, 40, 357–360.
[21] Luban NLC. (1998) Hemolytic disease of the newborn: progenitor cells and late effects. *N Engl J Med*, 338, 829–831.
[22] Nicolini U, Nicolaidis P, Tannirandorn Y, Fisk NM, Nasrat H, Rodeck CH.(1991) Fetal liver dysfunction in Rh alloimmunisation. *Br J Obstet Gynaecol,* 98, 287–293.
[23] Bartha JL, Illanes S, Abdel-Fattah SA, Harrison G, Soothill PW. (2006) Maternal serum alpha-fetoprotein, fetal middle cerebral artery blood flow velocity and fetal haemoglobin in pregnancies at risk of fetal anaemia. *Prenat Diagn*, 26, 101-4.
[24] Komischke K, Searle K, Enders G.(1997) Maternal serum alpha-fetoprotein and human chorionic gonadotropin in pregnant women with acute parvovirus B19 infection withand without fetal complications. *Prenat Diagn*, 7, 1039-46.
[25] Simms RA, Liebling RE, Patel RR, Denbow ML, Abdel-Fattah SA, Soothill PW, Overton TG. (2009) Management and outcome of pregnancies with parvovirus B19 infection over seven years in a tertiary fetal medicine unit. *Fetal Diagn Ther*, 25, 373-8.
[26] Tongprasert F, Wanapirak C, Tongsong T. (2001) Maternal serum hCG, PAPP-A and AFP as predictors of hemoglobin Bart disease at mid-pregnancy. *Prenat Diagn*, 31, 430-3.
[27] Tongprasert F, Srisupundit K, Luewan S, Tongsong T. (2014) Second trimester maternal serum alpha-fetoprotein (MSAFP) as predictor of fetal hemoglobin Bart's disease. *Prenat Diagn*, 34, 1277-82.

In: Alpha-Fetoprotein: Functions and Clinical Applications ISBN: 978-1-63484-875-6
Editors: Nisha Lakhi and Michael Moretti

Chapter 11

ALPHA-FETOPROTEIN AND POOR PREGNANCY OUTCOMES: OBSERVED PATTERNS, PATHOPHYSIOLOGY, AND CLINICAL UTILITY

Laura L. Jelliffe-Pawlowski[1,*], PhD, and Rebecca J. Baer[2], MPH

[1]Department of Epidemiology and Biostatistics,
University of California San Francisco School of Medicine, San Francisco, CA, US
[2]Department of Pediatrics, University of California San Diego, CA, US

ABSTRACT

During pregnancy, maternal serum alpha-fetoprotein (MSAFP) is commonly used as a second trimester prenatal screening biomarker for chromosomal and structural birth defects. In the absence of birth defects, abnormal levels of mid-pregnancy MSAFP are associated with adverse pregnancy outcomes including fetal demise and preterm birth. Pregnancies ending in fetal demise or preterm birth often have unusually lower or higher mid-pregnancy MSAFP levels than pregnancies that do not. The same is true for pregnancies with specific conditions that are closely related to fetal demise and preterm birth including preeclampsia, conditions of abnormal placentation (e.g., previa, abruption, accreta), and pregnancies carrying a fetus with intrauterine growth restriction. In this chapter, we review specific studies looking at the relationship between MSAFP and fetal demise or preterm birth. In addition, we will review associations between MSAFP and conditions with close links to fetal demise and preterm birth including preeclampsia, placenta previa, placental abruption, placenta accreta, and intrauterine growth restriction. We also provide a brief review of the pathophysiological literature that underscores these relationships and we suggest next steps with respect to research and clinical use of MSAFP data.

---

* Correspondence to: laura.jelliffe@ucsf.edu.

INTRODUCTION

Alpha-fetoprotein (AFP) is a tumor-associated fetal glycoprotein produced by the yolk sac, fetal liver, and fetal gastrointestinal tract [1]. During pregnancy, maternal serum AFP levels (MSAFP) are dependent on gestational age and are often used in screening for chromosomal and structural birth defects. Low levels of MSAFP are used to screen for aneuploidies [2] and high levels are used to screen for neural tube and other related structural birth defects. [3-5] In the absence of chromosomal or structural birth defects, abnormally low and high MSAFP levels have been increasingly shown to be associated with poor pregnancy outcomes – most notably fetal demise and preterm birth. [6-18] Abnormal MSAFP levels are consistently shown to be associated with conditions with close ties to fetal demise and preterm birth including preeclampsia [10, 11, 19, 20], conditions of abnormal placentation (including placenta previa, placental abruption, and morbidly adherent placenta) [11, 21-26], and intrauterine growth restriction [8, 11, 14, 15, 18].

This chapter reviews a number of studies that have examined the relationship between mid-pregnancy levels of MSAFP, fetal demise, preterm birth, and conditions with close links to these outcomes including preeclampsia, placenta previa, placental abruption, placental accreta, and intrauterine growth restriction. Further, this chapter provides a brief overview of the pathophysiological literature that informs these relationships and provides suggestions for future areas of research and clinical use of MSAFP levels in the management of women at increased risk for fetal demise, preterm birth, and other closely related diagnoses.

ALPHA-FETOPROTEIN AND FETAL DEMISE

Fetal demise, which is also referred to as fetal death or fetal loss, is defined by the World Health Organization as "death prior to the complete expulsion or extraction from its mother of a product of conception, irrespective of the duration of pregnancy; the death is indicated by the fact that after such separation a fetus does not breathe or show any other evidence of life." [27] While great controversies exist with respect to classification of fetal demise and stillbirth, in general, fetal demise is distinguished from stillbirth by death in utero (fetal demise) versus the birth of fetus who is deceased (stillbirth) [28].

A number of studies have examined the association between mid-pregnancy MSAFP levels and fetal demise. Given that AFP is typically drawn between the 15th and 20th week of gestation for aneuploidy and neural tube defect screening purposes, most studies examining the relationship between fetal demise and MSAFP levels have studied associations in gestations of at least 15 weeks. In general, investigators have examined patterns using MSAFP multiple of the medians (MoMs) and associated low and high cut points (e.g., < 0.5 and at or > 2.5 MoMs), keeping with prenatal screening methodologies and allowing for adjustment of values by gestational age at draw, maternal weight, and other factors determined to independently influence MSAFP values in the specific population from which MoMs were derived (e.g., race/ethnicity, smoking, diabetic status) [29].

Most studies conducted to date have observed a relationship between abnormal MSAFP levels and fetal demise. While most often the observed relationship has been between abnormally high MSAFP levels (generally defined by MoMs > 2.5) [6-11], some studies have observed a relationship between unusually low levels of MSAFP and demise (e.g., MoM < 0.5) [6, 8]. As noted in Table 1, studies have been conducted across Western populations and have generally found consistent results – particularly in studies that have considered demise across all gestational ages or have focused specifically on fetal demise before 20 or 24 weeks. One exception are the results from the First and Second Trimester Evaluation of Risk (FaSTER) study which observed a relationship between elevated MSAFP and fetal demise before 24 weeks gestation but did not observe an association with fetal demise at or after 24 weeks [9].

Although it is unclear why the FaSTER group did not find a relationship between elevated mid-pregnancy MSAFP and fetal demise where others have, it should be noted that our recent study of nearly 40,000 women undergoing first and second trimester prenatal screening in California, found that women who had a MSAFP MoM of 2.5 or more were at more than a 10-fold increased risk of fetal demise before 20 weeks gestation (2.5% of those with a MSAFP MoM at or above 2.5 versus 0.2% of those with a MSAFP MoM < 2.5 and no other marker abnormality leading to a prenatal screening positive result) and more than a 7-fold increased risk of fetal demise at 20 weeks gestation or more (3.8% of those with a MSAFP MoM at or above 2.5 versus 0.4% of those with a MSAFP MoM < 2.5 and no other marker abnormality leading to a prenatal screening positive result) (see Figure 1) [11].

Table 1. Sample studies examining the association between mid-pregnancy MSAFP levels and fetal demise

Reference	Sample (location, years, n =)	MSAFP Metric Used	Finding
Burton, 1988 [6]	North Carolina, USA, 1978-1981 and 1983-1986, n = 15,390	MSAFP MoM < 0.25 or ≥ 2.5	↑Fetal demise (< 20 weeks) with ↑MSAFP and ↑fetal demise (all weeks) with ↓MSAFP
Robinson et al., 1989 [7]	California, USA, 1980-1982 and 1986-1987, n = 560 (with MSAFP ≥ 2.5)	MSAFP MoM 2.5-2.9, 3.0-3.9, ≥ 4.0	↑Fetal demise (all weeks combined) with ↑MSAFP by category
Krause et al., 2001 [8]	Denmark, 1980-1993, n = 77,149	MSAFP MoMs by 0.25 intervals <0 .25 to > 2.5	↑Fetal demise (15 to 28 weeks) where MSAFP < 0.51 or > 1.25 (risks higher as MSAFP increases)
Dugoff et al., 2008 [9]	Multiple sites across United States, 1999-2002, n = 36,014	MSAFP MoM by value (rather than cut point)	↑MSAFP levels higher in pregnancies with fetal demise (< 24 weeks), no association with fetal demise ≥ 24 weeks
Huang et al., 2010 [10]	Ontario, Canada, 2000-2003, n = 141, 698	MSAFP MoM by value (rather than cut point)	↑MSAFP levels higher in pregnancies with fetal demise (all weeks combined)
Baer et al., 2014 [11]	California, USA, 2009-2012, n = 39,979	MSAFP MoM ≥ 2.5	↑Fetal demise < 20 weeks and ↑fetal demise ≥ 20 weeks

Of note for the purposes of this discussion – especially with respect to pathophysiological underpinnings – is the finding by investigators in California and Denmark showing that the risk of fetal demise increased at very low and very high MSAFP levels [6, 8]. These data suggest that optimal levels of MSAFP are required for normal pregnancy progression and levels below and above increase the risk of a poor pregnancy outcome.

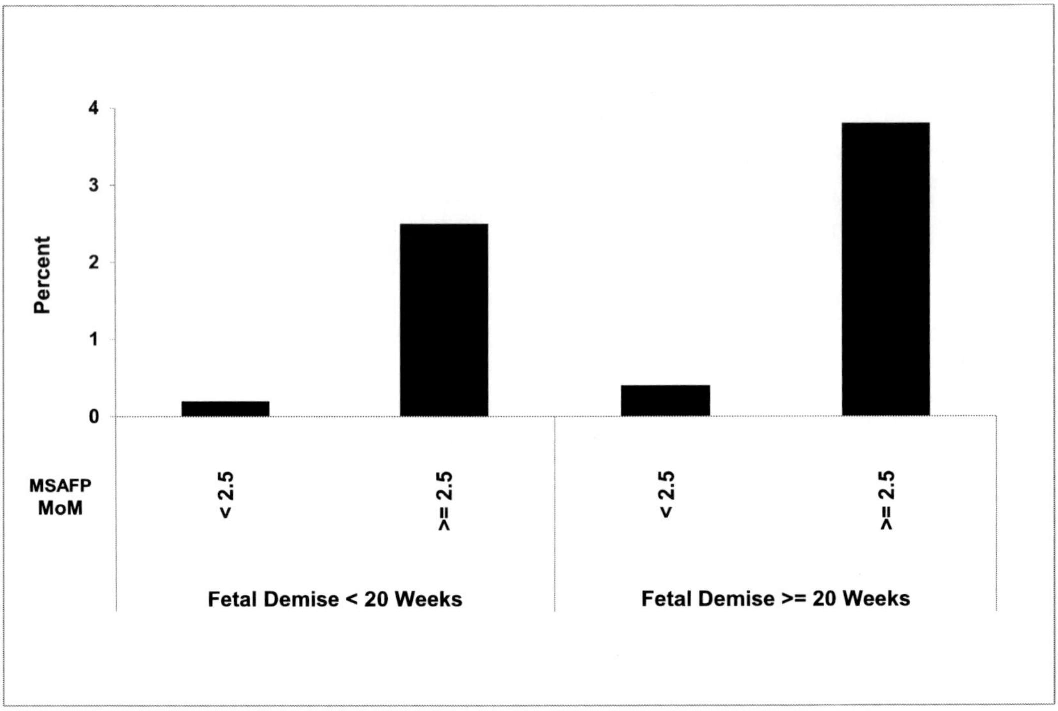

Figure 1. Percent of pregnancies ending in fetal demise before or after 20 weeks of gestation by MSAFP multiple of the median (MoM) above or below 2.5 (adapted from Baer, 2014 [11]).

Alpha-Fetoprotein and Preterm Birth

Preterm birth, defined as birth before 37 completed weeks of gestation, is associated with increased risks of infant mortality and morbidity [30, 31]. While mortality and morbidity risks are highest among babies born at earliest gestational ages (e.g., < 32 weeks), risks extend across all ages and indications of preterm birth, which historically, includes those resulting from early spontaneous delivery due to preterm premature rupture of membranes (PPROM), premature labor and those resulting from delivery by induction or cesarean section due to maternal or fetal indication (medically indicated, where there is no PPROM or premature labor) [30-33].

There exists a large volume of studies that have examined the association between MSAFP and preterm birth with near universal agreement in findings of elevated mid-pregnancy MSAFP increasing the risk of preterm birth. In 2011, Beta and colleagues provided a brief review of 21 studies examining the relationship between elevated mid-pregnancy MSAFP levels (generally defined by MoMs at or above 2.0) and preterm birth and found that 17 of the 21 studies reviewed observed an increased risk [34]. The four studies that found no significant increased risk between elevated mid-pregnancy MSAFP levels and preterm birth were those with case sample sizes < 50 (see Table 2 for a sample of studies conducted from 1999 – 2015).

Table 2. Sample studies examining the association between mid-pregnancy MSAFP levels and preterm birth

Reference	Sample (location, years, n =)	MSAFP Metric Used	Finding
Wenstrom et al., 1996 [12]	Alabama,USA 1988-1994, n = 6,221	MSAFP MoM ≥ 2.0	↑Preterm birth < 37 weeks (combined spontaneous/ medically indicated)
Van Rijn et al., 1999 [13]	Utrecht, Netherlands, 1991-1997, n = 7,380	MSAFP > 90th %ile	↑Spontaneous preterm birth < 35 weeks
Chandra et al., 2003 [14]	Nova Scotia, 1994-2000, n = 5,789	MSAFP MoM ≥ 2.0	↑Spontaneous and medically indicated preterm birth (examined separately) < 37 weeks
Dugoff et al., 2005 [15]	Multiple sites across United States, 1999-2002, n = 33,145	MSAFP MoM ≥ 2.0	↑Preterm birth ≤ 32 weeks (combined spontaneous/ medically indicated)
Alleman et al., 2013 [16]	Iowa, USA, 2009-2010, n = 2,509	MSAFP MoM by 0.5 groups from < 1.0 to ≥ 2.5	↑Preterm birth < 37 weeks (combined spontaneous/ medically indicated)
Jelliffe-Pawlowski et al., 2015 [17]	California, USA, 2007-2012, n = 125,202	MSAFP MoM ≥ 95th %ile	↑Spontaneous and medically indicated preterm birth (examined separately) < 32 and 32-36 weeks
Puntachai et al., 2015 [18]	Chiang Mai, Thailand, 2010-2013, n = 5,486	MSAFP MoM > 2.0	↑Preterm birth < 37 weeks (combined spontaneous/ medically indicated

Although the relationship between elevated MSAFP levels and preterm birth was previously thought to be driven by co-occurrence with other marker abnormalities associated with poor placental function (e.g., by elevated second trimester human chorionic gonadotrophin (hCG) [35], larger, more recent investigations have demonstrated that elevated MSAFP occurring in isolation (where other prenatal screening biomarkers are normal) does, in fact, portend an often substantial, preterm birth risk [17, 36, 37].

Increased risk of preterm birth with mid-pregnancy MSAFP MoMs of 2.0 and higher have been described as 2- to 7-fold depending on the referent group and the statistic used [11, 16], while the top 95th percentile for populations (MoMs of approximately 1.7) has been shown to confer at 2- to 4.5-fold increased risk even after adjustment for other common premature birth risks (e.g., hypertension, diabetes) [37, 38]. It is critical to note that investigators have also found that as mid-pregnancy MSAFP measurements increase, so does the risk of preterm birth [16]. Also, similar to what has been observed with fetal demise and other adverse outcomes, it has been shown that women with more than one abnormal prenatal screening biomarker value (e.g., elevated MSAFP and hCG) are at a much higher risk of preterm birth [9, 11, 36-38].

As research has progressed in the study of MSAFP and preterm birth, investigations have tended to narrow their focus into characterizing the relationship between mid-pregnancy

MSAFP and the timing of preterm birth (e.g., before 32 weeks and 32 to 36 weeks), and indication for premature delivery. Studies evaluating birth before 32 weeks gestation have observed a particularly potent relationship between elevated mid-pregnancy MSAFP and early preterm birth. In a 2010 publication [36], our group in California found that women with a mid-pregnancy MSAFP MoM of 2.0 or more were at 5.4-fold greater risk of a delivery at 32 weeks or earlier than those with MSAFP MoMs between 0.6 and 1.9. The relative risk was 2.6 for gestations between 33 and 36 weeks. In our investigation of risk factors for birth before 30 weeks gestation, women with elevated MSAFP had a 2.8-fold increased risk of spontaneous preterm birth and 8.3-fold increased risk of a medically indicated birth before 30 weeks [37]. In 2015 [17], we investigated risk factors for spontaneous or medically indicated preterm birth before 32 weeks and demonstrated that elevated MSAFP (MoM of 1.68 or more) had a relative risk of 3.2 for PPROM, 3.5 for spontaneous labor without PPROM, and 4.0 for medically indicated preterm birth. In contrast, those with elevated MSAFP were at 2.9-fold increased risk of having PPROM and delivering from 32 to 36 weeks, 2.2-fold increased risk of having spontaneous labor from 32 to 36 weeks, and 1.7-fold risk for medically indicated preterm birth from 32 to 36 weeks (Figure 2).

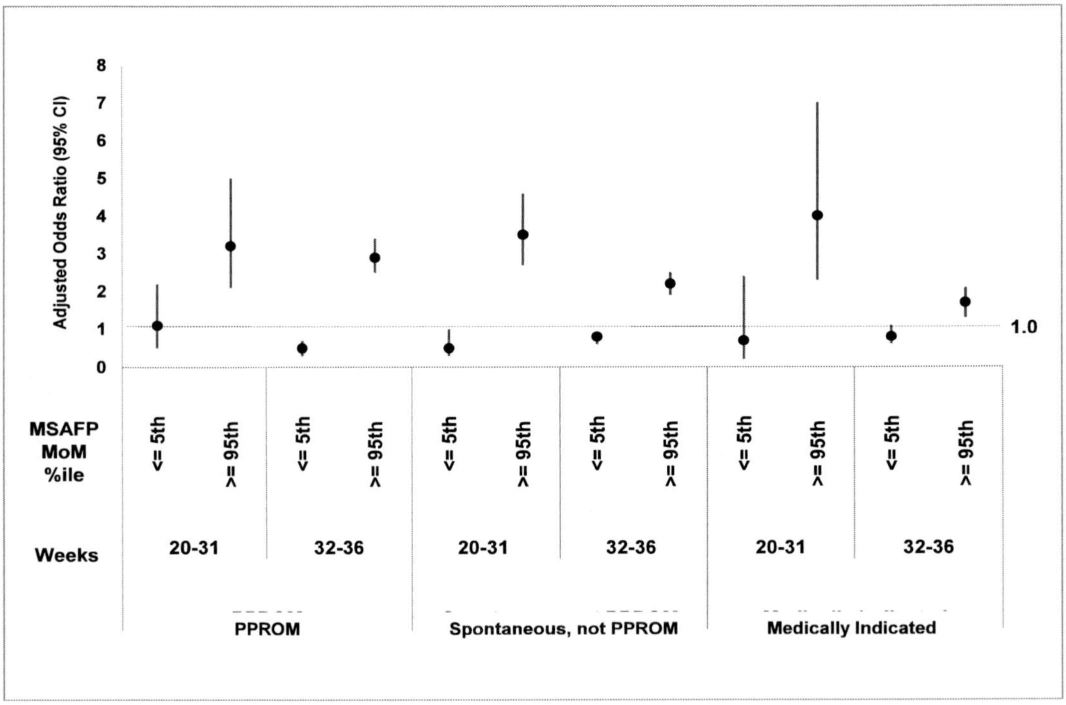

Figure 2. Risk of preterm birth (by timing and indication) by low and high mid-pregnancy MSAFP multiple of the median (MoM, by <= 5th and >= 95th compared to 6th to 94th percentile) (adapted from Jelliffe-Pawlowski, 2015 [17]).

While the vast majority of studies have shown an association between elevated MSAFP and preterm birth, it is also of note that some studies have demonstrated a decreased risk of preterm birth among women with low levels of MSAFP. Using a MoM cutoff of 0.76, Puntachai and colleagues [18] found a reduced risk of preterm birth with lower versus normal MSAFP levels. Recently our group found a protective effect of MSAFP at or below the 5th

percentile (0.59 MoMs) for preterm birth (PPROM and spontaneous labor without PPROM) at 32 to 36 weeks (see Figure 1) [17].

Alpha-Fetoprotein and Other Conditions with Close Links to Fetal Demise and Preterm Birth: Preeclampsia, Conditions of Abnormal Placentation, and Intrauterine Growth Restriction

Abnormal levels of mid-pregnancy MSAFP, particularly high MSAFP levels, have consistently been shown to be associated with other conditions with close links to fetal demise and preterm birth including preeclampsia [10, 11, 19, 20], conditions of abnormal placentation (including placenta previa, placental abruption, and morbidly adherent placenta) [11, 21-26], and intrauterine growth restriction [8, 11, 14, 15, 18].

Preeclampsia

Preeclampsia, which occurs in 2 to 8% of all pregnancies, is a syndrome generally diagnosed in the second half of pregnancy. Preeclampsia is a multisystem disorder whose hallmarks include maternal hypertension and proteinuria where the presence of additional maternal symptoms such as headache, visual changes, and abdominal pain often reflect the severity of the disease [39-41]. Severe preeclampsia is defined by persistently elevated blood pressure or by the presence of several other criteria including cerebral or visual symptoms such as severe headache, presence of pulmonary edema or cyanosis, oliguria, epigastric pain, thrombocytopenia, or impaired liver function. Early-onset severe preeclampsia is generally defined as severe symptoms before 34 weeks gestation [42-44]. Although occurring in only 0.3 to 1.5% of all pregnancies, early-onset severe preeclampsia contributes to approximately 15% of all preterm births. Early-onset severe preeclampsia is associated with both short- and long-term morbidity and mortality risks in the mother, fetus, and newborn [40, 45].

Elevated MSAFP levels have consistently been shown to be associated with preeclampsia [10, 11, 13, 14, 18, 20, 46]. This association has been demonstrated for both early and late onset preeclampsia, with the strongest association with onset before 34 weeks gestation (Figure 3) [20, 47]. The higher the MSAFP levels, the higher the likelihood of developing early preeclampsia [19]. In a meta-analysis of cohort studies evaluating second trimester prenatal screening markers, Morris and colleagues [48] suggested that a MSAFP MoM of 2.0 or more resulted in the most effective screening threshold for preeclampsia, with a positive likelihood ratio of 2.36 and negative likelihood ratio of 0.96.

Conditions of Poor Placentation: Placenta Previa, Placental Abruption and Morbidly Adherent Placenta

While there are other conditions that are indicative of poor placentation (e.g., velamentous cord insertion), for the purposes of this discussion we focus specifically on placenta previa, placental abruption, and morbidly adherent placenta, as they are not only associated with preterm birth and fetal demise [49-51], but have also been studied in the context of MSAFP (Table 3) [11, 21-26]. All of these conditions reflect placental abnormalities. The most serious of these conditions is morbidly adherent placenta including placenta accreta, increta and percreta where parts of the placenta grow into the uterine wall (placenta accreta), into the uterine muscles (placenta increta), or through the uterine muscles (placenta percreta) [52]. Placenta previa is characterized by the placenta partially or totally covering the opening of the cervix [53]. Placental abruption is defined by the premature separation of implanted placenta before delivery [54].

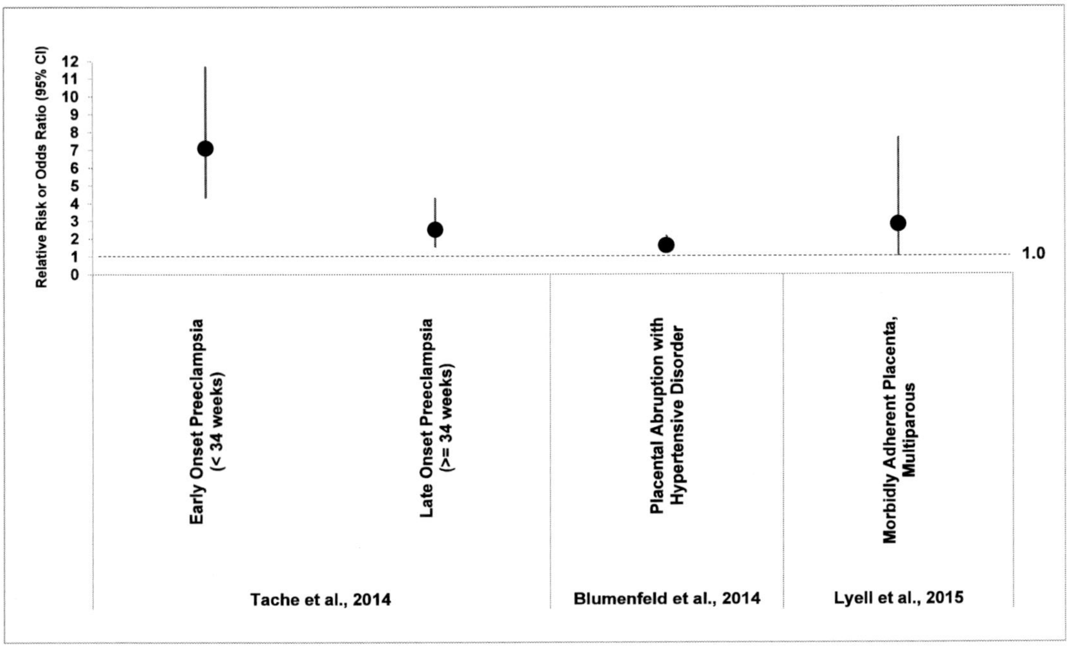

Figure 3. Risk of preeclampsia, placental abruption, and morbidly adherent placenta by mid-pregnancy MSAFP multiple of the median >= 95th (adapted from Tache et al. [20], Blumenfeld et al. [23], and Lyell et al. [25]).

In comparison to women with negative first and second trimester prenatal screening results, in a large population-based study of California births, women with a MSAFP MoM of 2.5 or greater were 4-times as likely to develop placental abruption [11]. In a population of low risk women (excluding women with a preexisting medical condition, previous poor obstetric outcome, and women over 34 years), those with a MSAFP MoM or 2.0 or greater were at more than 6-fold increased risk of developing placental abruption [14]. Our group in California also demonstrated that women with or without hypertension and MSAFP levels in the 95th percentile or more (MoM of 1.79 for the population) were at risk of placental abruption (Figure 3) [23].

Placenta previa has also been associated with elevated MSAFP [11, 13]. Koster and colleagues [21] showed an association between increasing MSAFP values and increasing likelihood of placenta previa persisting until delivery, while women with previa and MSAFP MoMs below 1.0 were at reduced risk of the previa persisting until birth. Among women with previa, those with elevated MSAFP are at increased risk of bleeding in the third trimester and preterm birth [55]. Elevated MSAFP has consistently been shown to be a risk factor for abnormal placental adherence, including placenta accreta [25, 26, 56, 57].

Table 3. Sample studies examining the association between mid-pregnancy MSAFP levels and conditions closely linked to preterm birth and fetal demise: Preeclampsia, and conditions of abnormal placentation (placenta previa, placental abruption, morbidly adherent placenta)

Reference	Sample (location, years, n =)	MSAFP Metric Used	Finding
Preeclampsia			
Huang et al., 2010 [10]	Ontario, Canada, 2000-2003, n = 141,698	MSAFP MoM by value (rather than cut point)	↑MSAFP levels higher in pregnancies with preeclampsia (all combined)
Olsen et al., 2012 [19]	San Diego, California, USA, 1995-2010, n = 7,767	MSAFP MoM ≥ 2.0, ≥ 2.5, ≥ 3.0	↑Early (< 34 weeks) and late preeclampsia
Baer et al., 2014 [11]	California, USA, 2009-2012, n = 39,979	MSAFP MoM ≥ 2.5	↑Preeclampsia (all combined)
Tache et al., 2014 [20]	California, USA, 2006-2008, n = 136,139	MSAFP MoM ≥ 95th %ile	↑Early (< 34 weeks) and late severe preeclampsia
Placenta Previa			
Koster et al., 2004 [21]	Dallas, Texas, USA, 1991-2000, n = 275	MSAFP MoM 0-0.99, 1.0-1.99, 2.0-2.99, ≥ 3.0	↑Placenta previa
Baer et al., 2014 [11]	California, USA, 2009-2012, n = 39,979	MSAFP MoM ≥ 2.5	↑Placenta previa
Placental Abruption			
Tikkanen et al., 2007 [22]	Finland, 1997-1999, n = 165 (58 cases, 108 controls)	MSAFP MoM ≥ 1.5	↑Placental abruption
Baer et al., 2014 [11]	California, USA, 2009-2012, n = 39,979	MSAFP MoM ≥ 2.5	↑Placental abruption
Blumenfeld et al., 2014 [23]	California, USA, 2009-2010, n = 137,915	MSAFP MoM ≥ 95th %ile	↑Placental abruption
Morbidly Adherent Placenta (MAP)			
Kupferminc et al., 1993 [24]	Chicago, Illinois, USA, 1985-1992, n = 20 (Placental accreta cases)	MSAFP MoM ≥ 2.5	↑Placental accreta
Hung et al., 1999 [26]	Taipei, Taiwan 1994-1999, n = 10,672	MSAFP MoM > 2.5	↑ Placental accreta
Lyell et al., 2015 [25]	California, USA, 2009-2010, n = 236,714	MSAFP MoM ≥ 2.5	↑MAP

Table 4. Sample studies examining the association between mid-pregnancy MSAFP levels and conditions closely linked to preterm birth and fetal demise: Intrauterine growth restriction (IUGR)

Reference	Sample (location, years, n =)	MSAFP Metric Used	Finding
Krause et al., 2001 [8]	Denmark, 1980-1993, n = 77,149	MSAFP MoM by 0.25 intervals (< 0.25 to > 2.5)	↑IUGR where MSAFP > 1.25 (risks higher as MSAFP increases)
Chandra et al., 2003 [14]	Nova Scotia, 1994-2000, n = 14,374	MSAFP MoM ≥ 2.0,	↑IUGR
Dugoff et al., 2005 [15]	Multiple sites across United States, 1999-2002, n = 33,145	MSAFP MoM ≥ 2.0	↑IUGR
Baer et al., 2014 [11]	California, USA, 2009-2012, n = 39,979	MSAFP MoM ≥ 2.5	↑IUGR
Puntachai et al., 2015 [18]	Chiang Mai, Thailand, 2010-2013, n = 5,486	MSAFP MoM > 2.0	↑IUGR

Intrauterine Growth Restriction

Intrauterine growth restriction in a fetus is generally defined as weight for gestational week at measurement below the 10th percentile. MSAFP has been associated with intrauterine growth restriction (Table 4) [8, 10, 11, 13-15, 18, 46, 58]. Among women with hyperemesis gravidum, those with MSAFP MoMs of 1.55 or higher were more likely to experience intrauterine growth restriction [59]. Among women with preeclampsia, those with abnormally high MSAFP levels were more likely to have newborns with intrauterine growth restriction [60].

Pathophysiological Underpinnings

While a detailed review of the pathophysiological underpinnings related to the observed association between MSAFP levels, fetal demise, preterm birth, and associated conditions is beyond the scope of this chapter and is addressed elsewhere in this book and other publications [61-63], for the purposes of this review, it is useful to note that much remains unknown with respect to underlying pathophysiology and abnormal MSAFP levels and poor outcomes of pregnancy.

Data suggest that the underpinnings of abnormally high or low MSAFP levels and poor outcomes of pregnancy may differ by condition and even, by mother-infant pair. In some instances it may be, for example, that high MSAFP levels indicate abnormalities in placentation allowing excess leakage of AFP produced by the yolk sac, fetal liver, or fetal gastrointestinal tract into maternal circulation [64-66]. Studies of placentas after birth in pregnancies with mid-pregnancy elevated MSAFP levels compared to those without elevation support this theory given increased chronic villitis and vascular lesions [64].

Other placental dysfunction and abnormalities have also been implicated in observed elevated mid-pregnancy MSAFP levels. Investigators have found that there is increased placental ischemia and inflammation in pregnancies with higher MSAFP levels [64, 67] and have also found that inadequate placentation related to insufficient angiogenesis is more common in pregnancies with higher MSAFP levels [65, 66]. Further support for elevated MSAFP levels having a direct link to compromised placental function comes from the extensive data showing that elevated MSAFP levels commonly co-occur with abnormalities in other markers produced by the placenta that also have close links with poor pregnancy outcomes including pregnancy-associated plasma protein A, human chorionic gonadotrophin and inhibin [9, 11, 20, 36-38, 68]. Data show that when MSAFP levels are elevated and there are abnormalities in these other markers related to placental function, the risk of abnormal outcomes such as preterm birth are dramatically increased (Figure 4) [37].

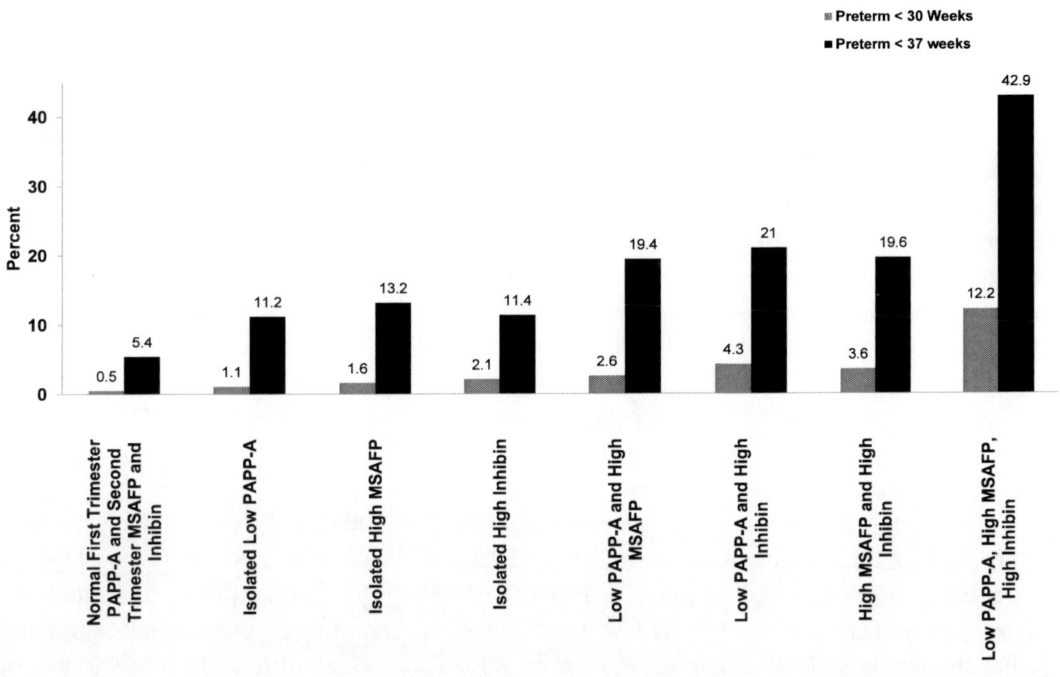

Figure 4. Risk of preterm birth by isolated and combined elevated MSAFP with other placental markers (adapted from Jelliffe-Pawlowski et al., 2013 [37]).

Elevated MSAFP levels have also been shown to have close links to infection and inflammation. Of particular note are findings which show that large number of women with elevated MSAFP levels also have abnormal levels of acute inflammatory factors (e.g., intracellular adhesion molecule-1) [69]. Investigators have also found that MSAFP levels can vary with viral load in some instances (e.g., with human immunodeficiency) and that medications aimed at reducing viral load can also lead to decreases in MSAFP levels [63, 70, 71]. When combined with other inflammatory markers, such as interleukin-6, levels of AFP have also been shown to identify pregnancies with microbial invasion of the amniotic cavity [72].

While elevated MSAFP levels have been associated with poor pregnancy outcomes, low MSAFP levels have consistently been associated with fetal demise. These low MSAFP levels are thought to be an indication of a nonviable fetus and may have some relationship to undiagnosed chromosomal defects [6, 73].

Clinical Implications and Next Steps

Despite extensive data demonstrating that women with abnormal MSAFP levels are at increased risk for poor pregnancy outcomes, (Tables 1-3, Figures 1-3) there are no universal protocols or guidelines for managing these pregnancies in the United States and in most other countries. While in 2008, the Society of Obstetricians and Gynaecologists of Canada (SOGC) Genetics Committee published their recommendations for the management of women with MSAFP MoMs > 2.5 (as well as the management of other biomarker abnormalities identified as part of routine prenatal screening) [74], we are not aware of any other published recommendations.

Specific recommendations of the SOGC when a women has a mid-pregnancy MSAFP MoM > 2.5 included: Ultrasound and a magnetic resonance imagining scan of the placental-uterine interface if there is also placenta previa in the second or third trimester; Consideration of uterine artery Doppler; Development of a fetal surveillance plan (with consideration of increased patient education for signs and symptoms of common complications, increased frequency of antenatal visits, increased ultrasound for monitoring of fetal growth and amniotic fluid levels, increased fetal surveillance of biophysical profile and arterial and venous Doppler); And, increased cervical measurement [74]. This group also recommended that further studies be conducted to assess the benefits of increased monitoring and the potential benefit of other therapeutic options such as administration of low-dose aspirin. Given low sensitivity and high false positive rates for specific outcomes, the SOGC did not recommend use of MSAFP or other routinely collected biomarkers to screen for poor pregnancy outcomes.

While the SOGC recommendations were published in 2008, few studies have evaluated whether any of their proposed strategies would lead to better outcomes in babies born to women with elevated mid-pregnancy MSAFP levels. One exception is a study by investigators in Tehran, Iran, in 2014 who conducted a randomized control trial examining whether the administration of low-dose aspirin (80 mg/day) beginning at 15 to 18 weeks of gestation in women with MSAFP MoMs at or above 2.5 would improve outcomes. These investigators found that women receiving low-dose aspirin had fewer overall adverse pregnancy outcomes and had longer gestational durations [75]. While this study only had 65 cases and 68 controls, these findings suggest benefits of low-dose aspirin when mid-pregnancy MSAFP levels are elevated and as such, seems to point to an area of promise for future study.

Given the extensive available data on poor outcomes in women with elevated mid-pregnancy MSAFP levels, it is surprising that so little work has been conducted to evaluate management strategies. Although it is unclear why little management data exist, one possible explanation may be a misperception of the burden of outcomes in women with elevated MSAFP levels. Given that most studies conducted have evaluated one outcome at a time, it may be that clinicians and researchers are unaware of the percentage of women with elevated MSAFP levels that have any poor outcome which is arguably the most relevant statistic from a management point of view. Recently our group examined the rate of "any adverse outcome" (e.g., placenta previa, placental abruption, preeclampsia, fetal demise, preterm birth, intrauterine growth restriction, structural birth defect) and found that more than one in two pregnancies with a mid-pregnancy MSAFP MoM at or above 2.5 had an adverse outcome (50.6%) (Figure 5) [11].

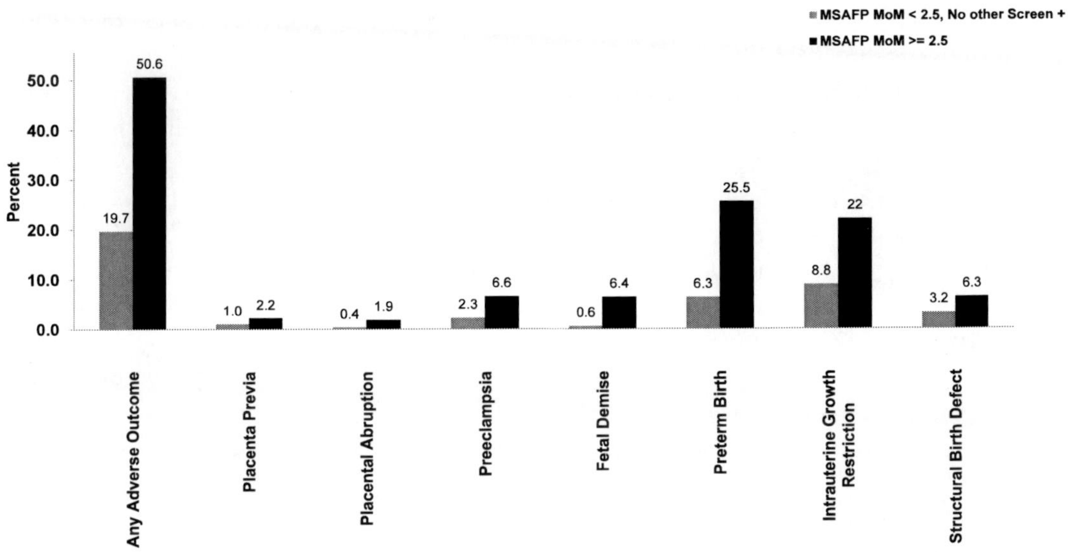

Figure 5. Outcomes in pregnancies with MSAFP MoMs at or above 2.5 (adapted from Baer et al., 2014 [11]).

For the purposes of this review, we contend that motivation of the field towards translating association findings with respect to elevated mid-pregnancy MSAFP levels and poor outcomes into studies and management protocols will require a focus on understanding the overall burden of poor outcomes in women with abnormally high MSAFP levels. We suspect that such efforts will also benefit from a greater focus on evaluating whether existing or novel markers might improve screening performance for "any poor pregnancy outcomes" including improved sensitivity and reduced false positives. This effort may further benefit from the consideration of testing MSAFP levels at multiple time points throughout pregnancy given recent findings showing that prediction is improved if there is a posttest after an elevated MSAFP [76].

References

[1] Sell S, Skelly H. (1976) Tissue sites of alpha fetoprotein synthesis by the rat during pregnancy and hepatoma growth. *J Natl Cancer Inst*, 56, 645-648.

[2] Leek AE, Ruoss CF, Kitau MJ, Chard T. (1973) Raised alpha-fetoprotein in maternal serum with anencephalic pregnancy. *Lancet*, 2, 385.

[3] Merkatz IR, Nitowsky HM, Macri JN, Johnson WE. (1984)An association between low maternal serum alpha-fetoprotein and fetal chromosomal abnormalities. *Am J Obstet Gynecol*, 148, 886-894.

[4] Wald NJ, Cuckle HS, Sneddon J, Haddow JE, Palomaki GE. (1989) Screening for Down syndrome. *Am J Hum Genet*, 44, 586-590.

[5] Canick JA, Palomaki GE, Osathanondh R. (1990) Prenatal screening for trisomy 18 in the second trimester. *Prenat Diagn*, 10, 546-548.

[6] Burton BK. (1988) Outcome of pregnancy in patients with unexplained elevated or low levels of maternal serum alpha-fetoprotein. *Obstet Gynecol*, 72, 709-713.

[7] Robinson L, Grau P, Crandall BF. (1989) Pregnancy outcomes after increasing maternal serum alpha-fetoprotein levels. *Obstet Gynecol*, 74, 17-20.

[8] Krause TG, Christens P, Wohlfahrt J, Lei U, Westergaard T, Norgaard-Pedersen B, Melbye M. (2001) Second-trimester maternal serum alpha-fetoprotein and risk of adverse pregnancy outcome. *Obstet Gynecol,* 97, 277-282.

[9] Dugoff L, Cuckle HS, Hobbins JC, Malone FD, Belfort MA, Nyberg DA, Comstock CH, Saade GR, Eddleman KA, Dar P, Craigo SD, Timor-Tritsch IE, Carr SR, Wolfe HM, D'Alton ME. (2008) Prediction of patient-specific risk for fetal loss using maternal characteristics and first- and second-trimester maternal serum Down syndrome markers. *Am J Obstet Gynecol*, 199, 290-296.

[10] Huang T, Hoffman B, Meschino W, Kingdom J, Okun N. (2010) Prediction of adverse pregnancy outcomes by combinations of first and second trimester biochemistry markers used in the routine prenatal screening of Down syndrome. *Prenat Diagn*, 30, 471-477.

[11] Baer RJ, Currier RJ, Norton ME, Flessel MC, Goldman S, Towner D, Jelliffe-Pawlowski LL. (2014) Obstetric, perinatal, and fetal outcomes in pregnancies with false-positive integrated screening results. *Obstet Gynecol*, 123, 603-609.

[12] Wenstrom KD, Owen J, Davis RO, Brumfield CG. (1996) Prognostic significance of unexplained elevated amniotic fluid alpha-fetoprotein. *Obstet Gynecol*, 87, 213-216.

[13] van Rijn M, van der Schouw YT, Hagenaars AM, Visser GH, Christiaens GC. (1999) Adverse obstetric outcome in low- and high- risk pregnancies: Predictive value of maternal serum screening. *Obstet Gynecol*, 94, 929-934.

[14] Chandra S, Scott H, Dodds L, Watts C, Blight C, Van Den HM. Unexplained elevated maternal serum alpha-fetoprotein and/or human chorionic gonadotropin and the risk of adverse outcomes. (2003) *Am J Obstet Gynecol*, 189, 775-781.

[15] Dugoff L, Hobbins JC, Malone FD, Vidaver J, Sullivan L, Canick JA, Lambert-Messerlian GM, Porter TF, Luthy DA, Comstock CH, Saade G, Eddleman K, Merkatz IR, Craigo SD, Timor-Tritsch IE, Carr SR, Wolfe HM, D'Alton ME. (2005) Quad screen as a predictor of adverse pregnancy outcome. *Obstet Gynecol*, 106, 260-267.

[16] Alleman BW, Smith AR, Byers HM, Bedell B, Ryckman KK, Murray JC, Borowski KS. (2013) A proposed method to predict preterm birth using clinical data, standard maternal serum screening, and cholesterol. *Am J Obstet Gynecol*, 208, 472 e471-411.

[17] Jelliffe-Pawlowski LL, Baer RJ, Blumenfeld YJ, Ryckman KK, O'Brodovich HM, Gould JB, Druzin ML, El-Sayed YY, Lyell DJ, Stevenson DK, Shaw GM, Currier RJ. (2015) Maternal characteristics and mid-pregnancy serum biomarkers as risk factors for subtypes of preterm birth. *BJOG*, 122, 1484-1493.

[18] Puntachai P, Wanapirak C, Sirichotiyakul S, Tongprasert F, Srisupundit K, Luewan S, Traisrisilp K, Tongsong T. (2015) Associations between pregnancy outcomes and unexplained high and low maternal serum alpha-fetoprotein levels. *Arch Gynecol Obstet*, 292, 81-85.

[19] Olsen RN, Woelkers D, Dunsmoor-Su R, Lacoursiere DY. (2012) Abnormal second-trimester serum analytes are more predictive of preterm preeclampsia. *Am J Obstet Gynecol*, 207, 228 e221-227.

[20] Tache V, Baer RJ, Currier RJ, Li CS, Towner D, Waetjen LE, Jelliffe-Pawlowski LL.(2014) Population-based biomarker screening and the development of severe preeclampsia in California. *Am J Obstet Gynecol*, 211, 377 e371-378.

[21] Koster EL, Dashe JS, McIntire DD, Ramus RM. (2004) Association of maternal serum alpha-fetoprotein with persistent placenta previa. *J Matern Fetal Neonatal Med*, 16, 3-7.

[22] Tikkanen M, Hamalainen E, Nuutila M, Paavonen J, Ylikorkala O, Hiilesmaa V.(2007) Elevated maternal second-trimester serum alpha-fetoprotein as a risk factor for placental abruption. *Prenat Diagn*, 27, 240-243.

[23] Blumenfeld YJ, Baer RJ, Druzin ML, El-Sayed YY, Lyell DJ, Faucett AM, Shaw GM, Currier RJ, Jelliffe-Pawlowski LL. (2014) Association between maternal characteristics, abnormal serum aneuploidy analytes, and placental abruption. *Am J Obstet Gynecol*, 211, 144 e141-149.

[24] Kupferminc MJ, Tamura RK, Wigton TR, Glassenberg R, Socol ML. (1993) Placenta accreta is associated with elevated maternal serum alpha-fetoprotein. *Obstet Gynecol*, 82, 266-269.

[25] Lyell DJ, Faucett AM, Baer RJ, Blumenfeld YJ, Druzin ML, El-Sayed YY, Shaw GM, Currier RJ, Jelliffe-Pawlowski LL. (2015) Maternal serum markers, characteristics and morbidly adherent placenta in women with previa. *J Perinatol*, 35, 570-574.

[26] Hung TH, Shau WY, Hsieh CC, Chiu TH, Hsu JJ, Hsieh TT. (1999) Risk factors for placenta accreta. *Obstet Gynecol*, 93, 545-550.

[27] World Health Organization: International statistical classification of diseases, and related health problems, 10th revision (ICD-10), 2nd ed. (2004) Geneva (Switzerland), World Health Organization.

[28] Joseph KS, Kinniburgh B, Hutcheon JA, Mehrabadi A, Dahlgren L, Basso M, Davies C, Lee L. (2015) Rationalizing definitions and procedures for optimizing clinical care and public health in fetal death and stillbirth. *Obstet Gynecol*, 125, 784-788.

[29] Wald NJ, Hackshaw AK, George LM. (2000) Assay precision of serum alpha fetoprotein in antenatal screening for neural tube defects and Down's syndrome. *J Med Screenl* 7, 74-77.

[30] Vohr BR, Wright LL, Poole WK, McDonald SA. (2005) Neurodevelopmental outcomes of extremely low birth weight infants <32 weeks' gestation between 1993 and 1998. *Pediatrics*, 116, 635-643.

[31] Tomashek KM, Shapiro-Mendoza CK, Davidoff MJ, Petrini JR. (2007) Differences in mortality between late-preterm and term singleton infants in the United States, 1995-2002. *J Pediatr*, 151, 450-456, 456 e451.

[32] Morken NH, Kallen K, Jacobsson B. (2007) Outcomes of preterm children according to type of delivery onset: A nationwide population-based study. *Paediatr Perinat Epidemiol*, 21, 458-464.

[33] Bastek JA, Srinivas SK, Sammel MD, Elovitz MA. (2010) Do neonatal outcomes differ depending on the cause of preterm birth? A comparison between spontaneous birth and iatrogenic delivery for preeclampsia. *Am J Perinatol*, 27, 163-169.

[34] Beta J, Bredaki FE, Rodriguez Calvo J, Akolekar R, Nicolaides KH. (2011) Maternal serum alpha-fetoprotein at 11-13 weeks' gestation in spontaneous early preterm delivery. *Fetal Diagn Ther*, 30, 88-93.
[35] Yuan W, Chen L, Bernal AL. (2009) Is elevated maternal serum alpha-fetoprotein in the second trimester of pregnancy associated with increased preterm birth risk? A systematic review and meta-analysis. *Eur J Obstet Gynecol Reprod Biol*, 145, 57-64.
[36] Jelliffe-Pawlowski LL, Baer RJ, Currier RJ. (2010) Second trimester serum predictors of preterm birth in a population-based sample of low-risk pregnancies. *Prenat Diagn*, 30, 727-733.
[37] Jelliffe-Pawlowski LL, Shaw GM, Currier RJ, Stevenson DK, Baer RJ, O'Brodovich HM, Gould JB.(2013) Association of early-preterm birth with abnormal levels of routinely collected first- and second-trimester biomarkers. *Am J Obstet Gynecol*, 208, 492 e491-411.
[38] Smith GC, Shah I, Crossley JA, Aitken DA, Pell JP, Nelson SM, Cameron AD, Connor MJ, Dobbie R. (2006) Pregnancy-associated plasma protein A and alpha-fetoprotein and prediction of adverse perinatal outcome. *Obstet Gyneco*l, 107, 161-166.
[39] Rudra CB, Williams MA. (2005) Monthly variation in preeclampsia prevalence: Washington state, 1987-2001. *J Matern Fetal Neonatal Med*, 18, 319-324.
[40] Khan KS, Wojdyla D, Say L, Gulmezoglu AM, Van Look PF. (2006) WHO analysis of causes of maternal death: A systematic review. *Lancet*, 367, 1066-1074.
[41] Duley L. (2009) The global impact of pre-eclampsia and eclampsia. *Semin Perinatol*, 33, 130-137.
[42] Sibai BM. (2011) Evaluation and management of severe preeclampsia before 34 weeks' gestation. *Am J Obstet Gynecol*, 205, 191-198.
[43] ACOG practice bulletin. (2002) Diagnosis and management of preeclampsia and eclampsia. Number 33, January 2002. American College of Obstetricians and Gynecologists. *Int J Gynaecol Obstet*, 77, 67-75.
[44] Steegers EA, von Dadelszen P, Duvekot JJ, Pijnenborg R. (2010) Pre-eclampsia. *Lancet*, 376, 631-644.
[45] von Dadelszen P, Magee LA, Roberts JM. (2003) Subclassification of preeclampsia. *Hypertens Pregnancy*, 22, 143-148.
[46] Metcalfe A, Langlois S, Macfarlane J, Vallance H, Joseph KS. (2014) Prediction of obstetrical risk using maternal serum markers and clinical risk factors. *Prenat Diagn*, 34, 172-179.
[47] Shenhav S, Gemer O, Sassoon E, Volodarsky M, Peled R, Segal S. (2002) Mid-trimester triple test levels in early and late onset severe pre-eclampsia. *Prenat Diagn*, 22, 579-582.
[48] Morris RK, Cnossen JS, Langejans M, Robson SC, Kleijnen J, Ter Riet G, Mol BW, van der Post JA, Khan KS. (2008) Serum screening with Down's syndrome markers to predict pre-eclampsia and small for gestational age: Systematic review and meta-analysis. *BMC Pregnancy Childbirth*, 8, 33.
[49] Vahanian SA, Lavery JA, Ananth CV, Vintzileos A.(2015) Placental implantation abnormalities and risk of preterm delivery: A systematic review and metaanalysis. *Am J Obstet Gynecol*, 213, S78-90.
[50] Delorme P, Goffinet F, Ancel PY, Foix-L'Helias L, Langer B, Lebeaux C, Marchand LM, Zeitlin J, Ego A, Arnaud C, Vayssiere C, Lorthe E, Durrmeyer X, Sentilhes L,

Subtil D, Debillon T, Winer N, Kaminski M, D'Ercole C, Dreyfus M, Carbonne B, Kayem G. (2016) Cause of preterm birth as a prognostic factor for mortality. *Obstet Gynecol*, 127, 40-8.

[51] Reddy UM, Rice MM, Grobman WA, Bailit JL, Wapner RJ, Varner MW, Thorp JM, Jr., Leveno KJ, Caritis SN, Prasad M, Tita AT, Saade GR, Sorokin Y, Rouse DJ, Blackwell SC, Tolosa JE. (2015) Serious maternal complications after early preterm delivery (24-33 weeks' gestation). *Am J Obstet Gynecol*, 213, 538 e531-539.

[52] Silver RM, Barbour KD. (2015) Placenta accreta spectrum: Accreta, increta, and percreta. *Obstet Gynecol Clin North Am*, 42, 381-402.

[53] Lal AK, Hibbard JU. (2015) Placenta previa: An outcome-based cohort study in a contemporary obstetric population. *Arch Gynecol Obstet*, 292, 299-305.

[54] Ananth CV, Lavery JA, Vintzileos AM, Skupski DW, Varner M, Saade G, Biggio J, Williams MA, Wapner RJ, Wright JD. (2015) Severe placental abruption: Clinical definition and associations with maternal complications. *Am J Obstet Gynecol.*

[55] Butler EL, Dashe JS, Ramus RM. (2011) Association between maternal serum alpha-fetoprotein and adverse outcomes in pregnancies with placenta previa. *Obstet Gynecol*, 97, 35-38.

[56] Dreux S, Salomon LJ, Muller F, Goffinet F, Oury JF, Sentilhes L.(2012) Second-trimester maternal serum markers and placenta accreta. *Prenat Diagn*, 32, 1010-1012.

[57] Zelop C, Nadel A, Frigoletto FD, Jr., Pauker S, MacMillan M, Benacerraf BR. (1992) Placenta accreta/percreta/increta: A cause of elevated maternal serum alpha-fetoprotein. *Obstet Gynecol*, 80, 693-694.

[58] Proctor LK, Toal M, Keating S, Chitayat D, Okun N, Windrim RC, Smith GC, Kingdom JC. (2009) Placental size and the prediction of severe early-onset intrauterine growth restriction in women with low pregnancy-associated plasma protein-A. *Ultrasound Obstet Gynecol*, 34, 274-282.

[59] Ozkaya E, Cakir E, Cinar M, Altay M, Gelisen O, Kara F. (2011) Second trimester serum alpha-fetoprotein level is a significant positive predictor for intrauterine growth restriction in pregnant women with hyperemesis gravidarum. *J Turk Ger Gynecol Assoc*, 12, 220-224.

[60] Critchfield AS, Paulus JK, Farez R, Urato AC. (2013) Abnormal analyte preeclampsia: Do the second-trimester maternal serum analytes help differentiate preeclampsia subtypes? *J Perinatol*, 33, 754-758.

[61] Mizejewski GJ. (2003) Levels of alpha-fetoprotein during pregnancy and early infancy in normal and disease states. *Obstet Gynecol Surv,* 58, 804-826.

[62] Mizejewski GJ. (2004) Biological roles of alpha-fetoprotein during pregnancy and perinatal development. *Exp Biol Med (Maywood)*, 229, 439-463.

[63] Mizejewski GJ. (2007) Physiology of alpha-fetoprotein as a biomarker for perinatal distress: Relevance to adverse pregnancy outcome. *Exp Biol Med (Maywood)*, 232, 993-1004.

[64] Salafia CM, Silberman L, Herrera NE, Mahoney MJ. (1988) Placental pathology at term associated with elevated midtrimester maternal serum alpha-fetoprotein concentration. *Am J Obstet Gynecol*, 158, 1064-1066.

[65] Greenwold N, Jauniaux E, Gulbis B, Hempstock J, Gervy C, Burton GJ. (2003) Relationship among maternal serum endocrinology, placental karyotype, and intervillous circulation in early pregnancy failure. *Fertil Steril*, 79, 1373-1379.

[66] Kutluer G, Cicek NM, Moraloglu O, Ertargin P, Sarikaya E, Artar I, Erdem O. (2012) Low VEGF expression in conceptus material and maternal serum AFP and beta-hCG levels as indicators of defective angiogenesis in first-trimester miscarriages. *J Turk Ger Gynecol Assoc*, 13, 111-117.

[67] Spong CY, Ghidini A, Walker CN, Ossandon M, Pezzullo JC. (1997) Elevated maternal serum midtrimester alpha-fetoprotein levels are associated with fetoplacental ischemia. *Am J Obstet Gynecol*, 177, 1085-1087.

[68] Jelliffe-Pawlowski L, Baer R, Moon-Grady AJ, Currier RJ. (2011) Second trimester serum predictors of congenital heart defects in pregnancies without chromosomal or neural tube defects. *Prenat Diagn,* 31, 466-472.

[69] Salafia CM, DeVore GR, Mainolfi E, Kelly J, Pezzullo JC, Rothlein R. (1993) Circulating intercellular adhesion molecule-1 in amniotic fluid, maternal serum alpha-fetoprotein levels, and intrauterine growth retardation. *Am J Obstet Gynecol*, 169, 830-834.

[70] Gross S, Castillo W, Crane M, Espinosa B, Carter S, DeVeaux R, Salafia C. (2003) Maternal serum alpha-fetoprotein and human chorionic gonadotropin levels in women with human immunodeficiency virus. *Am J Obstet Gyneco*l, 188, 1052-1056.

[71] Einstein FH, Wright RL, Trentacoste S, Gross S, Merkatz IR, Bernstein PS. (2004) The impact of protease inhibitors on maternal serum screening analyte levels in pregnant women who are HIV positive. *Am J Obstet Gynecol*, 191, 1004-1008.

[72] Combs CA, Garite TJ, Lapidus JA, Lapointe JP, Gravett M, Rael J, Amon E, Baxter JK, Brady K, Clewell W, Eddleman KA, Fortunato S, Franco A, Haas DM, Heyborne K, Hickok DE, How HY, Luthy D, Miller H, Nageotte M, Pereira L, Porreco R, Robilio PA, Simhan H, Sullivan SA, Trofatter K, Westover T. (2015) Detection of microbial invasion of the amniotic cavity by analysis of cervicovaginal proteins in women with preterm labor and intact membranes. *Am J Obstet Gynecol*, 212, 482 e481-482 e412.

[73] Muller F, Dreux S, Sault C, Galland A, Puissant H, Couplet G, Lemay C, Larcher ME, Renom G. (2003) Very low alpha-fetoprotein in down syndrome maternal serum screening. *Prenat Diagn*, 23, 584-587.

[74] Gagnon A, Wilson RD, Audibert F, Allen VM, Blight C, Brock JA, Desilets VA, Johnson JA, Langlois S, Summers A, Wyatt P. (2008) Obstetrical complications associated with abnormal maternal serum markers analytes. *J Obstet Gynaecol Can*, 30, 918-949.

[75] Khazardoost S, Mousavi S, Borna S, Hantoushzadeh S, Alavi A, Khezerlou N. (2014) Effect of aspirin in prevention of adverse pregnancy outcome in women with elevated alpha-fetoprotein. *J Matern Fetal Neonatal Med*, 27, 561-565.

[76] Spaggiari E, Ruas M, Dreux S, Valat AS, Czerkiewicz I, Guimiot F, Schmitz T, Delezoide AL, Muller F. (2013) Management strategy in pregnancies with elevated second-trimester maternal serum alpha-fetoprotein based on a second assay. *Am J Obstet Gynecol*, 208, 303 e301-307.

In: Alpha-Fetoprotein: Functions and Clinical Applications
ISBN: 978-1-63484-875-6
Editors: Nisha Lakhi and Michael Moretti

Chapter 12

ALPHA-FETOPROTEIN AND DISORDERS OF PLACENTATION: PATHOGENESIS AND CLINICAL IMPLICATIONS

Nisha Lakhi*[*], *Joanna Pessolano*, *Jane M. Ponterio* and *Michael Moretti

Department of Obstetrics and Gynecology,
Richmond University Medical Center, Staten Island, NY, US

ABSTRACT

It is well established in the literature that an abnormal elevation of maternal serum alpha-fetoprotein (MSAFP) that cannot be explained by either fetal aneuploidy, multiple gestation, incorrect dates or structural defects, is associated with a multitude of adverse pregnancy outcomes. These outcomes include pre-eclampsia, fetal growth restriction, preterm delivery, fetal demise, placental abruption, persistent placenta previa, and placenta accreta. The most likely unifying etiology is an inherent defect in placentation, probably occurring early in the pregnancy, that ultimately affects the distal stem villi of the placenta. Damage to the distal villi can originate from either an ischemic insult or a chronic inflammatory state. This is supported by pathological findings in placentas from pregnancies complicated by unexplained elevations of MSAFP. Common pathological lesions in these placentas include evidence of feto-maternal hemorrhage, increased fibrin deposition, decidual vasculopathy, and chronic villitis. These changes can damage the maternal-fetal-placental interface. This disruption of the placental interface is believed to allow increased diffusion of fetal AFP into the maternal serum, accounting for the unexplained elevations. This chapter strives to explain the pathological mechanisms accounting for unexplained elevated MSAFP, specifically focusing on disorders of placentation and implantation.

[*] Corresponding author: nlakhi@yahoo.com.

Introduction

The ability to use analytes from the maternal serum to detect structural malformation and chromosomal aneuploidy in the fetus has evolved over the past forty years. Starting in the 1970s, Wald and Cuckle first began using maternal serum alpha-fetoprotein (MSAFP) to screen for open neural tube defects [1]. Later, in 1984, Merkatz discovered a relationship between low levels of MSAFP and Down syndrome [2]. Today MSAFP is a component of second trimester screening which is routinely offered at 15-20 weeks gestation. It is used to identify women who may be at increased risk of delivering a fetus with chromosomal aneuploidy such as trisomy 18 and 21, or those with structural defects [3].

Elevated MSAFP levels are associated with pregnancies complicated by open neural tube defects, abdominal wall defects, cystic hygromas, and renal anomalies [4]. Elevated MSAFP can occur after fetal death, in multiple gestations, and when the fetal age is underestimated [4]. An unexplained elevation of MSAFP is typically defined as an AFP level greater than 2.5 multiple of the median (MoM), in the absence of multiple gestation, fetal aneuploidy, structural anomalies, or any of the above described situations that could justify the elevated level [3, 5].

Theoretical mechanisms accounting for unexplained elevations of MSAFP include placental vascular damage, feto-maternal hemorrhage, placental ischemia, and chronic inflammatory processes affecting the villous architecture. These pathological processes ultimately lead to disruption of the maternal-placental-fetal interface [6, 7]. The disruption of the placental interface allows for increased diffusion of AFP from the fetal circulation into the maternal serum, causing elevated levels [3, 7]. The relationship between unexplained second trimester elevations of MSAFP and adverse pregnancy outcomes have been well described. Adverse obstetrical outcomes that are associated with unexplained elevations of AFP in singleton gestations include pre-eclampsia, low birth weight, preterm birth, intrauterine growth restriction, placental abruption, and fetal or neonatal demise [8, 9, 10]. Abnormalities in placentation, including placenta accreta and placenta previa have been associated with elevated MSAFP [11, 12, 13].

Alpha Fetoprotein in Pregnancy

AFP is a 69 kDa single polypeptide fetal glycoprotein containing 3-5% carbohydrate and is associated with both ontogenic and oncogenic growth [3]. Synthesis of AFP in the yolk sack of the human embryo can be detected as early as 29 days post conception [14]. After the yolk sac involutes around the 9th gestational week, the fetal liver becomes the principal source of AFP production during development [15]. AFP synthesis by the fetal hepatocytes continues to increase until the 20th week of gestation when the level plateaus and remains fairly constant until the 32nd gestational week. After the 32nd week there is a decline in synthesis [16, 17]. Specifically, between the seventh and tenth week of gestation, fetal serum AFP levels increase 30-fold from 70 to 2000 micrograms per ml. They reach a maximum concentration in the fetal serum of approximately 3000 micrograms per ml by the 14th gestational week. From the 14th to the 32nd week of pregnancy, fetal AFP serum concentrations decline exponentially due to the dilutional effects of a rapid increase in fetal size [14]. The fetal AFP concentration continues to decline and at term reaches 20-120 micrograms per ml [14].

Serum fetal AFP is excreted into the fetal urine and enters the amniotic fluid. It is then transported to the maternal serum through trans-placental passage or by diffusion across fetal membranes [16, 17]. Amniotic fluid AFP levels are highest during the 14th-18th weeks of gestation, and then fall in parallel with the decline in serum fetal AFP concentration. Despite the decrease in fetal serum AFP concentration, the maternal serum AFP concentration continues to rise, reaching a peak at the 32nd gestational week, and then declining until parturition [17]. Between the 16th to the 18th gestational week, the normal MSAFP level is approximately 50 ng per ml, rising to 100 to 300 ng per ml at peak levels [14].

Unexplained Elevated Alpha-Fetoprotein and Adverse Pregnancy Outcomes

Approximately 1% of patients have an elevated maternal serum level that cannot be accounted for by incorrect dates, structural or chromosomal abnormalities, or multiple gestation [18]. Several studies have shown a relationship between unexplained elevated MSAFP levels in the second trimester and adverse pregnancy outcomes [18-24]. The reported complications include gestational hypertension, pre-eclampsia, intrauterine growth restriction (IUGR), fetal loss, pre-term delivery, placental abruption, placenta previa, and abnormal placental adherence, including placenta accreta, increta, and percreta [22].

These risks have been demonstrated by several large population based studies. Krause and colleagues analysed data of 77,149 women from a Danish registry, and found that women with MSAFP levels greater than or equal to 2.5 MoM had an increased risk of spontaneous abortion (RR 12.5; 95% CI 9.7,16.1), preterm birth (RR 4.8; 95% CI 4.1, 5.5), small for gestational age (RR 2.8; 95% CI 2.4, 3.2), low birth weight (RR 5.8; 95% CI 5.0, 6.6), and infant death (RR 1.9; 95% CI 1.2, 2.8) [20]. Cusick, et al. were able to demonstrate that the rate of an adverse pregnancy outcome is increased with higher elevations of MSAFP [23]. Stratified by MSAFP elevations of 2.0-2.49 MoM, 2.5-2.99 MoM and > or = 3.0 MoM, the rates of adverse pregnancy outcome were: (1) preterm birth: 14.3%, 15.6%, 20.3%; (2) small for gestational age at birth: 7.4%, 11.1%, 22.2%; and (3) perinatal deaths (neonatal and fetal): 2.6%, 3.3%, 5.6% respectively [23]. Similarly, in a population of 6,927 women in Birmingham, Alabama, Maher et al., demonstrated that as MSAFP levels increased, there was also an increase in the fetal death rate: (MoM less than 2.0, 11 per 1000; MoM 2-2.49, 29 per 1000; MoM 2.5 or greater, 95 per 1000; $P < .001$) [24].

Although the adverse obstetrical outcomes described seem to be varied, an underlying defect in placentation could be a unifying etiology. Therefore the understanding of placental structure and the trans-placental transport of AFP is critical.

Structure of the Placenta

The mature placenta is roughly a discoid organ, weighing approximately 500 g. Its surfaces are the chorionic plate that faces the fetus and to which the umbilical cord is attached, and the basal plate that abuts the maternal endometrium [25]. Between these plates is a cavity, the intervillous space, in which 30– 40 elaborately branched fetal villous trees project [25]. Each villous tree arises from a stem villus attached to the surface of the chorionic plate, and branches repeatedly to create a lobule 1–3 cm in diameter [25]. The center of a lobule is located over the opening of a maternal spiral artery originating from the basal plate. Maternal blood released at these openings percolates between the villous branches before draining into the uterine veins and exiting the placenta. Each lobule thus represents an independent maternal–fetal exchange unit. The final branches of the villous tree are the terminal villi, which are formed primarily from the 20th week of gestation onwards [25].

Villous cytotrophoblast stem cells differentiate along two possible pathways, forming either syncytiotrophoblasts or cytotrophoblasts. The cytotrophoblasts comprise the inner layer of the trophoblastic cells which are located interior to the syncytiotrophoblasts. Invasive cytotrophoblast, found in anchoring villi, break through the syncytiotrophoblast to form extravillous trophoblast columns that invade into the maternal decidua, eventually reaching the maternal spiral arteries [25]. Invasion of the spiral arteries by extravillous trophoblasts leads to transformation of these vessels into large bore conduits which are necessary to allow the increased maternal blood flow. The epithelial covering of the villous tree is the syncytiotrophoblast, which forms a true multinucleated syncytium. The syncytiotrophoblast shows regional variations in thickness around the villous surface, being very thin and devoid of organelles at the site of vasculo-syncytial membranes and thicker over non-vascular parts of the villous surface [25].

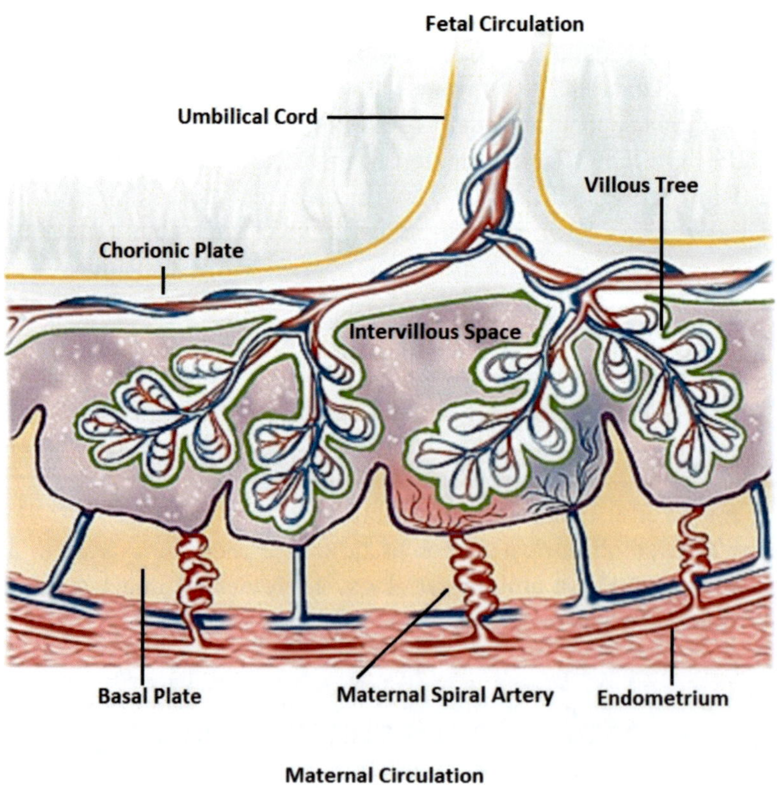

Figure 1. Placental Structure.

ALPHA-FETOPROTEIN AND PLACENTAL TRANSFER

There are four anatomical barriers for proteins traversing between the maternal and fetal circulations of the human placenta: the syncytiotrophoblasts, the cytotrophoblasts, the maternal capillary basement membrane, and the fetal capillary endothelium [26]. Although the absence of intercellular junctions in the syncytiotrophoblast layer suggests that exchange must take place through the apical and basal plasma membranes, there are exceptions. Both *in vivo* and *in vitro* studies have shown that the permeability of the human placenta to inert, hydrophilic tracers is proportional to their coefficients of free diffusion in water [27, 28], thereby demonstrating that transfer by paracellular diffusion can occur for a molecule such as AFP [29]. The anatomical correlate for the paracellular route of diffusion in the human placental villus is controversial since the syncytiotrophoblast is a true syncytium. However, discontinuities in the syncytiotrophoblast layer have been shown to exist on placental villi from uncomplicated pregnancies. These discontinuities are associated with fibrin containing fibrinoid deposits [30]. These areas are postulated to be a portal for transport of AFP from amniotic fluid to the maternal serum.

Brownbill, et al., investigated the mechanisms of AFP transfer across the human placenta by correlating AFP transfer with cytochemical localization of AFP immune reactivity in placentas obtained from uncomplicated pregnancies [26]. Using a monoclonal antibody, they localized AFP in two areas of the placenta: (1) fibrinoid deposits in regions of villi with discontinuities of the syncytiotrophoblast and (2) trophoblast cells in the decidua. From these cytochemical findings, the authors concluded that in uncomplicated pregnancies, two pathways for transfer of AFP into the maternal circulation exist: (1) Fetal AFP can pass from the fetal circulation into the villous core and traverse fibrinoid deposits at discontinuities in the syncytiotrophoblast layer to enter the maternal circulation, or (2) AFP present in the decidua basalis could enter blood vessels that traverse the basal plate and gain entry into the maternal circulation [26].

PLACENTAL FINDINGS ASSOCIATED WITH UNEXPLAINED ELEVATED MSAFP

1. Feto-Maternal Hemorrhage

Breakdown of the fetal-maternal-placental interface has been associated with elevated MSAFP in otherwise normal pregnancies [6]. Due to the high concentration gradient of AFP from the fetal compartment to the maternal compartment in early pregnancy, leakage of small amounts of fetal serum into maternal circulation can significantly raise the maternal serum level of AFP. Los, et al., used the Kleihauer-Betke test (test used to detect and quantify the presence of fetal hemoglobin in maternal serum), and demonstrated that feto-maternal transplacental hemorrhage was significantly more common in pregnant women with elevated serum MSAFP compared with pregnancies with normal MSAFP levels [31].

Placental hemorrhage has also been demonstrated by ultrasound in placentas with elevated MSAFP despite normal anatomic appearance of the fetus [33]. In a study of 25 pregnancies with unexplained elevated MSAFP, the incidence of sonographically demonstrated placental hemorrhage in women with elevated AFP was 64% (16/25), compared to 3.6% in a control group of 112 patients with normal serum AFP [33]. Of the sixteen patients with elevated AFP with sonographically evident hemorrhage, 8 were retroplacental and 8 were subchorionic [33]. The elevated maternal serum level of AFP may be attributed to feto-maternal admixture occurring after retroplacental hemorrhage near the basal plate or subchorionic hemorrhage immediately beneath the chorionic plate. However, since not all pregnancies associated with sonographically demonstrated hemorrhage are associated with elevated MSAFP, a more complex mechanism is involved. In fact, Lachmen, et al., reported negative Kleihauer-Betke test in patients with sonographically demonstrated retroplacental hemorrhage [33]. Therefore, it is more likely that an anatomical disruption in either the basal or chorionic plate may contribute to increased diffusion of AFP from the fetal to maternal compartment, resulting in an elevated concentration.

This hypothesis was further supported by a prospective cohort study of 95 singleton pregnancies complicated with unexplained elevated MSAFP [34]. In this study, women with elevated MSAFP had an increased incidence of placental sonolucencies, positive second trimester Kleihauer-Betke test, pre-term delivery, fetal growth restriction, and fetal death compared to the control group. Interestingly, in patients with elevated MSAFP levels, neither the presence of sonolucencies nor a positive Kleihauer-Betke test correlated with a history of first trimester vaginal bleeding or the quantitative MSAFP level. This study also demonstrated that in the sub-group of women with elevated MSAFP, a positive Kleihauer-Betke test or the presence of placental sonolucencies did not further increase the risk of adverse pregnancy complications [34]. These findings are again suggestive that a breakdown in the placental interface, rather than a simple admixture of fetal blood into the maternal serum resulting from placental hemorrhage, is responsible for unexplained elevations in MSAFP associated with adverse obstetrical outcomes.

2. Vascular/Ischemic Lesions

Unexplained elevations of MSAFP were correlated to placental ischemia and neovascularization in a study by Spong, et al. [7]. This study measured angiogenin, a marker of tissue ischemia and a potent inducer of neovascularization, in the amniotic fluid of patients with an unexplained elevated second trimester MSAFP. Significantly elevated levels of amniotic fluid angiogenin were noted in pregnancies that had elevated MSAFP as compared to normal controls. The authors attributed the elevation of angiogenin secondary to neovascularization induced by an ischemic insult early in the pregnancy.

Placental malperfusion, secondary to abnormal spiral artery development and flow, can lead to ischemic pathological changes. Developmental abnormalities of the maternal stromal-vascular compartment of the placenta involve defects in appropriate differentiation and expansion of trophoblasts [35]. Inappropriate development of the extravillous population of trophoblastic cells can lead to shallow implantation and lack of maternal spiral artery remodelling [35]. Maternal placental malperfusion that develops as a result of abnormal spiral artery flow is classified into two categories.

The first type is global, and is secondary to decreased maternal perfusion that globally affects the entire placenta. Accelerated villous maturation is the term used to describe all histological changes seen in areas of the villous tree deprived of the low velocity, high volume maternal flow which is characteristic of normal placentas [35]. Specific findings include increased syncytial knots, intravillous fibrin deposition, and villous paucity. The second form of maternal placental malperfusion segmentally affects the placenta and is due to a complete villous infarct that involves a portion of the placenta, leading to ischemic necrosis overlying an occluded spiral artery [35]. Pathological changes associated with both forms have been noted in placentas from patients with elevated levels of MSAFP [36].

Alkazaleh, et al., pathologically analysed placentas from 50 singleton pregnancies with both elevated AFP (>3.5 MoM) and human chorionic gonadotrophin (>5.3 MoM) and normal fetal anatomical ultrasound findings [36]. Signs of utero-placental insufficiency were identified in 88% of the placentas, with only 12% considered normal. Decidual vasculopathy and accelerated villous maturation were noted in 53% and 44% of placentas respectively. Decidual vasculopathy describes pathological findings seen in the spiral arteries in pre-eclampsia. Morphologically, decidual vasculopathy is characterized by fibrinoid necrosis and foamy macrophages within the vessel walls. Other lesions consistent with thrombo-occlusive disease, including focal or massive perivillous fibrin deposition, intervillous thrombosis, and villous infarction were present. Excessive perivillous fibrin deposition was the most common finding in 63% of the placentas [36].

Fibrinoid deposits occur at discontinuities in syncytiotrophoblasts and are a common histologic feature of villi in term human placentas [37, 38]. Fibrin deposits have been noted in placentas from pregnancy-induced hypertension and pre-eclampsia patients. This was demonstrated in a study by Knafer, et al., that compared term placenta of pre-eclamptic and hypertensive patients to those from uncomplicated pregnancies [38]. The authors noted that the percentage of villi with fibrinoid deposits was higher in the pre-eclamptic/hypertensive placentas compared with the controls: 13.2% versus 6.75% ($p < 0.001$), as well as the total amount of perivillous fibrinoid deposits, 4.8 versus 1.5 deposits ($p = 0.04$) [38]. Similar placental pathology findings were noted in a more recent study by Orabona, et al., [39]. Both studies did not attempt to correlate findings with MSAFP levels. However as demonstrated by the Brownbill study described earlier, areas of fibrin/fibrinoid deposition may be one paracellular route for diffusion of hydrophilic molecules, such as AFP, from the maternal to fetal circulation [26].

3. Chronic Villitis

Villitis of unknown etiology (VUE) is a T-cell mediated inflammatory disorder affecting the distal villous tree. Histologically, it represents a destructive inflammatory lesion characterized by the infiltration of maternal T cells into the chorionic villi [40]. VUE is caused by CD-8 positive maternal T lymphocytes that inappropriately gain access into the villous stroma and activate fetal antigen-presenting cells (Hofbauer cells). Activation of Hofbauer cells leading to the production of various cytokines has been implicated in the destruction of the villous architecture. VUE is thought to represent a breach of immune tolerance, resembling a maternal graft-versus-host type response to fetal antigens in the placenta [35].

VUE can be found affecting 5% to 15% of term placentas, and is often low grade and sub-clinical in nature [35, 40]. However, extensive involvement of the placenta by VUE has been associated with fetal growth restriction, spontaneous preterm delivery, small-for-gestational-age fetuses, pre-eclampsia and fetal death [40]. Studies have reported unexplained elevations of MSAFP associated with this chronic villitis [41, 42].

Salafia, et al., found that MSAFP elevations occurred with greater frequency in association with VUE than in placentas without pathological findings. (10.1% versus 3.4%, p=0.02) [41]. The placentas with VUE were examined for the grade of villitis, which is a measure of the index of severity of the inflammatory process. Sixty of 80 (75%) placentas with VUE and normal MSAFP demonstrated grade 1 or mild villitis [41]. Of the nine placentas with VUE and elevated MSAFP, five had grade 2 or greater changes ($p = 0.075$) [41]. This suggests that more severe villous inflammation may be associated with elevations in MSAFP. Since either an ischemic insult to the placenta or VUE can lead to damage of villus architecture of the placenta, the mechanism accounting for MSAFP elevation may be similar.

Defects of Placental Implantation Associated with Elevated MSAFP

1. Placental Abruption

Placental abruption is defined as the premature separation of the placenta before the delivery of the fetus. The loss of maternal vascular integrity can occur secondary to maldevelopment of the maternal spiral arteries due to ischemia-reperfusion or atherosis [35]. This can lead to high pressure flow through the maternal spiral arteries, indentation of the basal plate, and extension into the intervillous space [35]. Although placental abruption is a third trimester diagnosis, it is likely to result from an underlying defect in placentation occurring in early pregnancy. It is possible that prior to the acute phase of abruption, poorly implanted spiral arteries may rupture and allow for small leakage of fetal serum into the maternal compartment, accounting for the elevated AFP in the maternal serum. Additionally a poorly perfused placenta with ischemic injury could lead to disruption of the placental interface.

Tikkanen, et al., conducted a retrospective case-control review demonstrating that MSAFP remained an independent risk factor for placental abruption when adjusting for other risk factors [8]. MSAFP $\geq$ 1.5 MoMS had a sensitivity of 29% and a false-positive rate of 10% to predict abruption. In a large population based study by Blumenfeld, et al., elevated second trimester MSAFP greater than the 95% percentile was independently associated with placental abruption, even in the absence of hypertensive disorders or other risk factors [43].

2. Placenta Accreta

The term placenta accreta characterizes a spectrum of disorders in which the trophoblasts invade the myometrium without intervening decidua [44]. Myometrial invasion by the trophoblasts is considered to be pathologic. Under normal circumstances the myometrium is not reached because the trophoblasts stop invading the uterus when Nitabuch's layer is reached. Nitabuch's layer is made up of eosinophilic fibrinoid material that separate the cytotrophoblasts from the compact endometrium, and serves as a plane of cleavage at the time of placental separation [45]. The term placenta accreta is a general term used to describe the entire spectrum of conditions, including increta, percreta, and morbidly adherent placenta [45]. If the trophoblasts invade the myometrium it is called placenta increta, and if the trophoblasts invade through the myometrium and serosa, it is termed placenta percreta. The exact mechanism for elevated MSAFP secondary to placenta accreta is not fully understood. It is possible that pathological myometrial invasion by the trophoblasts might compromise the placental interface, lead to ischemic changes, or allow for passage of fetal serum to enter the maternal circulation.

Several studies have demonstrated an association of elevated MSAFP and placenta accreta [12, 13, 46]. Both Kupfermincm et al., and Zelop et al., retrospectively analysed serum MSAFP levels in patients that underwent cesarean hysterectomies for placenta accreta, percreta, or increta and matched them to controls that did not have any evidence of adherent placenta [12, 13]. Both studies indicated a significant association between elevated MSAFP and placenta accreta. Lyell and colleagues analyzed 37 cases of placenta accreta and found that women with MSAFP greater than the 95th percentile (1.79 MoM), had a nearly 3-fold increased risk for placenta accreta [46].

3. Placenta Previa

Placenta previa is a condition in which the placenta overlies or is in close proximity to the internal cervical os. The incidence is approximately 1 in 200 pregnancies at term. It is unclear why some placentas implant in the lower uterine segment rather than the fundus. However, it is believed that uterine scarring maybe a predisposing factor [44]. With the progression of pregnancy, more than 90% of low lying placentas will appear to move away from the cervix towards the fundus [45]. The apparent movement of the placenta towards the fundus is through a process known as trophotropism [45]. That is, the placenta migrates towards the better vascularized fundus, and the portion overlying the less vascularized cervix and lower uterine segment undergoes atrophy. As there is impaired blood supply to the lower uterine segment compared to the uterine fundus, it is possible that this may lead to deficient decidualization in low-lying placentas [45]. This may be a factor leading to disruption of the placental interface and subsequent elevation in MSAFP.

Studies have shown that MSAFP is more likely to be elevated in cases of persistent placenta previa. Koster et al., showed an association between increasing MSAFP values and greater likelihood of persistent placenta praevia at term [11]. This study analyzed 275 women with placenta previa at 15-20 weeks gestation and correlated with MSAFP levels. Women with a normal AFP (<1.0 MoM) had a 65% reduction in the risk of persistence compared with all cases of placenta previa that were diagnosed. In contrast, women with AFP > 1.0 MoM were 16% more likely to have persistent placenta previa at term [11].

A study by Butler and colleagues that compared a cohort of women with placenta previa and elevated MSAFP to a control group with placenta previa but normal MSAFP showed an increased risk of antepartum bleeding before 30 weeks gestation (50% vs 15%, $p = 0.002$), delivery before 30 weeks gestation (29% vs 5%, $p = 0.004$), and pregnancy induced hypertension (14% vs 0%, $p < 0.001$) in the cohort with elevated MSAFP [47]. Sixty-seven percent of patients with placenta previa and MSAFP greater than 2.0 MoM had at least one of those outcomes, and when 3.0 MoM was used as a cutoff, 100% of patients had at least one of the outcomes [47]. Other studies have also demonstrated a significantly increased risk of placenta accreta if both placenta previa and elevated MSAFP were present [13, 46].

CONCLUSION

The adverse obstetrical outcomes associated with unexplained elevated MSAFP, although diverse in their presentation, share a common etiology: a defect in placentation. Disruption to the placental interface, by either defective implantation, ischemic insult, or chronic inflammation, allows for increased diffusion of AFP from the fetal compartment to the maternal serum. This is the most probable mechanism accounting for an unexplained elevation of MSAFP.

Whether the elevated MSAFP can be used as a population-based screening test to detect pregnancy complications is still questionable as the sensitivity is low, and the false positive rate is high. No surveillance protocol has been established and proven effective to alter outcomes [3, 9].

Further research to investigate prediction models that incorporate MSAFP in combination with other serum analytes, angiogenesis markers, or markers associated with oxidative stress, requires further study. Addition of other modalities such as Doppler and ultrasound assessment may improve detection rates in pregnancies with unexplained MSAFP that are destined for adverse obstetrical outcomes [3].

REFERENCES

[1] Wald, N. J., Cuckle, H. S., Boreham, J., Brett, R., Stirrat, G. M., Bennett, M. J., Turnbull, A. C., Solymar, M., Jones, N., Bobrow, M. and Evans, C. J. (1979). Antenatal screening in Oxford for fetal neural tube defects. *Br J Obstet Gynaecol*, *86*, 91-100.

[2] Merkatz, I. R., Nitowsky, H. M., Macri, J. N., et al., (1984). An association between low maternal serum alpha-fetoprotein and fetal chromosomal abnormalities. *Am J Obstet Gynecol*, *148*, 886-894.

[3] Lakhi, N., Govind, A., Moretti, M. and Jones, J. (2012). Abnormal Maternal Serum Analytes as a Marker for Adverse Obstetrical Outcome. *The Obstetrician and Gynaeclogist, Official Publication of Royal College of Obstetricians and Gynaecologists*, Fall 20-26.

[4] Hong, S., Berkowitz, G., Wang, W., Stone, J. and Ainbender, E. (1996). Unexplained elevated maternal serum alpha-fetoprotein levels and pregnancy outcome in twins. *Obstet Gynecol*, *88*, 337-42.

[5] Mizejewski, G. J. (2007). Physiology of Alpha-Fetoprotein as a Biomarker for perinatal Distress: Relevance to Adverse Pregnancy Outcome. *Exp Biol Med (Maywood)*, *232*, 993-1004.

[6] Berkeley, A. S., Killackey, M. A. and Cederqvist, L. L. (1983). Elevated maternal serum alpha-fetoprotein levels associated with breakdown in fetal-maternal-placental barrier. *Am J Obstet Gynecol*, *146*, 859-61.

[7] Spong, C. Y., Ghidini, A., Walker, C. N., Ossandon, M. and Pezzullo, J. C. (1997). Elevated maternal serum midtrimester alpha-fetoprotein levels are associated with fetoplacental ischemia. *Am J Obstet Gynecol*, *177*, 1085-7.

[8] Tikkanen, M., Hämäläinen, E., Nuutila, M., Paavonen, J., Ylikorkala, O. and Hiilesmaa, V. (2007). Elevated maternal second-trimester serum alpha-fetoprotein as a risk factor for placental abruption. *Prenat Diagn*, *27*, 240-3.

[9] Dugoff, L., Hobbins, J. C., Malone, F. D., Vidaver, J., Sullivan, L., Canick, J. A., Lambert-Messerlian, G. M., Porter, T. F., Luthy, D. A., Comstock, C. H., Saade, G., Eddleman, K., Merkatz, I. R., Craigo, S. D., Timor-Tritsch, I. E., Carr, S. R., Wolfe, H. M. and D' Alton, M. E. (2005). Quad screen as a predictor of adverse pregnancy outcome. *Obstet Gynecol*, *106*, 260-7.

[10] Gagnon, A., Wilson, R. D., Audibert, F., Allen, V. M., Blight, C., Brock, J. A., Désilets, V. A., Johnson, J. A., Langlois, S., Summers, A. and Wyatt, P. (2008). Society of Obstetricians and Gynaecologists of Canada Genetics Committee. Obstetrical complications associated with abnormal maternal serum markers analytes. *J Obstet Gynaecol Can*, *30*, 918-49.

[11] Koster, E. L., Dashe, J. S., McIntire, D. D. and Ramus, R. M. (2004). Association of maternal serum alpha-fetoprotein with persistent placenta previa. *J Matern Fetal Neonatal Med*, *16*, 3-7.

[12] Kupferminc, M. J., Tamura, R. K., Wigton, T. R., Glassenberg, R. and Socol, M. L. (1993), Placenta accrete is associated with elevated maternal serum alpha-fetoprotein. *Obstet Gynecol*, *82*, 266-9.

[13] Zelop, C., Nadel, A., Frigoletto, F. D., Jr. Pauker, S., MacMillan, M. and Benacerraf, B. R. (1992). Placenta accreta/percreta/increta: a cause of elevated maternal serum alpha-fetoprotein. *Obstet Gynecol*, *80*, 693-4.

[14] Yachnin, S. (1978). The clinical significance of human alpha-fetoprotein. *Ann Clin Lab Sci*, *8*, 84-90.

[15] Gitlin, D., Perricelli, A. and Gitlin, G. M. (1972). Synthesis of -fetoprotein by liver, yolk sac, and gastrointestinal tract of the human conceptus. *Cancer Res*, *32*, 979–82.

[16] Jones, E. A., Clement-Jones, M., James, O. F. and Wilson, D. I. (2001). Differences between human and mouse alpha-fetoprotein expression during early development. *J Anat*, *198*, 555–9.

[17] Mizejewski, G. J. (2003). Levels of alpha-fetoprotein during pregnancy and early infancy in normal and disease states. *Obstet Gynecol Surv*, *58*, 804–26.

[18] Chandra, S., Scott, H., Dodds, L., Watts, C., Blight, C. and Van Den, H. M. (2003). Unexplained elevated maternal serum alpha-fetoprotein and/or human chorionic gonadotropin and the risk of adverse outcomes. *Am J Obstet Gynecol*, *189*, 775–81.

[19] van Rijn, M., van Der Schouw, Y. T., Hagenaars, A. M., Visser, G. H. and Christiaens, G. C. (1999). Adverse obstetric outcome in low- and high-risk pregnancies: predictive value of maternal serum screening. *Obstet Gynecol*, *94*, 929–34.

[20] Krause, T. G., Christens, P., Wohlfahrt, J., Lei, U., Westergaard, T., Norgaard-Pedersen, B., et al., (2001). Second-trimester maternal serum alpha-fetoprotein and risk of adverse pregnancy outcome. *Obstet Gynecol*, *97*, 277–82.

[21] Spencer, K. (2000). Second-trimester prenatal screening for Down syndrome and the relationship of maternal serum biochemical markers to pregnancy complications with adverse outcome. *PrenatDiagn*, *20*, 652–6.

[22] Dug off, L. (2010). First- and second-trimester maternal serum markers for aneuploidy and adverse obstetric outcomes. *Obstet Gynecol*, *115*, 52-61.
[23] Cusick, W., Rodis, J. F., Vintzileos, A. M., Albini, S. M., McMahon, M. and Campbell, W. A. (1996). Predicting pregnancy outcome from the degree of maternal serum alpha-fetoprotein elevation. *J Reprod Med*, *41*(5), 327-32.
[24] Maher, J. E., Davis, R. O., Goldenberg, R. L., Boots, L. R. and Du Bard, M. B. (1994). Unexplained elevation in maternal serum alpha-fetoprotein and subsequent fetal loss. *Obstet Gynecol*, *83*, 138-41.
[25] Burton, G. J. and Fowden, A. L. (2015). The placenta: a multifaceted, transient organ. *Philos Trans R Soc Lond B BiolSci*, *370*, 1663.
[26] Brownbill, P., Edwards, D., Jones, C., Mahendran, D., Owen, D., Sibley, C., Johnson, R., Swanson, P. and Nelson, D. M. (1995). Mechanisms of alphafetoprotein transfer in the perfused human placental cotyledon from uncomplicated pregnancy. *J Clin Invest*, *96*, 2220-6.
[27] Bain, M. D. D. K., Copas, M. J. and Landon Stacey, T. E. (1988). In vivo permeability of the human placenta to inulin and mannitol. *J. Physiol*, *399*, 313-319.
[28] Schneider, H., Sodha, R. J., Progler, M. and Young, M. P. A. (1985). Permeability of the human placenta for hydrophilic substances studied in the isolated dually in vitro perfused lobe. *Contrib. Gynecol. Obstet*, *13*, 98-103.
[29] Edwards, D., Jones, C. J. P., Sibley, C. P. and Nelson, D. M. (1993). Paracellular permeability pathways in the human placenta: a quantitative and morphological study of maternal fetal transfer of horseradish peroxidase. *Placenta*, *14*, 63-74.
[30] Nelson, D. M., Crouch, E. C., Curran, E. M. and Farmer, D. R. (1990). Trophoblast interaction with fibrin matrix. Epithelialization of perivillous fibrin deposits as a mechanism for villous repair in the human placenta. *Am. J. Pathol*, *136*, 855-865.
[31] Los, F. J., de Wolf, B. T. and Huisjes, H. J. (1979). Raised maternal serum-alpha-fetoprotein levels and spontaneous fetomaternal transfusion. *Lancet*, *2*, 1210-2.
[32] Fleischer, A. C., Kurtz, A. B., Wapner, R. J., Ruch, D., Sacks, G. A., Jeanty, P., Shah, D. M. and Boehm, F. H. (1988). Elevated alpha-fetoprotein and a normal fetal sonogram: association with placental abnormalities. *AJR Am J Roentgenol*, *150*, 881-3.
[33] Lachman, E., Hingley, S. M., Bates, G., Ward, A. M., Stewart, C. R. and Duncan, S. L. (1977). Detection and measurement of fetomaternal haemorrhage: serum alpha-fetoprotein and the Kleihauer technique. *Br Med J*, *28*, 1377-9.
[34] Bernstein, I. M., Barth, R. A., Miller, R. and Capeless, E. L. (1992). Elevated maternal serum alpha-fetoprotein: association with placental sonolucencies, fetomaternal hemorrhage, vaginal bleeding, and pregnancy outcome in the absence of fetal anomalies. *Obstet Gynecol*, *79*, 71-4.
[35] Redline, R. W. (2015). Classification of placental lesions. *Am J Obstet Gynecol*, *213*(4 Suppl), S21-8.
[36] Alkazaleh, F., Chaddha, V., Viero, S., Malik, A., Anastasiades, C., Sroka, H., Chitayat, D., Toi, A., Windrim, R. C. and Kingdom, J. C. (2006). Second-trimester prediction of severe placental complications in women with combined elevations in alpha-fetoprotein and human chorionic gonadotrophin. *Am J Obstet Gynecol*, *194*, 821-7.
[37] Stevens, D. U., Al-Nasiry, S., Bulten, J. and Spaanderman, M. E. Decidual vasculopathy and adverse perinatal outcome in preeclamptic pregnancy. *Placenta*, *33*, 630-3.

[38] Kanfer, A., Bruch, J. F., Nguyen, G., He, C. J., Delarue, F., Flahault, A., Nessmann, C. and Uzan, S. (1996). Increased placental antifibrinolytic potential and fibrin deposits in pregnancy-induced hypertension and preeclampsia. *Lab Invest*, *74*, 253-8.

[39] Orabona, R., Donzelli, C. M., Falchetti, M., Santoro, A., Valcamonico, A. and Frusca, T. (2015). Placental histological patterns and uterine artery Doppler velocimetry in pregnancies complicated by early and late pre-eclampsia. *Ultrasound Obstet Gynecol*, 2015.

[40] Kim, C. J., Romero, R., Chaemsaithong, P. and Kim, J. S. (2015). Chronic inflammation of the placenta: definition, classification, pathogenesis, and clinical significance. *Am J Obstet Gynecol*, *213*, S53-69.

[41] Salafia, C. M., Silberman, L., Herrera, N. E. and Mahoney, M. J. (1988). Placental pathology at term associated with elevated midtrimester maternal serum alpha-fetoprotein concentration. *Am J Obstet Gynecol*, *158*, 1064–1066.

[42] Ho, M., Faye-Petersen, O. M., Goldenberg, R. L., Carlo, W. A., Cliver, S. P. and Andrews, W. W. (2012). Elevated midtrimester α-fetoprotein and delivery markers of inflammation in a preterm population. *J Matern Fetal Neonatal Med*, *25*, 2424-7.

[43] Blumenfeld, Y. J., Baer, R. J., Druzin, M. L., El-Sayed, Y. Y., Lyell, D. J., Faucett, A. M., Shaw, G. M., Currier, R. J. and Jelliffe-Pawlowski, L. L. (2014). Association between maternal characteristics, abnormal serum aneuploidy analytes, and placental abruption. *Am J Obstet Gynecol*, *211*, 144, e1-9.

[44] Oyelese, Y. and Smulian, J. C. (2006). Placenta previa, placenta accreta, and vasa previa. *Obstet Gynecol*, *107*, 927-41.

[45] Silver, R. M. (2015). Abnormal Placentation: Placenta Previa, Vasa Previa, and Placenta *Accreta. Obstet Gynecol*, *126*, 654-68.

[46] Lyell, D. J., Faucett, A. M., Baer, R. J., Blumenfeld, Y. J., Druzin, M. L., El-Sayed, Y. Y., Shaw, G. M., Currier, R. J. and Jelliffe-Pawlowski, L. L. (2015). Maternal serum markers, characteristics and morbidly adherent placenta in women with previa. *J Perinatol*, *35*, 570-4.

[47] Butler, E. L., Dashe, J. S. and Ramus, R. M. (2001). Association between maternal serum alpha-fetoprotein and adverse outcomes in pregnancies with placenta previa. *Obstet Gynecol*, *97*, 35-8.

SECTION 4: ALPHA-FETOPROTEIN: LIGAND BINDING, RECEPTORS, AND DRUGS

In: Alpha-Fetoprotein: Functions and Clinical Applications ISBN: 978-1-63484-875-6
Editors: Nisha Lakhi and Michael Moretti

Chapter 13

ALPHA-FETOPROTEIN STRUCTURES, FUNCTIONS AND BINDING CHEMICALS

Sugunadevi Sakkiah, PhD, Hui Wen Ng, PhD, Hao Ye, PhD, Heng Luo, PhD, Weigong Ge, MSc, Weida Tong, PhD, and Huixiao Hong*, PhD

National Center for Toxicological Research,
US Food and Drug Administration, Jefferson, AK, US

ABSTRACT

Alpha-fetoprotein (AFP), a well-known oncofetal protein, is a member of the albuminoid gene superfamily and is very similar to the albumin protein. It functions as a serum binding and transport protein of chemicals. AFP is one of the essential proteins in the mammalian development and is produced in the yolk sac and the liver of the developing fetus. AFP has a U-shaped structure and consists of three domains. AFP has served as a serum biomarker for liver, gonads, and gastrointestinal tract cancers. AFP was investigated as an activator for cell surface receptors. It acts as a regulator of cytoplasmic transcription factors in signaling pathways. AFP can bind chemicals with diverse structures. In this chapter, the structure of AFPs will be reviewed; the functions and clinical applications will be discussed and the chemicals known for AFP binding will be summarized.

1. INTRODUCTION

Alpha-fetoprotein (AFP), one of the most widely studied oncofetal proteins [1], was discovered by Bergstrand and Czar in serum which migrated between albumin and α1-globulin [2]. AFP is produced during the early development in mammals at various stages of

* Correspondence should be addressed to Dr. Huixiao Hong at huixiao.hong@fda.hhs.gov.
Address: Building 5C Room 109, National Center for Toxicological Research, U.S. Food and Drug Administration, 3900 NCTR Road, Jefferson, AR 72079, USA. Tel: +1 870 543 7296. Fax: +1 870 543 7854.

ontogenesis like embryonic, fetal, and neoplastic growth [3]. During evolution the main site of AFP synthesis has shifted from fetal stomach in sharks to the yolk sac in birds and to the liver in mammals [4, 5]. AFP was identified as an embryonic protein [6]. It was discovered in the amniotic fluid, liver tissue extracts, and blood serum of mouse embryos [6, 7]. It was reported that the patients with liver cancer contain elevated level of α-globulin in blood [8]. In the early 1970s, Brock and Sutcliffe correlated the elevated levels of AFP in human amniotic fluids and maternal serum with the neural tube defects in the fetus [9, 10]. AFP was abundantly identified and characterized in rabbit, mouse, rat, guinea, pig, cow, sheep, goats, pigs and monkeys [11]. In 1972, Uriel et al. elucidated the binding of steroidal chemical properties of rat AFP and proved that the rat serum from embryos, neonatal and pregnant rat plasma contains a feto-specific protein that is capable of binding to estrogen [12]. Human AFP gene was employed as a carrier for anticancer drugs and can act as a carrier for a multitude of ligands such as fatty acids, steroids, retinoids, heavy metals, drugs and dyes [13, 14]. AFP can bind to the key proteins involved in the growth or apoptosis signal pathways and act as an intracellular signaling molecule [13, 15].

AFP is a member of the albuminoid gene family that consists of four members: albumin (ALB), vitamin D-binding (Gc) protein (DBP), AFP and α-ALB (αALB) [16, 17]. The cysteine residues in AFP play a vital role to maintain the U-shaped molecular structure by forming a disulfide bridge resulting in a triplet domain [11]. The ALB gene family members display structural similarities, homologous amino acid sequence and similar cysteine disulfide bridge cluster. DBP alone is truncated in the third domain which results in a protein with a smaller molecular mass [18]. The αALB was discovered in both rodent and human [19]. All gene members are mainly capable of ligand/carrier transport function, but display a vast array of other functions, including chemotaxis, oxygen free radical scavenging, esterase activity, leukocyte adherence, copper stimulated and lipid peroxidation [20, 21].

AFP is similar to albumin and functions as a serum binding/transport protein of fatty acids, steroids, retinoic acid, heavy metals and various drugs [22]. AFP and albumin are produced during different phases of cell cycle and synthesized during G1 and S phase and from mid-S through G2, respectively [22]. AFP, distinguished from albumin and other blood protein carrier/transport molecule, binds with the estrogen and plays a role in sexual differentiation of the brain by protecting the fetus from the effects of circulating estrogen [23]. AFP has been assumed to be essential for mammalian development due to its high level of expression during embryonic development which decreases dramatically in the adult [24, 25]. In adults, the human AFP gene is silenced by methylation processes and reappears only in instances of hepatic damage/regeneration, hepatomas and germ cell cancers [26]. Only the adult hepatic oval cells surrounding bile ducts and blood vessels secrete human AFP; in contrast, liver parenchymal cells secrete little if any of the fetal protein. High serum level of AFP was detected with hepatoma, germ cell tumors, neoplastic and non-neoplastic patients [4, 22, 27].

AFP acts as a dual regulator of growth, enhancement and inhibition, in cancer and in fetal cells [28]. Extensive studies have been conducted to elucidate the AFP molecular functions such as in fetus development, as a serum marker in cancer and for detection of fetal and congenital defects [1]. AFP has attracted the attention of many scientists and researchers. More recent studies have demonstrated that AFP is capable of both oncogenic and ontogenetic growth regulation, and that such modulation might be attributed to transcriptional enhancement and/or suppression activities [29, 30]. This book chapter will briefly describe

AFP structure and function and the role of small molecules that can bind with AFP to alter its functions.

2. AFP STRUCTURE

2.1. Primary Structure

AFP is present in the fetal cells of various species like human [31], rat [32], mouse [33], chicken [34] and baboon [35]. The primary structure of inactive human AFP consists of 609 amino acids. Removal of the 18 amino acid signal peptide from the inactive form of AFP yields the mature protein (591 amino acids) [36-38].

AFP is a major fetal transport protein with a molecular mass ranging from 66 to 72 kDa [39]. It consists of three homologous domains (Domain I: amino acids 2-187, Domain II: amino acids 194-379, and Domain III: amino acids 386-577) as shown in Figure 1. Each domain consists of two globular subdomains (IA, IB, IIA, IIB, IIIA, and IIIB) which are linked by 15 regularly distributed disulfide bonds [6].

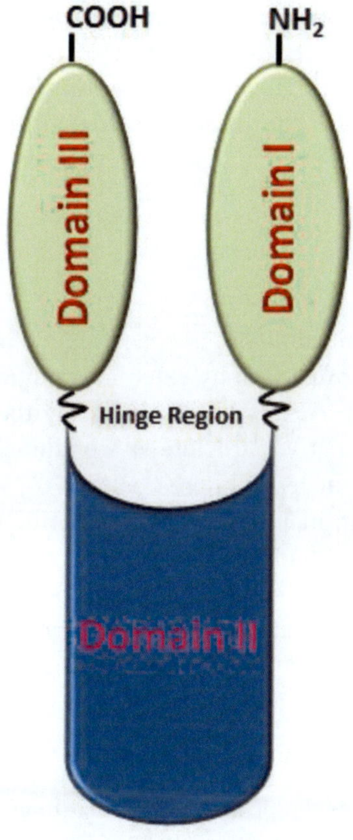

Figure 1. Diagrammatic representation of three homologous domains of AFP.

AFP in different species share structural similarity, but have distinct difference in their structure and biological and physiological functions. Human, mouse and rat AFPs possess different estrogen-binding interfaces but share a high sequence similarity [37, 40, 41]. The human and mouse AFPs share 66% identity in the primary structures [37]. The sequence identity of Domain III (72%) was the highest, while the identity of Domain I (59%) was the lowest between AFPs of different species. Interestingly, the sequence identity between AFP and albumin from the same species for some domains were lower than the sequence identity of the same domains between the AFPs or between the albumins of different species [42, 43]. AFP and albumin shared only 50% identity at bilirubin and unsaturated fatty acids binding sites. The Domain III and subdomain IA show the highest (48%) and lowest (16%) sequence identity between human AFP and albumin, respectively [44, 45]. Human AFP contains only one glycosylation site (Asn233), while mouse and rat AFPs include three glycosylation sites (Mouse: Asn232, Asn310 and Thr483; Rat: Asn232, Ser96 and Asn310) [46].

2.2. Secondary Structure

The AFP molecule was perceived as a U-shaped structure: one on the top and the other two on the periphery of the molecule. This U-shaped structure provides the cavity for the estrogen or other small molecules to bind so that AFP undergoes conformational changes to promote the interactions [47]. The tertiary structure of Domains I and III are tightly packed and joined by a flexible Domain II. Hence, the C-terminal part of Domain II was considered as a "hinge region" and responsible for the conformational flexibility of AFP which facilitates the binding or release of various potential ligands and proteins.

2.3. Tertiary Structure

Proteins possess a remarkable ability to adopt different conformations due to the substrate or small molecule binding and influenced by external conditions such as temperature, pH and the composition of solvent [48]. AFP is characterized by the relatively high conformational stability and possesses a rigid native structure in conditions such as within the pH range between 10.5 and 4.5 [43, 49], at temperatures below 70°C [49-51], at urea concentrations below 7.5 M [46, 50-52] and at guanidinium hydrochloride concentrations below 2.0 M [6, 52]. A large number of disulfide bonds (15 S-S bridges) allow AFP to maintain an extremely high conformational stability. Some natural ligands also increase AFP stability. The protein can undergo an irreversible conformational transition from an inactive to an active form due to the chemical condition. The structural changes in AFP may be induced by the release of ligands which might be important for functioning as an immunoregulator. AFP represents a two-receptor model for AFP endocytosis and fatty acid transfer into a target cell [53]. Binding of saturated AFP ligand to the corresponding AFP cell surface receptors produces a signal for endocytosis and is responsible for the protein conformational changes which are accompanied by the dissociation of AFP-fatty acid complex. Binding of high concentration of estrogens and fatty acids or any other hydrophobic ligands can induce a rapid and reversible conformational change in AFP tertiary structure [10, 54].

The decrease in pH and effective dielectric constant can potentially denature the significant interactions between AFP and cellular membranes. Membrane induced transformation of proteins into the molten globule state is of functional importance to facilitate the transport of transmembrane protein and toxin penetration into cells. Some organic solvents like methanol and trifluroethanol can induce conformational changes in AFP. The rigid tertiary structure of AFP can be completely disrupted at high concentrations of organic solvents. Structural transformation of the rigid native protein structure to molten globule state increases the structural flexibility to facilitate the binding and release of the hydrophobic ligands [48].

3. BINDING SITES OF AFP

3.1. Functionally Important Sites of AFP

Experiments proved that the stretch of 15 amino acids in the primary structure of AFP were responsible for the vital functions of AFP [11]. Some functionally important sites of binding for AFP are immunoregulation and peptide epitopes, extracellular matrix binding and cell adhesion, enzyme-related modulation, transcription factor/homeodomains, apoptosis-associated activities, hydrophobic, ion and metal ligand binding, hormone and growth factors, cell surface receptors, dimerization and nuclear localization signals, amino acid (AA) rich repeat sequences, cytoskeletal proteins, filaments and microtubules, and multiple chemokine sites. AFP is similar to albumin and acts as a binding and transport protein for a variety of ligands such as bilirubin, fatty acids, steroids, heavy metals, retinoids, flavonoids, dyes, dioxin, phytoestrogens, and several drugs [44].

3.2. Hydrophobic Ligand Binding Sites

AFP can bind with different hydrophobic ligands and enhances transport of the metal ions, estrogens, bilirubin, retinoids, flavonoids, phytoestrogens, exotoxins, various dyes and drugs into the cells [55]. AFP requires the increased energy supply and intermediates of fatty acid metabolism to transport the fatty acids into actively proliferating cells. One of the important regulatory mechanisms of AFP is its estrogen binding activity which prevents the degradation of estrogens and transports them into the organs and tissues. AFP binds to both the free and immobilized hormones. It was reported that the estrogen binding site, an amino acid stretch of the sequence ELIDLTGKMVSIAST (amino acids 424-438), was present in Domain III of AFP [55]. The P149b (GIP-14), AADIIIGHLCIRHE (amino acids 458-471) hydrophobic fragment of growth inhibitory peptide (GIP), had shown a low affinity at the estrogen binding site due to the hidden conformation of P149b and becomes susceptible only in the presence of estrogens. The Lys224 present in Domain II (amino acids 209-228, MKNFGTRTFQAITVTKLSQK) plays a vital role for the binding of unsaturated fatty acids [56]. Butanol decouples the AFP-estrogen and AFP-fatty acid complexes that results in the conformational changes which increases the accessibility of hydrophobic estrogen binding sites for various hormones. Still, further investigation is needed to provide insight on the molecular mechanism of AFP binding to hydrophobic ligands.

3.3. Fatty Acid Binding Site

The presence of six fatty acids (palmitic, stearic, oleic, linoleic, arachidonic (AA), and docosahexaenoic (DHA) acids) on human AFP has been reported [10]. The poly-unsaturated fatty acids (PUFA's) are intracellularly delivered into the developing cells by AFP. During endocytosis of a fatty acid, AFP binds to its cell surface receptors and then transfers into cells by a specific fatty acid binding protein [10]. Subsequently the fatty acid is released and the AFP is recycled back across the cell surface. Lysine plays an important role in the process and contains a free amino group which can form an ionic bond with the carboxyl group of the fatty acid. Hence, Lys223 is crucial for the fatty acid to bind in a suitable orientation in AFP.

3.4. Effect of Natural Ligands

AFP was isolated from human fetal tissues based on the two isoelectric points (4.7 and 5.3). The values of isoelectric points were altered according to the natural ligands. For example, the pI values were 4.7 and pI 5.3 for AFP bound with fatty acids and for fatty acid free AFP, respectively. The ampholyte displacement chromatography [57] and isoelectric focusing [40] methods revealed that there are seven differently charged forms of AFP can be isolated from human amniotic fluid or umbilical cord blood and nine AFP variants in the pH 4.5-5.2 range, respectively [6]. AFP isolated from fetal tissues or amniotic fluids has shown less charge heterogeneity than from different hepatoma sources. The heterogeneity of various AFP variants depends on amino acid sequence, glycosylation or ligands bound [58].

3.5. Epitope Sites

The eleven antigenic clusters of AFP are grouped into six distinct major epitopes (A-F) that were derived from 30 monoclonal antibodies (*MoAbs*) to human AFP [59]. These MoAbs to human AFP exhibit four types of reactivity: full competition; partial competition; full independence, and enhancement of binding. MoAbs react similarly with AFP major epitopes from umbilical cord serum, placental extracts, or tumor tissue. One of the hidden forms of minor AFP epitopes was identified and recognized only after partial denaturation by adsorption on a nitrocellulose membrane (NCM) [60, 61]. The knowledge about the major and minor epitopes of AFP leads to the rational design of monoclonal polyclonal antibodies. Such design might be helpful to determine whether it is related to disease state and/or cancer stages. Overall, it is important to understand the AFP antigenic structure for several functions.

3.6. Estrogen Binding Site

The binding of estrogens in the C-terminal of AFP suggests that it can act as an estrogen carrier in blood to compete with estrogen receptor (ER) for estrogens [40, 62-64]. There are two estrogen binding site in human AFP. Site 1 is for the estrogen binding and ligation and Site 2 is the docking site for proteins like heat shock protein family. Both sites overlap with the fatty acid binding site.

In Site 1, the residues from 424-439 have shown a high affinity for estrogens. Glycine 425, methionine 427, isoleucine 430, alanine 432 and threonine 433 are five crucial amino acids in rodent AFP. Three out of the five amino acids were substituted in human AFP: glycine to arginine, isoleucine to threonine, and serine to alanine. The estrogen binding site in the human ER (amino acid 419-431) was compared with human and rodent AFPs. Interestingly, four out of the five residues of rodent AFP were matched and aligned with ER estrogen binding site. Most of the vital amino acids in rodent AFP and ER were replaced by alanine in human AFP.

Site 2, amino acids from 545-468 in human AFP, has a low binding affinity to estrogens. Glycine and cysteine which are absent in ER but are present commonly in human and rodent AFPs. This site is also involved in protein folding and unfolding process.

3.7. Other Putative Functionally Important Sites

Human AFP and serum albumin share a high homology in the homo and hetero-dimerization motifs. Thyroid receptor hetero-dimerization motif was found in AFP Domain III. AFP acts as a transcription cofactor to form a heterodimer with nuclear receptors and stabilizing their binding to hormone responsive elements of DNA. AFP also regulates the gene transcription by promoting or blocking of certain genes on the transcription factor dimerization which results in either stimulation or inhibition of cell proliferation and tissue growth.

Primary sequence comparison of AFP and extracellular matrix proteins reveal that AFP Domain II contains short amino acid stretches (LRE, 195_197; LDV, 242_244; RGD, 253_255; DGEK, 262_265; ILRVAK, 352_356, etc.) similar to the signal adhesive sites in laminin, collagen, and fibronectin [11, 46]. In addition, AFP contains cell adhesion motifs which may have some functions typical for extracellular matrix proteins. Identification of these motifs support that AFP regulates various processes such as proliferation, differentiation, migration, inhibition of growth and metastases of tumors [65].

Many short segments of amino acids of AFP have shown similarity with various receptor proteins like insulin receptor, tumor necrosis factor receptors, growth factor β1 etc. AFP and cytokines are involved in the regulation of apoptosis and the sequence analysis reveal that the amino acid stretch in AFP domains I, II, and III show a 71% identity with tumor necrosis factor receptors (TNFRα), Fas and Bcl_2 proteins, and also of α-interferon and interleukins binding sites [28]. Some amino acid segments of AFP share a 33% identity with insulin α-chain and also has similarity with transforming growth factor β1 (TGFβ1). In addition, a short amino acid segment of AFP has shown similarity with the inhibitor of cyclin dependent kinase, RxL [66]. Most of the functionally important sites require the conformational changes to occur during protein–protein, protein–ligand or protein–membrane interactions.

4. Function of Cytoplasmic AFP

AFP is one of the well-defined regulatory factors in the growth of tumor cells [14]. It is involved in pleiotropic activities and affects the processes of growth regulation, cell differentiation and tumorigenesis [67-69]. Both tumor and normal cells contain AFP and the elevated level of AFP was found to be related to human hepatocellular carcinoma presence [13, 15, 70]. AFP binds with retinoic acid receptors-b (RAR-b) to block the RAR signaling to inhibit the translocation of cytoplasmic RAR-b into the nucleus [68, 70, 71]. It is responsible for retinoid resistance in tumor chemotherapy and impacts the expression of growth arrest and DNA damage-inducible 153 gene (GADD153) in human hepatoma cell lines [13, 72]. The apoptotic signal was disrupted by cytoplasmic AFP by blocking the caspase-3 cascade process [68, 73]. AFP may function as a regulator by interacting with PTEN to promote the PI3K/AKT pathway which leads to aberrant growth of hepatocellular carcinoma cells [13]. AFP suppresses the mitogenic response and functions as a regulator by interacting with phosphatase and tensin homolog deleted on chromosome ten protein and stimulating cell growth through the PI3K/AKT signal pathway [74]. In the brain, AFP plays a vital role in the neuronal development and differentiation. The presence of AFP in choroids plexus hinted that the fetal protein was transudate from the blood to the cerebrospinal fluid via a cellular route across the choroids plexus epithelial layers. Mainly, AFP was found in the pyramidal cell dendritic processes and also at the axonal pole of the cell to differentiate the neurons in the cytoplasm. The presence of the AFP fetal protein in the developing nervous system indicates the binding and transport of polyunsaturated fatty acids (PUFAs) by AFP. Myelin sheath of the neuronal cells requires PUFAs for development [10]. AFP secreted from the oviduct may serve to mediate sperm survival and motility by decreasing the response of the acrosomal reaction so as to prolong sperm viability and function [10].

5. Biological Activities

5.1. Regulation of Cell Proliferation and Tumor Growth

The mechanism of inhibition of cell proliferation and tumor growth is unclear. Several researchers have reported that AFP can stimulate proliferation and differentiation of various cell types including lymphoid and epidermal cells, fibroblasts, hepatocytes, and ovarian and uterine cells [75]. AFP is synthesized during embryonic development and the high concentration of AFP found in fetal serum suggests that it may stimulate tissue growth. The recombinant human AFP shows an inhibitory effect on growth of estrogen-dependent tumor cells but not in estrogen-independent tumor cells. The binding of AFP to its receptor on the surface of normal and tumor cells is required to penetrate into cells. It was reported that the AFP may inhibit or induce programmed cell death [76]. However, it was demonstrated that AFP leads to the cytotoxic effect of tumor necrosis factor [77].

5.2. AFP Peptide Epitopes and the MHC Proteins

The major histocompatibility complex (MHC) gene consists of three major classes of proteins. Class I addresses the adaptive and innate immune systems employing CD8 cytotoxic T-cells (CTL). Class II is involved in immunological recognition as the MHC restricting CD4 helper (Th) cells. Class III relates to the complement cleavage (lysis) cascade in the inflammatory response. Most of the T-cells recognize peptides bound to Class I or Class II cell surface proteins. The Class II proteins serve to present antigenic peptides to CD4(+) helper T-cells [78]. The effector T-cells differentiate in three types of CD4 helper cells that can respond to foreign antigens. T-helper type1 (Th1) clears the intracellular pathogens. T-helper type2 (Th2) controls the parasitic infections [79]. T-helper cell (Th17) is induced by interleukin17 and responds to autoimmune tissue injury [80]. The MHC Class I molecules are subdivided into two groups (Ia and Ib) which are essential for the functions of adoptive and innate immune response systems [81]. The Class Ia molecules are glycoproteins and also known as human leukocyte antigen (HLA)-A, HLA-B, HLA-C and HLA-DR and play a seminal role in adaptive immunity. The Class Ib molecules possess a larger number of primary anchor sites that are common to some Class Ia molecules and display more limited forms and regulate innate immunity encompassing pathogens and tumors.

5.3. Extracellular Matrix

Components of the Extracellular matrix (ECM) and its receptors virtually interact with all types of cells in embryonic and adult organisms. Most of these molecules inhibit the neurite outgrowth, cell migration during embryogenesis, modulation of synaptic transmission, anti-adhesive activities, growth factor interaction, ECM anchoring, binding of proteases and/or protease inhibitors, initiation of intracellular signaling cascades, and cytoskeletal signal transduction. The adhesion molecules and/or receptors influence many biological activities including cell proliferation, differentiation, cell junction formation and polarity by targeting cell surface adhesion to specific ECM proteins/ligands on adjacent cells. The AFP Domain II and Domain I are associated with ECM activities such as hemostasis, basement membrane architecture, fibril scaffolding and cross-banding, cytoskeletal attachment, and signal transduction which was confirmed by the sequence identify between AFP and ECM. The Domain II had shown a high identity with ECM. Hence it that AFP was distinguished from other members of albumin family due to the presence of the short peptide sequences similar to ECM.

5.4. Fibrin Binding

Human tissue type plasminogen activator (TPA) is responsible for the dissolving fibrinogen in blood clots. The plasminogen and TPA enzymes facilitate the enzyme binding with substrate by interacting with fibrin polymer in clots. The epidermal growth factor motif in AFP overlaps with a portion of TPA. The border of Domain I and II reveals an identity site with TPA. The Domain III of AFP resembles the fibrin binding finger domains for the serine protease/substrate contact which are present in the amino-terminal of TPA. Interestingly, the estrogen binding site of AFP may overlap with a portion of the substrate binding site for serine protease [40].

5.5. Oncogene Transcription

Some of the transcription factors, tumor suppressors and homeodomain protein have shown sequence homologies with AFP. Oncoproteins are the products of the various protooncogenes such as ras, rel, myc, myb, fos, jun, ski, ets, cbl and erb A [82, 83]. Most oncoproteins are predominantly present in the nucleus, can perceive the incoming signals and modulate the transcription of specific genes in the plasma membrane or cytoplasm [84-86]. In mammals, during early embryogenesis POU domain genes are expressed in many regions of the developing brain and required for homo and hetero-dimerization of the POU proteins.

5.6. Transport of Small Molecules

The major function of AFP is the binding with ligands and the transport of the ligands during the fetal development. It enhances the embryotoxic and teratogenic effects by facilitating the dioxin (2,3,7,8-tetrachlorodibenzo-p-dioxin) transport into embryonic tissues. If AFP is similar to ALB in this regard, then it is probable that one fatty acid binding site exists on each of the three domains. AFP domain I site had shown similarity to fatty acid synthetase for folic acid binding and also another fatty acid protein like stretch of a Vitamin D binding site. On domain II, a fatty acid denaturase homolog site was identified. The estrogen binding interface of AFP was formed by α-helix segment lying adjacent to a β-sheet/turn structure. In humans, a high affinity estradiol (E2)-binding region displays overlapping binding sites for fatty acids, diethylstibesterol (DES), protease inhibitors/substrates, retinoids, warfarin, coumarin, phenylbutazone, pyrazolic drugs and anthranilic acid. Multiple binding sites have been reported for the binding of heavy metals likes Cu^{2+}, Zn^{2+}, Pb^{2+}, etc. The Domain II contains three histidine residues, the histidine rich region, and acts as a binding region for $Cu^{2+}$and/or Zn^{2+}. The Cu^{2+} transporting ATPase was identified in domain I. All the above evidence suggests that Domain I and II of AFP have a well-defined heavy metal binding site.

5.7. Chemokine-Like Peptide Sites

Chemokines attracted the attention of many researchers only after the discovery of the presence of HIV binding sites in their receptors. The chemokine family is classified into four members based on the N-terminal positioning of their cysteine residues [84]. The four members are further divided into four groups. Group I contains a single cysteine in the amino terminal. Group II has two cysteine residues. Group III has the CXC configuration. Group IV has the CX3C pattern where X is any residues. The cytokines were attracted by chemokines, essential to the immune response and inflammatory reactions, to mediate the migration of cells in tissue. The chemokines (ligands) binds to heptahelical receptors, which weave through the cell membrane seven times as transmembrane (TM) domains. Chemokines and their receptors are highly conserved and share high similarity in across various species. Chemokines are produced locally in tissues and direct the emigration and immigration of cells from the bloodstream into sites of inflammation, infection and cell proliferation. Chemokines directly influence many physiological processes such as angiogenesis, degranulation, autoimmunity, HIV infection, tumor growth, parasite infection and leukocyte trafficking [85]. The chemokines function as regulatory molecules in cell maturation, trafficking, homing and the development of various tissues. The chemokines family contains nearly 50 ligands and 20 receptors, which are remarkably homogeneous with properties similar to IL8. Many chemokines are secreted as a result of pathological conditions, some fulfill regulatory roles, and others function in cell migration during histo and organogenesis and in cancer metastasis. Fetal AFP has positive and negative correlation with fetal hemoglobin and red blood cell levels, hematocrit, and erythropoietin/transferring concentrations, respectively. Thus, during pregnancy, AFP has long been involved in erythropoiesis myelopoiesis (bone marrow) and lymphopoiesis and these developmental states are regulated by the chemokines. It was proposed that AFP has the property of chemokine or mimics its properties by comparing the sequence of AFP with chemokine. By sequence comparison, it was revealed that AFP resembles the segment of cytokine mainly the Group II, GROα, MiP1B, Eotaxin, Rantes and MCP1 chemokine. Full length human AFP can bind to the CCR5 receptor [90] and suppresses the growth factor in prostate and breast cancers as well as HIV cell fusion and infection which proved that AFP mimics the CCR5 ligand. All the above evidences prove that AFP mimics a chemokine and binds to a chemokine receptor and also induces the receptor conformational change and tubulin polymerization. In addition, AFP has an advantage compared to chemokine related inhibitors, the advantage of AFP is that it acts as a decoy ligand but other chemokines are not. It was speculated that AFP could serve as a novel type of chemokine inhibitor to retain the growth regulatory properties, but still be able to induce some of the early functions of a true chemokine ligand. In summary, AFP might be beneficial in pathological situations such as cancer therapeutics, HIV infection and transmission, inflammatory diseases, angiogenesis, autoimmune diseases, atherosclerosis, multiple sclerosis, diabetes, lung infections, malaria, endometriosis, graft rejection and wound healing.

5.8. Growth Inhibitory Peptide

Mizejewski's group identified a growth inhibitory motif from AFP which is responsible for the tumor growth inhibition, metastasis and suppression of estrogen dependent tumors [1, 86, 87]. This 34-mer amino acids stretch was known as a growth inhibitory peptide (GIP) and is termed as P149 (or P447) [1]. Most of this segment is buried in a molecular crevice and, hence, the antibodies are not able to interact with the native state AFP but are able to interact tightly with the AFP with the conformational changes caused by the extreme solvent concentration and stress/shock conditions during the fetal development [11, 14] when exposed to the GIP segment [86]. Chemical modification of cysteine residues followed by formation of S-methylcysteine or S(2-aminoethyl)cysteine have shown to be active in the inhibition of estrogen dependent proliferation of mouse uterine cells. Various GIP analogs were synthesized by replacing the two cysteine residues with alanine, glycine and serine. The alanine containing GIP analog have shown a higher biological activity towards the inhibition of estrogen dependent proliferation. The glycine and serine analogs showed two times less activity than alanine analog and failed to show any activity towards estrogen dependent proliferation. P149 acts as a negative regulator of both ontogenic and oncogenic growth and is composed of three bioactive fragments: (i) 12 amino acids in amino terminal (GIP-12 or P149a), (ii) hydrophobic mid piece consisting of 14 amino acids (GIP-14 or P149b) and (iii) 8 amino acids peptide located near the carboxyl terminus (GIP-8 or P149c) (Figure 2). It has been demonstrated that GIP-8 and GIP-34 have an effective anti-cancer growth activity [1, 88] against different types of cancers and appears to function largely in estrogen (E)-dependent cancers and inhibit both E-dependent and non-E-dependent (basal) cancer growth. GIP-8 and GIP-34 adopt the different signal transduction cascades to inhibit the tumor cell growth and proliferation in the cytoplasm. They simultaneously interpose the signaling transduction cascade which eventually results in (1) blockage of K^+ ion channels opened by estradiol and epidermal growth factor (EGF); (2) protection of p53 from inactivation by phosphorylation; (3) prevention of cyclin inhibitor degradation; and (4) cell cycle S-phase/G2-phase arrest. The GIPs could potentially aid in relieving the negative side effects of: (1) Tamoxifen resistance, uterine hyperplasia/cancer, and blood clotting; (2) Herceptin antibody resistance and cardiac (arrest) arrhythmias; and (3) Doxorubicin's bystander cell toxicity. Therefore, these peptides might be ideal candidates for identification of novel and potent lead compounds in development of anticancer drugs.

Figure 2. Amino acid sequence of GIP of AFP.

5.8.1. GIP-34

GIP-34 consists of 45% β-sheets turns, 45% of random coil (disorder) and 10% of alpha-helix [89-91]. The 8-mer peptide from the C-terminal of GIP-34 [92] has a reverse β-turn which accommodates the β-turn of a ligand into the cell surface receptor binding site to enhance its biological activity [91]. GIP-34 can enter into cancer cells by two pathways: receptor-mediated endocystosis and pore-forming/cell penetration. The biological activity of GIP-34 is dependent on its molar concentration and its oligomeric state [93]. GIP-34 disturbs the cell surface activities such as tumor cell adhesion, cell pseudopodial extensions, platelet aggregation and cell agglutination [93]. Thus it acts as a cell membrane disruptive agent. Many researchers have reported that GIP-34 and its sub-fragments can regulate and invoke an immune response of the cell.

denote that no competitive binding was observed or were extrapolated in the assay, and thus the corresponding chemicals were coded as non-binders. "NA" indicates that a chemical was not tested in either the AR or ER (expand) binding assay. The 125 chemicals are grouped into 15 chemical classes as shown in the first column. The reference chemicals for logRBA calculations are testosterone for SHBG, estrone for AFP, 17β-estradiol for ER and R1881 for AR

Class	Name	CAS	Binding Affinity (logRBA)			
			SHBG	AFP	ER	AR
Steroid	Testosterone	53-16-7	2			1.285
	17β-Estradiol	50-28-2	1.33	1.734	2	-0.121
	6α-OH-Estradiol	1229-24-9	-0.514	1.007	-0.15	NA
	17-Desoxyestradiol	53-63-4	2	0.813	1.14	-2.131
	3-OH-Estra-1,3,5(10)-trien-16-one	3601-97-6	1.193	0	-0.29	NA
	Estriol	50-27-1	-1.286	-0.266	0.99	-3.147
	17α-Estradiol	57-91-0	0.796	-0.886	0.49	-2.4
	3-Deoxyestrone	53-45-2	0.29		NA	NA
	3-deoxyestradiol	2529-64-8	-0.125		NA	0.538
	Ethinyl estradiol	57-63-6	-0.722	-0.301	2.28	-1.416
	Mestranol	72-33-3		-1.187	0.35	NA
	Moxestrol	34816-55-2	-1.548		1.14	
	ICI 164-384	98007-99-9			1.16	NA
	Fulvestrant	129453-61-8			1.57	
	Dihydrotestosterone	521-18-6	2.796		-2.89	2.14
	Progesterone	57-83-0	-0.467			-0.703
	Estrone	58-22-0	1	2	0.86	
	3α-Androstanediol	1852-53-5	1.897		-2.67	-0.809
	3β-Androstanediol	571-20-0	2.796		-0.92	0.357
	Corticosterone	50-22-6	-0.467			-1.875
	Norethynodrel	68-23-5	0.176		-0.67	-0.703

Table 1. (Continued)

Class	Name	CAS	Binding Affinity (logRBA)			
			SHBG	AFP	ER	AR
Flavone	Flavone	525-82-6				-2.397
	Quercetin	117-39-5	-2.623	-0.266		NA
	Kaempferol	520-18-3		-0.301	-1.61	NA
	Apigenin	520-36-5	-2.186	-0.868	-1.55	NA
	Myricetin	529-44-2		-1.026	-2.75	NA
	Fisetin	528-48-3		-1.266	-2.35	NA
	3,6,4'-Trihydroxyflavone	253195-19-6		-1.442	-0.35	NA
	Chrysin	480-40-0	-2.415	-1.53		NA
	Baicalein	491-67-8		-1.649	-3.05	NA
	Rutin	153-18-4			-4.09	NA
	6-Hydroxyflavone	6665-83-4			-3.41	-2.768
	7-Hydroxyflavone	6665-86-7				
	6-Hydroxy-2'-methoxyflavone	61546-59-6				NA
	Morin	480-16-0			-3.09	NA
	4',6-Dihydroxyflavone	63046-09-3			-0.82	NA
Flavanone	Flavanone	487-26-3	-2.415	-1.678		-2.249
	Naringenin	480-41-1	-1.103	-0.488	-2.13	NA
	6-Hydroxyflavanone	4250-77-5	-2.523	-1.778	-3.05	-1.78
	7-Hydroxyflavanone	6515-36-2	-2.415	-1.649	-3.73	NA
	4'-hydroxyflavanone	6515-37-3	-2.125	-1.085	-2.65	-2.479
	Taxifolin	480-18-2		-1.266		NA
	3'-hydroxyflavanone	92496-65-6	-1.903	-1.363	-2.78	NA
	Hesperetin	520-33-2	-1.903			NA
	Catechin	154-23-4				NA
	Naringin	10236-47-2				

Class	Name	CAS	Binding Affinity (logRBA)			
			SHBG	AFP	ER	AR
Isoflavone	Genistein	446-72-0	-1.865		-0.36	-2.44
	Daidzein	486-66-8			-1.65	NA
	Formononetin	485-72-3			-2.98	NA
	Genistin	529-59-9				NA
	Biochanin A	491-80-5			-2.37	NA
	3',4',7-Trihydroxy isoflavone	485-63-2			-2.35	NA
	4',6,7-Trihydroxy isoflavone	17817-31-1				NA
	Prunetin	552-59-0			-2.74	NA
	Equol	531-95-3	-1.699	-1.391	-0.82	-2.388
Chalcone	Chalcone	94-41-7	-3.215	-2.266	-2.82	-2.316
	Phloretin	60-82-2	-1.514	-1.53	-1.16	NA
	4-Hydroxychalcone	20426-12-4	-2.644	-2.151	-2.55	-2.194
	4'-Hydroxychalcone	2657-25-2	-3.664	-2.333	-2.43	-2.273
	2',4,4'-Trihydroxychalcone	961-29-5			-1.26	NA
Coumarin	Coumestrol	479-13-0	-1.054	-0.919	-0.05	
	4-Ethyl-7-OH-3-(p-methoxyphenyl)coumarin	5219-17-0		-1.141	-0.05	NA
	7-Hydroxycoumarin	93-35-6			NA	NA
	6-Hydroxy-4-methylcoumarin	2373-31-1			NA	NA
	6,7-Dihydroxy-4-methylcoumarin	529-84-0			NA	NA
	7,8-Dihydroxycoumarin	486-35-1			NA	NA
Phthalate	Diethyl phthalate	84-66-2	-3.079	-2.778		-3.437
	Dimethyl phthalate	131-11-3		-2.958		NA
	Bis(2-ethylhexyl)phthalate	117-81-7				
	Butyl benzyl phthalate	85-68-7	-1.824			-2.069
	di-n-Butyl phthalate	84-74-2	-3.699			-1.952
	Bis(n-octyl) phthalate	117-84-0	-4.437			-2.225
	di-i-Butyl phthalate	84-69-5				-3.277
	Diisononylphthalate	28553-12-0				-3.562

Table 1. (Continued)

Class	Name	CAS	Binding Affinity (logRBA)			
			SHBG	AFP	ER	AR
Paraben	Heptyl p-hydroxybenzoate	1085-12-7	-1.754	-0.919	-2.09	NA
	2-Ethylhexyl-4-hydroxybenzoate	5153-25-3	-1.125	-1.16	-1.74	NA
	Benzyl 4-hydroxybenzoate	94-18-8	-1.865	-1.266	-2.54	NA
	Methylparaben	99-76-3	-3.028		-3.44	
	n-Propyl 4-hydroxybenzoate	94-13-3	-2.548		-3.22	-3.004
	Ethyl 4-hydroxybenzoate	120-47-8	-2.721		-3.22	NA
Benzophenone	Dioxybenzone	131-53-3	-2.028	-1.266	NA	NA
	o-Benzoylphenol	117-99-7	-2.186	-1.391	NA	NA
	2,2'-dihydroxy benzophenone	835-11-0	-1.903	-1.549	NA	NA
	2,4-Hydroxybenzophenone	131-56-6	-1.329		-2.61	-2.534
	4,4'-Dihydoxy-benzophenone	611-99-4	-1.166		-2.46	-2.674
Diphenylethylene	Aurin	603-45-2	-1.772		-1.5	-1.703
	Dihydroxymethoxychlor olefin	14868-03-2	-0.222	-0.187	0.42	-1.312
	Clomiphene	911-45-5	-2.67		-0.14	-1.643
	(2,2-Dichloro-1-phenylvinyl)benzene	2779-69-3	-0.768		NA	NA
	Nafoxidine	1845-11-0			-0.14	-1.627
Diphenylmethane	Bisphenol A	80-05-7	-2		-2.11	-2.388
	Bisphenol B	77-40-7	-1.824		-1.07	-2.087
	p,p'-Methoxychlor	72-43-5				-1.944
	o,p'-DDD	53-19-0	0.607			-1.524
	o,p'-DDT	789-02-6	-1.523		-2.85	-1.689
	Phenolphthalein	77-09-8	-1.903		-1.87	NA

Class	Name	CAS	Binding Affinity (logRBA)			
			SHBG	AFP	ER	AR
Bibenzyl	DL-Hexestrol	5776-72-7	-0.824	-1.09	NA	NA
	Meso-Hexestrol	84-16-2	-0.368	-1.743	2.48	NA
	2,6-Dimethyl hexestro		-0.204	-1.8	0.95	NA
	3-(p-Phenol)-4-(p-tolyl)-hexane		-1.777	-1.877	0.6	NA
	Monomethyl ether hexestrol	13026-26-1	-1.365		0.97	-1.627
	Dienestrol	84-17-3	-0.721	-1.09	1.57	NA
	2,2',4,4'-Tetrahydroxybenzil	5394-98-9	-2.301		-0.68	NA
Stilbene	Diethylstilbestrol (DES)	56-53-1	-1.286	-0.301	2.6	-1.659
	Monomethyl ether diethylstilbestrol	633-60-3			1.31	NA
	Dimethylstilbestrol	552-80-7	-1.467		1.16	-1.659
	4,4'-Dihydroxystilbene	659-22-3	-2.301		-0.55	-2.437
	Toremifene	89778-26-7	-0.824		0.14	NA
	Droloxifene	82413-20-5	-1.342		1.18	NA
	4-OH-Tamoxifen	68047-06-3	-2.166		2.24	-1.489
	Tamoxifen	10540-29-1	-2.467		0.21	-1.592
	Triphenylethylene	58-72-0	-1.523		-2.78	-1.976
Mycoestrogens	α-Zearalanol	55331-29-8	-1.271	-0.767	1.48	NA
	β-Zearalanol	42422-68-4	-1.67	-1.602	-0.19	-1.717
	α-Zearalenol	36455-72-8	-1.079	-2.079	1.63	-1.643
	β-Zearalenol	71030-11-0	-1.079		-0.69	-2.093
	Zearalanone	5975-78-0	-1.355		0.32	-2.139
Phenol	2,3,4,5-Tetrachloro-4'-biphenylol	67651-34-7	-1.355	1.47	-0.64	NA
	2,5-Dichloro-4'-biphenylol	53905-28-5	-0.103	-1.266	-1.44	NA
	4-n-Nonylphenol	104-40-5	-0.77	-0.062	-1.53	NA
	4-n-Octylphenol	1806-26-4	-1.125	-0.602	-2.31	-1.803
	Nordihydroguaiaretic acid	500-38-9	-1.772	-1.056	-1.51	-2.284
	4-Heptyloxyphenol	13037-86-0	-1.903	-1.678	-2.88	-1.689
	2-sec-Butylphenol	89-72-5	-2.623	-2.391	-3.54	-2.522
	4-(Benzyloxyl)phenol	103-16-2	-1.609	-2.911	-3.44	-2.893

Table 2. Comparison between rat AFP and human SHBG and their binding for different classes of chemicals

Name	Chemicals	Rat AFP	Human SHBG
Steroid	21	Beneze ring and the 3-OH are important for binding. 9 compounds exhibited binding capability (strong?)	High binding affinity. Out of 21 compounds, 18 bind well and 3 failed.
Flavone	15	OH group and position are important 9 binders and 6 non-binders	12 compounds failed to bind. Quercetin, apigenin and chrysin bound with weak affinity.
Flavanone	10	S-configuration of carbon atom-2 is a key factor for binding	7 compounds were able to bind and 3 failed.
Iso-flavones	9	All compounds failed to bind 7 binders and 2 non-binders	Mostly did not bind. 7 compounds failed and 2 showed moderate activity (define what is moderate?)
Chalcone	5	4 weak binders Trihydroxychalcone failed	4 weak binders and 1 (Trihydroxychalcone failed) failed
Coumarin	6	Size of the courmarin is the key factor to bind rat AFP	Most of the derivatives failed. 5 compounds failed and one showed moderate activity
Phthalate	8	Binding affinity increased with molecular size Diethyl phthalate and dimethyl phthalate showed low binding activity	4 compounds showed weak activity and 4 compounds failed to bind
Paraben	6	Similar to coumarin derivaties but contrast to phthalates, parabens Large hydrophobic groups enable the binding	All binds to SHBG
Benzophenone	5	OH substitution at 1 or 1' are necessary for binding OH substitution at 3 or 3' diminishes binding affinity	All compounds binds
Diphenylethylene	5	Only dihydroxymethoxychlor olefin Larger side chain R and R' on the double bound inhibit receptor binding	Mostly bind
Diphenylmethane	6	All compounds failed to bind	Mostly bind
Bibenzyl	7	Hydrogen bonding and van der Waals interactions are important for binding	All compounds binds
Stilbene	9	Molecular shape of the stilbene is not favorable	8 compounds bind and one compound failed
Mycoestrogens	5	R-configuration of the chiral-2 atom having highest binding affinity.	5 compounds showed moderate activity
Phenol	8	Structural indicator for binding	Structural indicator for binding

5.8.2. GIP-8

Mizejewski first reported the anti-cancer and anti-growth fragment derived from GIP-34 [14, 94]. GIP-34 peptide was digested by trypsin to yield an 8–9 amino acid fragment that was found to retain the E-dependent anti-cancer growth properties from the original GIP-34 peptide. It has been confirmed that the GIP-8 is able to prevent the growth and suppress proliferation of E-dependent breast cancer. Mesfin et al. proved that the cyclized GIP-8, adding the Gln and Asn amino acids in the terminal region to form a 9-mer peptide, maintained the similar E-dependent anti-cancer growth properties as the linear 8-mer [95, 96]. The growth of tamoxifen-resistant human breast cancer in mice was inhibited by cyclized GIP-8 that suppressed the uterotrophic hyperplasia produced by tamoxifen in the immature mice [88, 97]. Among 8 amino acids in GIP-8, some (Glu, Val, Asn, Pro and Gly) are essential for the inhibition which was proved by Defreest et al. using pharmacophore modeling studies [98, 99]. Shields et al., discovered that the secondary structure of GIP-8 consists of a random coil which contains a reverse β-turn, a structure known to bind and activate the cell surface receptors linked to G-proteins [91, 92]. Interestingly, the β-turn was present in a tetra-amino acid of GIP-8, namely, Thr-Pro-Val-Asn (TPVN) which is a known motif for peptide binding to PDZ-containing proteins and the transient receptor potential (TRP), but not in the Pro-Val-Asn-Pro (PVNP) sequence which is involved in the activation of serine/threonine Src-kinases. Overall, the linear and cyclic GIP-8 plays vital role in the growth suppression of different type of cancers. GIP-8 might serve as a plasma membrane interactive agent to block the ion channels and interfere with cell surface-induced signal transduction pathways involved in tumor growth, progression and metastasis.

6. COMPOUNDS WHICH BIND TO RAT AFP

In 2012, Hong et al. developed a competitive binding assay to measure the binding affinity towards rat AFP of 125 structurally diverse chemicals spanning 15 chemical classes [100]. The 15 classes of chemicals (Table 1) are steroid, flavone, flavanone, isoflavone, chalcone, coumarin, phthalates, parabens, benzophenone derivative, diphenylethylene, diphenylmethane, bibenzyl, stilbene, mycoestrogens and phenol. The same compounds have been tested for their binding affinity to the human sex hormone binding globulin (SHBG) [101], estrogen receptor (ER) [102] and androgen receptor (AR) [103] (Table 2). The binding data have been included in the FDA's endocrine disruptors knowledgebase (EDKB) [47, 104-106] that has been used in the development of the *in silico* prediction models using machine learning methods such as Decision Forest [107-118] and Mold2 molecular descriptors [119, 120] that are calculated using the algorithm developed for the expert systems of structure elucidation [121-126].

Among the 21 steroidal compounds (Figure 3A and B), compounds 9 and 17 have shown a good binding affinity towards rat AFP and human SHBG, respectively. The planar and the non-planar structure of the benzene ring A in steroid has shown a good activity and lack of binding affinity towards AFP, respectively. However, the planar ring along with the 3-hydroxyl group is more important for the tight binding of steroid to rat AFP and the ethyl or any other group at position-7 in the steroid inhibits the binding. The steroidal chemicals show greater binding affinity towards human SHBG than rat AFP. Out of the 15 flavones (Figure

3C), 9 were bound to rat AFP and 3 bound to human SHBG. The substitution of the larger group at position-2'or 6' diminished the binding affinity of flavones towards rat AFP. The hydroxyl group at 5, 6, 7, and 4' are favored positions for flavones to bind with rat AFP. Another class of flavonoids named as flavanones have estrogenic activity [100] and are derived from flavones by changing the double bond. In contrast to steroids the non-planar flavanone (Figure 3D) fits well in the binding site but the planar flavanone failed to show binding affinity towards rat AFP. Evidently, the chiral carbon S-configuration at atom 2 is the key motif for binding the flavonoids in the ligand binding site of rat AFP. Isoflavones (Figure 3E) that failed to bind to both rat AFP and SHBG are structurally different from flavones in the attachment of the phenyl group. The structure of isoflavones is similar to E2, hence it can compete with estrogen in the binding site of ER. The reason for the flavones binding and why isoflavones fail to bind with rat AFP might be that the carbonyl and phenyl groups are in the same plane at close proximity and it is difficult to form a hydrogen bond due to the presence of the phenyl group in isoflavones [100]. Chalcones (Figure 3F) are associated with several biological activities [127] such as binding to ER [102]. Only the trihydrochalcone showed a weak binding affinity towards the rat AFP and human SHBG. The size of Coumarin (Figure 3G) derivatives plays an important role in binding with rat AFP and ER [102]. Adding a large chemical group instead of small groups helps coumarin to occupy a full binding site present in rat AFP. In contrast with coumarin, the phthalates (Figure 3H) binding affinity decrease with molecular size. Parabens, esters of p-hydroxybenzoic acid (Figure 3I), show a weak estrogenic activity [102, 128]. Parabens show a similar binding affinity like coumarin towards rat AFP and are in contrast to phthalic acids. Parabens and coumarin bind well in the ligand binding site of rat AFP in presence of large hydrophobic groups instead of small groups. The hydroxyl group in benzophenone (Figure 3J) at position 1 or 1' and at position 3 or 3' is vital for the tight and diminished binding affinity with rat AFP, respectively. All the above observations suggest that different types of chemicals have the diffident binding orientations at the binding site of rat AFP.

It was reported that the diphenylethylene (Figure 3K) and diphenylmethane (Figure 3L) derivatives had shown an estrogenic activity. Interestingly, among 5 diphenylethylene derivatives only dihydroxymethoxychlor olefin showed an IC_{50} value of 0.001 μM and the remaining compounds fails to bind with rat AFP, suggesting that the hydroxyl group present in the 3 and 3' of the benzene ring and the chlorine atoms allowed the hydroxyl group to form hydrogen bonds with rat AFP. The remaining four diphenylethylene derivatives have larger side chains which inhibit the binding with rat AFP. Surprisingly, all six diphenylmethane derivatives failed to produce the binding affinity towards the rat AFP.

Previously, it was observed that chemicals with the bibenzyl scaffold (Figure 3M) have high binding affinity towards ER. Out of the seven bibenzyl derivatives, five had shown a binding affinity towards both ER and rat AFP. The reason for the failure of the other two molecules might be the lack of van der Waals interactions with rat AFP. The presence of 3 and 3'–OH group and the hydrophobic group in the middle of bibenzyl derivatives, forms a hydrogen bond and van der Waals interactions, plays a vital role in binding of the compounds with rat AFP. The two dimensional structure of stilbene (Figure 3N) is similar to bibenzyl except that the bond which connects the two benzene rings, is a double bond in stilbene and a single bond in bibenzyl. The rigid stilbene planar shape was not able to form a suitable ligand conformation to occupy the AFP binding pocket. These type of scaffold should have an –OH groups in the 3 and 3' position of the benzene ring to form the hydrogen bonds with the receptor and larger hydrophobic groups are required to bind with rat AFP. The R-configuration of mycoestrogen (Figure 3O) had shown a greater binding affinity towards rat AFP, due to the presence of chiral carbon atom-2 which enables the –OH groups to form a tight hydrogen bond with rat AFP.

7. Clinical Significance of AFP

Of clinical significance, AFP has been used as a serum biomarker for diagnosis of chronic hepatitis C. Recently, AFP attracted a significant interest due to the increasing evidence of the role of AFP in biological processes such as promoting cell growth function. Elevated serum AFP has been associated with various chronic liver diseases and hepatic regeneration and acute exacerbation of chronic hepatitis B in association with elevated transaminases. In addition, significant and mild elevation of AFP was occasionally found in CHC patients but not in patients with HCC.

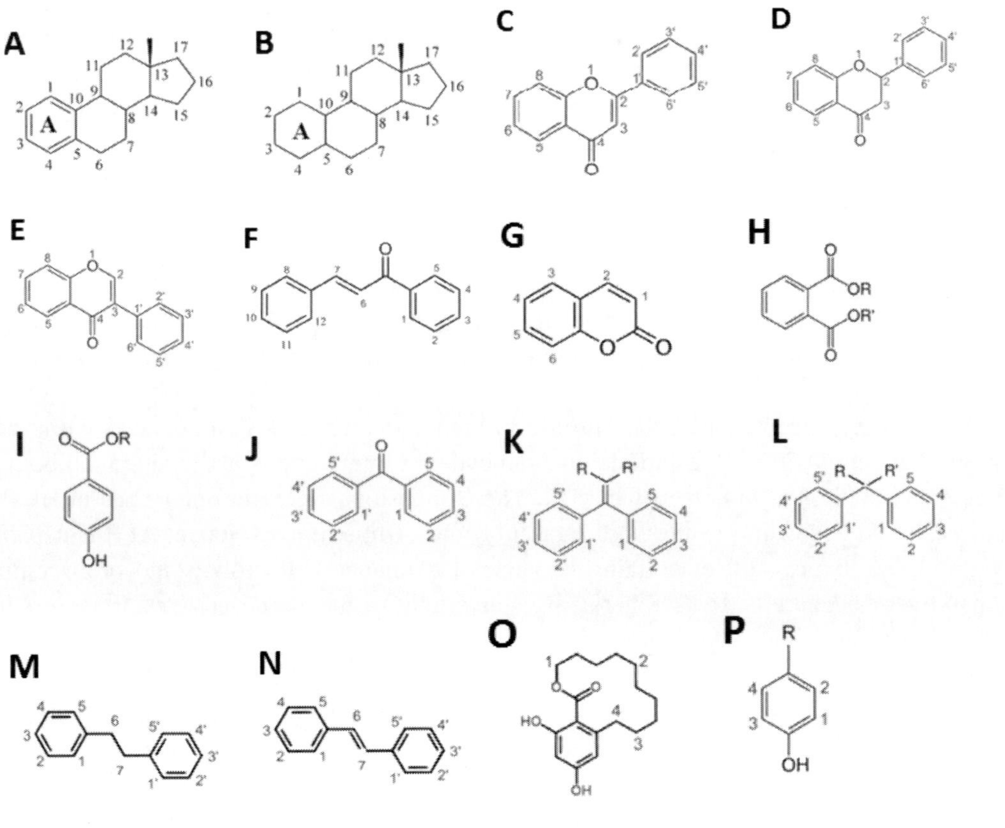

Figure 3. Chemical scaffolds tested for binding to rat AFP: chemicals with steroid structural backbones (A and B); flavones (C); flavanones (D); isoflavones (E); chalcones (F); coumarin derivatives (G); phthalates (H); parabens (I); benzophenone derivatives (J); diphenylethylene derivatives (K); diphenylmethane derivatives (L); bibenzyl derivatives (M); stilbene derivatives (N); mycoestrogens (O); chemicals with a phenol structural backbone (P).

During pregnancy, AFP acts as a positive and a negative marker of acute inflammation in postnatal and adulthood periods, respectively. In the second trimester, elevated amniotic fluid AFP showed the elevated levels of intracellular Adhesion Molecule-1 (ICAM-1), a known marker of acute inflammation, were correlated with pre-term birth, intrauterine growth retardation (IUGR), and prematurity. Elevated levels of AFP are associated with multiple pregnancy, congenital nephrosis, intrauterine fetal death, hydrocephalus, hemolytic disease secondary to Rh isoimmunization, omphalocele, duodenal or esophageal atresia, threatened or missed abortion, placental separation and open neural tube defects such as anencephaly, women with chronic villitis, meningocele, low birth weight offspring and myelocele [22, 27] in maternal blood and in amniotic fluid.

AFP has long been implicated with autoimmune disorders, including those of the central nervous system (CNS). AFP is one of the major factors produced during pregnancy. During pregnancy, AFP plays a protective role against autoimmunity and is associated with myasthenia gravis (MG). AFP inhibits antibody binding to acetylcholine receptor and has a potential to serve as a biomarker for MG and other autoimmune diseases [129]. AFP could be well tolerated and decreases various aspects of neuroinflammation, including disease severity, axonal loss and damage, T-cell reactivity and antigen presentation. AFP was found elevated in suppression of induced autoimmune arthritis and could be used as an immunosuppressive agent to improve the development of this inflammatory autoimmune disease.

Depending on the ontogeny states, AFP can serve as an acute and a chronic phase reactant. During pregnancy, AFP functions as a positive and negative acute phase protein in the embryo, fetus, and the placenta and also in postnatal and adulthood periods. The proteolytic processing of proinflammatory cytokine IL-1b was activated by the AFP-caspase complex, which is elevated in some forms of preterm birth and maternal metabolic disorders [73, 130]. Various states of inflammation and undesirable growth during histogenesis and organogenesis are modulated by AFP. Depending on the developmental stage of ontogeny, AFP can serve as an acute phase protein of inflammation in both a positive and a negative manner.

It is widely accepted that the chronic inflammatory process is associated with various types of cancers. AFP is a gestational age-dependent protein and is associated with both fetal defects and malignant tumor growth [11]. The expression of certain oncogenes in HCC was induced by AFP, leading to the high mortality rate of this tumor. Hence, AFP is capable of promoting the expression of certain oncogenes to enhance cell growth and proliferation of human liver cancer cells. Researchers have shown that women exposed to AFP are less likely to develop breast cancer. AFP showed a close association with HCC carcinogenesis [28, 89]. AFP was used as a vector for specific delivery of anticancer drug conjugates to target cells. Elevated AFP levels were identified in a few cases of malignant melanoma and a patient with basal cell carcinoma. Using immunofluorescence techniques, AFP was detected in Bowen's disease.

Future Perspectives

AFP can be used as a serum marker in cancer and fetal defects. It can be used in the detection of congenital defects. The measurement of serum AFP levels has been found to be useful in the diagnosis and prognosis of patients with various types of diseases such as cancer, inflammation and fetal disorders. AFP should be considered as a candidate target for preventing and/or treating congenital defects. Various studies are presently under progress to find novel and potent inhibitors for AFP. More research is required to determine which clinical applications and doses of small molecules will help to inhibit the function of AFP in a way to accomplish the desired therapeutic outcomes.

ACKNOWLEDGMENT

This research was supported in part by an appointment to the Research Participation Program at the National Center for Toxicological Research (Sugunadevi Sakkiah, Hui Wen Ng and Hao Ye) administered by the Oak Ridge Institute for Science and Education through an interagency agreement between the U.S. Department of Energy and the U.S. Food and Drug Administration.

DISCLAIMER

The content is solely the responsibility of the authors and does not necessarily represent the official views of the Food and Drugs Administration.

NOTE

The findings and conclusions in this article have not been formally disseminated by the US Food and Drug Administration (FDA) and should not be construed to represent the FDA determination or policy.

REFERENCES

[1] Mizejewski, G. J. and MacColl, R. (2003) Alpha-fetoprotein growth inhibitory peptides: potential leads for cancer therapeutics, *Mol Cancer Ther.* 2, 1243-55.
[2] Bergstrand, C. G. and Czar, B. (1956) Demonstration of a new protein fraction in serum from the human fetus, *Scand J Clin Lab Invest.* 8, 174.
[3] Mizejewski, G. J. (1995) Alpha-fetoprotein signal sequences: a proposed mechanism for subcellular localization and organelle targeting, *Journal of Theoretical Biology.* 176, 103-113.
[4] Soltani, K. (1979) Alpha-Fetoprotein: A Review, *J Investig Dermatol.* 72, 211-213.
[5] Gitlin, D., Perricelli, A. and Gitlin, G. M. (1972) Synthesis of -fetoprotein by liver, yolk sac, and gastrointestinal tract of the human conceptus, *Cancer Res.* 32, 979-82.
[6] Terentiev, A. A. and Moldogazieva, N. T. (2006) Structural and functional mapping of alpha-fetoprotein, *Biochemistry (Mosc).* 71, 120-32.

[7] Abelev, G. I., Perova, S. D., Khramkova, N. I., Postnikova, Z. A. and Irlin, I. S. (1963) Production of embryonal alpha-globulin by transplantable mouse hepatomas, *Transplantation.* 1, 174-80.

[8] S, T. Y. (1963) in I All-Union Biochem. Cong. Book of Abstr. in, Academy of Sciences of the USSR, Moscow-Leningrad.

[9] Brock, D. J. and Sutcliffe, R. G. (1972) Alpha-fetoprotein in the antenatal diagnosis of anencephaly and spina bifida, *Lancet.* 2, 197-9.

[10] Mizejewski, G. J. (2004) Biological roles of alpha-fetoprotein during pregnancy and perinatal development, *Exp Biol Med (Maywood).* 229, 439-63.

[11] Mizejewski, G. J. (2001) Alpha-fetoprotein structure and function: relevance to isoforms, epitopes, and conformational variants, *Exp Biol Med (Maywood).* 226, 377-408.

[12] Uriel, J., de Nechaud, B. and Dupiers, M. (1972) Estrogen-binding properties of rat, mouse and man fetospecific serum proteins. Demonstration by immuno-autoradiographic methods, *Biochem Biophys Res Commun.* 46, 1175-80.

[13] Li, M., Li, H., Li, C., Wang, S., Jiang, W., Liu, Z., Zhou, S., Liu, X., McNutt, M. A. and Li, G. (2011) Alpha-fetoprotein: A new member of intracellular signal molecules in regulation of the PI3K/AKT signaling in human hepatoma cell lines, *International Journal of Cancer.* 128, 524-532.

[14] Mizejewski, G. J., Dias, J. A., Hauer, C. R., Henrikson, K. P. and Gierthy, J. (1996) Alpha-fetoprotein derived synthetic peptides: assay of an estrogen-modifying regulatory segment, *Molecular and Cellular Endocrinology.* 118, 15-23.

[15] Wang, S., Jiang, W., Chen, X., Zhang, C., Li, H., Hou, W., Liu, Z., McNutt, M. A., Lu, F. and Li, G. (2012) Alpha-fetoprotein acts as a novel signal molecule and mediates transcription of Fn14 in human hepatocellular carcinoma, *J Hepatol.* 57, 322-9.

[16] McLeod, J. F. and Cooke, N. E. (1989) The vitamin D-binding protein, alpha-fetoprotein, albumin multigene family: detection of transcripts in multiple tissues, *Journal of Biological Chemistry.* 264, 21760-21769.

[17] Lichenstein, H. S., Lyons, D. E., Wurfel, M. M., Johnson, D. A., McGinley, M. D., Leidli, J. C., Trollinger, D. B., Mayer, J. P., Wright, S. D. and Zukowski, M. M. (1994) Afamin is a new member of the albumin, alpha-fetoprotein, and vitamin D-binding protein gene family, *Journal of Biological Chemistry.* 269, 18149-18154.

[18] Ray, K., Wang, X. K., Zhao, M. and Cooke, N. E. (1991) The rat vitamin D binding protein (Gc-globulin) gene. Structural analysis, functional and evolutionary correlations, *Journal of Biological Chemistry.* 266, 6221-6229.

[19] Bélanger, L., Roy, S. and Allard, D. (1994) New albumin gene 3' adjacent to the alpha 1-fetoprotein locus, *Journal of Biological Chemistry.* 269, 5481-5484.

[20] Gutteridge, J. M. C. (1986) Antioxidant properties of the proteins caeruloplasmin, albumin and transferrin. A study of their activity in serum and synovial fluid from patients with rheumatoid arthritis, *Biochimica et Biophysica Acta (BBA) - Protein Structure and Molecular Enzymology.* 869, 119-127.

[21] Nathan, C., Xie, Q. W., Halbwachs-Mecarelli, L. and Jin, W. W. (1993) Albumin inhibits neutrophil spreading and hydrogen peroxide release by blocking the shedding of CD43 (sialophorin, leukosialin), *J Cell Biol.* 122, 243-56.

[22] Tomasi, T. B., Jr. (1977) Structure and function of alpha-fetoprotein, *Annu Rev Med.* 28, 453-65.

[23] MacLusky, N. J. and Naftolin, F. (1981) Sexual differentiation of the central nervous system, *Science.* 211, 1294-302.

[24] Leighton, P. C., Kitau, M. J., Gordon, Y. B., Leek, A. E. and Chard, T. (1975) Levels of Alpha-Fetoprotein in, aternal blood as a screening test for fetal neural-tube defect, *The Lancet.* 306, 1012-1015.

[25] Cuckle, H. S., Wald, N. J. and Lindenbaum, R. H. (1984) Maternal serum alpha-fetoprotein measurement: a screening test for Down syndrome, *Lancet.* 1, 926-9.

[26] Kuo, M. T., Iyer, B., Wu, J. R., Lapeyre, J. N. and Becker, F. F. (1984) Methylation of the alpha-fetoprotein gene in productive and nonproductive rat hepatocellular carcinomas, *Cancer Res.* 44, 1642-7.

[27] Yachnin, S. (1978) The clinical significance of human alpha-fetoprotein, *Annals of Clinical and Laboratory Science.* 8, 84-90.

[28] Mizejewski, G. J. (2002) Biological role of alpha-fetoprotein in cancer: prospects for anticancer therapy, *Expert Rev Anticancer Ther.* 2, 709-35.

[29] Mizejewski, G. J. and Warner, A. S. (1989) Alpha-fetoprotein can regulate growth in the uterus of the immature and adult ovariectomized mouse, *Journal of Reproduction and Fertility.* 85, 177-185.

[30] Mizejewski, G. J., Keenan, J. F. and Setty, R. P. (1990) Separation of the estrogen-activated growth-regulatory forms of alpha-fetoprotein in mouse amniotic fluid, *Biology of Reproduction.* 42, 887-898.

[31] Laborda, J., Naval, J., Allouche, M., Calvo, M., Georgoulias, V., Mishal, Z. and Uriel, J. (1987) Specific uptake of alpha-fetoprotein by malignant human lymphoid cells, *Int J Cancer.* 40, 314-8.

[32] Alava, M. A., Iturralde, M., Lampreave, F. and Pineiro, A. (1999) Specific uptake of alpha-fetoprotein and albumin by rat Morris 7777 hepatoma cells, *Tumour Biol.* 20, 52-64.

[33] Uriel, J., Faivre-Bauman, A., Trojan, J. and Foiret, D. (1981) Immunocytochemical demonstration of alpha-fetoprotein uptake by primary cultures of fetal hemisphere cells from mouse brain, *Neurosci Lett.* 27, 171-5.

[34] Hajeri-Germond, M., Trojan, J., Uriel, J. and Hauw, J. J. (1983) In vitro uptake of exogenous alpha-fetoprotein by chicken dorsal root ganglia, *Dev Neurosci.* 6, 111-7.

[35] Uriel, J., Trojan, J., Dubouch, P. and Pineiro, A. (1982) Intracellular alpha-fetoprotein and albumin in the developing nervous system of the baboon, *Pathol Biol (Paris).* 30, 79-83.

[36] Terentiev, A. A., Moldogazieva, N. T., Levtsova, O. V., Maximenko, D. M., Borozdenko, D. A. and Shaitan, K. V. (2012) Modeling of three dimensional structure of human Alpha-Fetoprotein complexed with diethylstilbestrol: Docking and molecular dynamics simualtion study, *Journal of Bioinformatics and Computational Biology.* 10, 1241012.

[37] 37. Morinaga, T., Sakai, M., Wegmann, T. G. and Tamaoki, T. (1983) Primary structures of human alpha-fetoprotein and its mRNA, *Proceedings of the National Academy of Sciences of the United States of America.* 80, 4604-4608.

[38] 38. Pucci, P., Siciliano, R., Malorni, A., Marino, G., Tecce, M. F., Ceccarini, C. and Terrana, B. (1991) Human alpha-fetoprotein primary structure: a mass spectrometric study, *Biochemistry.* 30, 5061-6.

[39] Bergstrand, C. G. and Czar, B. (1957) Paper electrophoretic study of human fetal serum proteins with demonstration of a new protein fraction, *Scand J Clin Lab Invest.* 9, 277-86.

[40] Nishi, S., Matsue, H., Yoshida, H., Yamaoto, R. and Sakai, M. (1991) Localization of the estrogen-binding site of alpha-fetoprotein in the chimeric human-rat proteins, *Proc Natl Acad Sci U S A.* 88, 3102-5.

[41] Hérve, F., Gentin, M., Rajkowski, K. M., Wong, L. T., Hsia, C. J. C. and Cittanova, N. (1990) Estrogen-binding properties of rat serum α1-fetoprotein and its isoforms. Investigation of the apparent non-integrality of sites on the unfractionated protein, *Journal of Steroid Biochemistry.* 36, 319-324.

[42] Baker, M. E. (1988) Evolution of alpha-fetoprotein: sequence comparisons among AFP species and with albumin species, *Tumour Biol.* 9, 123-36.

[43] Gorin, M. B., Cooper, D. L., Eiferman, F., van de Rijn, P. and Tilghman, S. M. (1981) The evolution of alpha-fetoprotein and albumin. I. A comparison of the primary amino acid sequences of mammalian alpha-fetoprotein and albumin, *J Biol Chem.* 256, 1954-9.

[44] Deutsch, H. F. (1991) Chemistry and biology of alpha-fetoprotein, *Adv Cancer Res.* 56, 253-312.

[45] Ruoslahti, E. and Terry, W. D. (1976) alpha foetoprotein and serum albumin show sequence homology, *Nature.* 260, 804-5.

[46] Mizejewski, G. J. (1997) alpha-fetoprotein as a biologic response modifier: relevance to domain and subdomain structure, *Proc Soc Exp Biol Med.* 215, 333-62.

[47] Shen, J., Zhang, W., Fang, H., Perkins, R., Tong, W. and Hong, H. (2013) Homology modeling, molecular docking, and molecular dynamics simulations elucidated α-fetoprotein binding modes, *BMC Bioinformatics.* 14, S6-S6.

[48] Gillespie, J. R. and Uversky, V. N. (2000) Structure and function of alpha-fetoprotein: a biophysical overview, *Biochim Biophys Acta.* 1480, 41-56.

[49] Breborowicz, J. (1988) Microheterogeneity of Human Alphafetoprotein, *Tumor Biology.* 9, 3-14.

[50] Zorin, N. A., Krivonosov, S. K., Zorin, R. M., Terentiev, A. A. and Ionova, N. K. (1986) Competitive inhibition of interactions between phytohemaglutinin A and blood serum glycoproteins by carbohydrates, *Vopr Med Khim.* 32, 96-98.

[51] Krivonosov, S. K., Zorin, R. A., Ionova, N. K., Terentiev, A. A. and Tatarinov, Y. S. (1985) Interaction of pregnancy-associated proteins with hemagglutinin and concanavalin A, *Byul Eksp Biol Med.* 5, 568-571.

[52] Taketa, K. (1998) Characterization of sugar chain structures of human alpha-fetoprotein by lectin affinity electrophoresis. ibarahp@oka.urban.ne.jp, *Electrophoresis.* 19, 2595-602.

[53] Uriel, J., Laborda, J., Nadal, A., Gabant, P., Mizejewski, G. J. and Jacobson, H. I. (1989) *Biological Activities of alpha1-fetoprotein*, CRC Press, Boca Raton.

[54] Mizejewski, G. J. (1995) The Phylogeny of Alpha-Fetoprotein in Vertebrates: Survey of Biochemical and Physiological Data. 5, 281-316.

[55] Ruiz-Gutierrez, V., Moreno, R., Moreda, W., Copado, M. A. and Rodriguez-Burgos, A. (2001) Detection of squalene in alpha-fetoprotein and fetal serum albumin from bovine, *J Protein Chem.* 20, 19-23.

[56] Nishihira, J., Koyama, Y., Sakai, M. and Nishi, S. (1993) The fatty acid binding site of human alpha-fetoprotein, *Biochem Biophys Res Commun.* 196, 1049-57.
[57] Aoyagi, Y., Ikenaka, T. and Ichida, F. (1978) Copper(II)-binding ability of human alpha-fetoprotein, *Cancer Res.* 38, 3483-6.
[58] Ruoslahti, E. and Seppala, M. (1971) Studies of carcino-fetal proteins: physical and chemical properties of human alpha-fetoprotein, *Int J Cancer.* 7, 218-25.
[59] Yakimenko, E. F., Yazova, A. K., Goussev, A. I. and Abelev, G. I. (2001) Epitope mapping of human alpha-fetoprotein, *Biochemistry (Mosc).* 66, 524-30.
[60] Karamova, E. R., Yazova, A. K., Goussev, A. I., Yakimenko, E. F. and Abelev, G. I. (1998) Conformational Variants of Human Alpha-Fetoprotein, *Tumor Biology.* 19, 310-317.
[61] Yakimenko, E. F., Karamova, E. R., Goussev, A. I., Hilgers, J., Abelev, G. I. and Yazova, A. K. (1998) Epitope Mapping of Human Alpha-Fetoprotein, *Tumor Biology.* 19, 301-309.
[62] Savu, L., Benassayag, C., Vallette, G., Christeff, N. and Nunez, E. (1981) Mouse alpha 1-fetoprotein and albumin. A comparison of their binding properties with estrogen and fatty acid ligands, *J Biol Chem.* 256, 9414-8.
[63] Savu, L., Benassayag, C., Vallette, G., Nunez, E. and Jayle, M. F. (1977) Purification and estrogen binding properties of mouse alpha1-fetoprotein and of two forms of the protein with different affinities for concanavalin-A, *Biochimie.* 59, 323-8.
[64] Uriel, J., Bouillon, D., Aussel, C. and Dupiers, M. (1976) Alpha-fetoprotein: the major high-affinity estrogen binder in rat uterine cytosols, *Proc Natl Acad Sci U S A.* 73, 1452-6.
[65] Seppala, M. (1975) Fetal pathophysiology of human alpha-fetoprotein, *Ann N Y Acad Sci.* 259, 59-73.
[66] Vlach, J., Hennecke, S. and Amati, B. (1997) Phosphorylation-dependent degradation of the cyclin-dependent kinase inhibitor p27, *EMBO J.* 16, 5334-44.
[67] Mizejewski, G. J. (2011) Therapeutic use of human alpha-fetoprotein in clinical patients: Is a cancer risk involved?, *International Journal of Cancer.* 128, 239-242.
[68] Li, M., Li, H., Li, C., Guo, L., Liu, H., Zhou, S., Liu, X., Chen, Z., Shi, S., Wei, J., McNutt, M. A. and Li, G. (2009) Cytoplasmic alpha-fetoprotein functions as a co-repressor in RA-RAR signaling to promote the growth of human hepatoma Bel 7402 cells, *Cancer Lett.* 285, 190-9.
[69] Wang, L. and Levin, M. S. (2005) Suppression of FGF signaling: a putative mechanism for the chemopreventive effects of acyclic retinoid in hepatocellular carcinoma, *Gastroenterology.* 128, 228-31.
[70] Li, M. S., Li, P. F., Chen, Q., Du, G. G. and Li, G. (2004) Alpha-fetoprotein stimulated the expression of some oncogenes in human hepatocellular carcinoma Bel 7402 cells, *World J Gastroenterol.* 10, 819-24.
[71] Li, P., Wang, S. S., Liu, H., Li, N., McNutt, M. A., Li, G. and Ding, H. G. (2011) Elevated serum alpha fetoprotein levels promote pathological progression of hepatocellular carcinoma, *World J Gastroenterol.* 17, 4563-71.
[72] Li, C., Wang, S., Jiang, W., Li, H., Liu, Z., Zhang, C., McNutt, M. A. and Li, G. (2012) Impact of intracellular alpha fetoprotein on retinoic acid receptors-mediated expression of GADD153 in human hepatoma cell lines, *Int J Cancer.* 130, 754-64.

[73] Li, M., Li, H., Li, C., Zhou, S., Guo, L., Liu, H., Jiang, W., Liu, X., Li, P., McNutt, M. A. and Li, G. (2009) Alpha fetoprotein is a novel protein-binding partner for caspase-3 and blocks the apoptotic signaling pathway in human hepatoma cells, *Int J Cancer.* 124, 2845-54.

[74] Kaplan, M. J., Lewis, E. E., Shelden, E. A., Somers, E., Pavlic, R., McCune, W. J. and Richardson, B. C. (2002) The apoptotic ligands TRAIL, TWEAK, and Fas ligand mediate monocyte death induced by autologous lupus T cells, *J Immunol.* 169, 6020-9.

[75] Li, M. S., Li, P. F., Yang, F. Y., He, S. P., Du, G. G. and Li, G. (2002) The intracellular mechanism of alpha-fetoprotein promoting the proliferation of NIH 3T3 cells, *Cell Res.* 12, 151-6.

[76] Laderoute, M. P. and Pilarski, L. M. (1994) The inhibition of apoptosis by alpha-fetoprotein (AFP) and the role of AFP receptors in anti-cellular senescence, *Anticancer Res.* 14, 2429-38.

[77] Semenkova, L. N., Dudich, E. I., Dudich, I. V., Shingarova, L. N. and Korobko, V. G. (1997) Alpha-fetoprotein as a TNF resistance factor for the human hepatocarcinoma cell line HepG2, *Tumour Biol.* 18, 30-40.

[78] Kawai, T. and Akira, S. (2007) TLR signaling, *Semin Immunol.* 19, 24-32.

[79] Hanke, P., Rabe, C., Serwe, M., Bohm, S., Pagenstecher, C., Sauerbruch, T. and Caselmann, W. H. (2002) Cirrhotic patients with or without hepatocellular carcinoma harbour AFP-specific T-lymphocytes that can be activated in vitro by human alpha-fetoprotein, *Scand J Gastroenterol.* 37, 949-55.

[80] Ayaru, L., Pereira, S. P., Alisa, A., Pathan, A. A., Williams, R., Davidson, B., Burroughs, A. K., Meyer, T. and Behboudi, S. (2007) Unmasking of alpha-fetoprotein-specific CD4(+) T cell responses in hepatocellular carcinoma patients undergoing embolization, *J Immunol.* 178, 1914-22.

[81] Sherman, M. (2001) Alphafetoprotein: an obituary, *J Hepatol.* 34, 603-5.

[82] Luscher, B. and Eisenman, R. N. (1990) New light on Myc and Myb. Part I. Myc, *Genes Dev.* 4, 2025-35.

[83] Secombe, J., Pierce, S. B. and Eisenman, R. N. (2004) Myc: a weapon of mass destruction, *Cell.* 117, 153-6.

[84] Martin, K. J. (1991) The interactions of transcription factors and their adaptors, coactivators and accessory proteins, *Bioessays.* 13, 499-503.

[85] Robledo, M. M., Bartolome, R. A., Longo, N., Rodriguez-Frade, J. M., Mellado, M., Longo, I., van Muijen, G. N., Sanchez-Mateos, P. and Teixido, J. (2001) Expression of functional chemokine receptors CXCR3 and CXCR4 on human melanoma cells, *J Biol Chem.* 276, 45098-105.

[86] Vakharia, D. and Mizejewski, G. J. (2000) Human alpha-fetoprotein peptides bind estrogen receptor and estradiol, and suppress breast cancer, *Breast Cancer Res Treat.* 63, 41-52.

[87] Mizejewski, G. J. (2011) Mechanism of Cancer Growth Suppression of Alpha-Fetoprotein Derived Growth Inhibitory Peptides (GIP): Comparison of GIP-34 versus GIP-8 (AFPep). Updates and Prospects, *Cancers (Basel).* 3, 2709-33.

[88] Bennett, J. A., Mesfin, F. B., Andersen, T. T., Gierthy, J. F. and Jacobson, H. I. (2002) A peptide derived from alpha-fetoprotein prevents the growth of estrogen-dependent human breast cancers sensitive and resistant to tamoxifen, *Proc Natl Acad Sci U S A.* 99, 2211-5.

[89] Mizejewski, G. J., Eisele, L. and Maccoll, R. (2006) Anticancer versus antigrowth activities of three analogs of the growth-inhibitory peptide: relevance to physicochemical properties, *Anticancer Res.* 26, 3071-6.

[90] MacColl, R., Eisele, L. E., Stack, R. F., Hauer, C., Vakharia, D. D., Benno, A., Kelly, W. C. and Mizejewski, G. J. (2001) Interrelationships among biological activity, disulfide bonds, secondary structure, and metal ion binding for a chemically synthesized 34-amino-acid peptide derived from alpha-fetoprotein, *Biochim Biophys Acta.* 1528, 127-34.

[91] Kirschner, K. N., Lexa, K. W., Salisburg, A. M., Alser, K. A., Joseph, L., Andersen, T. T., Bennett, J. A., Jacobson, H. I. and Shields, G. C. (2007) Computational design and experimental discovery of an antiestrogenic peptide derived from alpha-fetoprotein, *J Am Chem Soc.* 129, 6263-8.

[92] 92. Shields, G. C. (2009) Computational approaches for the design of peptides with anti-breast cancer properties, *Future Med Chem.* 1, 201-12.

[93] Muehlemann, M., Miller, K. D., Dauphinee, M. and Mizejewski, G. J. (2005) Review of Growth Inhibitory Peptide as a biotherapeutic agent for tumor growth, adhesion, and metastasis, *Cancer Metastasis Rev.* 24, 441-67.

[94] Mizejewski, G. J. (2007) The alpha-fetoprotein-derived growth inhibitory peptide 8-mer fragment: review of a novel anticancer agent, *Cancer Biother Radiopharm.* 22, 73-98.

[95] Mesfin, F. B., Bennett, J. A., Jacobson, H. I., Zhu, S. and Andersen, T. T. (2000) Alpha-fetoprotein-derived antiestrotrophic octapeptide, *Biochim Biophys Acta.* 1501, 33-43.

[96] Mesfin, F. B., Andersen, T. T., Jacobson, H. I., Zhu, S. and Bennett, J. A. (2001) Development of a synthetic cyclized peptide derived from alpha-fetoprotein that prevents the growth of human breast cancer, *J Pept Res.* 58, 246-56.

[97] Bennett, J. A., DeFreest, L., Anaka, I., Saadati, H., Balulad, S., Jacobson, H. I. and Andersen, T. T. (2006) AFPep: an anti-breast cancer peptide that is orally active, *Breast Cancer Res Treat.* 98, 133-41.

[98] DeFreest, L. A., Mesfin, F. B., Joseph, L., McLeod, D. J., Stallmer, A., Reddy, S., Balulad, S. S., Jacobson, H. I., Andersen, T. T. and Bennett, J. A. (2004) Synthetic peptide derived from alpha-fetoprotein inhibits growth of human breast cancer: investigation of the pharmacophore and synthesis optimization, *J Pept Res.* 63, 409-19.

[99] Joseph, L. C., Bennett, J. A., Kirschner, K. N., Shields, G. C., Hughes, J., Lostritto, N., Jacobson, H. I. and Andersen, T. T. (2009) Antiestrogenic and anticancer activities of peptides derived from the active site of alpha-fetoprotein, *J Pept Sci.* 15, 319-25.

[100] Hong, H., Branham, W. S., Dial, S. L., Moland, C. L., Fang, H., Shen, J., Perkins, R., Sheehan, D. and Tong, W. (2012) Rat alpha-Fetoprotein binding affinities of a large set of structurally diverse chemicals elucidated the relationships between structures and binding affinities, *Chem Res Toxicol.* 25, 2553-66.

[101] Hong, H., Branham, W. S., Ng, H. W., Moland, C. L., Dial, S. L., Fang, H., Perkins, R., Sheehan, D. and Tong, W. (2015) Human sex hormone-binding globulin binding affinities of 125 structurally diverse chemicals and comparison with their binding to androgen receptor, estrogen receptor, and alpha-fetoprotein, *Toxicol Sci.* 143, 333-48.

[102] Blair, R. M., Fang, H., Branham, W. S., Hass, B. S., Dial, S. L., Moland, C. L., Tong, W., Shi, L., Perkins, R. and Sheehan, D. M. (2000) The estrogen receptor relative

binding affinities of 188 natural and xenochemicals: structural diversity of ligands, *Toxicol Sci.* 54, 138-53.

[103] Fang, H., Tong, W., Branham, W. S., Moland, C. L., Dial, S. L., Hong, H., Xie, Q., Perkins, R., Owens, W. and Sheehan, D. M. (2003) Study of 202 natural, synthetic, and environmental chemicals for binding to the androgen receptor, *Chem Res Toxicol.* 16, 1338-58.

[104] Ding, D., Xu, L., Fang, H., Hong, H., Perkins, R., Harris, S., Bearden, E. D., Shi, L. and Tong, W. (2010) The EDKB: an established knowledge base for endocrine disrupting chemicals, *BMC Bioinformatics.* 11, S5-S5.

[105] Shen, J., Xu, L., Fang, H., Richard, A. M., Bray, J. D., Judson, R. S., Zhou, G., Colatsky, T. J., Aungst, J. L., Teng, C., Harris, S. C., Ge, W., Dai, S. Y., Su, Z., Jacobs, A. C., Harrouk, W., Perkins, R., Tong, W. and Hong, H. (2013) EADB: an estrogenic activity database for assessing potential endocrine activity, *Toxicol Sci.* 135, 277-91.

[106] Ng, H. W., Perkins, R., Tong, W. and Hong, H. (2014) Versatility or promiscuity: the estrogen receptors, control of ligand selectivity and an update on subtype selective ligands, *Int J Environ Res Public Health.* 11, 8709-42.

[107] Tong, W., Hong, H., Fang, H., Xie, Q. and Perkins, R. (2003) Decision Forest: Combining the Predictions of Multiple Independent Decision Tree Models, *Journal of Chemical Information and Computer Sciences.* 43, 525-31.

[108] Hong, H., Tong, W., Perkins, R., Fang, H., Xie, Q. and Shi, L. (2004) Multiclass Decision Forest--a novel pattern recognition method for multiclass classification in microarray data analysis, *DNA Cell Biol.* 23, 685-94.

[109] Xie, Q., Ratnasinghe, L., Hong, H., Perkins, R., Tang, Z.-Z., Hu, N., Taylor, P. and Tong, W. (2005) Decision Forest Analysis of 61 Single Nucleotide Polymorphisms in a Case-Control Study of Esophageal Cancer; a novel method, *BMC Bioinformatics.* 6, S4.

[110] Hong, H., Tong, W., Xie, Q., Fang, H. and Perkins, R. (2005) An in silico ensemble method for lead discovery: decision forest, *SAR QSAR Environ Res.* 16, 339-47.

[111] Ng, H. W., Doughty, S. W., Luo, H., Ye, H., Ge, W., Tong, W. and Hong, H. (2015) Development and Validation of Decision Forest Model for Estrogen Receptor Binding Prediction of Chemicals Using Large Data Sets, *Chemical Research in Toxicology.* DOI: 10.1021/acs.chemrestox.5b00358.

[112] McPhail, B., Tie, Y., Hong, H., Pearce, B. A., Schnackenberg, L. K., Ge, W., Valerio, L. G., Fuscoe, J. C., Tong, W., Buzatu, D. A.; et al. (2012) Modeling chemical interaction profiles: I. Spectral data-activity relationship and structure-activity relationship models for inhibitors and non-inhibitors of cytochrome P450 CYP3A4 and CYP2D6 isozymes, *Molecules.* 17, 3283-306.

[113] Tong, W., Hong, H., Xie, Q., Shi, L., Fang, H., Perkins, R. (2005) Assessing QSAR limitations-a regulatory perspective, *Current Computer-Aided Drug Design.* 1, 195-205.

[114] Tie, Y., McPhail, B., Hong, H., Pearce, B. A., Schnackenberg, L. K., Ge, W., Buzatu, D. A., Wilkes, J. G., Fuscoe, J. C., Tong, W.; et al. (2012) Modeling chemical interaction profiles: II. Molecular docking, spectral data-activity relationship, and structure-activity relationship models for potent and weak inhibitors of cytochrome p450 cyp3A4 isozyme, Molecules. 17, 3407-60.

[115] Ng, H. W., Shu, M., Luo, H., Ye, H., Ge, W., Perkins, R., Tong, W. and Hong, H. (2015) Estrogenic Activity Data Extraction and in Silico Prediction Show the Endocrine Disruption Potential of Bisphenol a Replacement Compounds, *Chemical Research in Toxicology*. 28, 1784–1795.

[116] Tong, W., Xie, Q., Hong, H., Shi, L., Fang, H., Perkins, R. and Petricoin, E. F. (2004) Using Decision Forest to Classify Prostate Cancer Samples on the Basis of Seldi-Tof Ms Data: Assessing Chance Correlation and Prediction Confidence, *Environmental Health Perspectives*. 112, 1622-7.

[117] Hong, H., Tong, W., Fang, H., Shi, L., Xie, Q., Wu, J., Perkins, R., Walker, J. D., Branham, W. and Sheehan, D. M. (2002) Prediction of Estrogen Receptor Binding for 58,000 Chemicals Using an Integrated System of a Tree-Based Model with Structural Alerts, *Environmental Health Perspectives*. 110, 29-36.

[118] Hong, H., Fang, H., Xie, Q., Perkins, R., Sheehan, D. M. and Tong, W. (2003) Comparative Molecular Field Analysis (Comfa) Model Using a Large Diverse Set of Natural, Synthetic and Environmental Chemicals for Binding to the Androgen Receptor, *SAR and QSAR in Environmental Research*. 14, 373-88.

[119] Hong, H., Xie, Q., Ge, W., Qian, F., Fang, H., Shi, L., Su, Z., Perkins, R. and Tong, W. (2008) Mold2, Molecular Descriptors from 2D Structures for Chemoinformatics and Toxicoinformatics, *Journal of Chemical Information and Modeling*. 48, 1337-44.

[120] Hong, H., Slavov, S., Ge, W., Qian, F., Su, Z., Fang, H., Cheng, Y., Perkins, R., Shi, L. and Tong, W. (2012) Mold2 Molecular Descriptors for QSAR in *Statistical Modelling of Molecular Descriptors in QSAR/QSPR* pp. 65-109, Wiley-VCH Verlag GmbH and Co. KGaA.

[121] Hong, H., Xin, X. (1990) ESSESA: an expert system for structure elucidation from spectra analysis. 1. The knowledge base of infrared spectra and analysis and interpretation program, *J Chem Inf Comput Sci*. 30, 203-10.

[122] Hong, H., Xin, X. (1992) ESSESA: an expert system for structure elucidation from spectra analysis. 2. A novel algorithm of perception of the linear independent smallest set of smallest rings, *Analytica Chimica Acta*. 262, 179-91.

[123] Hong, H., Xin, X. (1992) ESSESA: an expert system for structure elucidation from spectra analysis. 3. LNSCS for chemical knowledge representation, *J Chem Inf Comput Sci*. 32, 116-20.

[124] Hong, H., Xin, X. (1994) ESSESA: an expert system for structure elucidation from spectra analysis. 4. Canonical representation of structures, *J Chem Inf Comput Sci*. 34, 730-4.

[125] Hong, H., Xin, X. (1994) ESSESA: an expert system for structure elucidation from spectra analysis. 5. Substructure constraints from from analysis of first-order 1H-NMR spectra, *J Chem Inf Comput Sci*. 34, 1259-66.

[126] Masui, H., Hong, H. (2006) Spec2D: a structure elucidation system based on 1H NMR and H-H COSY spectra in organic chemistry, *J Chem Inf Model*. 46, 775-87.

[127] Go, M. L., Wu, X. and Liu, X. L. (2005) Chalcones: an update on cytotoxic and chemoprotective properties, *Curr Med Chem*. 12, 481-99.

[128] Byford, J. R., Shaw, L. E., Drew, M. G., Pope, G. S., Sauer, M. J. and Darbre, P. D. (2002) Oestrogenic activity of parabens in MCF7 human breast cancer cells, *J Steroid Biochem Mol Biol*. 80, 49-60.

[129] Abramsky, O. and Brenner, T. (1979) Alpha-fetoprotein inhibits antibody binding to acetylcholine receptor. Significance for myasthenia gravis and other autoimmune diseases, *Isr J Med Sci.* 15, 943-5.

[130] Khan, R. N. and Hay, D. P. (2015) A clear and present danger: inflammasomes DAMPing down disorders of pregnancy, *Hum Reprod Update.* 21, 388-405.

In: Alpha-Fetoprotein: Functions and Clinical Applications
Editors: Nisha Lakhi and Michael Moretti
ISBN: 978-1-63484-875-6

Chapter 14

THE ALPHA-FETOPROTEIN RECEPTOR (RECAF): CHARACTERIZATION AND POTENTIAL USES FOR CANCER DIAGNOSIS AND THERAPY

Ricardo Moro*, MD

Program Chief and Director, Pacific Biosciences Research Centre
Richmond, British Columbia, Canada

ABSTRACT

Alpha-fetoprotein (AFP) was the first oncofetal antigen of clinical significance. Over 20000 papers have been published on this protein, covering from its physiology to its applications for cancer diagnosis and management. AFP has also been used as a tool to study cell differentiation and malignant transformation. During the course of those studies, AFP was found inside fetal and embryonic cells that do not synthesize it and therefore several groups started working on the hypothesis that AFP was internalized via a receptor mediated mechanism of endocytosis. This receptor for AFP (RECAF) would be expressed in most tissues and its expression would be related to the stage of cell differentiation.

Cancer cells are poorly differentiated and therefore, RECAF could be re-expressed in malignant cells, thus serving as a broad spectrum cancer marker. This was later confirmed, and since then, evidence has been accumulating supporting the use of RECAF in medical practice: (a) for detecting cancer cells on tissue sections, (b) as a circulating cancer marker for cancer diagnosis, (c) for scintigraphy, (d) for flow cytometry and (e) for cancer cell targeting. In this chapter, we review the literature on the AFP receptor and we present some original data on its biochemistry as well as on its potential uses in medicine.

Keywords: AFP, AFP fragment, AFP uptake, receptor, RECAF, binding, serum test, immunohistology, targeted therapy, cancer, tumor, malignant, leukemia, oncofetal, cell differentiation, monoclonal antibody, apoptosis, peptide, scintigraphy, immunoscintigraphy, imaging, glycan, glycoprotein

---

* Corresponding author: ricardoip@yahoo.com.

INTRODUCTION

In the literature, the term "AFP receptor" has been abbreviated in many different ways such as AFPR, AFPr, AFPRp, AFP-BP, etc. which are difficult to pronounce and lead to confusion with AFP itself. Thus, throughout this chapter, the receptor for AFP is abbreviated as "RECAF," a term used by the author in previous publications.

Fifty years ago, Abelev et al. reported the discovery of the first oncofetal antigen, alpha-fetoprotein (AFP) [1]. After peaking in serum concentration during fetal life, AFP almost disappears after birth, with normal adult serum leveling at concentrations below 50 ng/ml [2]. However, in hepatocarcinomas and teratocarcinomas, serum AFP increases dramatically, attaining concentrations 1000 times higher [3]. These findings drew the attention of both the medical community, who foresaw new means for diagnosing and monitoring cancer patients, and basic research groups interested in studying the physiology of this protein during fetal life as well as in using AFP as tool to explore the process of cell differentiation and malignization.

While the main sites of AFP synthesis are the liver, stomach and yolk sac, using anti-AFP antibodies and the immunoperoxidase technique, Benno and Williams described the distribution of AFP in the developing rat brain [4]. Soon after, a series of reports confirmed the presence of AFP (and other plasma proteins) within embryonic neuroblasts in several species including rodents [5], sheep [6], primates [7], humans [8] and birds [9]. Since neuroblasts do not synthesize a significant amount of AFP, in the latter reference we proposed the existence of a receptor to explain AFP uptake by embryonic neural cells, and we suggested that the expression of this receptor could be related to the degree of cell differentiation[1].

At the time, it was noticed that as fetal development progresses, the AFP (and serum Albumin) staining in neural tissues follows a similar pattern from one species to another [10, 11]: Very immature neuroblasts in a given neural structure are initially AFP and Albumin negative. Then, rather abruptly, the staining becomes positive for both proteins, even within the same cell [12]. After some time, the staining, which is always restricted to the cytoplasm, fades both in intensity and in number of positive cells, first for AFP, and then for SA. This is not due to a decrease of AFP in serum because the AFP fading takes place before AFP reaches its peak in serum [13]. Nor is it due to the closing of the blood-brain barrier, which would prevent access of serum proteins to cells, since spinal ganglia neuroblasts turn immunohistologically negative for AFP while they remain positive for Albumin [10], which has a similar molecular weight. Neither protein is found in fully mature nervous structures. Other serum proteins, such as IgG or Ovalbumin in chicken embryos, are never present in neuroblasts despite being conspicuously present in the cerebrospinal fluid [13].

[1] This reference is difficult to find. The text reads: "*As an alternative to passive diffusion, the selectivity of the protein uptake could be due to specific receptors for AFP and for SA. These receptors would be only present at some intermediary stage of neuronal differentiation*."

AFP Uptake by Embryonic and Fetal Cells

A question arising from these initial observations was whether AFP and serum Albumin are incorporated from extracellular sources, thus validating the receptor hypothesis, or synthesized by neuroblasts. While it is not clear whether neural cells are capable of synthesizing these plasma proteins [14-17] it has been well established, both *in vitro* [18, 19] and *in vivo* [20, 21] that neuroblasts internalize AFP and serum Albumin from extracellular sources.

Injecting ^{125}I-AFP into pregnant rats [20] confirmed the accumulation of the labeled protein in neuroblasts but also in other fetal tissues that do not synthesize AFP and yet are positively stained for that protein by immunohistology [22]. Autoradiographies from these experiments also confirmed the exclusive cytoplasmic distribution of AFP.

Some of the initial *in vivo* uptake experiments were carried out with heterologous proteins [21]: When newborn rat serum containing a large amount of AFP was injected into the mesencephalic cavity of chicken embryos, the neuroblasts staining for rat AFP and rat Albumin was positive, with a color gradient fading from the lumen toward the peripheral layers. This staining pattern was identical to the one seen for the corresponding native proteins. On the other hand, the injection of rat IgG (MW ~150,000) or Ovalbumin (MW ~43,000) in the same cavity resulted in no uptake even when Ovalbumin was injected at 30 times its physiological concentration, which, at that embryonic stage, was two times the normal molar concentration of AFP in the cerebrospinal fluid [13]. These results indicate that (a) the uptake of AFP and Albumin is protein selective and (b) rat AFP and chicken AFP are similar enough to be internalized across species.

At the time, it was not clear whether the progressive disappearance of intracellular AFP at the end of gestation was a cell regulated process or just a consequence of the dropping concentrations of circulating AFP. The first possibility was proven correct since it was demonstrated, first in chicken, as mentioned above [10], and then in human embryos [23], that neural cells within spinal ganglia become AFP negative while the serum concentration of that protein is still rising.

AFP Uptake by Cancer Cells

The data at that point supported the following conclusions: (a) Fetal and embryonic cells from different tissues take-up AFP, (b) the uptake is protein-specific, (c) the uptake is related to the degree of cell differentiation and (d) embryonic cells from one species take up AFP from another species. All of these elements were consistent with the hypothesis of a receptor-mediated mechanism of AFP endocytosis whose expression would depend on the stage of cell differentiation [9, 21]. The uptake of heterologous AFP suggested that the structure of the binding sites on the receptor and the ligand are highly conserved across species.

Cancer cells and fetal/embryonic cells share a number of common biochemical and antigenic features [24]. Hence, we hypothesized that cancer cells (from tissues that incorporate AFP during fetal life) might regain the ability to internalize AFP and that the internalization would be carried out via its receptor, which would therefore qualify as a new oncofetal antigen. To verify this hypothesis, a number of studies, both *in vitro* and *in vivo* were undertaken: *In vitro* experiments evidenced the uptake of AFP by a variety of cancer cell lines such as MCF-7 (human breast cancer) [25], a nickel induced rat rhabdomyosarcoma [26], the C1300 mouse neuroblastoma [27] and several human leukemias [28]. *In vivo* studies carried out in mice showed a significant accumulation of injected mouse ^{125}I-AFP in spontaneous mammary carcinomas when compared to normal tissues [29].

Tumor Imaging Using Radiolabeled AFP

The injection of ^{125}I-AFP in 19 C3H/Bi mice bearing 31 breast spontaneous tumors resulted in a 3.6 tumor/liver ratio of radioactivity per mg of tissue [30]. Following injection with ^{131}I-AFP, two mammary carcinomas could be imaged by scintigraphy [30]. A more extensive study showed that after the injection of ^{131}I-AFP, eleven out of twelve spontaneous breast mouse tumors were clearly visualized despite using a standard gamma camera designed for ^{99m}Tc instead of ^{131}I [31]. Tumors as small as 3 mm in diameter could be seen, as shown in Figure 1. In yet another study, the C1300 mouse neuroblastoma tumor could be imaged when transplanted into mice [32]. Similar results were obtained by Line et al. [33, 34].

All of these findings, along with the high quality of the images obtained in animals, prompted us to scan patients after being injected with human ^{131}I-AFP. Since poorly differentiated cells take up AFP, we were concerned that gonad cells, bone marrow and normal stem cells could be killed or mutated by the accumulation of radioactivity. In order to address this possibility, we injected twelve female and six male mice with a dose of 0.1 mCi of ^{131}I-AFP per animal, which is the equivalent, by weight, to 250 mCi for an adult human being. No abnormalities were detected in any of these mice, which had a normal life span. When mated, fertility was normal, and the two generations of their offspring that were kept under observation were also normal.

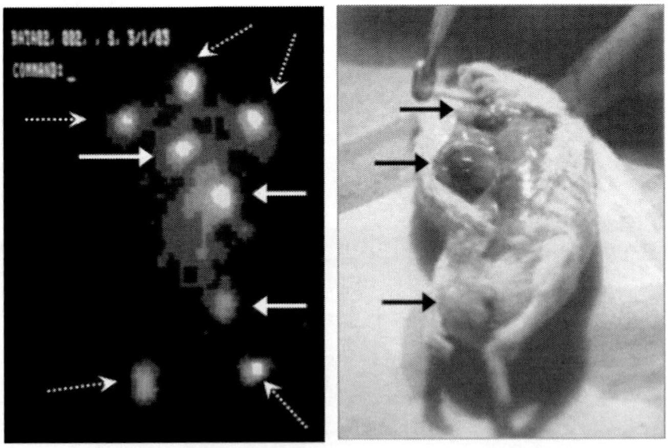

Figure 1. Scintigraphy of a mouse bearing 3 spontaneous tumors (histologically diagnosed as a primary mammary carcinoma and 2 metastases). On the left, the image obtained after injection of 72 uCi of ^{131}I-AFP. The dotted arrows point at radioactive markers used to locate the nose and extremities of the animal under the gamma-camera. The solid arrows show three positive lesions. The upper spot was initially thought to be the thyroid, even though the animals were given KI to drink for a week before injection. The autopsy on the right revealed that the spot corresponded to a metastasis 3 mm in diameter.

Thus, two healthy individuals (including the author) and three cancer patients were injected with 300-600 ug of ^{131}I-HuAFP (~0.6 mCi) and scanned 2 to 7 days later. In the normal patients, who were injected first, most of the radioactivity was localized in the liver, spleen and cardiac area. The first cancer patient (male, bearing a large abdominal tumor diagnosed as stomach cancer) showed a clear accumulation of radioactivity corresponding to the palpable mass. The second cancer patient was a female patient bearing bone metastases of a breast carcinoma. The original tumor had been surgically removed 2 years earlier. This patient was first injected with ^{99m}Tc-Pyrophosphate, which accumulates in regenerating bone tissue, and then was imaged according to routine procedures. As depicted in Figure 2 (left), the scintigraphy showed bone metastases in the lumbar vertebrae and the iliac crests. To track the location of the lesions, a radioactive pen was moved over the patient's back until the pen's spot observable on the gamma camera overlapped with the lesions spots. Their location on the patient's back was marked for future reference using a non-toxic marker. A week later, the patient was imaged again, fitting this time the gamma camera with a ^{131}I collimator. As a control, and before injecting the radioactive AFP, a new image was taken, which, as expected, showed no visible radioactivity given the short half life of ^{99m}Tc (~6 hours). The patient was then injected with human ^{131}I-AFP and scintigraphies were taken at 24 or 48-hour intervals. Figure 2 (right) shows the image obtained on the 6th day after injection. Repeating the lesion tracking process with the radioactive pen we were able to demonstrate that the 3 lesions pointed at by the solid arrows in both scintigraphies had the same location and therefore, that the lesions detected by ^{131}I-AFP were the same as the ones detected with ^{99m}Tc-Pyrophosphate. ^{99m}Tc-Pyrophosphate images show bone synthesis and therefore they are restricted to bone metastases, whereas ^{131}I-AFP images denote the presence of malignant cells and therefore they can be detected in other locations.

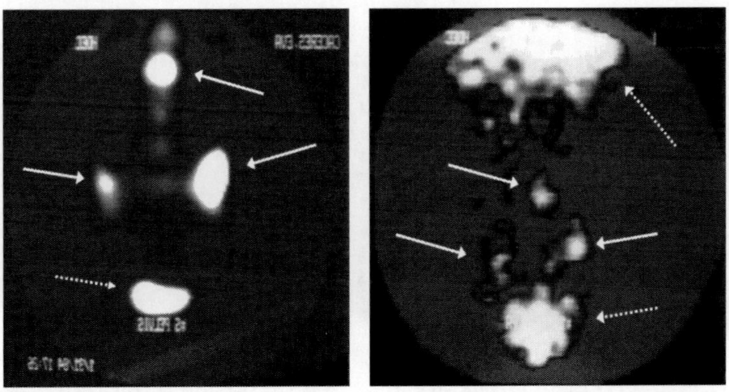

Figure 2. Scintigraphy of bone metastases of a human breast carcinoma. The patient was first injected with 99mTc-Pyrophosphate and then imaged according to routine procedures (left picture). The solid arrows show the areas of bone regeneration produced by metastases in a lumbar vertebra and the iliac crests. The dashed arrows show the liver area and the urinary bladder. Seven days later, the patient was injected with human 131I-AFP and scintigraphies were taken at 24-48 hour intervals. On the right, an image taken 6 days after the injection or radioactive AFP. The difference in magnification is due to the use of a different collimator for each isotope. The 3 lesions pointed at by the solid arrows coincided with those on the left picture.

The third patient was a 73-year old female who had had bone metastases of a breast carcinoma. Following chemotherapy and hormone therapy, the patient had been in remission for 18 months, at which time she was imaged as described above. Bone regeneration is a slow process and ^{99m}Tc-Pyrophosphate positive images can remain as such for many months after remission thus masking a complete remission. In effect, the ^{99m}Tc Pyrophosphate scintigraphy was still positive, whereas, as expected, no radioactive spots could be detected with ^{131}I-AFP.

These results, taken together with the images obtained in animals, strongly suggest that AFP might prove useful for tumor imaging in humans. Systematic studies involving a larger number of patients are necessary to fully ascertain the actual merits of this promising technique and the doses, as well as the time between radioisotope injection and imaging need to be optimized.

Using radiolabeled human AFP offers the advantage, over monoclonal antibodies, that it is a homologous molecule, thus virtually precluding anaphylactic reactions when used repeatedly. Also, only one radiopharmaceutical would be required to detect a wide variety of tumors, a practical consideration to be taken into account.

The discovery of the binding site of RECAF on the AFP molecule (see below) opens the possibility of using a radiolabeled AFP peptide for imaging. This offers several advantages over using whole AFP or a large fragment thereof: (a) Lower cost of production, (b) low risk of infection, and (c), the ability to tailor the peptide for: (i) radiolabeling, (ii) optimal tissue penetration, (iii) cell uptake and (iv) elimination/secretion, thus improving the image quality while reducing the radioactivity doses as well as the time between the injection and the image capture. This, in turn, could allow the use of ^{99m}Tc instead of ^{131}I or exotic isotopes such as ^{111}In or ^{123}I, etc. Using a peptide labeled with ^{99m}Tc would be practical since the latter is readily available in any nuclear medicine department, it would significantly reduce radiation exposure and it might enable the entire process, from injection to imaging, to be carried out in one session.

Something to be kept in mind when considering this type of scintigraphy (as well as in the evaluation of serum RECAF test results), is that AFP is accumulated in some acute inflammatory lesions [35, 36]. This does not seem to be the case in chronic inflammatory processes.

RECAF Expression and Detection

Even though the expression of a receptor to explain AFP uptake was first proposed in 1981 [9], indirect evidence of this receptor-ligand system can be traced to earlier papers [37, 38]. The first of these articles concluded that 18% of normal peripheral T-cells bind AFP. However, later articles showed that peripheral T-lymphocytes do not bind AFP unless they are previously stimulated with Phytohaemagglutinin (PHA) [39]. The second article [38] reported the presence of two AFP rich ultracentrifugation fractions in immature rat uterine cytosols: A 4S fraction, detectable by an anti-AFP antiserum and an 8S fraction where AFP was not detectable. However, treatment of the 8S fraction with 0.4 M KCl, converted it to a 4S fraction, in which AFP could then be detected. It is likely that the 8S fraction corresponded to a receptor-AFP complex, which was dissociated at high KCl concentration. This is peculiar because in those experiments, AFP was detected using a polyclonal antiserum and AFP is known to have at least 6 epitopes [40]. Thus, the AFP conformational changes required to void all of the epitopes when it binds to RECAF need to be very substantial. A similar phenomenon was observed by Smalley and Sarcione, who reported that the uterus of immature rats synthesizes AFP in a form that was not recognized by anti-AFP antibodies unless it was treated with 0.4 M KCl [41]. Sarcione et al. also reported the presence of complexed AFP in cancer cell extracts, in a form that could be dissociated by KCl [42]. In addition, Sarcione's team demonstrated that AFP is synthesized by the MCF-7 human breast cancer cell line as a complex that is not recognized by anti-AFP antibodies unless it is first treated with a high concentration of KCl [43].

The first direct evidence of a receptor for AFP was described on MCF-7 cells by Villacampa et al. [44] who found results consistent with the presence of a two-site receptor model exhibiting positive binding cooperativity. The higher affinity site showed a Kd of 1.5×10^{-9} M and a number of 2000 sites per cell. The lower affinity site, with a number of 320000 sites per cell exhibited a Kd of 2.2×10^{-7} M. Similar AFP receptor systems were subsequently detected on the surface of the YAC-1 mouse T-lymphoma [45] (but not on normal adult mouse T-cells) and on the human U937 and THP-1 cell lines [46]. The number of sites/cell as well as the binding affinity vary from one cell line to another. The binding assays were not restricted to intact cells but also included MCF-7 and primary breast cancer cytosols thus evidencing a soluble RECAF fraction [47].

Monoclonal Antibodies against RECAF

Monoclonal antibodies (Mabs) are more practical, consistent and inexpensive for detecting RECAF than AFP. Thus, we screened the supernatants of a fusion made against pooled human mammary carcinoma membranes and found 2 clones that inhibited the binding of AFP to malignant cells. Conversely, these Mabs were inhibited from binding to RECAF by an excess of AFP [48]. The Mabs also reacted with RECAF from human cord serum that was purified by AFP affinity. Paraffin sections stained with these Mabs showed strongly positive labeling of human fetal muscle and mammary carcinomas whereas benign breast tumors were negative.

Using one of these Mabs on placenta tissue sections, an independent study concluded that RECAF plays a role in moving AFP from the fetal circulation to the mother's blood stream [49], which is consistent with the earlier notion that AFP transports fatty acids to fetal (and cancer) cells [50-53].

These anti-RECAF Mabs also recognized mouse and dog cancer cells [54]. This cross reactivity is in agreement with the previous finding that AFP from one species is taken-up by cells from other species [21, 45] and it validates the use of syngeneic mouse tumor models for *in vivo* experimentation thus eliminating the need for xenografts in nude mice, which is a very artificial model.

More recently, we became interested in detecting circulating RECAF and therefore we made Mabs against the soluble RECAF fraction in MCF-7 cell extracts. One of these Mabs, named 1.4G11, complied with the conditions necessary to ascertain that it is directed against

RECAF[2]. As expected, this Mab also recognized mouse and dog RECAF. It is worth mentioning that all the anti-RECAF Mabs we have produced so far belong to the IgM class.

An AFP Peptide that Binds to RECAF

Both AFP and IgM Mabs are large molecules that are costly to produce and, given their size, are impaired when it comes to tissue penetration and secretion, an important point when considering targeted therapy. Moreover, as biologicals, they present some risks when injected to patients. These drawbacks could be overcome if the AFP binding site to its receptor consisted of a short peptide that could be synthesized chemically. To search for such a peptide, we digested human AFP with proteases following Dudich et al. methods [55] and we monitored the AFP binding of the fragments by binding inhibition of labeled AFP to microtiter plates coated with cancer cell extracts. Once the smallest RECAF binding fraction was identified, we applied a brute force approach: Starting from the C-terminal of the AFP molecule, we synthesized overlapping peptides covering the fragment[3] which we then tested for AFP binding inhibition,. We found an inhibiting peptide that we then trimmed until we established the shortest sequence that exhibited AFP binding inhibition activity [56, 57]. That sequence is KQEFLIN, starting at amino acid 549 of human AFP. For solubility reasons this sequence was extended to LQTMKQEFLIN (henceforth referred to as F4). It should be noted that this sequence does not correspond to other AFP peptides reported to have biological activity [58-64].

RECAF Purification and Characterization

RECAF has been purified by several methods and from different sources: It has been purified from cord serum [48] and from the pleural effusion of a patient bearing lung metastases of a mammary carcinoma [54]. In both these cases the purification relied on the presence of AFP-RECAF complexes in the starting material, which is consistent with several studies showing the synthesis and release of these complexes in different types of fetal and cancer cells [41-43, 65, 66]. A more direct approach was taken by Severin et al. [67] who used AFP covalently coupled to Sepharose® to bind RECAF, and KCl at high concentration to elute it from the gel. Their results were virtually identical to ours using cord serum and Con-A Sepharose. Another method involved the use of AFP conjugated to a photoactivatable thio-cleavable cross-linker that was activated once the AFP was attached to RECAF on U937 cells (a human monocytic malignant cell line) [46].

The fractions purified with all of those methods exhibited a molecular weight of approximately 65 kDa, which is consistent with the molecular weight of the AFP-RECAF complex as measured by gel filtration after incubating radiolabeled AFP with materials rich in

---

[2] These conditions are: (a) an excess of the Mab must inhibit the binding of AFP to RECAF, (b) an excess of AFP must inhibit the binding of the Mab to RECAF and (c) on a Western blot of a cancer cell extract, both AFP and the Mab must detect the same bands.

[3] The peptides were 20 amino acids in length and the overlap was 10 amino acids. To test them, the peptides were incubated together with labeled AFP onto microtiter plate wells coated with RECAF containing material. Positive peptides inhibited the binding of whole AFP.

RECAF [68]. The molecular weight of the complex was found to be ~130 kDa, of which 65 kDa could be accounted for by AFP, with the other 65 kDa corresponding to RECAF. A molecular weight of ~65 kDa is also consistent with the detection, using anti-RECAF Mabs, of two adjacent bands weighing 62 and 67 kDa respectively, on SDS polyacrylamide transblots of human breast cancer extracts [48, 54].

Heavier RECAF fractions weighing 180 kDa [54] and 250 kDa [67, 69] have also been reported. The heavier fractions could correspond to complexes with anti-receptor auto-antibodies [70] or other cellular proteins.

Torres et al., using AFP immobilized on nitrocellulose membranes, reported the isolation of two smaller (31 kDa and 18 kDa) AFP binding glycoproteins from Raji cells (a human B-lymphoma) [71]. The 18 kDa fraction was also found in PHA stimulated human peripheral blood lymphocytes. No binding fractions were isolated from resting lymphocytes. Interestingly, the pictures in this article show a fainter band exhibiting a molecular weight of ~65 kDa, which is never discussed throughout the article. The authors stated: *"These observations strongly suggest that the isolated 31 kDa and 18 kDa glycoproteins are probably AFP receptors previously demonstrated in several neoplastic and normal cells undergoing growth and/or differentiation; indeed, they were identical to albumin-binding proteins described by others."* This is important because some previous studies also showed that Albumin inhibited the binding of AFP to its receptor, albeit to a lesser extent than AFP itself [44, 45]. However, this is inconsistent with the above mentioned scintigraphy results since the amount of injected radiolabeled AFP was minuscule compared to the concentration of Albumin in serum, which would have competed out any binding or internalization of the former into cancer cells. Also, Suzuki et al. [46] reported that serum Albumin did not compete with AFP for binding to RECAF. Thus, to elucidate whether Albumin competes with AFP for binding to RECAF, we carried out a series of experiments and we found that neither human Albumin nor bovine Albumin competed with AFP for binding to a RECAF rich MCF-7 extract (Figure 3). In previous experiments we had shown that neither Albumin nor Ovalbumin competed with our anti-RECAF Mabs [48].

An explanation to Torres et al. finding that Albumin and AFP bind to the same bands on polyacrylamide gels might be that the AFP binding activity is located on a glycan that could be attached to several proteins with different molecular weight. It is also possible that within the same band, some molecules are glycosylated with the AFP binding glycan while others carry a glycan that binds Albumin. This could reconcile the apparent binding cross reactivity on the gel bands with the lack of competition found in other experiments.

Yet, it remains to be explained why some experiments show AFP binding competition by Albumin while others do not. A possible explanation could be connected to the fact that different lots of AFP show different RECAF binding activity, which could, in turn, be explained by the proposed hypothesis that one of the functions of AFP is to carry essential fatty acids to fetal cells [17, 72, 73]. Once the AFP-fatty acid complex is internalized, the fatty acid is retained and the "empty" AFP molecule is released into the extracellular fluid intact [74], ready to fetch and transport another fatty acid molecule. It is reasonable to assume that fatty acid depleted AFP molecules are not internalized; otherwise, they would be secreted and taken up again in a vicious circle that would be detrimental to the system. The AFP lot binding differences could then be explained by differences in their fatty acid content. To test this hypothesis, we precipitated AFP with cold acetone, which also extracts the fatty acids from the protein. Next, we re-lipidated aliquots with a mixture of fatty acids as described for native human AFP [75]. We then incubated the initial AFP, the delipidated AFP and the re-lipidated AFP with MCF-7 cells in culture. The results are shown in Figure 4, which clearly shows that the delipidated AFP is not taken up and that re-lipidation not only reverts the results of removing the lipids, but can also enhance AFP uptake depending on the amount of fatty acid added. These results have both practical and theoretical implications: From a practical point of view, the findings suggest that experiments involving AFP uptake by fetal or cancer cells could benefit from lipidating the protein with fatty acids. From a theoretical point of view, these results further support the notion that AFP functions as a carrier molecule for fatty acids and they could help explain the apparent competition between AFP and Albumin: Most commercial preparations of Albumin consist of Cohn fraction V, which is prepared using ethanol, a fatty acid solvent. The resulting de-fatted Albumin, when used in large concentrations, could sequester the fatty acids from the co-incubated AFP, thus greatly reducing the ability of the latter to bind to its receptor. In our competition experiments (Figure 3), we used pre-lipidated AFP (for both labeled and non-labeled AFP), and we added only 100 fold excess of Albumin (as compared to 5,000 fold excess in a previous paper [44]) since that amount was sufficient to show ~90% competition by unlabeled AFP.

It is worth noting that whole AFP requires fatty acids to bind to its receptor but the F4 peptide, which is part of AFP, does not. Hence, it appears that fatty acids induce AFP conformational changes [76] that expose the F4 sequence to the receptor rather than having the AFP-fatty acid complex binding directly to the receptor. It should be mentioned that the fatty acid binding site and the receptor binding site are far away on the AFP molecule; the fatty acid binding site is on or nearby amino acid 223 [77] whereas, as mentioned above, the receptor binding site is located nearby amino acid 549.

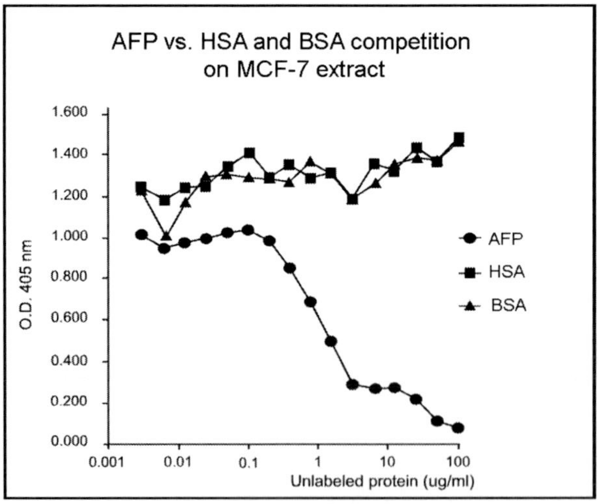

Figure 3. Binding competition between human serum Albumin (HSA), bovine serum Albumin (BSA) and AFP. The supernatant of an MCF-7 extract made without SDS (mainly 62 kDa RECAF) was coated onto microtiter wells. Biotin-AFP (1 ug/ml) was mixed with different concentrations of unlabeled AFP, HSA or BSA and incubated in the wells. After washing, Streptavidin-Peroxidase was added. The wells were washed, the reaction was developed with ABTS + H2O2 and the wells were read at 405 nm. Only unlabeled AFP competes with Biotin-AFP; there is no competition between Biotin AFP and HSA or BSA.

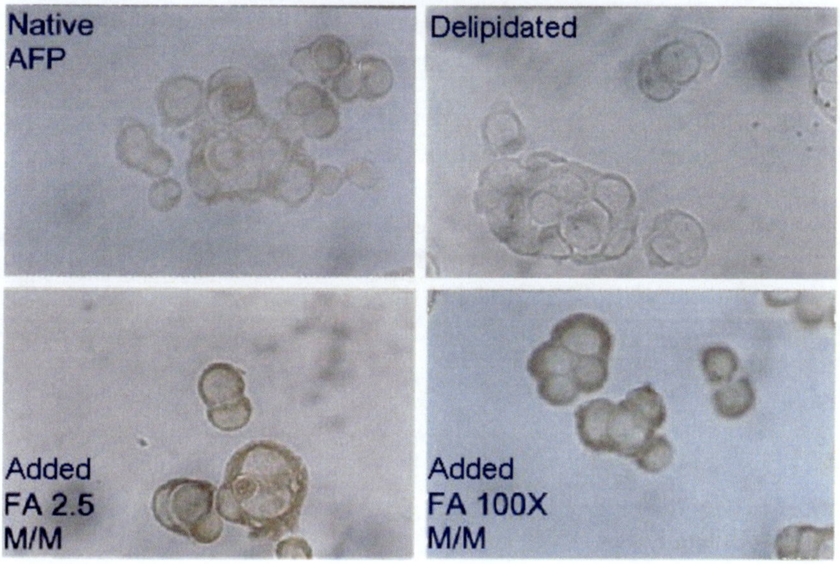

Figure 4. Native or treated AFP was incubated with MCF-7 cells in culture for 90 minutes at 37oC. After washing, the cells were fixed and immunostained using an anti-AFP antibody. AFP was delipidated by precipitation with cold acetone. To re-lipidate the delipidated AFP, we diluted a mixture of oleic, arachidonic and docohexanoic acids [75] in ether and then dried the solution on the wall of a rotating glass tube. The AFP was then incubated in the tube under rotation. The molar ratios (M/M) described in the figure correspond to the total amount of fatty acid dried on the glass tube wall over the total amount of AFP incubated in the tube. The delipidated AFP is not taken up, whereas the 2.2 M/M relipidated AFP exhibits stronger staining (and hence uptake) than the native AFP. Loading AFP with a fatty acids at 100X the molar ratio results in even stronger uptake.

In reference to RECAF being a carbohydrate, a previous report using glycan modification with sodium periodate or borohydrate showed that anti-RECAF Mabs recognize a structure that is compatible with an O-glycan [54]. Our own experiments show that AFP does not bind to an O-glycan but rather, it binds to an N-glycan on RECAF: We coated microtiter wells with K562 cell extracts and then we treated the wells with periodate (1-100 mM), followed by incubation with Biotinylated-AFP (B-AFP). As shown in Figure 5, the treatment with periodate results in a decrease of up to 90% of the binding activity and the decrease is proportional to the periodate concentration. As a control, wells coated with human albumin, which has no glycans, were treated in identical conditions and showed no decrease in binding to a polyclonal anti-HSA antibody.

To further demonstrate that only the glycan and not the RECAF peptidic chain is involved in the binding, we did the counter experiment: We terminally digested the RECAF containing material with pepsin and tested its binding inhibition activity. Since digesting the receptor coated onto a plate would result in detachment of peptidic fragments, we carried out the digestion in solution using 0.04% pepsin for 24 hours at 37°C, pH 2.5 (a control PAGE of the digested material showed no bands aside from the pepsin band). After bringing up the pH, we mixed a fixed concentration of B-AFP with different concentrations of the digested RECAF extract. The mixture was then added to wells coated with the same extract, but not digested. Thus, the higher the concentration of RECAF left undigested in solution, the more B-AFP it would complex and therefore the lower would be the amount of free B-AFP available to bind to the RECAF coating the plate. As a control, we subjected the same receptor preparation to the condition described above but without the pepsin. Another control was carried out using Albumin and an anti-Albumin antibody instead of the RECAF preparation and B-AFP respectively.

If the binding site were on a sugar, the digestion with pepsin would make no difference in the binding and therefore, the pepsin digested receptor and the control without pepsin should display similar binding curves for the receptor experiment.

Figure 6 shows that the inhibition curve obtained with the digested extracts was practically identical to that of non-digested extracts. This was not the case for the Albumin control, which showed a 5-fold difference in OD between digested and non-digested protein (data not shown).

Thus, the glucidic nature of RECAF has been established by two different methods; by destroying the glycan with periodate-borohydrate treatment and by destroying its peptidic backbone.

In addition, extracts from cells incubated with Tunicamycin, which specifically inhibits N-glycosylation [78], showed a ~90% decrease in AFP binding (Figure 7). Treating the cells with Galactosamide, which inhibits the linking of O-glycans, resulted in no change in binding.

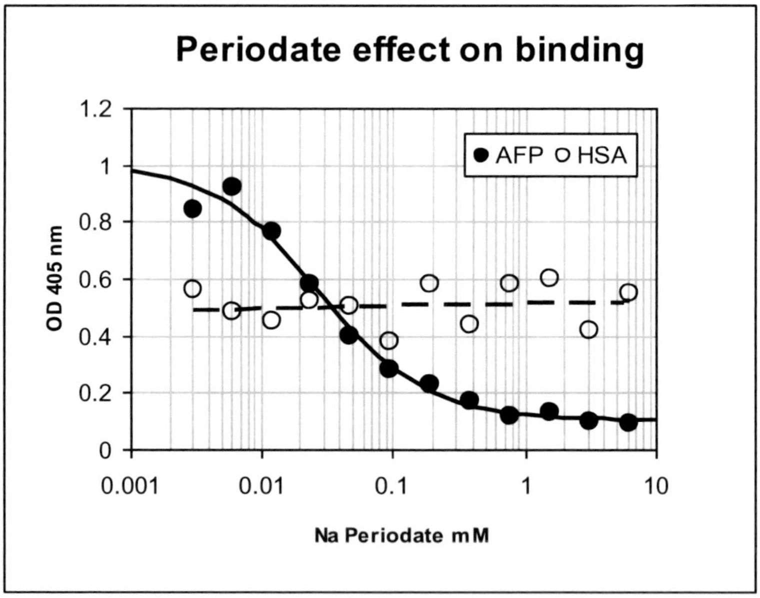

Figure 5. Sodium periodate treatment of RECAF demonstrating that RECAF is a sugar. Microtiter plates were coated with a K562 extract (black circles) or human serum Albumin (HSA, open circles) and then treated with different concentrations of periodate. The RECAF binding activity was tested using a constant concentration of Biotin-AFP. The HSA integrity was tested using an antibody. The Biotin-AFP and anti-HSA antibody were developed with Streptavidin-Peroxidase and anti-IgG-Peroxidase respectively. The Biotin-AFP binding decreases as the concentration of periodate increases, reaching a plateau above 10 mM. There is no effect of the periodate on HSA, indicating that the periodate is not affecting the RECAF peptidic chain but rather its glycans.

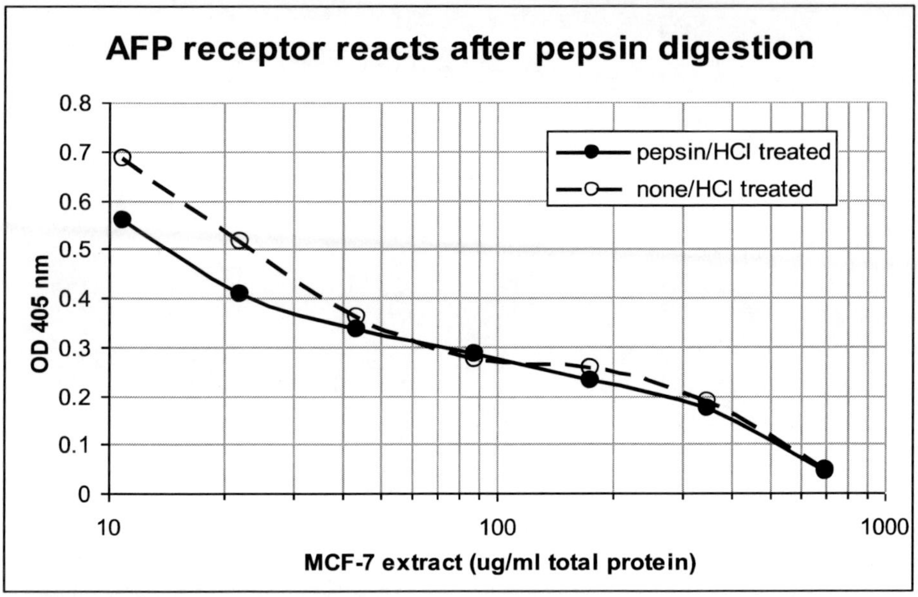

Figure 6. Pepsin treated RECAF behaves almost exactly like non pepsin treated RECAF. See text for the experiment's details.

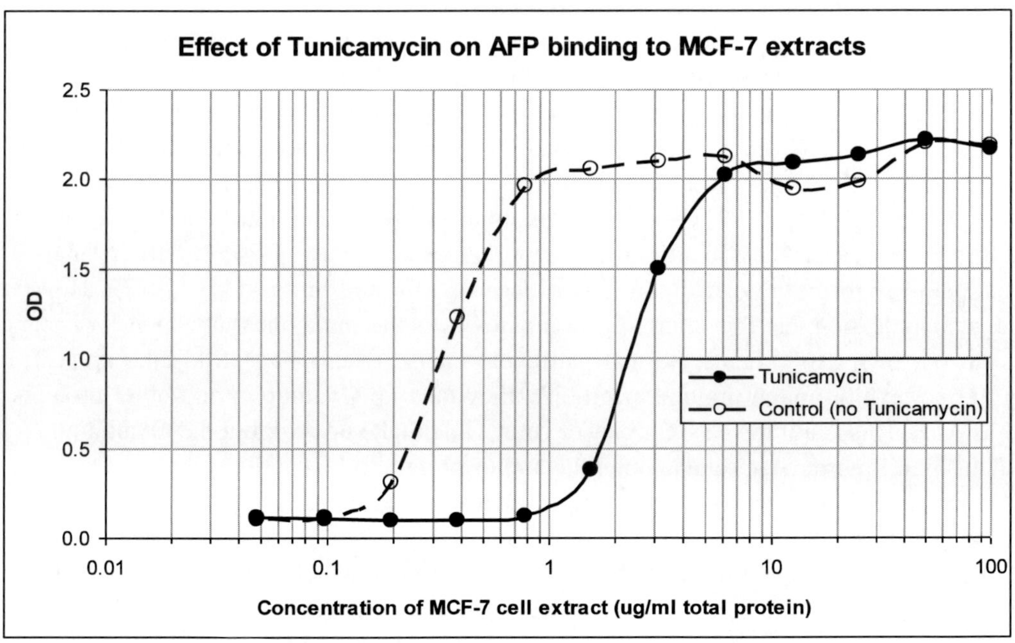

Figure 7. Binding of AFP to MCF-7 extracts from cells treated or not with Tunicamycin, a specific inhibitor of N-glycosylation. The cell extracts were coated at different concentrations onto ELISA plates, followed by incubation with 10 ug/ml AFP-Peroxidase. The AFP binding activity on Tunicamycin treated cells was significantly lower (~1/10) compared to not treated control cells, thus requiring a higher amount of MCF-7 extract coating to attain the same OD readings. These results indicate that RECAF is an N-glycan.

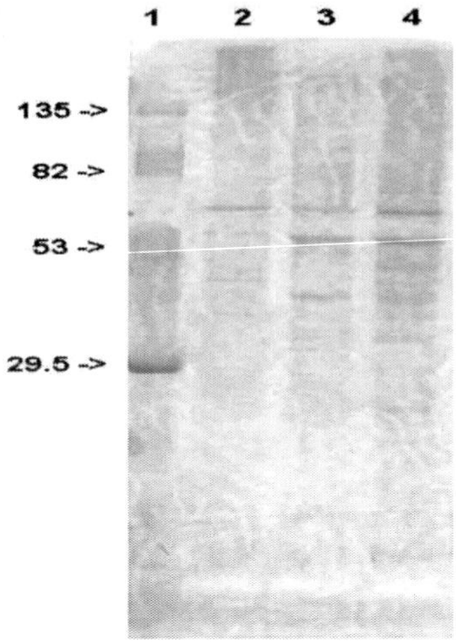

Figure 8. Different migration patterns of RECAF bands. Cell extracts were prepared in the absence of SDS. The supernatants were then treated with SDS which was also present in the gel. Lanes: (1) MW markers, (2 and 3) two K562 supernatants incubated and processed identically, (4) Ichikawa supernatant.

These results contradict the reports suggesting that the receptor is an O-glycan [54] or that the RECAF binding site is a widespread mucin [79, 80]. The author of the latter references also suggested that RECAF belongs to the scavenger receptor family [81]. The complex nature of this group of receptors and the general lack of information on the RECAF glycan structure makes this hypothesis difficult to ascertain. Perhaps once that structure is determined it might become apparent that RECAF belongs to some known family of receptors. Moreover, it is possible that there is more than one type of receptor involved: Pardee et al. have shown [82] that AFP is internalized by normal adult dendritic cells (DC), which is unexpected. They identified a mannose receptor (MR/CD206) as the primary DC uptake pathway for both normal cord blood-derived AFP and tumor-derived AFP. However, there is no indication that this internalization pathway is the same one that is present in cancer cells. In our own experiments, normal peritoneal mouse macrophages did take up AFP but they did not exhibit much protein specificity; they take up Ovalbumin and other proteins as well as AFP. This is not the case for cancer cells, which take up AFP but not Ovalbumin.

Thus, until more data are available, the author prefers to be cautious when speculating in a field in which information is scant. The evidence gathered so far supports the notion that RECAF is an N-glycan that is attached to a number of different proteins. Polyacrylamide gels of cell extracts show the ubiquitous 65/67 kDa doublet [48], which is water soluble and the 31 and 18 kDa bands, which require 0.5% SDS to be extracted [71]. In addition, we have observed two faint soluble bands in some of our Westerns; a 53 kDa and a 44 kDa band. The relative intensity of all these bands varies from one cell line to another. Moreover, extracts carried out in exactly the same conditions, from the same cells cultured, under identical conditions show variable RECAF bands patterns[4] as shown in Figure 8.

[4] Anecdotally, we observed that the receptor expression in cultured HL60 (a human acute promyelocytic leukemia), was modified by the plastic Petri dishes brand used for tissue culture.

Given our interest in a soluble band for cancer detection in serum, we concentrated our efforts on the 65/67 kDa band, which is consistently present in cancer cell extracts. While this fraction is soluble, it is also aggregated. When cancer cell extracts are passed through a gel filtration column, the AFP binding activity comes out in the void volume, even when using Bio-Gel A-5m, which has an exclusion size of 5 mega Daltons. Adding 0.5% SDS, or lowering the pH below 5.0 results in disaggregation with a single broad peak of AFP binding activity at ~70 kDA as measured by molecular weight calibrators.

POTENTIAL USES OF RECAF AS A TUMOR MARKER FOR *IN-VITRO* DIAGNOSIS

Aside from its potential for tumor scintigraphy, RECAF could prove useful as a widespread cancer marker to detect leukemia and malignancy on tissue sections, blood smears and bodily fluids.

RECAF on Leukemia Cells

Flow cytometry offers the possibility of monitoring residual disease in leukemia patients after treatment. The sensitivity of this method is far superior than detecting malignant cells under the microscope, thus prompting the administration of a new round of chemotherapy at an earlier recurrence stage. This technique showed strongly positive results when the 167H.4 Mab was used to stain HL60, Ramos and U-937 cells [54]. In those experiments, Molt-4 cells were not stained, which is consistent with the findings from other AFP uptake experiments [28, 83].

Immunohistology Using Anti-RECAF Mabs on Cancer Tissue Sections

Using an anti-RECAF Mab for immunofluorescence on paraffin sections, we reported the positive staining of 6 out of 6 human mammary carcinomas whereas 3/3 benign adenomas were negative [48]. Similar results were obtained with another antibody and the immunoperoxidase technique: We found positive staining in 21/23 mammary carcinomas, 14/15 lung cancers, 8/8 colon carcinomas and in approximately 90% of other malignancies. Tsuboi et al., using anti-RECAF antibodies found 34/47 positive stomach cancers and only 2 positive cases out of 61 benign specimens and non-malignant areas around cancer cells in specimens from cancer patients [84]. Interestingly, these authors found no statistical correlation between the staining intensity and the degree of cell differentiation. Also, they mention that a previous and unpublished study on colon cancer sections showed identical staining patterns between anti-RECAF Mabs and labeled AFP, which coincides with our own observations.

Figure 9 shows some examples of cancer and benign tissues stained by immunoperoxidase using an anti-RECAF Mab.

While paraffin sections stained with Hematoxylin and Eosin are usually clear enough for diagnosis purposes, there are some situations in which RECAF staining could prove helpful:

i. Frozen sections, where the sub-optimal quality of the specimen can lead to error.
ii. For detecting the presence of small metastatic foci within sentinel lymph nodes as currently done with cytokeratin immunostaining.
iii. Fine needle biopsies in which the lack of tissue architecture makes the diagnosis more difficult. Figure 10 shows the staining of fine needle biopsies from patients with benign and malignant breast lesions in which the difference between them becomes very clear after staining for RECAF.
iv. Smears, for the same reason as fine needle biopsies. To illustrate this point, we first spiked normal sputum or cervical smears with cancer cells and then we fixed them and stained them with an anti-RECAF Mab. The results are shown in Figure 11. Cervical smears are a special case because they are often examined with the aid of software. Some programs pinpoint the suspicious areas for the pathologist to focus on. Others carry out the whole diagnostic analysis, considering several parameters such as the nuclear/cellular surface ratio. In both cases, a great deal of microprocessing time is dedicated just to identifying cells and differentiating them from detritus and artifacts before analyzing whether they are malignant or not. On the other hand, detecting brown spots resulting from immunoperoxidase staining requires very little processing. Thus, pointing the pathologist's attention to the brown cells on a tissue section, or in fully automated systems, focusing processing time on the brown stained structures/cells should largely improve the speed and thus the efficiency of the system, even if the criteria used for diagnosis are kept unchanged and the positive staining is not considered as part of those criteria. Of course, the inclusion of positive RECAF staining as an additional criterion of malignancy should further improve the system's performance.

Serum RECAF and Its Use for Cancer Diagnosis

The soluble nature of some of the RECAF bands suggested that malignant cells could release the marker into the circulation, where it would be abnormally elevated. A first attempt to ascertain if circulating RECAF could be used for cancer detection compared sera from 17 cancer patients bearing assorted malignancies and 22 negative controls (normal donors or patients with hypertension or heart related conditions). RECAF was elevated in 16/17 cancer patients whereas 20/22 control sera were negative ($p < 2 \times 10^{-6}$ by t-test) [85]. Anecdotally, a female patient in the control group who consulted for severe hypertension had an extremely high level of circulating receptor. Given the statistical significance of the results, this patient was subjected to a CAT scan that revealed a previously unknown kidney tumor, which was soon after diagnosed as a hypernephroma.

Around that time, Sarcione et al., took a different and elegant approach for detecting RECAF in serum: Previous work [41-43] had shown that anti-AFP antibodies do no detect AFP bound to its receptor and it was well known that in breast cancer patients, AFP is not elevated [86, 87]. The authors hypothesized that the AFP in serum from cancer patients consisted of two fractions: Free AFP, which was detectable (and within normal values) and

RECAF bound AFP, which was elevated but undetectable. As a result, the *total* concentration of AFP would be higher in cancer patients than in normal individuals. To measure total AFP, they treated sera from both groups with 0.4M KCl to dissociate the RECAF-AFP complex[5] before measuring the AFP concentration, which had a mean of 26.4 ± 7.1 IU and a median of 28.0 IU in normal female serum. In the serum of breast cancer patients, the values were approximately twice as high; the mean AFP concentration was 53.0 ± 17.9 IU and the median was 50.0 IU [88]. It should be mentioned that the commercial kit used in this study was discontinued by the time we attempted to replicate those results. Thus, we tried using different antibodies and antisera, in both sandwich or competition assays without success; we found no difference between the serum AFP concentration in normal females and in breast cancer patients.

The development of Mabs first and RECAF binding peptides later allowed us to develop assays with which we have tested several thousands serum samples of the most prevalent types of cancer. We used radioimmunoassay [89], chemiluminescence [90, 91] and to a lesser extent ELISA [92]. We were also able to show proof of concept for lateral flow point-of-care tests [93]. The results obtained with the Mabs or the AFP peptides were similar [94].

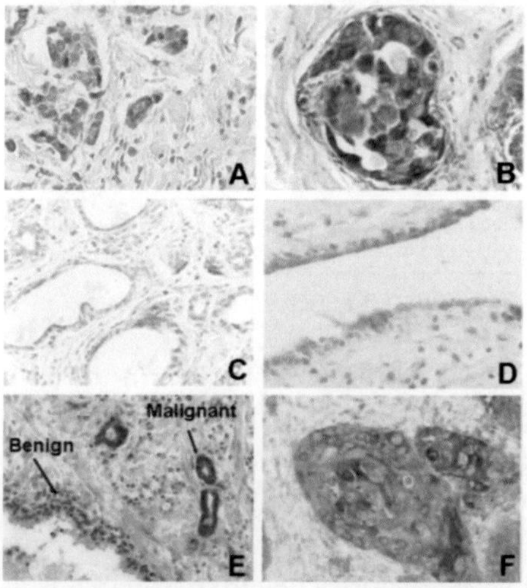

Figure 9. Immunoperoxidase of paraffin tissue sections using Mab 1.4G11. (A and B) Breast carcinomas, (C and D) negative breast benign fibroradenomas, (E) prostate carcinoma and (F) lung cancer. Staining is always cytoplasmic showing perinuclear accumulation. All but F are counterstained with Hematoxylin.

The preferred assay format consists of competing the RECAF in the sample with labeled RECAF for binding to the ligand (AFP peptide or Mab) coated onto a microtiter plate. This ensures that all proteins exhibiting the RECAF glycan are detected, which would not be the case with a sandwich assay in which one side of the sandwich would be the RECAF binding Mab and the other, a Mab against the peptidic chain of one of the proteins carrying the

[5] Chaotropic agents such as KCl at high concentration are known to split receptor-ligand bonds. The KCl needed to be diluted immediately before the assay because otherwise, the high salt concentration would prevent the assay's antibody from reacting with AFP.

RECAF glycan[6]. This is clearly illustrated in Figure 8: An assay in which the 2nd antibody were directed against the protein backbone of the 53 kDa band would detect samples 3 and 4 but not sample 2.

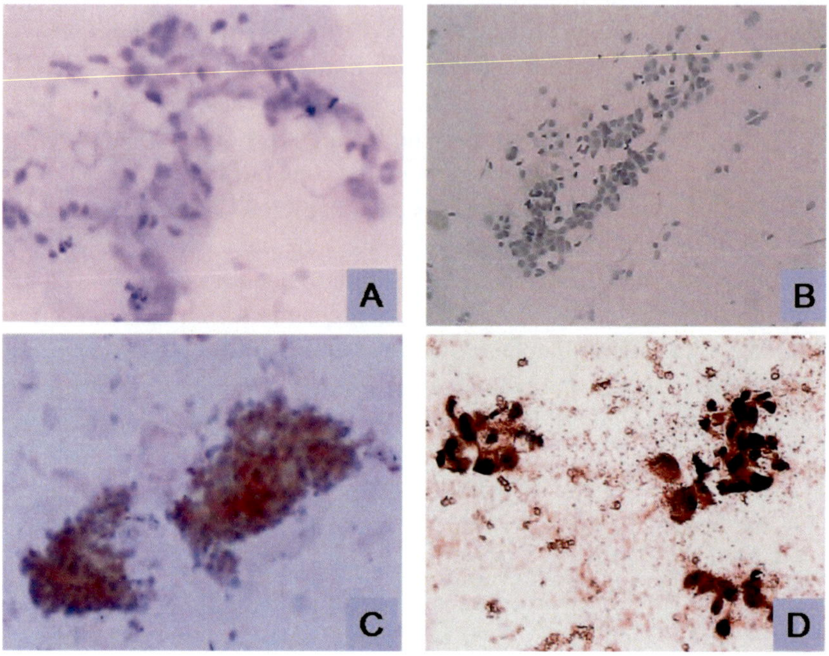

Figure 10. RECAF staining of breast fine needle biopsies. (A and B) negative benign lesions, (C and D) positive cancer cells. Counterstained with Hematoxylin.

Since we are measuring RECAF on several proteins with different molecular weights, the measurements cannot be expressed in ng/ml. As a result, the amount of RECAF is currently measured in arbitrary RECAF Units[7].

Figure 12 depicts some of our findings: At 95% specificity, sensitivity[8] values vary, from one cancer type to another, between 75% and 90%. Please note that benign lesions of the breast and prostate are in their vast majority negative. It remains to be determined whether the false positives in the benign groups are truly benign or if they correspond to premalignant lesions. In this regard, it is reasonable to consider that the biochemistry changes involved in the transformation process precede changes in the morphology and therefore a biochemical test might yield positive results before the pathologist can detect signs of malignancy under the microscope. This brings us to the issue of the "golden standards" used for evaluating the sensitivity and specificity of this kind of test. The so called "golden standards" consist of a set of serum samples from patients that were diagnosed as having cancer or benign lesions by a pathologist examining histological sections under the microscope. Yet, there is a percentage

[6] Our attempts at developing a sandwich immunoassay format resulted in a loss of 10-15% sensitivity (true positives) compared to the competitive assay. Using more than one Mab to capture all RECAF glycated proteins would be too noisy and impractical to manufacture and quality control.

[7] The 95 percentile of a population of normal donors used as cutoff value corresponds to 4,600 Units of RECAF.

[8] Sensitivity in this context does not refer to the amount of RECAF the test can measure but rather, it is the ability of the test to correctly identify patients with cancer. Specificity is the ability of the test to correctly identify individuals without the disease.

of cases in which pathologists err [95, 96]. This error is usually ignored because it is relatively small, but it becomes very significant when evaluating a test that approaches the highly sought 100% specificity and sensitivity necessary for screening. To yield those values, a blood test would have to replicate the pathologist's error, which is formally absurd.

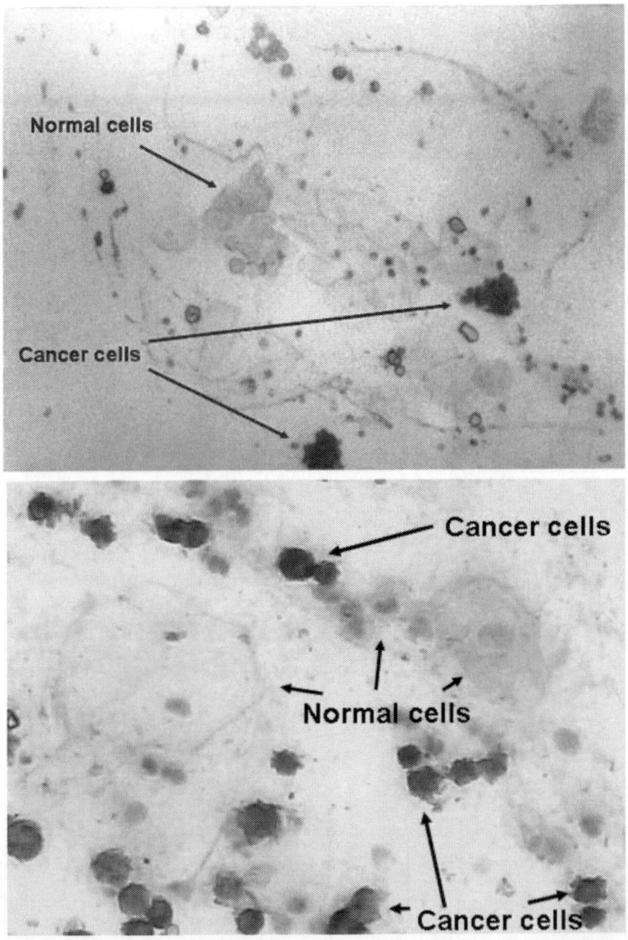

Figure 11. Sputum (top) and cervical (bottom) smears from normal individuals spiked with cultured lung cancer cells or cervical cancer cells respectively. After fixing, the slides were stained for RECAF. The cancer cells are notoriously stronger stained than the normal cells (including leukocytes in the bottom picture). Counterstained with Hematoxylin.

RECAF serum tests come close to the levels at which the pathologist error becomes relevant and this means that their actual sensitivity and specificity might be even higher. Also, it means that the test might be useful for cancer screening, which is very relevant given the fact that serum RECAF tests can detect a significant percentage of malignancies at early stages (I and II) [89, 91], when treatment is most effective. Furthermore, RECAF is not only elevated in the serum of cancer patients, but also in saliva, as shown in Figure 13. While the accuracy of the saliva test is lower than that of the serum test, saliva collection is remarkably simple and inexpensive. Thus, the saliva test should be set up with a cut-off value low enough to detect most of the cancer samples and then, on a second round, the saliva positive patients would get a RECAF serum test with a high cut-off value to eliminate most of the false positives.

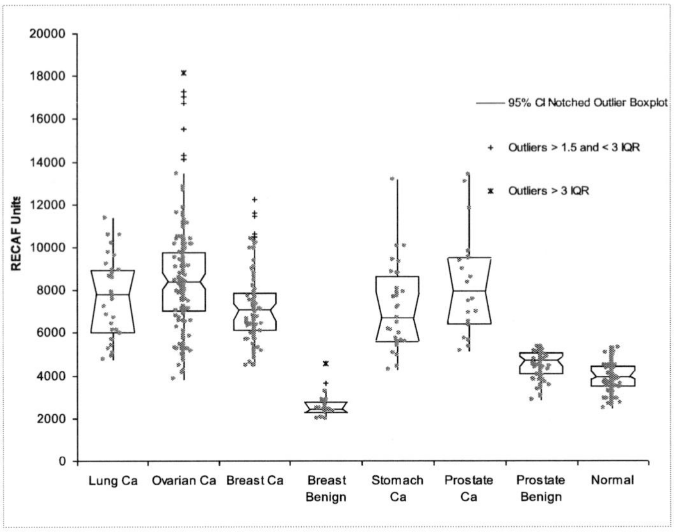

Figure 12. Serum RECAF concentration in patients with different types of cancer, benign lesions and normal donors. Ca = cancer. Circulating RECAF is elevated in cancer patients when compared to samples from patients with benign lesions or to serum from normal donors.

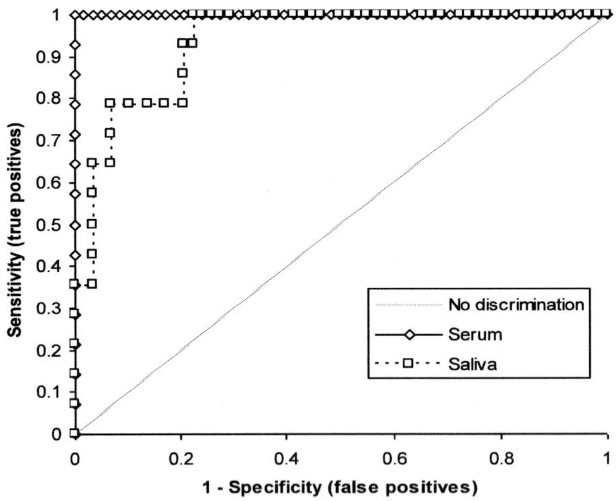

Figure 13. Receiver Operating Characteristic (ROC) curve of paired serum and saliva samples from 14 patients with breast cancer and 58 normal controls. The area under the curve (AUC) for serum = 1.0 and for saliva = 0.94.

The expression of RECAF in normal individuals is independent of age [89] and sex [unpublished results] and it seems to be unrelated to tumor mass. Moreover, there is no correlation in the expression of RECAF and CA125 [91] CEA [97] or PSA [98]. This allows combining the values of these common markers with the amount of RECAF thus increasing the overall sensitivity and specificity, in some cases, to the levels required for screening for early stages of ovarian cancer [91]. Combining PSA with RECAF is of particular interest because it brings together the RECAF specificity for malignancy with the organ specificity of PSA.

An intriguing possibility is to use serum RECAF as an indicator for prescribing chemotherapy to previously treated cancer patients with no clinical manifestation of cancer recurrence. It has been found that waiting for clinical or imaging evidence of recidivism results in poor outcome whereas guiding therapy with traditional markers before clinical manifestation improves prognosis [99]. The general idea is to administer chemotherapy when the cancer marker increases over a certain threshold even if the patient shows no evidence of recurrence. The broad spectrum of RECAF expression offers the enticing possibility of using this marker for guiding therapy in many different malignancies rather than the limited types of cancer that can be managed this way using current cancer markers.

In addition to the human RECAF tests, we developed a canine version using slightly modified AFP peptides. The test performance in terms of sensitivity and specificity was essentially the same as for the human test[9].

Potential Uses of RECAF for Cancer Therapy

Inhibition of Malignant Cell Proliferation by Anti-RECAF Mabs

Mouse Mabs directed against human RECAF recognize mouse malignant cells [48, 54]. The 167H.4 Mab reacts with EL-4 and YAC-1 (two mouse T-lymphomas), TA-3, (a mouse mammary carcinoma cell line) and P-388 (a mouse malignant cell line of dendritic origin). This is unusual but not totally unexpected since AFP from one species binds or is taken up by cells from another species [21, 45], which suggests that the binding sites on both AFP and RECAF are highly conserved and yet different enough to elicit an immune response. The cross reaction of anti-human RECAF Mabs with mouse malignant cells enables the use of simple syngeneic mouse tumor models for cancer targeting rather than transplanting human cancers into nude mice.

There is a plethora of literature ascribing biological activities to AFP, ranging from apoptosis to the stimulation of cell proliferation. The anti-RECAF 167H.4 Mab showed an inhibitory effect on the rate of cell division: P-388 cells incubated with the Mab showed a substantial inhibition in cell replication (Figure 14). The cells were not killed, as they continued to exclude Trypan Blue and would re-start replication after washing out the Mab. Complete DNA synthesis suppression was not achieved however, perhaps due to the presence of bovine AFP and bovine RECAF in the fetal calf serum required to obtain a measurable amount of cell replication in the controls. Bovine AFP binds to the mouse RECAF [45] and competes, to a certain extent, with the 167H.4 Mab [personal observations].

[9] While the overall sensitivity and specificity can be easily compared between humans and dogs, the comparison of the test performance within a given type of cancer is more difficult because the most prevalent types of cancer in dogs are different from the most prevalent cancers in humans: In dogs, the most common cancers are lymphomas, hemangiosarcomas, osteosarcomas and mast-cell-tumors. While breast cancer is also prevalent in dogs, it is less common than in humans.

A similar proliferation inhibition was obtained on LoVo cells (a human colon adenocarcinoma) [54]. In these experiments, it was shown that pure human AFP could reverse the inhibitory effect of the anti-receptor Mabs.

In addition, 167H.4 inhibited PHA transformation of normal human peripheral leukocytes.

All of these results point to a signal transduction pathway ending up on a cell proliferation control mechanism. In this context, it is worth mentioning that high concentrations of AFP can induce apoptosis [100, 101]. A great deal of work needs to be done in order to understand these control pathways, which could provide new strategies for cancer management.

These encouraging *in-vitro* results prompted a series of *in-vivo* experiments in which we inoculated C57 Black mice subcutaneously with 2 x 10^6 EL-4 cells followed, the same day, by the intravenous injection of 100 uL of either 167H.4 Mab ascites produced in Balb/c or normal Balb/c serum. In order to facilitate EL-4 cell grafting, the animals received a 300 Rads dose a week before injection. Four to five animals were included in each group and the experiments were repeated 6 times with consistent results: By the sixth day after injection, there was a clear difference between the control animals, which exhibited tumors 5-6 mm in size, and the treated animals, which had only a scar at the site of the injection. If allowed to progress, the tumors killed all of the control animals within 3 weeks. In 5 out of 6 experiments, all of the treated animals remained tumor free until euthanized. In one experiment we kept the treated mice for 8 months with no noticeable signs of disease or other abnormalities. Their offspring was normal. Figure 15 depicts 2 control and 3 treated animals from one of those experiments.

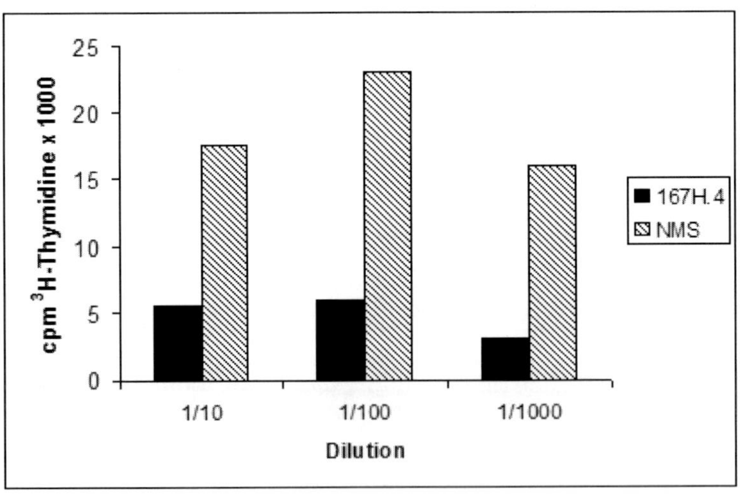

Figure 14 Inhibition of P-388 cell proliferation by the 167H.4 anti-RECAF Mab as compared to normal mouse serum (NMS). The incorporation of ^{3}H-Thymidine was reduced by 70%-75% in a 6 hour pulse chase experiment.

In the 6th and last experiment, the treated animals did developed tumors and eventually died of them, albeit they survived significantly longer than the controls (Figure 16). An assay conducted on the EL-4 cells in culture used for grafting these animals showed a drop to only 20% of specific ^{125}I-AFP binding. This unexplained and sudden drop in AFP binding, and hence in RECAF expression, has been previously noticed in other cell lines, such as MCF-7 and Ichikawa when they are cultured over many passages [personal observations], and could be related to the above mentioned RECAF gel bands polymorphism (Figure 8)[10].

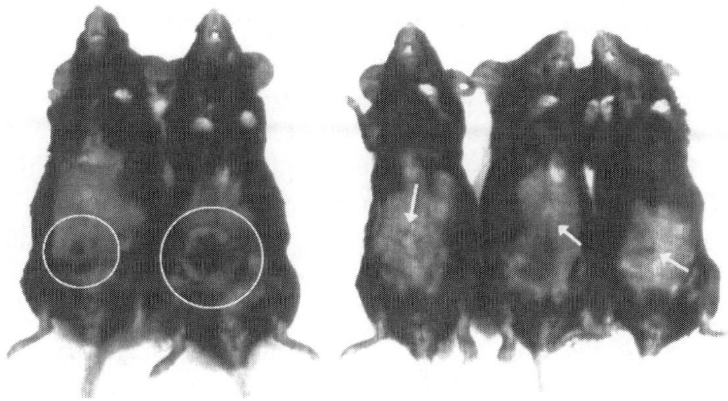

Figure 15. Tumor growth inhibition by the anti-RECAF 167H.4 Mab. Five mice on each group received an IV injection of either 100 uL of Mab ascites or normal mouse serum as a control, along with 2 x 10^6 EL-4 cells injected subcutaneously on the abdomen. After 2 weeks, all control animals exhibited large tumors (two of these animals are shown on the left, with a circle around the tumors). All treated animals were tumor free as shown on the three animals on the right. The arrows point to the small scars left where the EL-4 cells were injected.

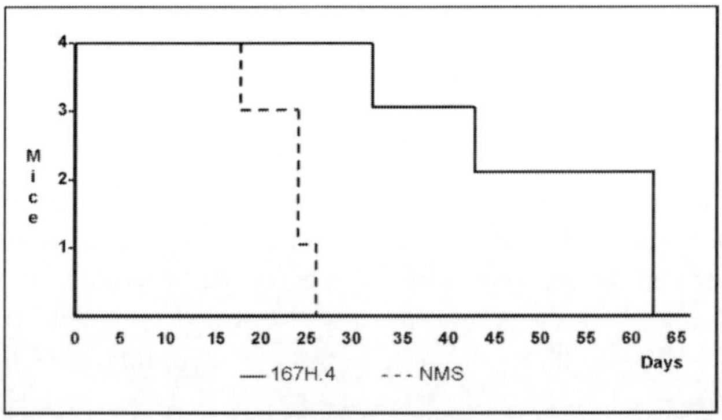

Figure 16. Tumor growth inhibition by the anti-RECAF 167H.4 Mab. This experiment was similar to the one depicted in Figure 15 with the difference that the specific AFP binding to the inoculated EL-4 cells had dropped by 80%. Nonetheless, the life span of the Mab treated animals was significantly longer than the control's.

These *in-vitro* and *in-vivo* results indicate that anti-RECAF antibodies might slow down or abrogate cell proliferation and makes our experience in carrying out over a dozen fusions to generate more anti-RECAF Mabs worth mentioning: After fusing the splenocytes of the immunized mouse with a myeloma partner[11], we allowed the resulting hybridoma colonies to grow in 96 well plates. Once they were visible to the naked eye, we screened the supernatants

[10] It must be stressed that this RECAF turn-off phenomenon is not related to selective pressure by the Mab or to *in-vivo* growing conditions since it takes place during cell culture under standard conditions and before the cells are injected into a mouse or exposed to Mabs, AFP or any other targeting molecule.

[11] SP2 or Fox-NY (the latter, was used for generating the previous round of anti-RECAF Mabs).

by ELISA against RECAF. We also screened for AFP binding inhibition as described elsewhere in this chapter. We usually found 5-15 positive clones in a 10 plate fusion. As time went by (and the concentration of Mab in the wells increased), the cells from those positive wells would stop growing, they would become picnotic and finally die. However, the cells could be kept alive by changing the medium every 48 hours. This 'suicidal' behavior could be explained by the presence of RECAF on the hybridoma cells, which would be carried over from the fusion partner. As the concentration of anti-RECAF secreted by the hybridoma increased, the cells would stop proliferating in a sort of autocrine negative feedback loop.

Another worth mentioning fact is that not all anti-RECAF Mabs inhibit binding. For example, Mab 1.4G11 has no *in-vitro* effect on cell proliferation despite (a) recognizing all the same bands that AFP recognizes on a Western (including the membrane associated 31 kDa and 18 kDa bands), (b) inhibiting the binding of AFP to its receptor and (c) being inhibited from binding to RECAF by an excess of AFP. These are the same criteria used for discovering 167H.4, which *did* inhibited cell proliferation. The main difference between the two Mabs is the immunogen used: 167H.4 was generated from a mouse immunized with a pool of human breast cancer biopsies solubilized with NP-40 and retained on Agarose-PNA (Peanut Agglutinin), which was then injected into a mouse [54]. 1.4G11 was produced from a mouse immunized against the soluble RECAF fraction from an MCF-7 extract.

It should be pointed out that a 1/300 dilution of the serum from the mouse used to generate the 167H.4 Mab also inhibited the binding of ^{125}I-AFP to Ichikawa cells by 71% [48] (as a reference, the supernatant of the 167H.4 clone inhibited ^{125}I-AFP binding by 73%). This is of practical interest: If a 1/300 dilution of the serum could inhibit cell proliferation *in-vitro*, it is likely that inside the animal the neat serum should inhibit cell proliferation even better, providing the mouse with some protection against cancer whether it was grafted of spontaneously generated. In essence, the animal would have been vaccinated against most cancer types given the broad spectrum of RECAF expression.

At present, the mechanisms underlying the proliferation inhibition are unknown but they must involve a message from the AFP receptor(s) on the cell surface to the nucleus. It seems reasonable to consider that the receptor molecules carry the message to the cytoplasm or even to the nuclear membrane, where immunohistology shows strong RECAF accumulation and capping. A thorough analysis of the possible pathways involving AFP binding proteins has been advanced by Mizejewski [102].

RECAF Targeted Chemotherapy

The discovery of cancer markers rekindled Paul Ehrlich's dream of a "magic bullet" which would target and kill cancer cells without destroying healthy cells, thus significantly improving standard chemotherapy.

The most common type of cancer-targeted compound consists of two parts, the delivery moiety (usually an antibody), which should be as cancer specific as possible, and the toxic moiety, which, once delivered into the cell, kills it. The latter can consist of chemotherapy agents such as Doxorubicin, Chlorambucil, Taxol, etc. or biological toxins such as Ricin, Abrine and others. Radioactive isotopes have also been used for this purpose [103]. The delivery moiety and the toxic moiety can be linked together in a variety of ways. The type of link is important because it must fulfill certain requisites: It should split only within the targeted cell and once inside, cancer cells must be able to break it down effectively to allow the toxic component to reach its target(s).

In general, targeted therapy has not delivered on its promise mainly because of the lack of suitable delivery vehicles: There are not that many cancer markers available and most – AFP is a typical example – move in the wrong direction, from within the cell outwards. The toxic moieties have also exhibited problems. For example, at tolerable doses, many chemotherapy agents do not achieve 100% cell killing, which leads to recurrence and in many cases, to selection of drug resistant cancer cells. Powerful cell toxins such as Ricin or Abrine are exceptions, since they exhibit phenomenal killing efficiency but their large size and immunogenicity are problematic.

RECAF offers several interesting features as a cancer cell target: Firstly, the receptor moves in the right direction; it carries AFP into the cell. This favors the intake of toxic payloads. Secondly, AFP, or a peptide thereof can be used as the delivery molecule with some obvious advantages over using antibodies such as: (i) no immune response triggered against the injected material (thus, no need to humanize Mabs), (ii) small size with faster tissue penetration and elimination if an AFP fragment or peptide is used and (iii) the possibility to tailor a peptide to optimize cell penetration and/or conjugation to the toxic moiety. The major difficulty associated with using RECAF as a target used to be the high cost and ethical issues related to purifying AFP of fetal origin in large quantities but this has been addressed by developing recombinant polypeptides [104, 105] or short peptides [56, 57] containing the AFP amino acid sequence that binds to RECAF.

As early as 1983, Deutsch et al. conjugated Daunomycin to arachidonic and docosahexaenoic acids and showed, both *in-vitro* and *in-vivo*, that the conjugates were effective against AFP producing hepatomas but ineffective against hepatomas that did not synthesize that protein. *In-vitro*, cell killing by Daunomycin-fatty acid conjugates was ~3 times higher than using Daunomycin alone. *In-vivo*, 100 days after grafting rats with an AFP producing hepatoma, all non-treated animals as well as the animals that received Daunomycin alone were dead, whereas 3 out of 5 animals injected with the conjugates were still alive [106]. These authors also found that the Daunomycin exhibited a "marked diminution" in toxicity when conjugated to fatty acids as compared to the free drug. They speculated that their results depended on the affinity of AFP for arachidonic acid, which would selectively bring the Daunomycin complex to the hepatoma cells.

A more direct approach was taken by Feldman et al., who conjugated Doxorubicin directly to AFP. This resulted in a 5-fold increase of the conjugate toxicity compared to the drug alone [107].

As mentioned above, the conjugation chemistry is important because AFP is known to enter and exit cancer cells unscathed [45] whereas the toxic payload must be retained and reach its cellular target. To address this issue, Yabbarov et al. used an acid-labile aconitil link between rAFP3D (a recombinant 3rd domain fragment of AFP) and Doxorubicin [108]. They found that while Doxorubicin alone was slightly more cytotoxic to SCOV3 cells (an ovarian cancer cell line sensitive to the drug) than the conjugate, the latter was significantly less toxic to normal lymphocytes than Doxorubicin alone, thus allowing the use of a higher concentration of the drug. Interestingly, the conjugate was far more effective on SKVLB ovarian cancer cells, which are resistant to Doxorubicin, than the drug alone. This suggests that the resistance mechanism might be related to poor cell internalization of the drug, which would be enhanced by the delivery mechanism provided by the AFP fragment. If this is representative of a more general situation, then AFP-drug conjugates might help treat some drug resistant cancers.

Similar results were obtained with Paclitaxel loaded nanoparticles attached to rAFP3D [109].

Besides drugs or toxins, RECAF could be used as a target for the intracellular delivery of interference RNA, as suggested by Glebova et al. [110].

The literature and our own experiments strongly support the idea that conjugating cytotoxic agents with AFP or fragments thereof moderately increases toxicity, but its main merit comes in the form of a substantial reduction of the drug cytotoxicity to normal cells. It is the cancer/normal cytotoxic ratio that is much improved. This indicates that the amount of conjugated drug could be greatly increased without generating the dreadful secondary effects of chemotherapy, thus rendering the treatment more effective.

A potential problem in using RECAF as a target is that its expression is related to the degree of cell differentiation and therefore, poorly differentiated normal cells, such as stem cells and reproductive cells, could be targeted as well. This, however, does not seem to be the case: As mentioned above, the administration of anti-RECAF Mabs completely abrogated the growth of tumors in mice but the animals did not seem to be affected by the treatment despite being followed up for several months. Also, mice injected with massive doses of ^{131}I-AFP were unaffected by the radiation and had normal litters indicating that the reproductive cells were not affected.

In a series of experiments *in-vitro*, we tried to circumvent the difficulties related to the drug-ligand[12] conjugation chemistry by coupling the anti-RECAF 1.4G11 Mab to Thimerosal®, a mercury based antimicrobial that attaches loosely to free thiols in proteins [111]. The conjugation method and cytotoxicity test were extremely simple: The Mab was incubated with Thimerosal® for 4-6 hours and the free antimicrobial was then removed by gel filtration on Sephadex® G-25. Immediately after, the conjugate was incubated for 48 hours with the target cells in serum free medium and the cell viability was determined by the MTT method [112]. As shown in Figure 17, the conjugate killed cancer cells very effectively while having little effect on MRC-5, which is considered as a normal human diploid lung cell line [113].

Figure 18 shows the effects of a 1.4G11-Thimerosal® conjugate and two irrelevant Mabs conjugated to Thimerosal® in identical manner, on JC mouse breast cancer cells [113]. Only the 1.4G11-Thimerosal® conjugate killed the cells, indicating that cell demise was not the result of Thimerosal® leaking from the conjugate into the medium but rather due to the intracellular delivery of the toxic payload by the anti-RECAF Mab.

It is unlikely that Thimerosal® will be used for *in-vivo* cancer targeting since its loose binding to proteins would result in serum proteins sequestering most of the injected chemical. Moreover, mercury is a toxic compound. However, these conjugates could be used *in-vitro* for purging bone marrow of leukemia cells prior to using it for auto-transplantation.

[12] The ligand in this case could be AFP, or an anti-RECAF Mab, an AFP peptide, etc.

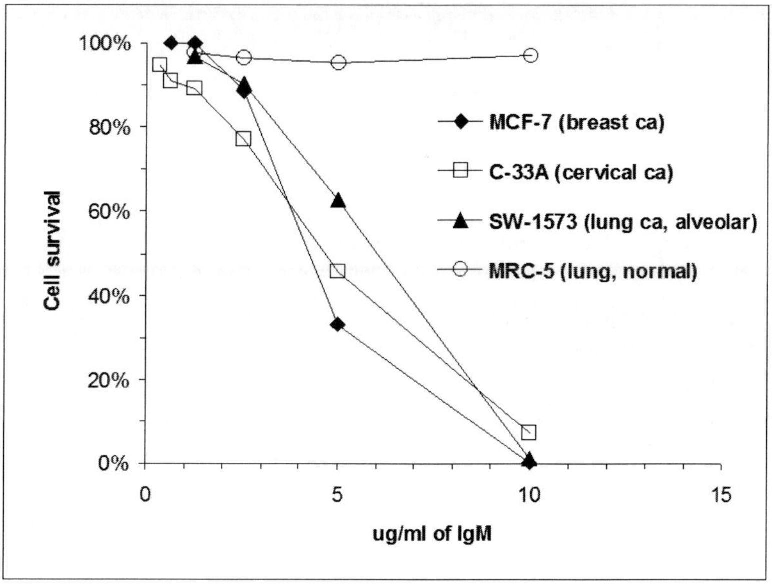

Figure 17. Incubation of cancer and normal human cell lines with different doses of the 1.4G11 anti-RECAF Mab conjugated to Thimerosal®. Cells were incubated with the conjugate for 48 hours in serum free medium. Cell killing was determined by the MTT assay. The conjugate's effect on MRC-5 (a normal cell line) is negligible compared to the effect on a lung cancer cell line (SW-1573).

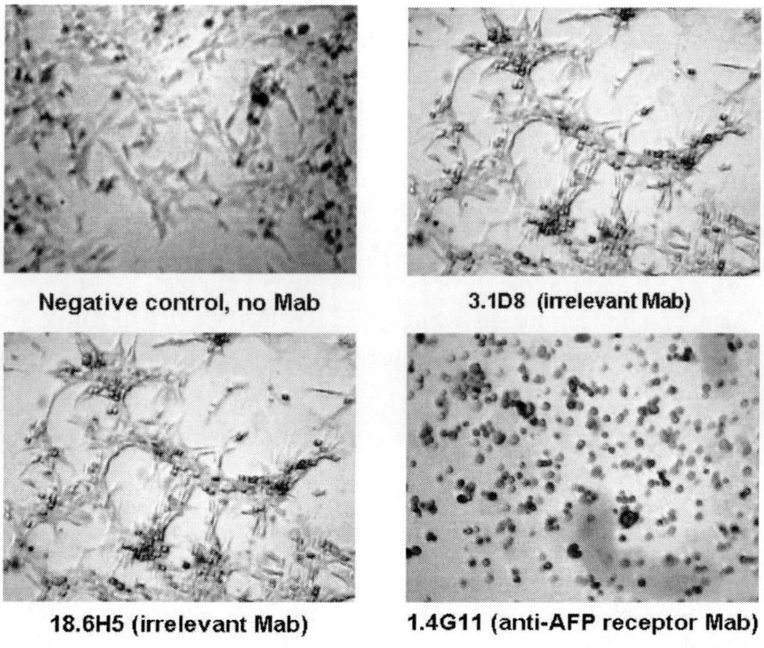

Figure 18. JC cells were incubated under identical conditions (48 hours, serum free medium and 10 ug/ml of IgM) with two irrelevant IgM Mabs and 1.4G11 (an IgM producing Mab against RECAF). The 3 Mabs were conjugated to Thimerosal® under identical conditions. Only the 1.4G11- Thimerosal® conjugate killed the cells, showing that cell demise is not due to leaking of conjugated Thimerosal® into the medium.

Conclusion

There is a substantial amount of research done on RECAF, ranging from its characterization and physiology, to its potential uses in medical practice.

On the characterization side, it is clear that a number of different proteins in cancer cell extracts bind to AFP. It is equally clear that the receptor for AFP is an N-glycan. Until proven otherwise, it seems logical to assume that the same glycan is attached to all these proteins. Thus we used, throughout this chapter and in previous articles, the acronym RECAF as a representation of the site, on that glycan, where AFP binds. This makes the term RECAF a functional descriptor rather than a protein name, which greatly simplifies things, particularly when working with complex mixtures such as cell extracts or serum where more than one protein may be binding to AFP. In addition, to denote a particular protein or gel band with RECAF activity, one can add the molecular weight at the end of RECAF. For example, the 65 kDa RECAF band would be RECAF-65.

The potential medical applications based on RECAF are particularly enticing: There is a body of evidence showing that RECAF could be used as a marker for detecting cancer cells on tissue sections and more importantly, on smears. Elevated serum RECAF has been shown to detect early stages of cancer, when conventional therapy yields 80-90% survival rates. The high sensitivity and specificity of the marker leaves open its potential for screening, particularly when used in combination with other cancer markers. On a more speculative basis, serum RECAF could be used for monitoring patients in remission, and perhaps prompting, if values increased over time, the administrating of therapy before recurrence became clinically evident, at which time the prognosis is usually ominous.

Several authors have demonstrated that RECAF has potential for scintigraphy, with which, small malignant tumors have been detected. This type of imaging could be used to confirm the malignant nature of a lesion in a suspicious mammography or to evaluate the presence of metastases in patients already diagnosed with cancer.

Finally, there is substantial evidence, both *in-vitro* and *in-vivo*, showing that cancer cells can be targeted and destroyed using anti-RECAF Mabs by themselves or coupled to cytotoxic agents, as well as by AFP and fragments or peptides thereof coupled to drugs.

Given the span of applications in which RECAF has yielded positive results and the medical significance of some of those applications, it appears as evident that this system warrants a great deal of future work.

References

[1] Abelev, G. I., Perova, S. D., Khramkova, N. I., Postnikova, Z. A. and Irlin, I. S. (1963). Production of embryonal alpha-globulin by transplantable mouse hepatomas. *Transplantation*, *1*, 174-180.

[2] Ruoslahti, E. and Seppala, M. (1971). Studies of carcinofetal proteins: III. Development of a radioimmunoassay for AFP. Demonstration of AFP in serum of healthy human adults. *Int J Cancer*, *8*, 374-378.

[3] Ruoslahti, E. and Seppala, M. (1979). AFP in cancer and fetal development. *Adv Cancer Res*, *29*, 275-310.

[4] Benno, R. H. and Williams, T. H. (1978). Evidence for intracellular localization of AFP in the developing rat brain. *Brain Res*, *142*, 182- 186.

[5] Trojan, J. and Uriel, J. (1980). Immunocytochemical localization of AFP in the developing rat brain. *Oncodevelop Biol Med*, *1*, 107-111.

[6] Dziegielewska, K. M., Evans, C. A. M., Lorscheider, F. L., Malinowska, D. H., Mollgard, K., Reynolds, M. L. and Saunders, N. R. (1981). Plasma proteins in fetal sheep brain: Blood-barrier and intracerebral distribution. *J Physiol*, *318*, 239-250.

[7] Uriel, J., Trojan, J. Dubouch, P. and Pineiro, A. (1982). Intracellular AFP and albumin in the developing nervous system of the baboon. *Path Biol*, *30*, 79-83.

[8] Mollgard, K., Jacobsen, M., Clausen, O. P. and Saunders, N. R. (1979). Immunohistochemical evidence for an intracellular localization of plasma proteins in human foetal choroid plexus and brain. *Neuroscience Letters*, *14*, 85-90.

[9] Moro, R. and Uriel, J. (1981). Early localization of AFP in the developing nervous system of the chicken. *Oncodevelop Biol Med*, *2*, 391-398.

[10] Moro, R. (1983). Selective localization of AFP and serum albumin within the sensory ganglia cells of developing chicken. *Neuroscience Letters*, *41*, 253-257.

[11] Uriel, J., Trojan, J., Moro, R. and Pineiro, A. (1983). Intracellular uptake of AFP: A marker of neural differentiation. *Ann NY Acad Sci*, *417*, 321-329.

[12] Torand-Allerand, C. D. (1980). Coexistence of AFP, albumin and transferrin immunoreactivity in neurons of the developing mouse brain. *Nature*, *286*, 733-734.

[13] Fielitz, W., Esteves, A. and Moro, R. (1984). Protein composition of the cerebrospinal fluid in the developing chick embryo. *Dev Brain Res*, *13*, 111-115.

[14] Ali, M., Balapure, K., Singh, D. R., Skulka, R. N. and Sahib, M. (1981). Ontogeny of AFP in human foetal brain. *Brain Res*, *207*, 459-464.

[15] Ali, M., Mujoo, K. and Sahib, M. (1983). Synthesis and secretion of AFP and albumin by newborn rat brain cells in culture. *Dev Brain Res*, *6*, 47- 55.

[16] Schachter, B. S. and Toran-Allerand, C. D. (1982). Intraneuronal AFP and albumin are not synthesized locally in developing brain. *Dev Brain Res*, *5*, 93-98.

[17] Pineiro, A., Calvo, M., Iguaz, F., Lampreave, F. and Naval, J. (1982). Characterization, origin and evolution of AFP and albumin in post-natal rat brain. *Int J Biochem*, *14*, 817-822.

[18] Uriel, J., Faivre-Bauman, A., Trojan, J. and Foiret, D. (1981). Immunocytochemical demonstration of AFP uptake by primary cultures of fetal hemisphere cells from mouse embryos. *Neurosci Lett*, *27*, 171.

[19] Hajeri-Germond, M., Trojan Uriel, J. and Hauw, J. J. (1984). *In-vitro* uptake of exogenous AFP by chicken dorsal root ganglia. *Dev Neurosci*, *6*, 111-117.

[20] Villacampa, M. J., Lampreave, F., Calvo, M., Naval, J., Pineiro, A. and Uriel, J. (1984). Incorporation of radiolabeled AFP in the brain and other tissues of the developing rat. *Dev Brain Res*, *12*, 77-82.

[21] Moro, R., Fielitz, W., Esteves, A., Grunberg, J. and Uriel, J. (1984). *In-vivo* uptake of heterologous AFP and serum albumin by ependymal cells of developing chicken embryos. *Int J Dev Neuroscience*, *2*, 143-148.

[22] Trojan, J. and Uriel, J. (1982). Immunocytochemical localization of AFP and serum albumin in ecto-, meso-, and endodermal tissue derivatives of the developing rat. *Oncodevelop Biol Med*, *3*, 13-22.

[23] Jacobsen, M., Lassen, L. C. and Mollgard, K. (1984). Immunohistochemical evidence for intracellular localization of plasma proteins in CNS and some neural crest derivatives in human embryos. *Tumor Biol*, *5*, 53-60.

[24] Uriel, J. (1979). Retrodifferentiation and the fetal patterns of gene expression in cancer. *Adv Cancer Res*, *29*, 127-174.

[25] Uriel, J., Failly-Crepin, C., Villacampa, M. J., Pineiro, A. and Geuskens, M. (1984). Incorporation of AFP by the MCF-7 human breast carcinoma cell line. *Tumor Biol*, *5*, 41-51.

[26] Uriel, J. Poupon, M. F. and Geuskens, M. (1983). AFP uptake by cloned cells lines derived from a nickel-induced rat rhabdomyosarcoma. *Br J Cancer*, *48*, 261-269.

[27] Hajeri-Germond, M., Naval, J., Trojan, J. and Uriel, J. (1985). The uptake of AFP by C1300 mouse neuroblastoma cells. *Br J Cancer*, *51*, 791-796.

[28] Calvo, M., Laborda, J., Naval, J., Georgoulias, V. and Uriel, J. (1985). Uptake of fluoresceinated AFP by human leukemic cells. XIII annual meeting of the International Society for Oncodevelopmental Biology and Medicine. Paris, France. *Tumour Biol*, *6*(4), 273-446.

[29] Uriel, J., Villacampa, M. J., Moro, R., Naval, J. and Failly-Crepin, C. H. (1983). Accumulation d'AFP radiomarque'e dans des tumeurs mammaires spontane'es de la souris. *C R Acad Sci Paris*, *297*, 589.

[30] Uriel, J., Villacampa, M. J., Moro, R., Naval, J. and Failly-Crepin, C. (1984). Uptake of radiolabeled AFP by mouse mammary carcinomas and its usefulness in tumor scintigraphy. *Cancer Res*, *44*, 5314-5319.

[31] Moro, R., Heuguerot, C., Vercelli-Retta, J., Fielitz, W., Lopez, J. J. and Roca, R. (1984). The use of radioiodinated AFP for the scintigraphic detection of mouse mammary carcinomas. *Nuclear Med Comm*, *5*, 5-12.

[32] Hajeri-Germond, M., Naval, J., Trojan, J. and Uriel, J. (1985). The uptake of AFP by C1300 mouse neuroblastoma cells. *Br J Cancer*, *51*, 791-796.

[33] Line, B. R. and ALBANY MEDICAL CENTER HOSPITAL NY. (1998). ^{99M}Tc Alpha-fetoprotein: A novel, specific agent for the detection of human breast cancer. *Defense Technical Information Center*, http:// www.dtic.mil/docs/citations/ADA357295.

[34] Line, B., Feustel, P., Festin, S., Andersen, T., Dansereau, R., Lukasiewicz, R. and Bennett, J. (1999). Scintigraphic detection of breast cancer xenografts with Tc-99m natural and recombinant human alpha-fetoprotein. *Cancer Biotherapy and Radiopharmaceuticals*, *14*(6), 485-494.

[35] Rieppi, G., Fielitz, W., Vercelli, J., Moro, R., Esteves, A. and Roca, R. (1985). ^{131}I-AFP uptake by experimentally induced inflamatory lesions. *R.I.R.C.S. Med Sci*, *13*, 515-516.

[36] Trojan, J., Naval, J., Jusforgues, H. and Uriel, J. (1989). Alpha-fetoprotein (afp) in granulomatous inflammation of the mouse. *British Journal of Experimental Pathology*, *70*(4), 469-478.

[37] Dattwyler, R. J., Murgita, R. A. and Tomasi, T. B. (1975). Binding of Alpha-Foetoprotein to murine T cells. *Nature*, *256*, 656-657.

[38] Uriel, J., Bouillon, D., Russel, C. and Dupiers, M. (1976). AFP: The major high affinity estrogen binder in rat uterine cytosols. *Proc Nat Acad Sci U.S.A.* *73*, 1452-1456.

[39] Torre, J. M., Laborda, J., Naval, J., Darracq, N., Calvo, M., Mishal, Z. and Uriel, J. (1989). Expression of alpha-fetoprotein receptors by human T-lymphocytes during blastic transformation. *Molecular Immunology*, *26*(9), 851-857.

[40] Yakimenko, E. F., Karamova, E. R., Goussev, A. I., Hilgers, J., Abelev, G. I. and Yazova, A. K. (1998). Epitope mapping of human alpha-fetoprotein. *Tumour Biol.*, *19*(4), 301-309.

[41] Smalley, J. R. and Sarcione, E. J. (1980). Synthesis of AFP by immature rat uterus. *Bioch Biophys Res Comm*, *92*, 1429-1434.

[42] Sarcione, E. J., Zloty, M., Delluomo, D. S., Mizejewski, G. and Jacobson, H. (1983). Detection and measurement of AFP in human breast cancer cytosol after treatment with 0.4 M KCl. *Cancer Res*, *43*, 3739-3741.

[43] Sarcione, E. J. and Hart, D. (1985). Biosynthesis of AFP by MCF-7 human breast cancer cells. *Int J Cancer*, *35*, 315-318.

[44] Villacampa, M. J., Moro, R., Naval, J., Failly-Crepin, C., Lampreave, F. and Uriel, J. (1984). RECAFs in a human breast cancer cell line. *Bioch Biophy Res Comm*, *122*, 1322-1327.

[45] Naval, J., Villacampa, M. J., Goguel, A. F. and Uriel, J. (1985). Cell type specific receptors for AFP in a mouse T-lymphoma cell line. *Proc Natl Acad Sci U.S.A.*, *82*, 3301-3305.

[46] Suzuki, Y., Carl, Q. Y. and Alpert, E. (1992). Isolation and partial characterization of a specific RECAF in human monocytes. *J Clin Invest*, *90*, 1530-1536.

[47] Biddle, W. and Sarcione, E. J. (1989). Cytoplasmic RECAFs in MCF-7 human breast cancer cells and primary breast cancer tissue from postmenopausial women. *Biological activities of Alpha1-Fetoprotein*, Vol. *II*, CRC Press, Boca Raton, FL, 129-138.

[48] Moro, R., Tamaoki, T., Wegmann, T. G., Longenecker, B. M. and Laderoute, M. P. (1993). Monoclonal antibodies directed against a widespread oncofetal antigen: The Alpha-Fetoprotein Receptor. *T Biol*, *14*, 116-130.

[49] Newby, D., Dalgliesh, G., Lyall, F. and Aitken, D. A. (2005). Alphafetoprotein and alphafetoprotein receptor expression in the normal human placenta at term. *Placenta*, *26*(2), 190-200.

[50] Benassayag, C., Vallette, G., Delorme, J., Savu, L. and Nunez, E. A. (1980). High affinity of nonesterified polyunsaturated fatty acids for rat alpha-fetoprotein (AFP). *Oncodev Biol Med*, *1*(1), 27-36.

[51] Aussel, C. and Masseyeff, R. (1983). Human alpha-fetoprotein-fatty acid interaction. *Bioch Biophys Res Comm*, *115*(1), 38-45.

[52] Aussel, C. and Masseyeff, R. (1983). On rat alpha-fetoprotein as a fatty acid carrier. *Bioch Biophys Acta (BBA)/Lipids and Lipid Metabolism*, *752*(2), 324-328.

[53] Torres, J. M., Anel, A. and Uriel, J. (1992). Alpha-fetoprotein-mediated uptake of fatty acids by human T lymphocytes. *J Cellular Physiology*, *150*(3), 456.

[54] Laderoute, M. P. (1991). The characterization of a novel, widespread, PNA-reactive tumor-associated antigen: The alpha-fetoprotein receptor/binding protein, ProQuest Dissertations Publishing.

[55] Dudich, I., Tokhtamysheva, N., Semenkova, L., Dudich, E., Hellman, J. and Korpela, T. (1999). Isolation and structural and functional characterization of two stable peptic fragments of human alpha-fetoprotein. *Biochemistry*, *38*(32), 10406-10414.

[56] Schmid, R., Moro, R. and Moro, R. J. (2011). AFP peptides that bind to RECAF (RECAF). *Tumor Biology*, *32*(S1), 111.

[57] Moro, R. J. and Schmid, R. H. (2011). Peptides that bind the alpha-fetoprotein (AFP) receptor and uses thereof. WIPO Patent application WO2011048503.

[58] Pang, X., Chen, M., Jia, W. and Zhou, X. (2008). Inhibitory effects of human AFP-derived peptide-pulsed dendritic cells on mouse hepatocellular carcinoma. *Chinese J of Cancer*, *27*(12), 1233.

[59] Butterfield, L. H., Ribas, A., Dissette, V. B., Lee, Y., Yang, J. Q. and Pilar De la Rocha Economou, J. S. (2006). A phase I/II trial testing immunization of hepatocellular carcinoma patients with dendritic cells pulsed with four α-fetoprotein peptides. *Clin Cancer Res*, *12*(9), 2817-2825.

[60] Mizukoshi, E., Nakamoto, Y., Tsuji, H., Yamashita, T. and Kaneko, S. (2006). Identification of α-fetoprotein-derived peptides recognized by cytotoxic T-lymphocytes in HLA-A24+ patients with hepatocellular carcinoma. *Int J Cancer*, *118*(5), 1194-1204.

[61] Vakharia, D. and Mizejewski, G. J. (2000). Human alpha – fetoprotein peptides bind estrogen receptor and estradiol, and suppress breast cancer. *Breast Cancer Res Treatment*, *63*(1), 41-52.

[62] Li, Z., Wang, X., Lin, H., Xu, B., Zhao, Q., Qi, B. and Wang, Z. (2015). Anti-tumor immunity elicited by cross-linking vaccine heat shock protein 72 and alpha-fetoprotein epitope peptide. *Neoplasma*, *62*(5), 713-721.

[63] DeFreest, L., Mesfin, F., Joseph, L., McLeod, D., Stallmer, A., Reddy, S. and Bennett, J. (2004). Synthetic peptide derived from alpha-fetoprotein inhibits growth of human breast cancer: Investigation of the pharmacophore and synthesis optimization. *J Peptide Res*, *63*(5), 409-419.

[64] Mesfin, F. B., Bennett, J. A., Jacobson, H. I., Zhu, S. and Andersen, T. T. (2000). Alpha-fetoprotein-derived antiestrotrophic octapeptide. *Bioch Biophys Acta - Molecular Basis of Disease*, *1501*(1), 33-43.

[65] Esteban, C., Geuskens, M. and Uriel, J. (1991). Activation of an Alpha-Fetoprotein (AFP) receptor autocrine loop in HT-29 human colo carcinoma cells. *Int J Cancer*, 49, 425-430.

[66] Esteban, C., Trojan, J., Macho, A., Mishal, Z., Lafarge-Frayssinet, C. and Uriel, J. (1993) Activation of an alpha-fetoprotein/receptor pathway in human normal and malignant peripheral blood mononuclear cells. *Leukemia*, 7, 1807-18016.

[67] Severin, S. E., Kanevski, V. Y., Sologub, V. K., Nakachian, R. and Severin, E. S. (1994). The purification of human alpha-fetoprotein receptor from fetal and cancerous tissues. *XXII Meeting of the ISOBM*, *Tumour Biol*, *16*(5), 297-344.

[68] Villacampa, M. J., Alava, M. A., Uriel, J. and Pineiro, A. (1985). Characterization of a membrane-receptor for Alpha-fetoprotein in rat fetal tissues. XIII Meeting of the ISOBM, Paris.

[69] Kanevsky, V., Pozdnyakova, L., Aksenova, O., Severin, S., Katukov, V. and Severin, E. (1997). Isolation and characterization of AFP-binding proteins from tumor and fetal human tissues. *Biochemistry and Molecular Biology International*, *41*(6), 1143-1151.

[70] Severin, E. S., Kanevsky, Y. V., Sologub, V. K., Luzhkov, Y. M., Koromyslova, I. A., Severin, S. E., Frias-Pena, J., Nakachian, R. and Andreani, J. (1996). The natural immunity against Alpha-Fetoprotein Receptor (AFPR). *XXIII Meeting of the ISOBM Montreal*, *Tumour Biol*, *17*, Suppl 1, 1-85.

[71] Torres, J., Darracq, N. and Uriel, J. (1992). Membrane proteins from lymphoblastoid cells showing cross-affinity for α-fetoprotein and albumin. isolation and characterization. *Bioch Biophys Acta (BBA)/Protein Structure and Molecular Enzymology*, *1159*(1), 60-66.

[72] Uriel, J., Naval, J. and Laborda, J. (1987). α-Fetoprotein-mediated Transfer of Arachidonic Acid into Cultured Cloned Cells Derived from a Rat Rhabdomyosarcoma. *J Biol Chem*, *262*(8), 3579-3585.

[73] Uriel, J., Torres, J. and Anel, A. (1994). Carrier-protein-mediated enhancement of fatty-acid binding and internalization in human T-lymphocytes. *Bioch Biophys Acta BBA - Molecular Cell Research*, *1220*(3), 231-240.

[74] Torres, J. M., Geuskens, M. and Uriel, J. (1991). Receptor-mediated endocytosis and recycling of alpha-fetoprotein in human B-lymphoma and T-leukemia cells. *Int J Cancer*, *47*(1), 110–117.

[75] Parmelee, D. C., Evenson, M. A. and Deutsch, H. F. (1978). The presence of fatty acids in human alpha-fetoprotein. *J Biol Chem*, *253*(7), 2114-2119.

[76] Vallette, G., Vranckx, R., Martin, M., Benassayag, C. and Nunez, E. (1989). Conformational changes in rodent and human alpha-fetoprotein - influence of fatty-acids. *Biochim Biophys Acta*, *997*(3), 302-312.

[77] Nishihira, J., Koyama, Y., Sakai, M. and Nishi, S. (1993). The fatty acid binding site of human α-fetoprotein. *Bioch Biophys Res Comm*, *196*(3), 1049-1057.

[78] Tkacz, J. S. and Lampen, J. O. (1975). Tunicamycin inhibition of polyisoprenyl N-acetylglucosaminyl pyrophosphate formation in calf-liver microsomes. *Biochem Biophys Res Commun*, *65*(1), 248-257.

[79] Mizejewski, G. J. (2013). Review of the adenocarcinoma cell surface receptor for human alpha-fetoprotein; proposed identification of a widespread mucin as the tumor cell receptor. *T Biol*, *34*(3), 1317-1336.

[80] Mizejewski, G. J. (2014). The adenocarcinoma cell surface mucin receptor for alpha-fetoprotein: Is the same receptor present on circulating monocytes and macrophages? A commentary. *T Biol*, *35*(8), 7397-7402.

[81] Mizejewski, G. (2015). The alpha-fetoprotein third domain receptor binding fragment: In search of scavenger and associated receptor targets. *J Drug Targeting*, *23*(6), 538-551.

[82] Pardee, A. D., Yano, H., Weinstein, A. M., Ponce, A. A. K., Ethridge, A. D., Normolle, D. P. and Butterfield, L. H. (2015). Route of antigen delivery impacts the immunostimulatory activity of dendritic cell-based vaccines for hepatocellular carcinoma. *J Immunotherapy of Cancer*, *3*(1), 32-48.

[83] Uriel, J., Laborda, J. Naval, J. and Gueskens, M. (1989). *RECAFs in malignant cells: An overview. Biological activities of Alpha1-Fetoprotein*, Vol. *II*. CRC Press, Boca Raton, FL, 103-117.

[84] Tsuboi, S., Taketa, K., Nouso, K., Fujikawa, T., Manabe, K., Ohmori, H. and Shiratori, Y. (2006). High level of expression of α-fetoprotein receptor in gastric cancers. *Tumor Biology*, *27*(6), 283-288.

[85] Moro, R. J. (filed: 1994). Detection of cancer using antibodies to the alphafeto protein receptor. US Patent US6514685.

[86] Lie-Injo, L. E., Caldwell, J., Ganesan, S. and Ganesan, J. (1976). Radioimmunoassay of serum alpha-fetoprotein in patients with different maliganant tumors. *Cancer*, *38*(1), 341-345.

[87] Vântu, A., Bălănescu, I., Stafidov, N. and Voiculeț, N. (1982). Variations of serum carcinoembryonic antigen, alpha-fetoprotein and immunoglobulin levels in patients with breast cancer. *Med Interne*, *20*(2), 151-155.

[88] Sarcione, E. J. and Biddle, W. (1987). Elevated serum alpha fetoprotein levels in postmenopausal women with primary breast carcinoma. *Disease Markers*, *5*(2), 75-79.

[89] Moro, R., Gulyaeva–Tcherkassova, J. and Stieber, P. (2012). Increased alpha-fetoprotein receptor in the serum of patients with early-stage breast cancer. *Current Oncology*, *19*(1), e1-e8.

[90] Dowell, B., Frost, S., Tcherkassova, J., Gerber, A., Moro, R. and Moro, R. J. (2007). Chemiluminescent assay (CIA) for the receptor of alpha fetoprotein (RECAF) to separate cancer from normal sera. *Tumor Biol*, *28*(suppl 1), 92.

[91] Tcherkassova, J., Abramovich, C., Moro R., Chen, C., Schmit, R., Gerber, A. and Moro, R. J. (2011). Combination of CA125 and RECAF biomarkers for early detection of ovarian cancer. *Tumour Biol*, *32*(4), 831-838.

[92] Ng, I., Tcherkassova, J., Lyubimova, N. and Moro, R. (2008). A New RECAF ELISA and Its Correlation with the Chemiluminescence RECAF Assay. *Tumor Biol*, *29*(suppl 1), 41.

[93] Tcherkassova, J., Schmid, R., Hu, X., Lyubimova, N. and Moro, R. (2007). Point-of-care serum test for cancer detection based on RECAF cancer marker. *Tumor Biol*, *28* (suppl 1), 101.

[94] Schmid, R., Moro, R., Tcherkassova, J., White, S. and Moro, R. J. (2011). RECAF serum test using an AFP derived peptide instead of antibody *Tumor Biology*, *32*(S1), 112.

[95] Ellis, I., Coleman, D., Wells, C., Kodikara, S., Paish, E., Moss, S. and Winder, R. (2006). Impact of a national external quality assessment scheme for breast pathology in the UK. *J Clinical Pathology*, *59*(2), 138-145.

[96] Palli, D., Galli, M., Bianchi, S., Bussolati, G., Di Palma, S., Eusebi, V. and Del Turco, M. R. (1996). Reproducibility of histological diagnosis of breast lesions: Results of a panel in Italy. *European Journal of Cancer*, *32*(4), 603-607.

[97] Moro, R., Tcherkassova, J. and Moro, R. J. (2009). Combination of CEA and the Receptor for AFP (RECAF) For Colorectal Cancer Screening. Presented at the 2009 ISOBM Meeting in Amsterdam.

[98] Tcherkassova, J. and Moro, R. (2011). RECAF as a replacement for free PSA in prostate cancer detection. *Tumor Biology*, *32*(S1), 71.

[99] Nicolini, A., Carpi, A., Michelassi, C., Spinelli, C., Conte, M., Miccoli, P., Fini, M. and Giardino, R. (2003). "Tumour marker guided" salvage treatment prolongs survival of breast cancer patients: final report of a 7-year study. *Biomedicine Pharmacotherapy*, *57*, 452–459.

[100] Dudich, E. I., Semenkova, L. N., Dudich, I. V., Nikolaeva, M. A., Gorbatova, E. A., Khromykh, L. M. and Sukhikh, G. T. (2000). α-fetoprotein-induced apoptosis of cancer cells. *Bulletin Exp Biol Med*, *130*(6), 1127-1133.

[101] Dudich, E., Semenkova, L., Dudich, I., Denesyuk, A., Tatulov, E. and Korpela, T. (2006). Alpha-fetoprotein antagonizes X-linked inhibitor of apoptosis protein anticaspase activity and disrupts XIAP-caspase interaction. *FEBS Journal*, *273*(16), 3837-3849.

[102] Mizejewski, G. J. (1995). Alpha-fetoprotein binding proteins: Implications for transmembrane passage and subcellular localization. *Life Sciences*, *56*, 1-9.

[103] Stigbrand, T. (2008). Targeted radionuclide tumor therapy. SpringerLink ebooks - Biomedical and Life Sciences, New York: Springer.

[104] Sharapova, O. A., Pozdnyakova, N. V., Laurinavichyute, D. K., Yurkova, M. S., Posypanova, G. A., Andronova, S. M. and Severin, E. S. (2010). Isolation and characterization of the recombinant human α-fetoprotein fragment corresponding to the C-terminal structural domain. *Russian Journal of Bioorganic Chemistry*, *36*(6), 696-703.

[105] Dudich, E., Semenkova, L., Dudich, I., Tatulov, E. (2015) Recombinant alpha-fetoprotein and compositions thereof. *US Patent application*, *8*, 932, 829.

[106] Deutsch, H. F., Tsukada, Y., Sasaki, T. and Hirai, H. (1983). Cytotoxic effects of daunomycin-fatty acid complexes on rat hepatoma cells. *Cancer Research*, *43*(6), 2668-2672.

[107] Feldman, N., Kiselev, S., Gukasova, N., Posypanova, G., Lutsenko, S. and Severin, S. (2000). Antitumor activity of alpha-fetoprotein conjugate with doxorubicin *in vitro* and *in vivo*. *Biochemistry-Moscow*, *65*(8), 967-971.

[108] Yabbarov, N., Posypanova, G., Vorontsov, E., Obydenny, S. and Severin, E. (2013). A new system for targeted delivery of doxorubicin into tumor cells. *J Controlled Release*, *168*(2), 135-141.

[109] Godovannyi, A. V., Vorontsov, E. A., Gukasova, N. V., Pozdnyakova, N. V., Vasilenko, E. A., Yabbarov, N. G. and Gnuchev, N. V. (2011). Targeted delivery of paclitaxel-loaded recombinant [alpha]-fetoprotein fragment-conjugated nanoparticles to tumor cells. *Doklady Biochemistry and Biophysics*, *439*(1), 158.

[110] Glebova, K. V., Marakhonov, A. V., Baranova, A. V. and Skoblov, M. Y. (2012). Therapeutic siRNAs and nonviral systems for their delivery. *Molecular Biology*, *46*(3), 335-348.

[111] Trumpler, S., Lohmann, W., Meermann, B., Buscher, W., Sperling, M. and Karst, U. (2009). Interaction of thimerosal with proteins-ethylmercury adduct formation of human serum albumin and beta-lactoglobulin A. *Metallomics*, *1*(1), 87-91.

[112] Mossman, T. (1983.) Rapid colorimetric assay for cellular growth and survival: Application to proliferation and cytotoxicity assays. *J Immunological Methods*, *65*, 55-63.

[113] Moro, R. J. and Moro, R. (2015). Cancer Imaging and Therapy Using the Receptor for AFP (RECAF) as a Target. *Tumor Biology*, *36*(S1), 1-88.

In: Alpha-Fetoprotein: Functions and Clinical Applications ISBN: 978-1-63484-875-6
Editors: Nisha Lakhi and Michael Moretti

Chapter 15

ALPHA-FETOPROTEIN AND RECOMBINANT ALPHA-FETOPROTEIN FRAGMENTS AS DRUG DELIVERY TOOLS

Galina Posypanova*[*], *PhD, and Sergey Severin, PhD, DSci
National Research Center "Kurchatov Institute," Moscow, Russia

ABSTRACT

This chapter deals with the problem of developing selectively acting antitumor agents in the form of bifunctional molecules and complexes. Such constructs comprise a targeting (vector) domain responsible for interaction with tumor cells, which is linked either directly to a therapeutic agent or to a polymeric carrier loaded with antitumor drugs.

Following a thorough search for an optimal vector molecule that would satisfy the principal requirements to targeting components of drug delivery systems, we concentrated our efforts on alpha fetoprotein (AFP). This oncofetal protein with a molecular weight of 68 KDa is structurally similar to serum albumin, but exhibits a number of dramatically different properties. AFP is highly conserved and is stable under physiological conditions. A considerable portion of human tumors express AFP-specific receptors on the cell surface. Therefore, it should be possible to employ AFP or its receptor-binding peptide fragments as drug delivery vectors to increase significantly the selectivity of antitumor therapy. The efficacy of such drug delivery systems has been demonstrated experimentally both in vitro and in vivo. A crucially important property of AFP-based drug delivery systems is their ability to reverse the multiple drug resistance tumor phenotype acquired due to ABC transporter activity.

In this chapter, we present the results of experimental studies demonstrating high antitumor activity and specificity of directed drug delivery systems employing AFP or its fragments.

[*] Corresponding author: galinapo@gmail.com.

INTRODUCTION

Insufficient efficacy and selectivity of conventional approaches to cancer treatment make the search for novel more effective and highly specific methods of diagnostics and therapy a problem of ultimate importance. The number of treatment variants that can be employed to eliminate malignant cells remaining in the patient's organism after surgical removal of the tumor bulk includes radiotherapy, chemotherapy, immunotherapy, etc. [1]. However, for metastatic tumors, the actual choice is limited, and chemotherapy remains the principal treatment option. At the same time, the efficacy of most pharmaceutical agents currently used for cancer therapy is insufficient due to several factors, such as their low therapeutic index, possible development of drug-resistant tumor cell populations, tumor heterogeneity, and metastasis. Because of their low bioavailability for tumor cells and nonselective action, conventional antitumor agents have to be administered in high doses, commonly causing severe side effects [2].

The selectivity of drug action can be improved by employing delivery systems that contain targeting agents, that is, molecules capable of binding to specific determinants on the tumor cell surface. Thus, by using a targeting component, it is possible to ensure that the drug delivery system interacts with strictly specific cells and tissues, e.g., with tumors. Promising candidate targeting agents (vectors) are physiological ligands of growth factor receptors or on cofetal protein receptors, i.e., growth factors and on cofetal proteins themselves. An antitumor agent can be linked to a vector either with a covalent bond (as a conjugate) or non covalently (as a stable complex). The binding of a targeting molecule to its specific receptor on the cell surface induces receptor-mediated end ocytosis; as a result, tumor cells accumulate the drug whose molecular target is located within the cell.

It should be pointed out that in some cases the use of targeted drug delivery represents the only way to make use of the therapeutic potential of some highly toxic antitumor antibiotics, such as enediynes. An appropriate example is Mylotarg, a drug comprising calicheamicin χ_1 conjugated with humanized mouse antibodies to CD33. Apart from that, the use of highly efficient delivery systems could significantly improve the therapeutic effect of antisense oligonucleotides, which unfortunately have not found use in clinical practice so far.

Successful realization of targeted drug delivery based on receptor mediated endocytosis depends first of all on the appropriate choice of the vector. Our choice of AFP as a vector was based on several reasons. First of all, the number of AFP receptors expressed on the surface of tumor cells is significantly higher than on normal proliferating cells, which serves for specific delivery of antitumor agents conjugated to AFP. Secondly, AFP is a physiological component of the blood serum, which diminishes the risk of immune reactions to the drug. Finally, an AFP molecule can be chemically modified by conjugation while largely maintaining the conformation of the receptor-binding center.

Expression of AFP Receptors on the Surface of Tumor Cells

AFP distribution in embryonic tissues was analyzed in different avian and mammalian species, and it was shown that embryonic cells selectively uptake AFP [3-6]. AFP accumulation depended on the cell differentiation grade and did not depend on their species.

When homologous ^{125}I–labeled AFP was administered to pregnant rats, radioactivity accumulated in the same fetal tissues where the fetal AFP was present. Similar results were obtained with heterologous AFP, which confirmed the similarity of AFP structures in birds and in different mammalian species [7]. Based on these data, it was speculated that cells accumulate AFP by endocytosis mediated by specialized AFP receptor, and that its intensity depends on the cell differentiation grade.

The first data confirming the existence of AFP receptors were obtained by Villacampa, et al., in a study in the human breast adenocarcinoma cell line MCF-7 [8]. Using ^{125}I-labeled AFP, the authors detected two types of AFP receptors present on MCF-7 cells: high affinity receptors with K_d of 1.5×10^{-9} M (2×10^3 per cell) and low affinity receptors with K_d of 2.2×10^{-7} M (320 x10^3 per cell). Experiments with the cell lines YAC-1, U937, and THP-1 produced similar results [9, 10]. Certainly, the fact that AFP receptors are widely present on different types of tumor cells evokes a natural question as to the origin of AFP "expected" to bind these receptors, since it is known that serum AFP levels are increased in only a few types of cancer (hepatocellular carcinoma, germ cell tumors, and keratocarcinomas). It was found that tumor cells synthesize small amounts of AFP which acts an autocrine trophic factor without affecting serum AFP concentrations [11]. Human T cells stimulated with phytohemagglutinin (PHA) were found to express a receptor that can bind both albumin and AFP. The curve describing the binding of iodine-labeled AFP by cell membranes exhibited two saturation plateaus, while the curve of serum albumin binding had only one plateau. A Scatchard plot analysis of these curves identified two types of AFP binding sites: one with K_d of 3.1×10^{-7} (25–30 × 10^3 receptors per cell) and the other one with K_d of 2.4×10^{-6} (110–200 × 10^3 receptors per cell), and a single type of albumin-binding sites with K_d of 2×10^{-6} (50 × 10^3 receptors per cell). The authors also described the differences in intracellular distribution and release of the two proteins. Albumin was nearly completely degraded within lysosomes, while intracellular AFP was detected in endosomes, multivesicular bodies, and tubulovesicular structures of the Golgi complex [12]. Most AFP was released in the intact form. Similar experiments on the binding of FITC-labeled AFP by human PHA-stimulated mononuclear leukocytes showed that the levels of AFP binding by mononuclear leukocytes, by the total T cell population, as well as by CD4+ and CD8+ subpopulations were similar. AFP accumulation increased rapidly for 72 h after PHA stimulation. A Scatchard plot analysis of the curves describing AFP binding by stimulated lymphocytes at 4°C identified a single group of binding sites with K_d of 3.03×10^{-7} (88 × 10^3 receptors per cell) [13].

Human histiocytic lymphoma cells of the U937 cell line were also found to possess two types of AFP binding sites on their cell membranes: 49 high-affinity sites with K_d of 5×10^{-11} M and 7.8×10^3 sites per cell with K_d of 2.5×10^{-7} M [10]. AFP receptor was isolated from these cells, and its molecular weight was determined as 65 KDa. Human peripheral blood monocytes were also capable of binding AFP, whereas resting T cell did not express AFP receptors [13, 14]. It was later shown that, apart from the membrane-bound form, there also exists a cytoplasmic form of AFP receptors [15].

There are currently known four types of membrane-bound AFP receptors with molecular weights of 18, 31, 60, and 62–67 kDa [16, 17]. The 18- and 31-kDa receptors are found in blood vessel endothelium and on epithelial cells; they are scavenger receptors that bind chemically modified or denatured AFP or serum albumin molecules. The 60-kDa receptor is an albumin-binding sialoglycoprotein present on endothelial cells. Finally, the 62–67 kDa protein, considered as the classical AFP receptor, was detected on the surface of monocytes, stimulated lymphocytes, cells of the reproductive system, and tumor cells. Experimental data suggest that tumor cells possess several AFP-binding proteins differing in molecular weight and K_d. The number of high-affinity AFP binding sites with K_d of 10^{-11}–10^{-10} M ranges usually from several dozens to several hundreds per cell, while the number of low-affinity sites with K_d of 10^{-7} to 10^{-6} M is several hundreds thousands per cell [8-10, 13, 18]. The function of AFP receptors is to bind AFP and to mediate its endocytosis. The currently available data indicate that AFP receptors expressed on most tumor and embryonic cells have a molecular weight of 62–67 KDa and constitute a part of a large protein complex whose components are glycosylated and surrounded with lipid molecules [19].

Quantitative characterization of AFP receptor expression on different types of cells is essential for evaluation of AFP potential as a vector for targeted tumor therapy. Table 1 lists the numbers of AFP receptors on certain human tumor cell lines, as detected with fluorescently labeled monoclonal antibodies to AFP receptor (clone 2B8, courteously provided by Dr. V.K. Sologub), and the number of AFP binding sites detected using fluorescently labeled AFP.

In MCF7 and HepG2 cells, the numbers of AFP receptors and AFP binding sites were similar, but in SKOV3, Jurkat, and Raji cells, the numbers of AFP binding sites were several times higher than the numbers of AFP receptors detected with MAb. Therefore, these tumor cell apparently carry other membrane proteins capable of AFP binding [20, 21], such as scavenger receptors present on Jurkat and Raji cells.

An analysis of AFP receptor expression in malignant and benign human tumors, as well as in normal tissues showed that this receptor was present only in malignant tumor cells [22, 23]. Immunohistochemical investigation of normal, benign and malignant tissue sections was performed using two different monoclonal antibodies: 2B8 [22] and 167H.4 [23]. AFP receptors were detected only in rapidly proliferating malignant tumor cells, but not in normal cells or benign tumors (Figure 1). Thus, it was experimentally confirmed that AFP receptor is a universal tumor-specific antigen that can be used as a unique target for specific delivery of antitumor drugs.

Table 1. Numbers of binding sites for monoclonal antibodies to AFP receptors and for AFP in different human tumor cell lines. Cells were incubated with AFP and antibodies for 1 h at 4°C, washed, and analyzed by flow cytometry [14]. The number of binding sites and K_d values were calculated using the Scatchard plot

Cell line	Number of AFP receptors on the cell surface (MAb)	Number of AFP binding sites (AFP)
HeLa	100 000	260 000 (K_d 1.2 × 10^{-7} M)
SKOV3	207 000	440 000 (K_d 6.4 × 10^{-7} M)
MCF7	300 000	310 000 (K_d 3 × 10^{-6} M)
Jurkat	175 000	580 000 (K_d 8.9 × 10^{-6} M)
Raji	180 000	420 000 (K_d 4.8 × 10^{-7} M)
HepG2	400 000	450 000 (K_d 4.9 × 10^{-7} M)

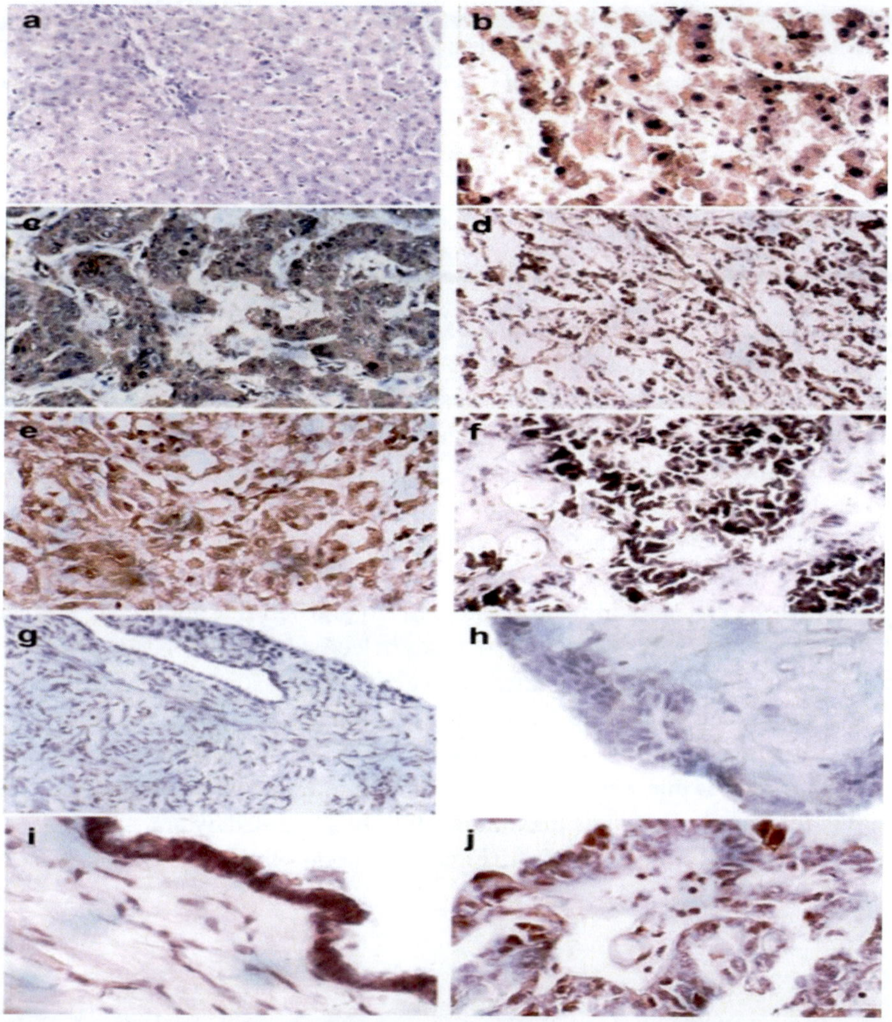

Figure 1. Expression of AFP receptors in normal and tumor tissues. Immunohistochemical staining of human tumor sections with monoclonal antibodies to AFP receptor, clone 2B8: *a*, intact liver tissue; *b*, hepatocellular carcinoma; *c*, cholangiocellular carcinoma; *d*, colon adenocarcinoma; *e*, scirrhous carcinoma of the stomach; *f*, breast adenocarcinoma; *g*, normal ovary; *h*, serous cystadenoma; *i*, ovarian papillary serous cystadenoma of borderline malignancy; *j*, ovarian serous papillary cystadenocarcinoma. Brown staining corresponds to positive reaction.

Upon binding to its specific receptor, AFP is internalized by the cell by receptor-mediated endocytosis. Moreover, some data obtained in proliferating human T cells and in human T and B lymphoma cell lines suggest that the levels of AFP endocytosis can exceed the levels of AFP binding as much as 100-fold. AFP is absorbed by the cells within clathrin vesicles and can be detected in endosomes, multivesicular bodies, and the Golgi complex [11, 24, 25]. A significant portion of internalized protein (up to 75%) is recycled and released from the cells in the intact form.

Figures 2 and 3 show the data on the intensity of AFP–FITC endocytosis by several tumor cell lines (Figure 2) and by mononuclear leukocytes (lymphocytes and monocytes) isolated from the peripheral blood of healthy volunteers (PBMCs). The level of endocytosis was evaluated by assessing the changes in cell fluorescence using flow cytometry. In cells incubated with AFP–FITC under endocytosis conditions (1 h, 37°C), the level of fluorescence was significantly higher than in those that were incubated at 4°C (binding-only conditions), which reflects intensive AFP uptake by tumor cells. AFP internalization rates varied among different cell lines and ranged from 23 × 10^3 to 117 × 10^3 molecules per hour per cell for the initial protein concentration of 500 nM.

A study of AFP accumulation in human PBMCs showed that, in isolated human lymphocytes, the levels of receptor-mediated AFP–FITC endocytosis were low. It should be noted that these experiments registered total endocytosis levels, including receptor-mediated and liquid-phase endocytosis. The contribution of liquid-phase endocytosis was evaluated in control experiments with FITC-labeled dextran (m.w. 70 000). It was found that the level of AFP–FITC endocytosis in lymphocytes was nearly the same as the level of dextran endocytosis, indicating that receptor-mediated AFP endocytosis did not occur in lymphocytes. At the same time, AFP–FITC endocytosis in monocytes was high, which confirmed the presence of AFP receptors on these cells [26].

Investigation of AFP distribution in intact animals and in animal cancer models is essential for the understanding of its therapeutic potential. Experimental studies in mouse models of breast cancer [27] and neuroblastoma [28] showed that mouse AFP labeled with iodine radionuclide (^{125}I–AFP and ^{131}I–AFP) was selectively accumulated in the tumor and could be detected there in 4 to 7 days after intravenous administration. An analysis of human ^{125}I–AFP pharmacokinetics showed that it could also accumulate selectively in tumor tissues [29]. In mice that received ^{125}I-labeled human AFP intravenously, tumor radioactivity was registered as soon as in 3 min after the injection. In 3 h after administration, the concentration of ^{125}I–AFP in the tumor reached its maximum, slightly decreased by 5 h, and was maintained at the level of 1.2–0.7% for 24 to 72 h (Figure 4), upon which the protein was gradually eliminated.

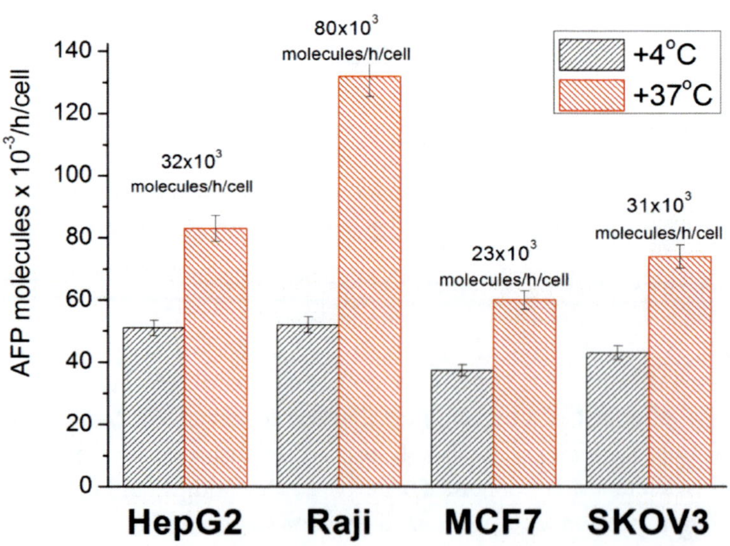

Figure 2. The binding (at 4°C) and uptake (at 37°C) of FITC-labeled AFP by HepG2, Raji, MCF-7, and SKOV3 human cancer cells.

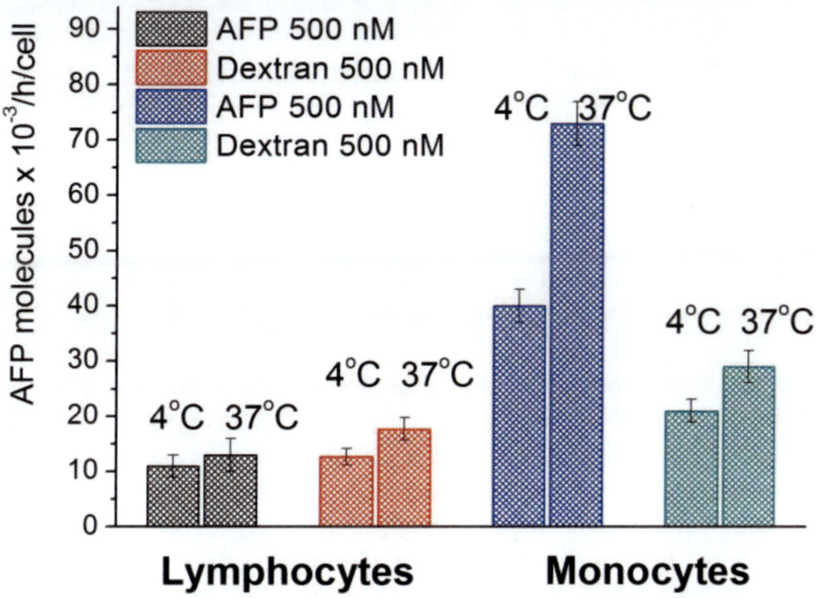

Figure 3. The binding and endocytosis of FITC-labeled AFP and dextrane 70 000 by lymphocytes and monocytes of human peripheral blood.

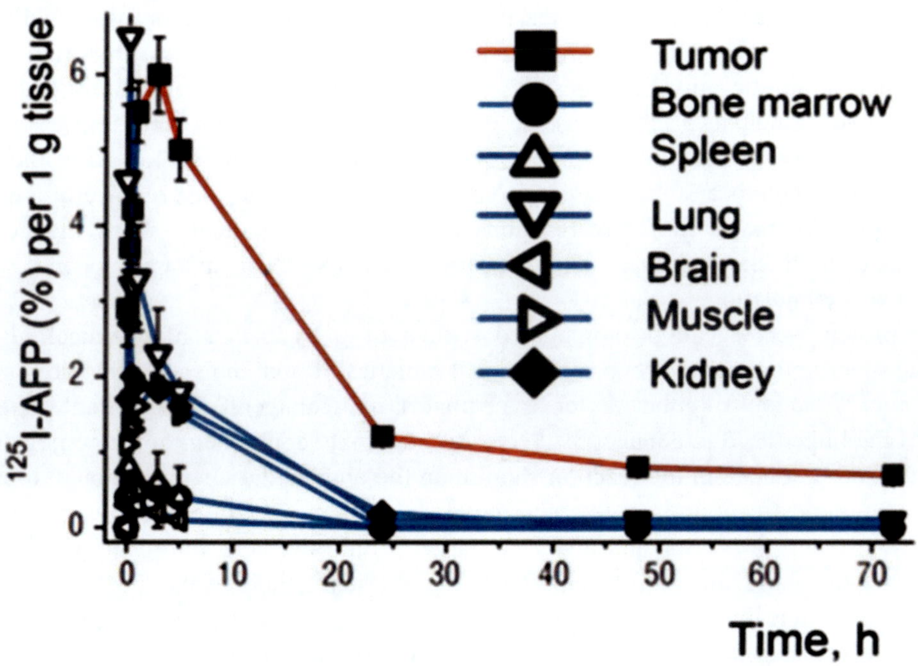

Figure 4. Pharmacokinetics of ^{125}I–AFP in DBA/2 mice carrying P388 murine leukemia transplants: ^{125}I–AFP accumulation in the bone marrow, spleen, lungs, brain, muscles, kidneys, and the tumor.

The maximal ^{125}I–AFP level reached 6% of the inoculated amount per 1 g tumor [29], which is considerably higher than the levels described for conventional radiopharmaceutic preparations. In normal tissues, the levels of labeled AFP were significantly lower and dropped to zero in 24 h, while in tumors considerable ^{125}I–AFP concentrations were detected for up to 2 or 3 days.

Thus, taking into account the experimental data on AFP receptor expression in tumor cells and the ability of AFP to enter such cells by receptor-mediated endocytosis, we supposed that the use of AFP as a vector molecule for directed drug delivery systems could significantly improve the selectivity of antitumor drug delivery.

Targeted Delivery of Antitumor Antibiotics, Cytostatic and Photosensibilizing Agents, and Antisense Oligonucleotides

Molecular targets of most antitumor drugs are located within the cell; therefore, the efficiency of the drug's cytostatic action depends on the efficiency of its internalization as a part of a drug delivery system. Apart from that, the cytostatic effect depends on the rate of the drug release from the delivery system, in particular, on the rate of degradation of the chemical bond between the cytostatic agent and the vector protein or peptide in a conjugate.

A comparison of delivery system variants designed as (protein vector)–linker–(drug) conjugates where human serum albumin (HSA), transferrin (Trf), and AFP were used as vector proteins showed that the latter had an advantage both in its uptake by tumor cells (Figure 5a) and in its cytotoxic effect on these cells (Figure 5b). The conjugates were constructed using 3-maleimidobenzoic acid hydrazide as the linker and doxorubicin (DOX) as the chemotherapeutic agent. This linker is acid-labile and can be hydrolyzed within late endosomes or lysosomes, thus releasing the drug from the delivery system. Doxorubicin is an anthracycline antibiotic, an efficient and widely used therapeutic agent. DOX causes cell death via two principal mechanisms: it impairs DNA replication and transcription by intercalating between neighboring nucleotides and strongly interacting with the DNA molecule and also binds and inhibits topoisomerase II [30]. Due to the presence of a quinone group, DOX participates in redox reactions producing free radicals which induce DNA damage and lipid peroxidation. The efficacy of DOX-based therapy depends on its intracellular accumulation.

The protein vector can be conjugated with a drug by means of chemical reactions involving specific functional groups or chemical moieties of both molecules that do not affect the biological functions of either vector or the pharmaceutical agent. An important issue is the choice of the linker used to connect the vector and the toxic component of the conjugate. The chemical bond produced in the reaction should, on the one hand, be strong enough to remain stable in the bloodstream and, on the other hand, be sufficiently labile to allow the antibiotic to disengage from the vector and attack its target within the cell. Introduction of a linker between the drug and the vector helps to overcome steric hindrances and to increase the ligand mobility, thus improving the efficacy of its binding to biological receptors. Linkers can be designed in such a way that they can additionally control the separation of the drug from the vector after the system has entered the cell by endocytosis. Such linkers serve to integrate the tasks of targeted delivery and efficacious intracellular localization of the system. The vector can also provide additional protection for the drug molecule during its transportation to the target. At later stages, bioavailability of the drug depends on the manner, rate, and efficacy of its separation from the linker [31]. The choice of the vector carrier is determined by the presence of specific ligands on the surface of tumor cells.

In vitro investigations of different AFP-based conjugates where the cytotoxic component was represented by various cytostatic agents and antitumor antibiotics (doxorubicin, daunomicin, calicheamicin, bleomycin, esperamicin, cisplatin, carboxyphosphamide, vinblastine), antimetabolites (metotrexate), photosensibilizing agents (phthalocyanines, chlorin), as well as antisense oligonucleotides showed that they possessed high and specific antitumor activity [32-36]. The conjugates caused the death of tumor cells, whereas their toxicity for control cells (PBMCs of healthy volunteers) was by one or two orders of magnitude lower than the toxicity of free cytostatic agents.

A covalent esperamicin A_{1b} (EsA) conjugate with AFP (AFP–EsA) was obtained by means of thiol–disulfide exchange reactions. *In vitro* experiments demonstrated its high cytotoxic activity against different human tumor cell lines (T and B cell lymphoma, breast adenocarcinoma, ovarian adenocarcinoma, etc.) [37]. AFP–EsA also exhibited significant efficacy in model *in vivo* experiments when used for treatment of mice with experimental tumors [37], suppressing the tumor growth and increasing the animal's survival time severalfold. Figure 6 shows the dynamics of tumor growth in mice treated with AFP–EsA and with EsA in the dose of 2.5 μg/kg body weight.

In mice treated with free EsA, tumors retained tendency to progression after the treatment was discontinued (experimental day 14). In animals that were treated with AFP–EsA, tumors were growing significantly slower in comparison both to the control group and to the group treated with free EsA, and complete tumor regression was observed in 30% of cases.

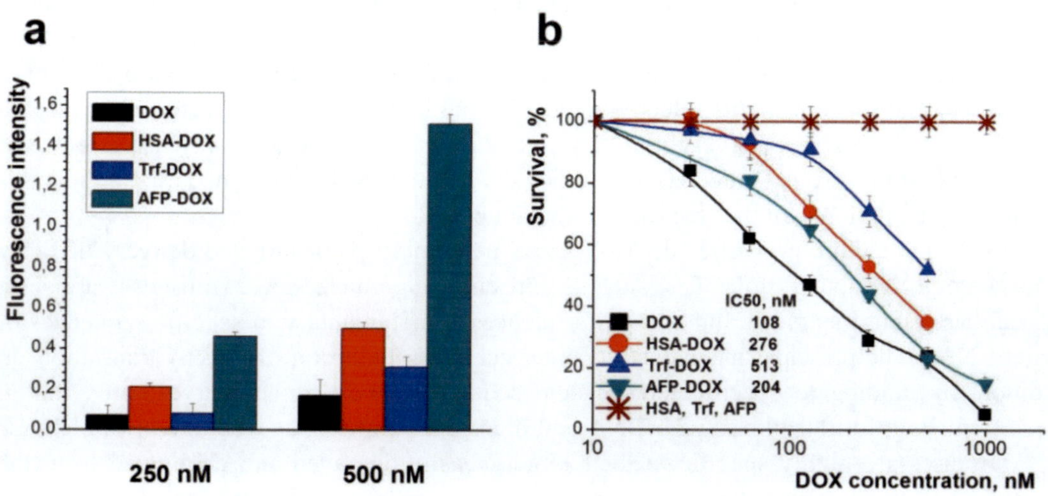

Figure 5. *In vitro* antitumor activity of DOX conjugates with serum albumin (HSA–DOX), transferrin (Trf–DOX), and AFP (AFP–DOX) against MCF-7 mouse breast adenocarcinoma cells. *a,* DOX accumulation in MCF-7 cells following 15-min incubation with DOX and DOX–protein conjugates evaluated by DOX fluorescence using flow cytometry; *b,* cytostatic activity of DOX–protein conjugates in comparison to DOX alone.

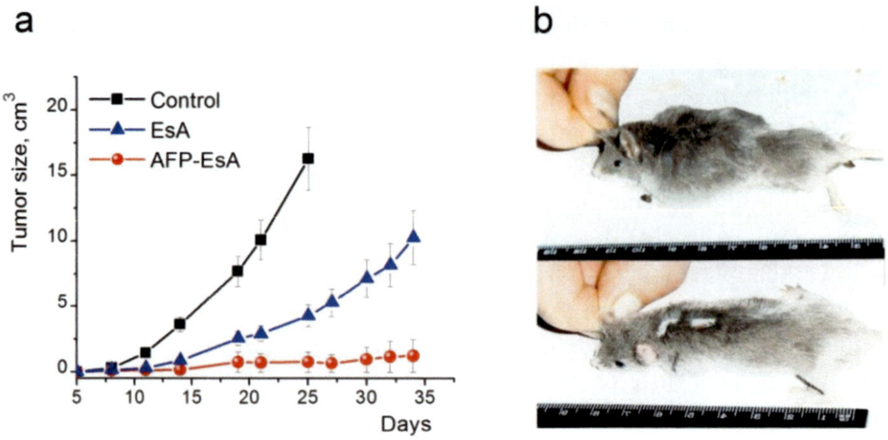

Figure 6. Antitumor activity of AFP–esperamicin A_{1b} conjugate. *a*, inhibition of P388 solid tumors in DBA/2 mice following treatment with AFP conjugates with esperamicin A_{1b} (AFP–EsA, 2.5 μg/kg EsA); *b,* photographs of control animals (top) and animals treated with AFP–EsA injections (bottom).

It should be noted that esperamicin A_{1b}, an enediyne antibiotic, is an extremely toxic compound. Enediyne antibiotics share a common structural element, commonly termed "warhead," a 10-atom enediyne cycle that contains two acetylene bonds in the α-position relative to the double bond and to the oxyrane ring. Under physiological conditions and following certain activation, this warhead undergoes transformation into a reactive diradical [38]. The cytotoxic effect of esperamicin is due to its ability to induce single- and double-stranded DNA breaks. This drug had to be withdrawn from clinical trials because of its high systemic toxicity. Increasing its selectivity by attaching it chemically to a vector molecule and thus ensuring its targeted delivery into tumor cells seems to be the only way to make use of the potential of this drug. Acute toxicity studies showed that AFP-conjugated esperamicin A_{1b} was less toxic in mice than the free form of the antibiotic.

Due to the unique properties of AFP, it can be employed for targeted delivery of fairly large biologically active molecules, such as antisense oligonucleotides. Antisenses are short (15–25 bases long) single-stranded DNA sequences complementary to specific sequences of target RNAs. The principal mechanism of their action is inhibition of mRNA translation. In addition, by binding to specific nucleic acid sequences, they can impair capping, inhibit mRNA splicing, form triplexes with DNA and block transcription, as well as activate RNAse H. Antisenses are highly specific to their targets, easily degraded and eliminated from the organism. Clinical applications and metabolic properties have been studied best for phosphorothioate oligonucleotides. They are similar to phosphodiester stands in their hybridization properties, but exhibit significantly higher tropism to cell membranes and are more resistant to nucleases. In 2004, a specific antisense complementary to the first six codons of *bcl-2* mRNA (Genasense, G3139) was approved by FDA for the treatment of melanoma. However, antisense drugs have to be administered in rather high doses, which unfortunately cause serious side effects, such as thrombocytopenia, leukopenia, fever, hypotension, and asthenia [39].

Table 2. Efficacy of AFP–antisense conjugates against human tumor cells. IC50 is the concentration causing 50% cell death

Antisense/target mRNA	Cell line	IC50 for antisense, nM	IC50 for AFP–antisense conjugate, nM
c-myc	Raji	1800	16
	Du145	2000	50
	MCF-7	>2000	140
	SKOV3	>2000	200
bcl-2	Raji	>2000	50
	MCF-7	>2000	200
Telomerase	MCF-7	>2000	132
	HeLa	2000	102

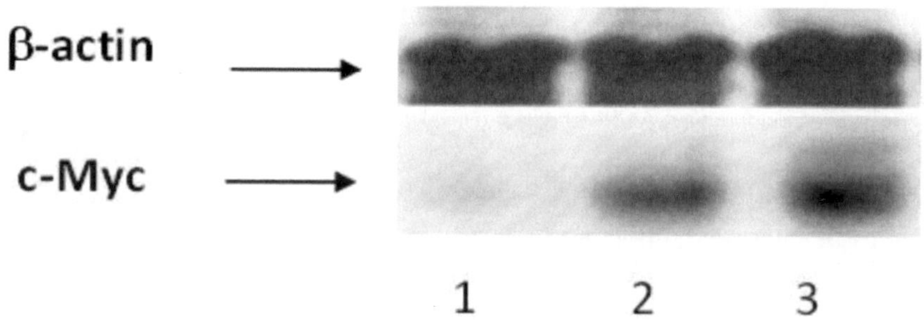

Figure 7. Inhibition of c-Myc synthesis in MCF-7 cells (Western blot). 1, 6 h of incubation with AFP–myc-antisense conjugate; *2,* 24 h of incubation; *3,* no treatment.

Some AFP–antisense conjugates exhibited significant biological activity. In particular, AFP conjugate with a *c-myc* antisense oligo substantially suppressed cell proliferation in Raji B cell lymphoma, Du145 prostatic carcinoma, and MCF-7 breast adenocarcinoma cell lines. Incubation of Raji cells with AFP–*bcl-2* antisense conjugates induced apoptosis and subsequent cell death. Although no such effect was observed in MCF-7 cells, preincubation with the same conjugates increased their sensitivity to doxorubicin. Free antisense oligos failed to induce a comparable effect even when they were used in tenfold higher concentrations (Table 2).

Targeted Drug Delivery Using Recombinant Receptor-Binding AFP Fragments

Experiments described above were performed with human AFP isolated from aborted fetuses and from retroplacental serum obtained from birthing mothers using a technique employing monoclonal anti-AFP antibodies and a combination of affinity, ion exchange, and gel filtration chromatography [33]. Obviously, the use of these natural AFP sources is complicated due both to technical problems and to ethical issues. Hence, to enable further development of this research direction, it was required to obtain recombinant AFP, not necessarily as a full-size molecule, but essentially as a one containing the receptor-binding site. Line, et al., [40] found that AFP domain III retained the ability of full-size human AFP to accumulate in tumor cells. Therefore, we supposed that the AFP receptor-binding site was located within protein domain III.

Recombinant C-terminal AFP fragments rAFP3D (amino acid residues 357 to 590) and rAFP-3BC (a shorter one, amino acid residues 454 to 577) were obtained using an *E. coli* expression system. Proteins were isolated from the producer biomass using metal chelate ion exchange and gel filtration chromatography [41-44]. Protein refolding was performed by slow renaturation.

A study of recombinant AFP binding and endocytosis by tumor cells showed that recombinant fragments were bound similarly to native AFP (Figure 8), were accumulated in the same manner as AFP, and competed with AFP for the common surface receptor on tumor cells.

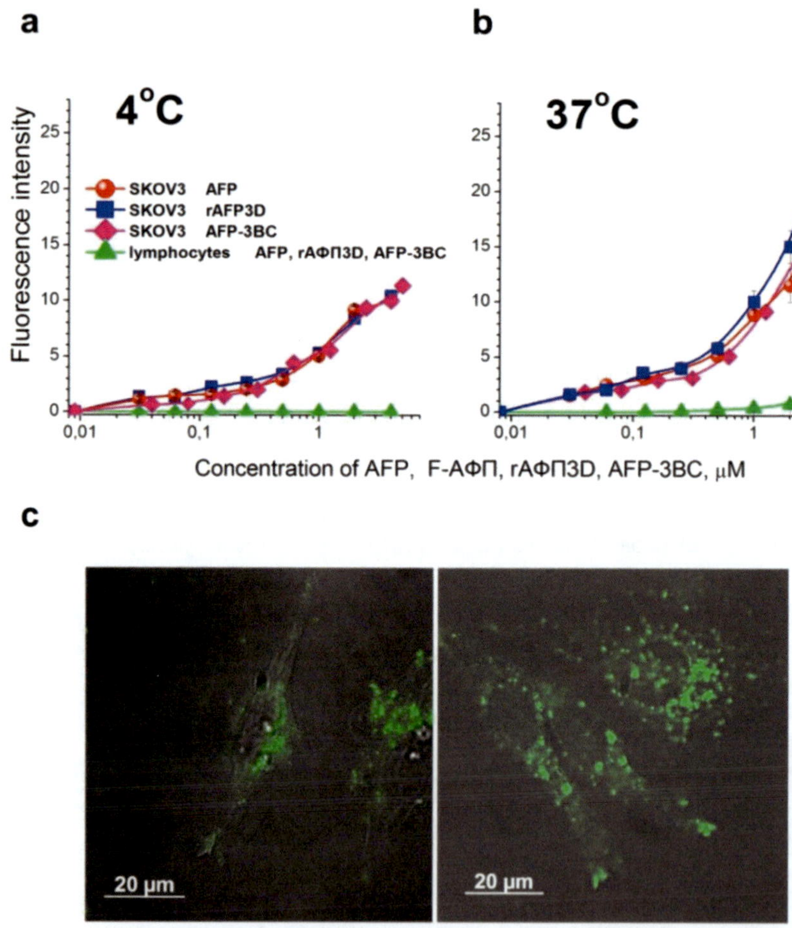

Figure 8. Binding and endocytosis of AFP and its recombinant C-terminal fragments, rAFP3D and AFP-3BC, assessed by cell fluorescence intensity following 1-h incubation with labeled proteins. *a,* Protein binding by SKOV3 human ovarian adenocarcinoma cells and peripheral blood leukocytes at 4°C; *b,* protein endocytosis at 37°C; *c,* intracellular distribution of FITC–AFP and FITC–rAFP3D in SKOV3 cells. Confocal microscopy, section at the mid-height.

Similarly to native AFP, its recombinant C-terminal fragments rAFP3D and the shorter rAFP-3BC variant were hardly bound or accumulated by PBMCs. The available experimental data suggest that the receptor-binding site of human AFP is located within its C-terminal fragment comprising amino acid residues 454 to 577 (rAFP-3BC); thus, the recombinant proteins rAFP3D and rAFP-3BC can be employed as selective transporters of pharmaceutical agents into tumor cells.

The rAFP-3BC and rAFP3D conjugates with the antitumor antibiotic doxorubicin was prepared by doxorubicin hydrazone condensation with thiolated proteins [42].

A confocal microscopy study of DOX–rAFP-3BC uptake by MCF-7 tumor cells, as detected by DOX fluorescence, showed that the conjugate was actively accumulated by the cells (Figure 9). Microscopic examination showed that the distribution of fluorescence signals was clearly different in cells treated with DOX–rAFP-3BC and those treated with free DOX. After 4 h of incubation, free DOX was predominantly localized in the nuclei (Figure 9*c, d*), whereas the DOX–rAFP-3BC conjugate was primarily found in lysosomes and endosomes in the cytoplasm (Figure 9*g--j*). However, after a 24-h incubation with DOX–rAFP-3BC, the red fluorescence of DOX was also observed in the cell nuclei.

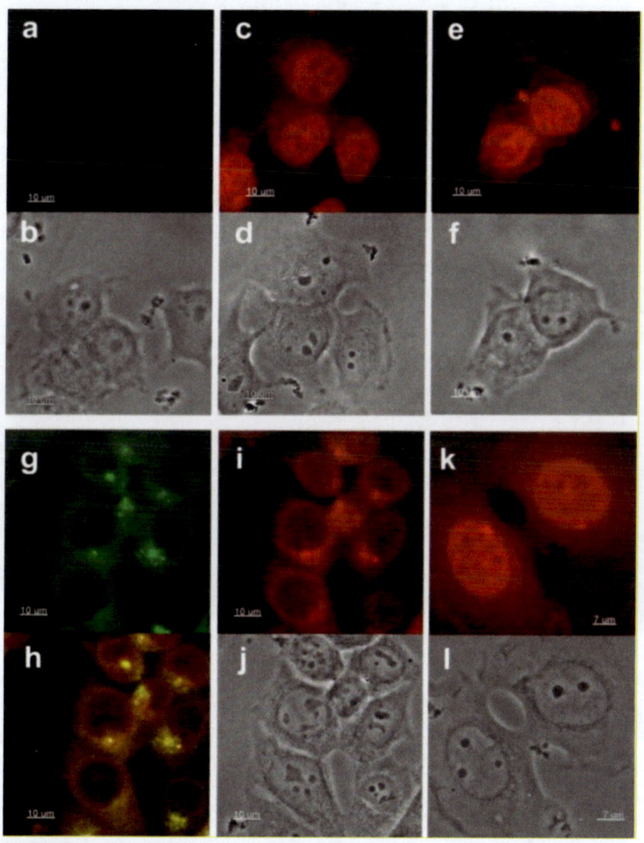

Figure 9. Localization of DOX within MCF7 cells after 4 h (*c, d, g–j*) and 24 h (*e, f, k, l*) of incubation with free DOX (*c–f*) and DOX–AFP-3BC conjugate (*g–l*); *a, b,* control cells; *a, c, e, g, h, i, k,* fluorescent image; *b, d, f, j, l,* phase contrast image; *g,* lysosome/ late endosome staining with FITC-labeled anti-LAMP2 mAb; *h,* superposition of *g* and *i.* Images were obtained by laser confocal microscopy at the mid-height of the cells.

As a result of the relatively slow release of DOX from recombinant protein conjugates and, accordingly, its delayed translocation into the nucleus (where its DNA target is located) in comparison to free DOX, the antitumor activity of these conjugates was two to four times lower than the activity of the free antibiotic. Moreover, an attempt to increase the doxorubicin load of recombinant C-terminal AFP fragments failed, because these proteins, in contrast to full-size AFP, lost their ability to bind the receptor under these conditions. Therefore, to increase the efficacy of targeted drug delivery, it was decided to employ polymer carriers loaded with the pharmaceutical agent. The use of vector-bound nanocontainers (liposomes, micelles, or nanoparticles of polymeric biodegradable compounds) in the development of optimized pharmaceutical preparations should make it possible to protect the drug from degradation and to improve its pharmacokinetics by enhancing drug accumulation in the tumor.

Tumor Targeted Nanoparticles and Dendrimers

A rAFP3D molecule can carry no more than two molecules of a cytostatic agent without losing its biological activity. To increase the payload, a vector molecule can be attached to a polymer possessing a large number of modifiable groups to which cytotoxic agents can be chemically attached. Dendrimers are a class of polymer carriers well-suited for such applications.

Dendrimers are strongly branched spherical polymer macromolecules, symmetrical or asymmetrical. They comprise the core, which serves as a branching center, individual branches, and terminal chemical groups [45]. Dendrimers are characterized with strictly controlled structure, definite molecular weight, monodispersity, and biocompatibility. In the recent years, dendrimers have been increasingly employed as carriers to transport biologically active compounds into cells and to ensure controlled release of pharmaceutical agents [46, 47].

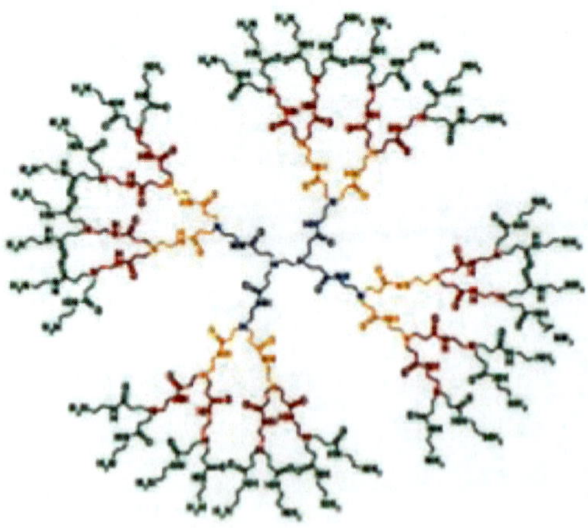

Figure 10. Third-generation PAMAM dendrimer. The ethylenediamine core is shown in blue.

A dendrimer molecule can carry a considerable number of drug molecules covalently bound to its surface groups. At the same time, by using linkers with specified properties, it is possible to control the drug release from the conjugate. Such linkers are cleaved under special conditions (e.g., at acidic pH values or in the presence of specific proteases); as a result, the drug molecule is rapidly released while retaining its initial structure.

In our experiments, we used second-generation polyamidoamine dendrimers (PAMAM G2) whose molecules possess 16 modifiable surface amino groups. Attachment of G2 to the rAFP3D vector did not affect its specific binding and endocytosis by SKOV3 ovarian adenocarcinoma cells. A rAFP3D–G2 conjugate contained two G2 molecules. After it was proven that rAFP3D conjugation to G2 did not cause any critical changes in the protein conformation that could affect rAFP3D binding to specific AFP receptors, we synthesized a rAFP3D–G2 conjugate with the antitumor antibiotic doxorubicin, rAFP3D–G2–DOX. It was obtained by condensation of DOX modified with cis-aconite anhydride and rAFP3D–G2 [48, 49]. The optimal rAFP3D-G2: DOX ratio was 1 : 6. Our experiments on DOX release from rAFP3D–G2–DOX showed that the resulting amide bond was stable at neutral pH values, but at pH < 7.4, the drug was released from the conjugate in the initial form. In particular, 75 to 93% of DOX was released from the conjugate after 24 h of incubation at pH 6.0, 5.5, and 5.0. Thus, using the acid-labile bonding between the antibiotic and the polymer carrier, it is possible to ensure that the conjugate is stable in the solution and releases DOX within cell compartments with acidic pH: late endosomes and lysosomes.

A study of rAFP3D–G2–DOX accumulation by tumor cells, as evaluated by intracellular DOX fluorescence, showed that rAFP3D–G2–DOX had a strong advantage in comparison to free DOX and G2–DOX conjugates (Figure 11). For longer incubation times, rAFP3D–G2–DOX accumulation by tumor cells was constantly growing (Figure 12). Fluorescence was detected both in the cytoplasm and in the cell nuclei, and in 24 h of incubation, it was localized predominantly in the nuclei. These observations suggest that DOX was gradually released from the rAFP3D–G2–DOX conjugate in the acidic environment of late endosomes and entered the cell nuclei by diffusion.

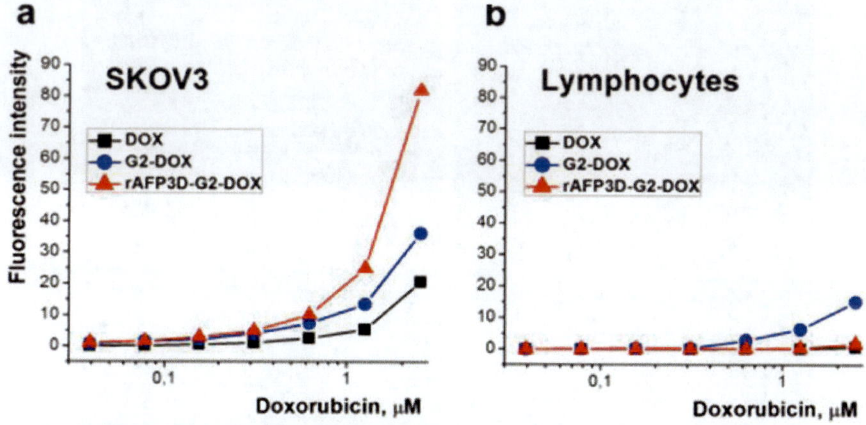

Figure 11. Uptake of DOX, G2-DOX and rAFP3D-G2-DOX by (a) SKOV3 cells, (b) human peripheral blood lymphocytes during 1-h incubation at 37°C.

The levels of rAFP3D–G2–DOX accumulation by control cells (mononuclear leukocytes) were very low: 50 times lower than the corresponding levels for SKOV3 tumor cells and more than 3 times lower than the uptake of vector-less G2-DOX conjugates by the same leukocytes. Apparently, monocytes are the PBMC fraction that is mainly responsible for rAFP3D–G2–DOX accumulation: although monocytes amount to only 5% of total mononuclear leukocyte population, they exhibited high levels of AFP [26] and rAFP3D uptake (unpublished data from our experiments). Our results demonstrate the selectivity of rAFP3D–G2–DOX action: the conjugate entered tumor cells expressing AFP receptors, but was not accumulated by nonstimulated lymphocytes, which do not express surface AFP receptors.

The data on DOX release were confirmed in a study of DOX localization following incubation of cells with rAFP3D–G2–DOX conjugate. In 4 h of incubation at pH 5.5 (typical for late endosomes), over 40% of DOX was released from the conjugate. DOX molecules released from the conjugate left endosomes or lysosomes and migrated into the nucleus, where their DNA target is located. During 24 h of incubation at pH 5.0–5.5, up to 90% of DOX was released from the conjugate and diffused into the nucleus. Free DOX entered the cell by diffusion and rapidly accumulated in the nucleus, the site of its cytotoxic activity.

Based on the data on DOX release from rAFP3D–G2–DOX and its intracellular distribution, it can be supposed that the conjugate will have a delayed cytotoxic effect in comparison to free DOX. Indeed, following 72 of incubation, free DOX exhibited two times higher activity in SKOV3 cells than rAFP3D–G2–DOX (as evaluated by IC50). However, it should be noted that the toxicity of the conjugate in control human PBMCs was much lower (by at least an order of magnitude) than the toxicity of either free DOX or the vector-less G2–DOX conjugate, and also lower than its own toxicity in SKOV3 tumor cells (Figure 13). Thus, the results of cytotoxicity studies also confirmed the selectivity of rAFP3D–G2–DOX action.

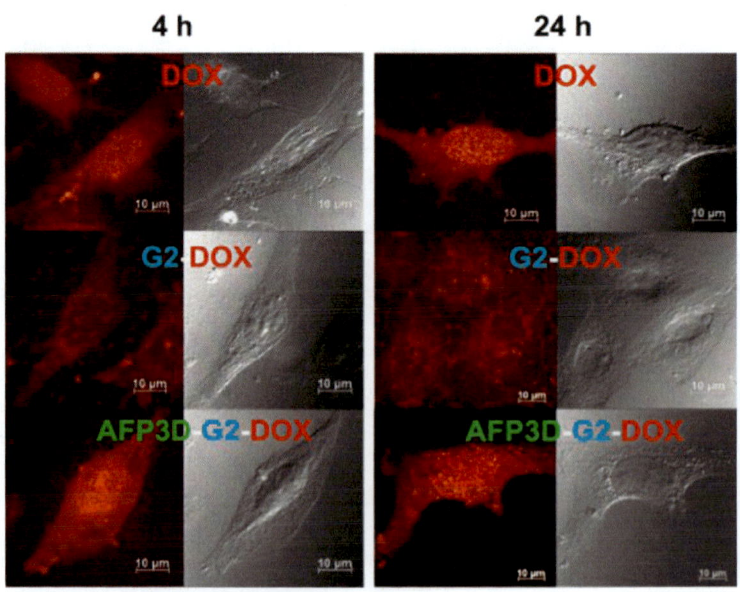

Figure 12. Intracellular distribution of DOX, G2-DOX and rAFP3D-G2-Dox in SKOV3 cells after 4 h and 24 h of incubation with free DOX and conjugates.

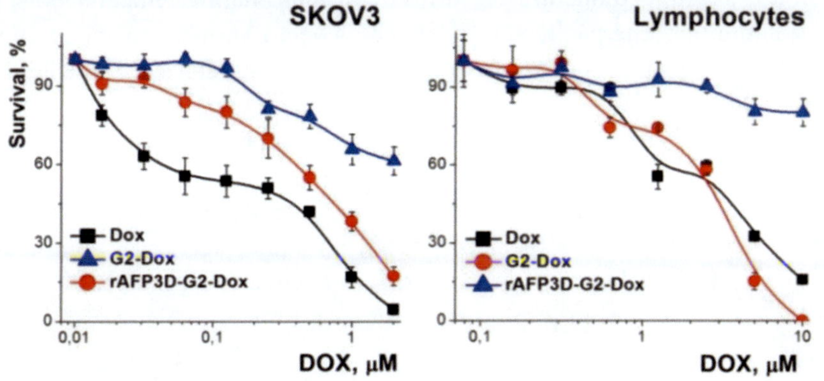

Figure 13. Cytotoxic activity of G2-DOX, rAFP3D-G2-Dox and free DOX against SKOV3 cells and human peripheral blood lymphocytes.

Certainly, the efficacy of dendrimer vectors can be increased by employing PAMAM dendrimers of higher generations that contain larger numbers of amino groups and thus can bind more antibiotic molecules. In our experiments, because of the hydrophobicity of DOX, no more than three antibiotic molecules could be attached to a single G2 molecule (accordingly, rAFP3D–G2–DOX comprised six DOX units). Higher loads caused precipitation of the conjugate.

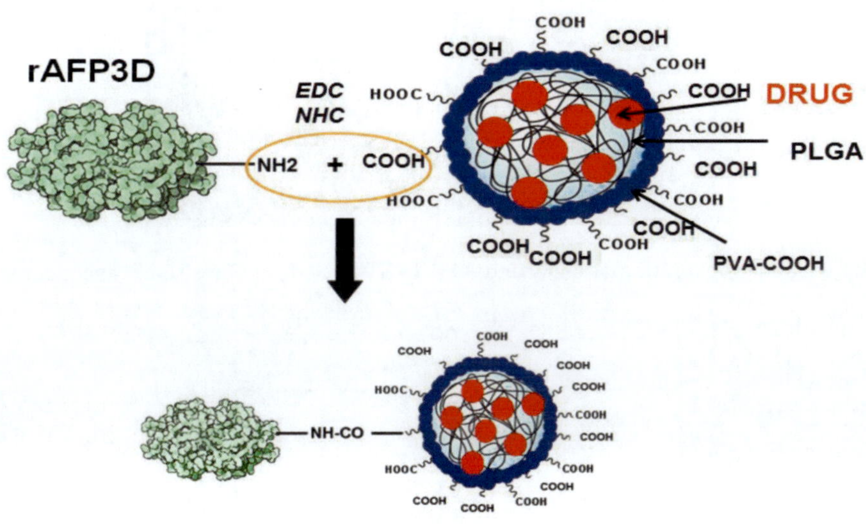

Figure 14. Scheme of obtaining rAFP3D conjugates with biodegradable poly(lactic-co-glycolic acid) (PLGA) nanoparticles. EDC, 1-ethyl-3-(3-dimethylaminopropyl) carbodiimide; NHS, N-hydroxysuccinimide ester; PVA, polyvinyl alcohol.

A further possibility to significantly increase the efficacy of targeted delivery of antitumor drugs is to construct nanoparticle vectors, which could contain a large number of cytostatic molecules and carry a vector molecule chemically attached to a reactive group on the surface of the nanoparticle (Figure 14). Currently, an ever more widely used polymer drug carriers are nanoparticles based on biodegradable poly(lactic-co-glycolic acid) (PLGA).

Using PLGA, it is possible to obtain nanoforms of various pharmaceutical agents, such as the poorly soluble antitumor drug paclitaxel. By attaching vector proteins to nanoparticles already carrying a cytostatic agent, it is possible to increase the antitumor effect of the drug both *in vitro* and *in vivo* [50].

Nanoparticles covalently bound to rAFP3D were obtained using an original novel technique: rAFP3D was immobilized on nanoparticles via modified polyvinyl alcohol associated with the polymer surface and serving as a stabilizer [51]. The efficacy of such targeted nanoparticles loaded with paclitaxel as the antitumor agent was significantly higher than the efficacy of free paclitaxel or of similar paclitaxel-carrying nanoparticles that did not contain rAFP3D.

Reversal of Multidrug Resistance

Development of multidrug resistance (MDR) by tumor cells is a major cause of low efficacy in cancer chemotherapy. A promising approach to overcoming the MDR problem is the development of high-molecular weight carriers for drug delivery into tumor cells. It is especially important that the macromolecules in question exhibit vector properties, since they provide the possibility to selectively eliminate malignant cells while minimizing the damage to normal cells.

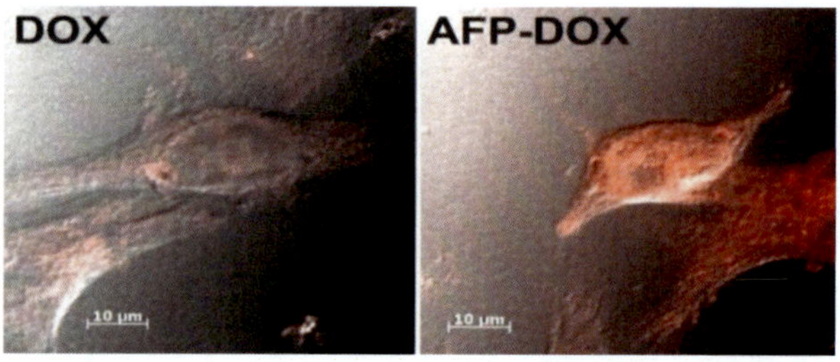

Figure 15. Intracellular localisation of DOX and AFP-DOX into SKVLB resistant cancer cells.

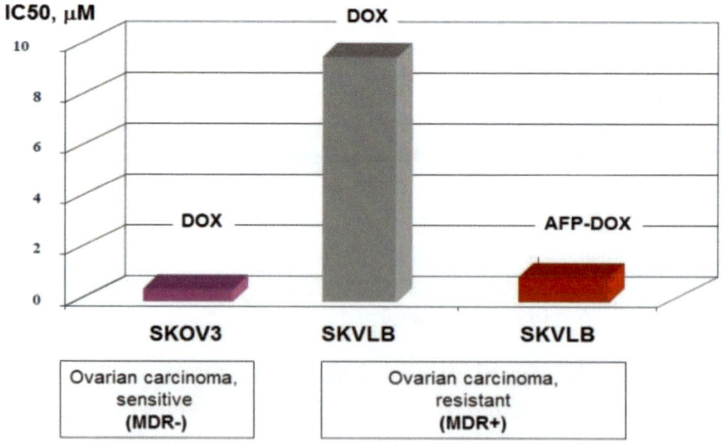

Figure 16. Cytotoxic activity of AFP-DOX conjugate against SKVLB resistant ovarian cancer cells.

An enormous advantage of nearly all drug delivery systems containing either full-size AFP or the recombinant receptor binding fragment rAFP3D as a vector is their ability to reverse the MDR tumor phenotype. This property is due to the fact that such systems enter the cell by receptor-mediated endocytosis rather than by transmembrane diffusion. This internalization pathway prevents the interaction of drugs with membrane proteins of the ABC transporter family. ABC transporter proteins are in fact membrane pumps that utilize the energy of ATP hydrolysis for transmembrane transportation of various substrates ranging from ions to macromolecules. They can act both to import and to export compounds, pumping toxins, pharmaceutical agents, and lipids out of the cell. Overexpression of ABC transporters interferes with drug uptake by tumor cells and thus leads to the development of the MDR phenotype.

Experiments in cell lines overexpressing the MDR1 (ABCB1) glycoprotein showed that doxorubicin contained in targeted drug delivery systems (direct AFP conjugates, targeted nanoparticles, or dendrimer) was efficiently accumulated by these resistant tumor cells [48, 52], in contrast to free DOX, whose accumulation was strongly suppressed (Figure 15). Resistant tumor cells accumulated DOX contained within targeted drug delivery systems similarly to initial drug-sensitive cell lines. DOX accumulation in resistant cells correlated with the cytotoxic activity of targeted drug delivery systems, which was comparable to their activity in drug-sensitive cell lines (Figure 16). Thus, the proposed drug delivery systems, both direct AFP conjugates and polymer-based delivery systems containing rAFP3D, were able to successfully overcome the MDR phenotype of tumor cells, which additionally confirms the selectivity of their action.

CONCLUSION

Targeted cancer therapy is a rapidly developing area of antitumor pharmaceutics whose purpose is create preparations with increased selectivity of drug action by employing vector molecules. This idea was first proposed in the early XXth century by Paul Ehrlich [53], who believed that a drug should act as a "magic bullet" selectively hitting its target, the vulnerable spot of a malignant cell.

Directed drug transportation by means of specially developed targeted delivery systems is a promising approach providing a possibility to increase the therapeutic efficacy of a drug and to decrease the risks associated with systemic toxicity of antitumor agents. Several targeted chemo- and radiotherapeutic agents based either on monoclonal antibodies (Mabthera, Bexar, Mylotarg) or on peptides ([(90)Y-DOTA(0), Tyr(3)] octreotide) have already been successfully introduced into clinical practice. The experimental data presented in this chapter demonstrate the high efficacy and selectivity of targeted drug delivery systems employing alpha fetoprotein or its recombinant receptor-binding fragment rAFP3D. Taking into account the high levels of AFP receptor expression on the surface of tumor cells of many different types, we firmly believe that both full-size AFP and its receptor-binding fragment are extremely promising vector agents for the development and implementation of novel targeted antitumor drugs.

REFERENCES

[1] Jain, R. K. (2001). Delivery of molecular and cellular medicine to solid tumors. *Adv Drug Deliv Rev*, *46*, 149-68.

[2] Jang, S. H., Wientjes, M. G., Lu, D. and Au, J. L. (2003). Drug delivery and transport to solid tumors. *Pharm Res*, *20*, 1337-50.

[3] Trojan, J. and Uriel, J. (1980). Immunocytochemical localization of alpha-fetoprotein in the developing rat brain. *Oncodevelopmental biology and medicine: the journal of the International Society for Oncodevelopmental Biology and Medicine*, *1*, 107-11.

[4] Dziegielewska, K. M., Evans, C. A., Lorscheider, F. L., Malinowska, D. H., Mollgard, K., Reynolds, M. L. and Saunders, N. R. (1981). Plasma proteins in fetal sheep brain: Blood-brain barrier and intracerebral distribution. *J Physiol*, *318*, 239-50.

[5] Uriel, J., Trojan, J., Moro, R. and Pineiro, A. (1983). Intracellular uptake of alpha-fetoprotein: A marker of neural differentiation. *Ann N Y Acad Sci*, *417*, 321-9.

[6] Uriel, J., Faivre-Bauman, A., Trojan, J. and Foiret, D. (1981). Immunocytochemical demonstration of alpha-fetoprotein uptake by primary cultures of fetal hemisphere cells from mouse brain. *Neurosci Lett*, *27*, 171-5.

[7] Moro, R., Fielitz, W., Esteves, A., Grunberg, J. and Uriel, J. (1984). *In vivo* uptake of heterologous alphafetoprotein and serum albumin by ependymal cells of developing chick embryos. *Int J Dev Neurosci*, *2*, 143-8.

[8] Villacampa, M. J., Moro, R., Naval, J., Failly-Crepin, C., Lampreave, F. and Uriel, J. (1984). Alpha-fetoprotein receptors in a human breast cancer cell line. *Biochem Biophys Res Commun*, *122*, 1322-7.

[9] Naval, J., Villacampa, M. J., Goguel, A. F. and Uriel, J. (1985). Cell-type-specific receptors for alpha-fetoprotein in a mouse t-lymphoma cell line. *Proc Natl Acad Sci, U S A*, *82*, 3301-5.

[10] Suzuki, Y., Zeng, C. Q. and Alpert, E. (1992). Isolation and partial characterization of a specific alpha-fetoprotein receptor on human monocytes. *J Clin Invest*, *90*, 1530-6.

[11] Esteban, C., Geuskens, M. and Uriel, J. (1991). Activation of an alpha-fetoprotein (afp)/receptor autocrine loop in ht-29 human colon carcinoma cells. *Int J Cancer*, *49*, 425-30.

[12] Geuskens, M., Torres, J. M., Esteban, C. and Uriel, J. (1989). Morphological characterization of the pathway of endocytosis and intracellular processing of transferrin and alpha-fetoprotein in human t lymphocytes stimulated with phytohemagglutinin (pha). *Eur J Cell Biol*, *50*, 418-27.

[13] Torres, J., M., Laborda, J., Naval, J., Darracq, N., Calvo, M., Mishal, Z. and Uriel, J. (1989). Expression of alpha-fetoprotein receptors by human t-lymphocytes during blastic transformation. *Mol Immunol*, *26*, 851-7.

[14] Nitsvetov, M. B., Rodina, A. V., Moskaleva, E. Y., Posypanova, G. A., Popova, O. N., Sologub, V. K., Koromyslova, I. A., Shmyrev, I. I., Sotnichenko, A. I., Zabolotnev, D. V., Karaulov, A. V. and Severin, S. E. (2001). Comparison of binding rates of anti-afpr monoclonal antibodies and afp during their interactions with human tumor cells. *Vopr Biol Med Pharm Chem (Rus)*, *3*, 19-25.

[15] Biddle, W. and Sarcione, E. J. (1987). Specific cytoplasmic alpha-fetoprotein binding protein in mcf-7 human breast cancer cells and primary breast cancer tissue. *Breast Cancer Res Treat*, *10*, 279-86.

[16] Mizejewski, G. J. (1995). Alpha-fetoprotein binding proteins: Implications for transmembrane passage and subcellular localization. *Life Sci*, *56*, 1-9.

[17] Mizejewski, G. J. (2001). Alpha-fetoprotein structure and function: Relevance to isoforms, epitopes, and conformational variants. *Exp Biol Med (Maywood)*, *226*, 377-408.

[18] Moro, R., Tamaoki, T., Wegmann, T. G., Longenecker, B. M. and Laderoute, M. P. (1993). Monoclonal antibodies directed against a widespread oncofetal antigen: The alpha-fetoprotein receptor. *Tumour Biol*, *14*, 116-30.

[19] Kanevsky, V., Pozdnyakova, L. P., Aksenova, O. A., Severin, S. E., Katukov, V. and Severin, E. S. (1997). Isolation and characterization of afp-binding proteins from tumor and fetal human tissues. *Biochem Mol Biol Int*, *41*, 1143-51.

[20] Mizejewski, G. J. (2011). Review of the putative cell-surface receptors for alpha-fetoprotein: Identification of a candidate receptor protein family. *Tumour Biol*, *32*, 241-58.

[21] Mizejewski, G. J. (2013). Review of the adenocarcinoma cell surface receptor for human alpha-fetoprotein; proposed identification of a widespread mucin as the tumor cell receptor. *Tumour Biol*, *34*, 1317-36.

[22] Nitsvetov, M. B., Moskaleva, E. Y., Posypanova, G. A., Makarova, O. V., Stepanov, V. A., Rogov, K. A., Koromyslova, I. A., Karaulov, A. V., Severin, S. E. and Severin, E. S. (2005). Immunohistochemical study of afp receptor expression in normal and malignant tumor human tissues. *Immunologiya (Rus)*, *26*, 122-25.

[23] Moro, R., Tcherkassova, J. and Song, E. (2005). A new broad-spectrum cancer marker. Ivd technology *IVD Technology* http://wwwivdtechnologycom/article/new-broad-spectrum-cancer-marker.

[24] Laborda, J., Naval, J., Allouche, M., Calvo, M., Georgoulias, V., Mishal, Z. and Uriel, J. (1987). Specific uptake of alpha-fetoprotein by malignant human lymphoid cells. *Int J Cancer*, *40*, 314-8.

[25] Alava, M. A., Iturralde, M., Lampreave, F. and Pineiro, A. (1999). Specific uptake of alpha-fetoprotein and albumin by rat morris 7777 hepatoma cells. *Tumour Biol*, *20*, 52-64.

[26] Esteban, C., Trojan, J., Macho, A., Mishal, Z., Lafarge-Frayssinet, C. and Uriel, J. (1993). Activation of an alpha-fetoprotein/receptor pathway in human normal and malignant peripheral blood mononuclear cells. *Leukemia*, *7*, 1807-16.

[27] Uriel, J., Villacampa, M. J., Moro, R., Naval, J. and Failly-Crepin, C. (1984). Uptake of radiolabeled alpha-fetoprotein by mouse mammary carcinomas and its usefulness in tumor scintigraphy. *Cancer Res*, *44*, 5314-9.

[28] Hajeri-Germond, M., Naval, J., Trojan, J. and Uriel, J. (1985). The uptake of alpha-foetoprotein by c-1300 mouse neuroblastoma cells. *Br J Cancer*, *51*, 791-7.

[29] Severin, S. E., Kulakov, V. N., Moskaleva, E., Severin, E. S., Slobodianik, II. and Klimova, T. P. (2012). The distribution of iodine-125 labeled alpha-fetoprotein in the animal organism and its accumulation in the tumor. *Vestn Ross Akad Med Nauk*, 11-5.

[30] Sinha, B. K. and Politi, P. M. (1990). Anthracyclines. *Cancer Chemother Biol Response Modif*, *11*, 45-57.

[31] Sanderson, R. J., Hering, M. A., James, S. F., Sun, M. M., Doronina, S. O., Siadak, A. W., Senter, P. D. and Wahl, A. F. (2005). *In vivo* drug-linker stability of an anti-cd30 dipeptide-linked auristatin immunoconjugate. *Clin Cancer Res*, *11*, 843-52.
[32] Severin, S. E., Moskaleva, E., Shmyrev, II., Posypanova, G. A., Belousova Yu, V., Sologub, V. K., Luzhkov Yu, M., Nakachian, R., Andreani, J. and Severin, E. S. (1995). Alpha-fetoprotein-mediated targeting of anti-cancer drugs to tumor cells *in vitro*. *Biochem Mol Biol Int*, *37*, 385-92.
[33] Moskaleva, E. Y., Posypanova, G. A., Koromyslova, I. A., Shmyrev, I. I., Krivonos, A. V., Myagkikh, I. V., Feldman, N. B., Finakova, G. V., Katukov, V. Y., Luzhkov, Y. M., Nakachian, R., Andreani, J., Severin, E. S. and Severin, S. E. (1996). *In vivo* antitumor activity of cytotoxic drugs conjugated with human alpha-fetoprotein. *Tumor Targeting*, *2*, 299-306.
[34] Severin, S. E., Posypanova, G. A., Katukov, V. Y., Shmyrev, I. I., Luzhkov, Y. M., Gerasimova, G. K., Zhukova, O. S., Vorozhtsov, G. N., Kaliya, O. L., Lukyanets, E. A. and Severin, E. S. (1997). Antitumor activity of conjugates of the oncofetal protein alpha-fetoprotein and phthalocyanines *in vitro*. *Biochem Mol Biol Int*, *43*, 1081-89.
[35] Severin, E. S., Moskaleva, E. Y. and Severin, S. E. (2002). Modern approaches to cancer therapy: Immunotherapy and targeted drug delivery. *Russian Journal of Immunology*, *6*, 345-56.
[36] Lutsenko, S. V., Feldman, N. B., Finakova, G. V., Gukasova, N. V., Petukhov, S. P., Posypanova, G. A., Skryabin, K. G. and Severin, S. E. (2000). Antitumor activity of alpha fetoprotein and epidermal growth factor conjugates *in vitro* and *in vivo*. *Tumour Biol*, *21*, 367-74.
[37] Severin, S. E., Posypanova, G. A., Sotnichenko, A. I., Moskaleva, E., Fel'dman, N. B., Grigor'ev, M. I., Severin, E. S. and Petrov, R. V. (1999). Antitumor activity of a covalent conjugate of the endiene antibiotic esperamicin a1 with human alpha-fetoprotein. *Dokl Akad Nauk (Rus)*, *366*, 561-4.
[38] Long, B. H., Golik, J., Forenza, S., Ward, B., Rehfuss, R., Dabrowiak, J. C., Catino, J. J., Musial, S. T., Brookshire, K. W. and Doyle, T. W. (1989). Esperamicins, a class of potent antitumor antibiotics: Mechanism of action. *Proc Natl Acad Sci U S A*, *86*, 2-6.
[39] Flaherty, K. T., Stevenson, J. P. and O'Dwyer, P. J. (2001). Antisense therapeutics: Lessons from early clinical trials. *Curr Opin Oncol*, *13*, 499-505.
[40] Line, B. R., Feustel, P. J., Festin, S. M., Andersen, T. T., Dansereau, R. N., Lukasiewicz, R. L., Zhu, S. and Bennett, J. A. (1999). Scintigraphic detection of breast cancer xenografts with tc-99m natural and recombinant human alpha-fetoprotein. *Cancer Biother Radiopharm*, *14*, 485-94.
[41] Posypanova, G. A., Gorokhovets, N. V., Makarov, V. A., Savvateeva, L. V., Kireeva, N. N., Severin, S. E. and Severin, E. S. (2008). Recombinant alpha-fetoprotein c-terminal fragment: The new recombinant vector for targeted delivery. *J Drug Target*, *16*, 321-8.
[42] Posypanova. G. A., Makarov. V. A., Savvateeva. M. V., Bereznikova, A. V. and Severin, E. S. (2013). The receptor binding fragment of alpha-fetoprotein is a promising new vector for the selective delivery of antineoplastic agents. *J Drug Target*, *21*, 458-65.
[43] Sharapova, O. A., Pozdniakova, N. V., Laurinavichiute, D. K., Iurkova, M. S., Posypanova, G. A., Andronova, S. M., Fedorov, A. N., Severin, S. E. and Severin, E. S.

(2010). Purification and characterization of recombinant human alpha-fetoprotein fragment, corresponding to the c-terminal structural domain. *Bioorg Khim*, *36*, 760-8.

[44] Sharapova, O. A., Pozdnykova, N. V., Laurinavichyute, D. K., Yurkova, M. S., Posypanova, G. A., Fedorov, A. N., Severin, S. E. and Severin, E. S. (2010). High-efficient expression, refolding and purification of functional recombinant c-terminal fragment of human alpha-fetoprotein. *Protein Expr Purif*, *73*, 31-5.

[45] Tomalia, D. A., Baker, H., Dewald, J., Hall, M., Kallos, G., Martin, S., Roeck, J., Ryder, J. and Smith, P. (1985). A new class of polymers: Starburst-dendritic macromolecules. *Polym J*, *17*, 117-32.

[46] Aulenta, F., Hayes, W. and Rannard, S. (2003). Dendrimers: A new class of nanoscopic containers and delivery devices. *European Polymer Journal*, *39*, 1741-71.

[47] Svenson, S. and Tomalia, D. A. (2005). Dendrimers in biomedical applications--reflections on the field. *Adv Drug Deliv Rev*, *57*, 2106-29.

[48] Yabbarov, N. G., Posypanova, G. A., Vorontsov, E. A., Obydenny, S. I. and Severin, E. S. (2013). A new system for targeted delivery of doxorubicin into tumor cells. *J Control Release*, *168*, 135-41.

[49] Yabbarov, N. G., Posypanova, G. A., Vorontsov, E. A., Popova, O. N. and Severin, E. S. (2013). Targeted delivery of doxorubicin: Drug delivery system based on pamam dendrimers. *Biochemistry*, *78*, 884-94.

[50] Van der Meel, R., Vehmeijer, L. J., Kok, R. J., Storm, G. and van Gaal, E. V. (2013). Ligand-targeted particulate nanomedicines undergoing clinical evaluation: Current status. *Adv Drug Deliv Rev*, *65*, 1284-98.

[51] Godovannyi, A. V., Vorontsov, E. A., Gukasova, N. V., Pozdnyakova, N. V., Vasilenko, E. A., Yabbarov, N. G., Dubovik, E. G., Severin, S. E., Severin, E. S. and Gnuchev, N. V. (2011). Targeted delivery of paclitaxel-loaded recombinant alpha-fetoprotein fragment-conjugated nanoparticles to tumor cells. *Doklady Biochemistry and biophysics*, *439*, 158-60.

[52] Moskaleva, E., Posypanova, G. A., Shmyrev, II., Rodina, A. V., Muizhnek, E. L., Severin, E. S., Katukov, V., Luzhkov, Y. M. and Severin, S. E. (1997). Alpha-fetoprotein-mediated targeting--a new strategy to overcome multidrug resistance of tumour cells *in vitro*. *Cell Biol Int*, *21*, 793-9.

[53] Ehrlich, P. (1954). The partial function of cells. (nobel prize address given on 11 december 1908 at stockholm). *Int Arch Allergy Appl Immunol*, *5*, 67-86.

In: Alpha-Fetoprotein: Functions and Clinical Applications
Editors: Nisha Lakhi and Michael Moretti
ISBN: 978-1-63484-875-6

Chapter 16

ANTI-BREAST CANCER DRUG DERIVED FROM ALPHA-FETOPROTEIN

James A. Bennett[1,*], PhD, Herbert I. Jacobson[1,2], PhD, and Thomas T. Andersen[3], PhD

[1]Center for Immunology and Microbial Diseases,
Albany Medical College, Albany, NY, US
[2]Department of Obstetrics and Gynecology,
Albany Medical College, Albany, NY, US
[3]Center for Cardiovascular Sciences,
Albany Medical College, Albany, NY, US

ABSTRACT

We will review the clinical and laboratory investigations that suggested the anti-estrogenicity of AFP, as well as the epidemiological studies that linked AFP to reduced breast cancer incidence, and we will articulate the concepts associated with using truncated segments of AFP for the control of breast cancer. In terms of drug development, we will review the generalizable means by which the anti-breast cancer activity of AFP was identified, isolated in the third domain of the molecule, and developed into an orally active cyclic nine amino acid peptide (AFPep). We illustrate the properties of AFPep, emphasizing its efficacy and lack of toxicity. Looking forward, we articulate the potential of AFPep for the treatment and prevention of breast cancer, as well as speculate about its utility for other clinical challenges. We conclude that the development of the homobiotic peptide AFPep may serve as a prototype for the development of other pharmaceutical molecules derived from molecules of human physiology.

* Corresponding author: James A. Bennett, CIMD MC-16, Albany Medical College, 47 New Scotland Avenue, Albany, NY 12208. Phone: 518 262-5247 E-mail: bennetj@mail.amc.edu.

Introduction

Many putative functions have been ascribed to α-fetoprotein (AFP) [1, 2], including fatty acid transport [3], immune regulation [4], and anti-estrogenicity [2]. The teleological concept of AFP possessing an anti-estrogenic effect to protect a fetus from maternal estrogen [5] rests on the recognition that the fetus is exposed to high levels of estrogen [6, 7] and has estrogen receptors [8] but does not respond to estrogen in the expected manner [9]. Few other components of the antenatal milieu could offer such protective activity. Further, the AFP gene is substantially down-regulated shortly after birth [2], coincident with the drop in estrogen levels in neonate tissues [7]. We have long been interested in the anti-estrogenic effects of AFP [10], with a particular goal of developing molecules that might be useful in clinical situations that call for anti-estrogens, notably including treatment of breast cancer [11]. The purpose of this contribution is to describe the rationale and strategies by which we developed an anti-breast cancer drug based on the structure of AFP and suggest this approach may be generalizable for some of the other putative functions of AFP (or other natural molecules).

The importance of developing pharmaceutical agents that are homobiotic can be summarized in the phrase "efficacy without toxicity" [12]. It may be that overuse of the word "natural" in both the scientific and lay literature has led to loss of meaning or even specificity. Taxols, for example, are natural products but their role in human physiology is certainly xenobiotic and might be expected to be accompanied by toxicity. On the other hand, homobiotic replacement agents such as hormones and hematopoeitic factors are generally effective and reasonably well tolerated. It makes sense that a molecule (or an analog of a molecule) that is indigenous to human physiology might be expected to be well-tolerated (i.e., exhibit no toxicity) and effective since it would utilize the normal signaling pathways. Furthermore, isolating from a natural molecule the structural component that possesses only one desired function of that molecule should improve selectivity and tolerability. We surmised that homobiotic analogs of AFP might be able to capture specific functions of AFP and generate the single desired response without the complications that would be associated with introduction of the intact AFP molecule. We describe here the implementation of that strategy to AFP and the subsequent development and characterization of a small homobiotic peptide which we call AFPep, and we illustrate why it is safe and effective for the treatment and prevention of breast cancer.

AFP Is Anti-Estrogenic

That AFP was involved in estrogen-modulated processes was suggested by clinical observations many years ago. AFP was found to be produced by the liver [13]. Amenorrhea was identified as an early symptom of hepatoma (over-production of AFP) [14], and partial hepatectomy of the tumor-involved liver lobe brought about reversal of both amenorrhea and galactorrhea [15]. Nerad and Skaunic reported a patient who became pregnant after liver lobectomy to remove hepatoma [16]. AFP produced by hepatoma or gastric cancer serves as a biomarker for those diseases [13, 17, 18], though, of course, there are no reports to suggest that AFP causes those diseases. Soto and Sonnenschein [19] reported on the role of AFP in controlling growth of estrogen-sensitive cells. Despite the presence of estrogen receptors in fetal, infant, and child tissues [8], fetal tissue does not respond to estrogen [7-9]. Serum AFP levels are high (mg/mL) during fetal life and decline rapidly after birth to ng/mL [20, 21]. These reports from the clinic suggest that AFP interferes with estrogen-induced responses, and this became important in the breast cancer arena because of the numerous reports that suggest overexposure to estrogen promotes development of breast cancer [22].

In 1979, laboratory studies began to report on the probable role of AFP in growth inhibition of estrogen-sensitive tumors in newborn rats [23]. Soto and Sonnenschein [19, 24, 25] showed that estrogen-independent tumors implanted into newborn or adult rats grew without delay and without respect to added estrogen. However, estrogen-dependent tumors grew promptly in adult animals but experienced a delay of several days when inoculated into newborn rats. Their interpretation was that AFP present only in newborn animals was responsible for the inhibition of tumor growth [26]. These authors also showed that purified AFP had growth-inhibiting effects against estrogen-sensitive cell lines in culture [19]. Other endpoints reflected similar suggestions for the anti-estrogenic activity of AFP, one example being that uterine wet weight and stimulation of uterine epithelial mucosa were significantly decreased in rats bearing an AFP-secreting hepatoma [27].

An elegant series of studies highlighted the role of AFP in breast cancer [19, 26]. Rats already bearing an estrogen receptor-positive breast cancer were implanted with a hepatoma cell line that secreted large quantities of AFP, or with a hepatoma cell line that did not secrete AFP. Mammary tumors actually decreased in size in animals receiving the AFP-secreting hepatoma cell line, but not in animals that received the control hepatoma cell line. Similar results were obtained by injecting partially purified preparations of AFP. AFP purified from cultured human hepatoma cells inhibited the growth of human breast cancer xenografts (including MCF-7 and T47D lines) growing in SCID mice [28], emphasizing the importance of continuing to develop AFP as a pharmaceutical entity to combat breast cancer. Historically, that development process included a progression in the purity of AFP, as technology allowed. Originally, studies used animals bearing AFP-secreting hepatomas [26, 27], then AFP preparations purified from rat fetuses [19] or from human umbilical cord blood [29], from rat amniotic fluid [10], or from culture media of human hepatoma cells [28], and eventually from a bacterial expression system [30]. As described below, further development included chemical synthesis of analogs of AFP, assuring that the anti-cancer activity is associated with AFP, not with any putative contaminant molecule.

A recurring observation in laboratory studies was that AFP could undergo a change in conformation in the presence of hydrophobic ligands such as estrogen [31] and fatty acids [3], and that this conformational change would increase its anti-estrogenic activity [31]. For example, difference spectroscopy studies indicated that AFP undergoes a conformation change in a molar excess of estrogen [31]. Coincidentally, AFP's antiuterotrophic activity is increased when it is in the presence of a molar excess of estrogen [10]. The suggestion was made [3, 31] that there was an active anti-estrogenic site in AFP, an 'epitope' which could become more available to growth regulating receptors depending on the molecular environment in which AFP found itself. This observation, too, encouraged further development of that 'epitope,' but it also led to the conclusion that the obvious mechanism of simple binding and sequestration of estrogen by AFP is insufficient to explain the anti-estrogenic activity of the protein, especially considering that adding a molar excess of estrogen could not overcome the growth inhibitory effect of AFP [10].

Epidemiology Studies Suggest AFP Modulates Breast Cancer

Epidemiological data suggest that AFP is one of the factors in pregnancy that confer on parous women their significant reduction in risk of breast cancer. As shown in Table 1, AFP is elevated in maternal serum during pregnancy [28]. Furthermore, there are various conditions associated with pregnancy (such as maternal race, weight, consumption of alcohol, hypertension, number of fetuses in utero, or neural tube defect in the fetus) in which the maternal serum AFP concentration (MSAFP) is substantially altered from normal pregnancy levels. In studying the literature, Jacobson noted [32] the consistent and striking correlation that in those pregnancy conditions associated with an elevated level of MSAFP (e.g., race and weight), there was a significant reduction in the lifetime risk to the mother of acquiring breast cancer. Conversely, in pregnancy conditions characterized by low MSAFP (e.g., alcohol use), subsequent breast cancer risk was elevated [28]. This led Jacobson and Janerich [33-35] to undertake epidemiological studies of three other pregnancy conditions (hypertension, twinning, and neural tube defects) that are associated with elevated MSAFP. For each of these conditions, a correlation between high maternal serum AFP and reduced breast cancer risk was observed [28]. Then, too, Ekbom et al. [36] published an epidemiological study which suggests that, at least in the case of hypertension during pregnancy, the reduction in breast cancer risk is also passed on to the fetus. All known reports that have failed to confirm this observation have employed small populations and lacked sufficient power to confirm or negate the observation [37-39], or included older populations of women in whom malignant transformations are likely to have occurred after their final pregnancy [38, 39]. On the other hand, a report by Albrektsen et al. [40] using a large population of Norwegian women stated explicitly that the "reduced risk of breast cancer observed among women with multiple births, compared to women with singletons only, is consistent with results reported by Jacobson et al." [33]. Our results supported this observation and added that the protective effect of twin pregnancy was even more pronounced when it was the last birth [33, 37].

Table 1. Association of High Maternal Serum [AFP] with Decreased Breast Cancer Risk

Maternal Condition 1	Maternal Condition 2	[MSAFP]	Maternal Breast Cancer Risk
Pregnant	Non-pregnant	1 > 2 (41)	1< 2 (42)
Pregnant, Black	Pregnant, White	1 > 2 (20)	1< 2 (43)
Pregnant, lean	Pregnant, obese	1 > 2 (20)	1< 2 (22)
Pregnant, consuming no alcohol	Pregnant, consuming alcohol	1 > 2 (44)	1< 2 (45)
Pregnant, hypertensive	Pregnant, normotensive	1 > 2 (46)	1< 2 (34)
Pregnant with multiple fetuses	Pregnant with a single fetus	1 > 2 (47)	1< 2 (33)
Pregnant, fetus with NTD	Pregnant, fetus without NTD	1 > 2 (41)	1< 2 (35)

NTD is Neural Tube Defect. Numbers in parentheses are references for the data. Redone and corrected from [28].

More recently, Richardson et al. [48] employed data from the University of California at Berkeley Child Health and Development Study to measure directly (i.e., in a single cohort) the association between maternal serum AFP concentrations and subsequent breast cancer incidence. They found that the concentration of AFP in maternal sera that had been cryogenically preserved was inversely correlated to the risk of breast cancer in these same mothers 20 to 30 years after their pregnancies. Richardson [48] concluded, "the results of this study agree with the protective effect reported by Jacobson et al. [33] and Thompson et al. [34] using surrogate indicators (multiple births and hypertension) for a high level of AFP during the index pregnancy." Melbye et al. [49] concluded "a high level of AFP in maternal serum during any pregnancy is associated with a low overall incidence of breast cancer, and, in particular, with a low incidence of advanced breast cancer. This association appears particularly strong for a pregnancy occurring at a young age." These studies suggested to us that AFP should be pursued as a modulator of breast cancer [32].

Other workers had proposed various hormones of pregnancy as the agent(s) responsible for the pregnancy-associated reduction in risk of breast cancer. Estriol [50-52], progesterone together with estrogen [53], or human chorionic gonadotropin [54] injected serially into carcinogen-exposed rats decreased the incidence of mammary cancer compared to carcinogen-exposed animals not treated with those hormones, but there was no understanding of why those various treatments would elicit a similar response. Jacobson et al. [55] proposed that these hormones were not the proximal modulators of cancer in the mammary tissue, but rather that they acted on the liver to stimulate production of AFP, and that AFP modulated breast cancer. These workers repeated the experiments involving treatment of carcinogen-exposed rats with progesterone plus estrogen, progesterone plus estriol, estriol alone, or chorionic gonadotropin and observed the previously identified reduction in mammary cancer incidence. However, they also measured serum AFP levels in the animals and showed that in every case, AFP levels were increased by the hormone treatment. An *in vitro* system using human HepG2 cells treated with the hormones of pregnancy elicited AFP, which when applied to MCF-7 cells inhibited their proliferation. Antibody directed against human AFP obviated that inhibition of proliferation, suggesting rather convincingly that, in either animal studies or in women who experience pregnancy, AFP is the proximal agent that inhibits breast cancer [55].

Summarizing, the observations are clear that breast cancer is driven by estrogen, that AFP is an anti-estrogen and is an anti-breast cancer agent. Therefore, it seemed logical to develop AFP as a pharmaceutical agent for the treatment of breast cancer. Of course, an intact protein with as many activities as has been reported for AFP would be undesirable as a pharmaceutical agent. A better approach would be to identify the anti-cancer active site of AFP and develop it into a small molecule, a process which is described below.

Development of AFPep

In overview, an extensive series of parsing studies summarized in [12], employing an anti-estrogenic activity assay [10] (i.e., inhibition of estrogen-stimulated growth of immature mouse uterus) for assessment of intended effect, was used to identify the anti-estrogenic active site of AFP [10, 12] which was then developed into a small, readily synthesized, stable, orally active cyclic 9-amino acid peptide referred to as AFPep.

Festin et al. [31] used a baculovirus system to express domains and subdomains of AFP and showed that the anti-estrogenic activity was contained primarily in the C-terminal third of the protein. Subsequent efforts to further define the active site employed peptide synthesis of large fragments of this domain [56-58], followed by synthesis of smaller and smaller peptides [59], each time guided by the anti-estrogenic uterine growth inhibition assay. An 8-amino acid peptide that retained essentially full anti-estrogenic activity of the intact AFP molecule was identified [59], whereas smaller fragments had diminished activity [60]. This 8-mer peptide had none of the other activities associated with full-length AFP. It was developed further through modeling and rational design approaches [59, 61-63], evolving eventually [12] into a small, readily synthesized, stable, orally active cyclic 9-amino acid peptide which we call AFPep (Figure 1) [31, 59, 61-66]. AFPep is the anti-estrogenic/anti-breast cancer site in the AFP molecule [57, 59, 61, 65-67], it inhibits estrogen-induced development and growth of experimental breast cancers [59, 61, 62, 64-66, 68]. AFPep has none of the other active sites and none of the side effects of AFP such as immune suppression or hepatic growth [4, 69, 70]. AFPep has been found to be active after oral administration in three different assays in two different species in both immature and adult animals for the prevention and therapy of breast cancer [66] (see below).

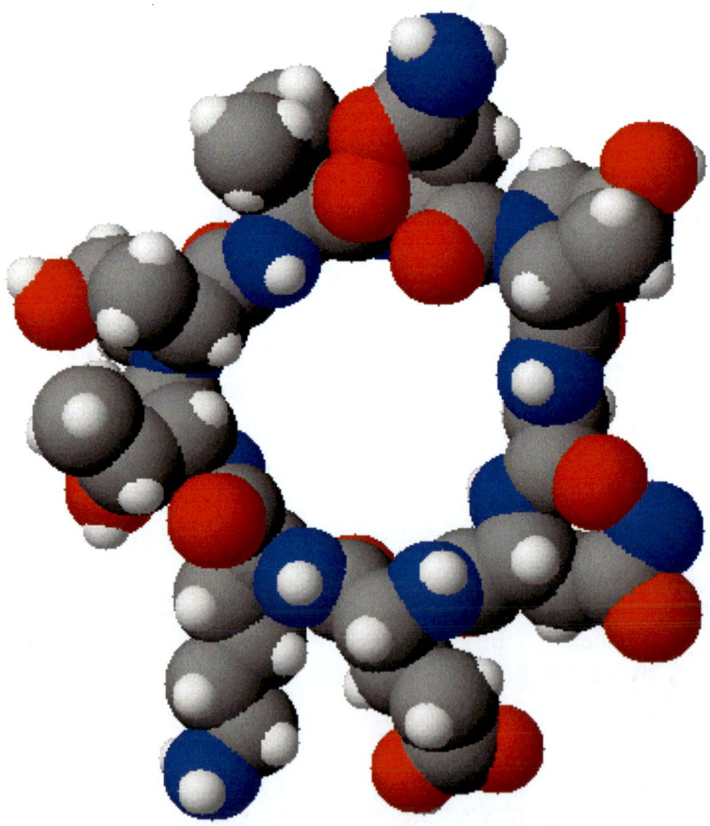

Figure 1. AFPep is *cyclo (EKTOVNOGN)*, the active site of AFP. Its metabolites are simple amino acids. *E* is at the six o'clock position; *O* is hydroxyproline.

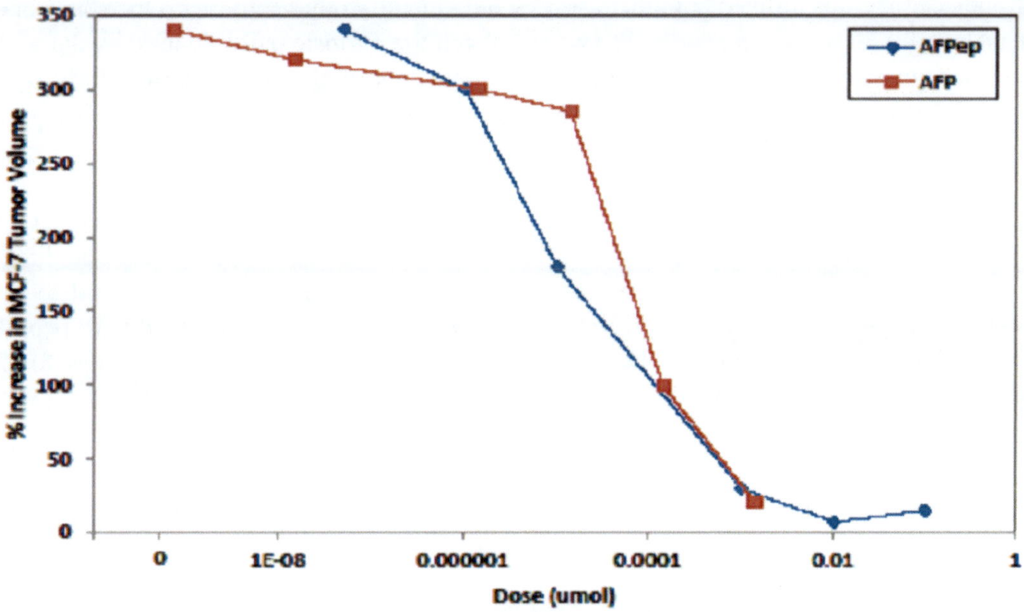

Figure 2. AFP and AFPep Inhibit Growth of MCF-7 Tumors. MCF7 tumor was implanted under the kidney capsule of SCID mice. Either AFP or AFPep were administered once daily i.p. for 30 days. Tumor volume was then measured and reported here as percent increase in tumor volume compared to Day 1. Each point represents the mean of three replicate mice.

Using a human breast cancer xenograft assay in mice [65], we found that either AFPep or its parent protein, AFP, given once daily, significantly inhibited the growth of human MCF-7 breast cancer xenografts [28, 65]. Dose response curves are shown for both entities in Figure 2. AFPep (m.w. 970) and AFP (m.w. 69,000) are similar in their potencies in that significant activity is apparent above a dose of 0.0001 μmoles of each agent. AFPep was evaluated against additional ER+ human breast cancer cell lines (T47D and ZR75-1 grown as xenografts) and was found to inhibit their growth [65, 71]. These important observations (together with those related to efficacy against tamoxifen-resistant cancers, see below) provided impetus for development of AFPep as a therapeutic agent.

As discussed in [12], several important aspects of peptide drug development were incorporated in the process of developing AFPep, but here we wish to emphasize that it is entirely likely that other activities associated with intact AFP could be developed into small molecules using similar approaches. Advantages of peptide analogs of AFP, be they cyclic or linear molecules, include generation of a single activity, utilization of intrinsic pathways, and low toxicity. Specificity of drugs may be more readily achieved by using peptides (as compared to non-peptidic organic molecules) because of the multitude of isosteres available to peptide chemistry, peptides can be exquisitely designed to interact with a specific receptor [72]. Moreover, peptides are widely accepted now as a good investment for development by the pharmaceutical industry, and there are more than 60 peptide drugs in current clinical use, capturing a significant market share of the pharmaceutical industry [73]. In the case of AFP, it should be possible, using the parsing techniques discussed herein, to identify individual 'epitopes' corresponding to the desired activity, and so remove all other active sites. As always, a sensitive assay is required to monitor the parsing process. In this instance, an anti-estrogenic assay was utilized but other assays would identify other activities. Once developed, an analog of AFP would be expected to work through the intrinsic pathway used by the parent AFP, and this is likely to minimize off-target effects and perhaps to prolong utility of the drug. Since such a molecule would be using native signaling pathways, it may be less likely to engender chemotherapeutic failure through development of drug resistance. While this concept has not yet been definitively demonstrated, it may be that AFPep will retain efficacy longer than tamoxifen or other agents (see below). Low toxicity is inherently associated with peptide drugs, due to target specificity as well as to metabolite tolerability. The metabolites of AFPep are simple amino acids, and are not toxic. Many non-peptidic drugs get oxidized and become more toxic than the original molecule, a process that is easy to avoid with peptides derived from parent proteins such as AFP. Thus, there are multiple advantages to this approach in drug development.

AFPEP IS USEFUL FOR TREATMENT AND PREVENTION OF BREAST CANCER

AFPep Can Be Added to, or Used in Place of, Tamoxifen

Tamoxifen has been the most widely used agent for the treatment of estrogen receptor-positive breast cancers and has provided significant benefits to women [74]. However, a vexatious problem encountered in its clinical use is that not all ER+ breast cancers are sensitive to tamoxifen. Moreover, it is not uncommon that women whose disease is being managed successfully by tamoxifen therapy will experience recurrence of cancer apparently because their tumor has acquired resistance to tamoxifen. We considered it important to determine whether AFPep was active against ER+ breast cancer that had become resistant to tamoxifen, so we developed an MCF-7 human breast cancer sub-line that was resistant to tamoxifen, grew it as xenografts, and showed that AFPep was equally effective against wild type and tamoxifen-resistant MCF-7 breast cancer [65]. In other studies, AFPep was found to be additive with tamoxifen in its inhibition of wild-type MCF-7 xenograft growth, and in its prevention of MNU-induced mammary cancer [64]. Thus, AFPep may be used as a replacement for tamoxifen, or at the very least, be used after tamoxifen has failed. But AFPep may be even better than that. First, an important observation is that AFPep actually inhibited the uterine hyperplasia induced by tamoxifen [64], which is a troublesome side effect of tamoxifen that may be a precursor to uterine malignancy [75]. Thus AFPep might advantageously be used in combination with tamoxifen to obviate tamoxifen's toxicity. Secondly, as discussed above, it is reasonable to expect that AFPep may not engender acquired resistance because it uses natural signaling pathways and will not be perceived as being foreign to the body because it closely mimics the native human sequence of AFP. Since the usual animal models are of short duration and have not lent themselves to assessment of acquired resistance, we plan to use the ACI rat model of breast cancer which allows for extended treatment duration. These studies may well suggest that acquired resistance does not occur with AFPep.

AFPep Has a Unique Mechanism of Action

AFPep is a multikinase inhibitor blocking the phosphorylation of c-Kit (Y703), FAK (S910), and MEK (S298, T386) and subsequently the phosphorylation of ERα (S118). Phosphorylation of S118 in ERα is needed for the transcription of growth signals in response to liganded ER [76]. AFPep is different from tamoxifen in that it does not block binding of estrogen to ER [65]. Instead, by blocking phosphorylation of ER it prevents formation of activated ER-E. In MCF-7 orthotopically xenografted in SCID mice and treated with AFPep, tumor biopsies indicated significant inhibition of phosphorylation of ERα, increases in ERβ, p21Cip1, Rb, and decreases in PCNA, suggesting that phospho-ERα inhibition is signaling a delay in the progression of tumor cells through the cell cycle, consistent with the significant reduction in tumor growth rate seen between days 28 and 35 of this experiment (**Tables 2-3**). As future preclinical and clinical studies are carried out, these biomarkers will be very useful in monitoring biological activity of AFPep since they will change well before tumor burden.

Table 2. Effect of AFPep on *Orthotopic* Breast Cancer Xenograft Growth
AFPep initially inhibited, then stopped, growth of palpable tumors

Days after Tumor Implantation	Tumor Volume (mm^3)	
	Control	AFPep
21	156 ± 22	156 ± 22
22	Begin: Vehicle or AFPep	
28	487 ± 68	308 ± 30
35	897 ± 71	313 ± 45

MCF-7 human breast cancer cells (5 x 10^6) were injected into the mammary fat pad of SCID mice. Tumors were on average 6.7 mm in diameter (156 mm^3) on day 21 after implantation. Either vehicle or AFPep (100 μg/mouse) was given in 0.5 ml once daily for 14 days by oral gavage, beginning 22 days after tumor implantation. Tumors were harvested for biomarker analysis (see Table 4) on day 35 after tumor implantation. Mean ± SE of 3 replicate mice/group are shown.

Table 3. Effect of AFPep on Biomarkers in Breast Cancer

Biomarker	Change from Control, %
pERα	- 48 ± 5
pERβ	110 ± 15
Rb	38 ± 4
P21	82 ± 9
PCNA	- 53 ± 4

Tumor tissue was harvested for biomarker determination 35 days after tumor transplantation and biomarkers were assayed by Western blot. Change in band density in AFPep-treated mice relative to control group is reported. Mean ± SE from three replicate mice per group. These biomarkers can also be assessed by immunohistochemistry.

AFPep Prevents Breast Cancer

We have shown in prevention studies [64, 68] that AFPep prevents the development of MNU-induced mammary cancers in rats (**Table 4**). Table 4 clearly indicates decreased incidence, multiplicity, and burden. The level of prevention was similar to that found with tamoxifen in similarly designed studies [64, 68, 77]. It is impressive that AFPep (or any other agent) can prevent breast cancer in the face of a potent carcinogen such as MNU. Still, the model is quite artificial, using a bolus dose of a carcinogen not normally encountered by women and generating mammary cancers that are not similar to most that occur in women [78]. Thus it is important to test AFPep against more realistic models of breast cancer such as that in the ACI rat model which is genetically primed to develop mammary cancer under the promotional influence of estrogen [78-86]. Nevertheless, there is a very important proof of principle here: AFPep has the potential to be used to *prevent* as well as to *treat* breast cancer.

Table 4. AFPep prevents MNU-induced breast cancer

Treatment	AFPep Dose		Tumor Incidence		Multiplicity	Burden	Latency
		µg/rat/day	%	p value	tumors/rat	cm^3	days
Saline	none	0	78	-	2.1	69	80 ± 31
AFPep	Sub-optimal	100	63	0.16	1.1*	39.1*	98 ± 33
AFPep	Optimal	270	40	0.02	0.5*	34.5*	88 ± 16

Incidence is defined as % of rats with one or more tumors at termination. Multiplicity is total number of tumors/number of rats in each group. Burden is the sum of the volume of tumors in each group. Latency is mean number of tumor-free days ± standard deviation. There were 30 rats in each group. * Statistically significant at $p < 0.05$ Data from [68].

Table 5. Summary of AFPep treatment studies that showed no toxicity

Species	Treatment Duration Days	Autopsy Date Days after Treatment	Dose, Route µg/animal/day	Number of Animals
Mouse	1	1	1 – 10,000 i.p. or p.o.	4000
Mouse	30	1	10 – 100 i.p. or p.o.	300
Rat	23	200	3 – 300 sc or p.o.	1000
Dog	1	None	10,000 i.v.	1
Mouse	5	1	2,000 i.v.	5
Mouse	1	5	10,000 i.v.	5

No toxicity was seen in the course of a variety of interventions, in three species. This lack of toxicity was not unexpected for four reasons: (1) AFPep is derived from a natural human protein; (2) AFPep has been designed/parsed/shaped to have only the anti-estrogenic, anti-breast cancer properties of AFP (59, 61, 62); (3) The active, anti-breast cancer concentration of AFPep *in vivo* is well below that of the parent protein concentration in the blood of the human fetus (20); and (4) The metabolites of AFPep are simple amino acids.

AFPep Is Non-Toxic

In studies to date no toxicity in mice and rats has been found with AFPep, even at doses as high as 1000 times the effective dose of 10 μg/mouse [66, 68]. Figures 6, 7, and 8 in reference [68] are typical of dozens of observations in which no side effects or toxicity has been seen when AFPep was administered to rats, mice, or dogs. Table 5 is a summary of the number of animals we have treated with AFPep in the past several years, none of which have exhibited side effects.

Treatment of rats with AFPep showed no disruption of the estrous cycle, no effect on fecundity or fertility, and no evidence of teratogencity [87]. While this bodes well for the potential of AFPep to serve as a breast cancer preventive agent, it also highlights the concepts of drug development that we discuss in this chapter. As noted in [87], the parent protein AFP disrupts reproductive tissues and function [14, 15, 27, 88, 89] but AFPep does not. This observation suggests that in isolating the anti-oncogenic site of AFP, the epitopes of that protein that adversely impact the reproductive system were effectively parsed out.

The favorable host toxicity profile cannot be over-emphasized since drug toxicity is one of the main reasons for failure of new drugs, and it is almost a requirement that an agent intended for use as a preventive be extremely well tolerated. We proffer that AFPep is efficacious without being toxic, which is true to date. The risk that there will be some side effects in humans being treated with AFPep for lengthy times is mitigated: by our extensive earlier studies in lower mammals, because AFPep is built on the human AFP sequence; because it is active at fetal physiological levels of AFP; and because the metabolites of AFPep are simple amino acids. At the very least, it is completely reasonable to expect that AFPep will be less toxic than cytotoxic agents such as doxorubicin and even SERMs such as tamoxifen, allowing use of AFPep as a therapeutic agent and probably as a preventive agent.

CONCLUSION

We have described the development of AFPep, a new peptide derived from AFP that has enormous potential as a well-tolerated, efficacious drug for the treatment and prevention of breast cancer. We see no reason why some of the other functions (such as immunosuppression) associated with AFP (or other macromolecules) could not be isolated and developed into therapeutic agents using an approach similar to that outlined here. The homobiotic nature of AFPep approaches the Holy Grail of drug development, namely efficacy without toxicity. Efficacy as well as longevity of the biological action of a homobiotic drug is presumably based on re-engaging the natural biochemical circuitry of the cell, rather than utilization of interdiction of pathways in unusual processes as most xenobiotic agents do. Lack of toxicity is based on AFPep's mimicry of its parent protein, as well as the fact that metabolites of AFPep are simple amino acids (unlike xenobiotic molecules which often get metabolized to become even more toxic).

Presumably, the anti-breast cancer activity of AFPep derives from its anti-estrogenic activity, although its exact mechanisms of action have not been completely delineated and, like many drugs, probably will not be until AFPep has experienced extensive clinical use. If the pre-clinical data continue to manifest in humans, AFPep may be useful for the prevention of breast cancer, as well as treatment, highlighting the innate potential of homobiotic molecules.

Although AFPep was developed with a narrow focus on breast cancer, the question always arises as to whether it would work for other cancers, and preliminary cell culture evidence suggests that AFPep should be tested against a variety of maladies. In the tumor arena, AFPep has inhibited growth of uterine leiomyoma, glioblastoma, prostate and thyroid cancers. Mechanistically, given its inhibition of c-kit, AFPep has potential against gastrointestinal stromal tumors. From an endocrinological perspective, it may be useful in situations that require the tamping down of estrogen effects without their complete eradication, such as in the treatment of uterine fibroids.

It seems likely that AFPep will have multiple future uses including treatment and prevention of breast cancer, due to its exquisite efficacy and tolerability through its re-engagement of natural growth regulatory circuits and will become a significant addition to the pharmacopeia as well as serve as an example for future development of other homobiotic agents.

REFERENCES

[1] Crandall, B. F. (1981). Alpha-fetoprotein: a review. *Crit Rev. Clin. Lab Sci*, *15*, 127-85.

[2] Abelev, G. I. (1989). Alpha-fetoprotein: 25 years of study. T*umour. Biol*, *10*, 63-74.

[3] Haourigui, M., Thobie, N., Martin, M. E., Benassayag, C. and Nunez, E. A. (1992). *In vivo* transient rise in plasma free fatty acids alters the functional properties of alpha-fetoprotein. *Biochim Biophys Acta*, *1125*, 157-65.

[4] Semeniuk, D. J., Boismenu, R., Tam, J., Weissenhofer, W. and Murgita, R. A. (1995). Evidence that immunosuppression is an intrinsic property of the alpha-fetoprotein molecule. *Adv Exp Med Biol*, *383*, 255-69.

[5] MacLusky, N. J. and Naftolin, F. (1981). Sexual differentiation of the central nervous system. *Science*, *211*, 1294-302.

[6] Mazor, M., Hershkowitz, R., Ghezzi, F., Cohen, J., Silber, A., Levy, J., Leiberman, J. R. and Glezerman, M. (1996). Maternal plasma and amniotic fluid 17 beta-estradiol, progesterone and cortisol concentrations in women with successfully and unsuccessfully treated preterm labor. *Arch Gynecol Obstet*, *258*, 89-96.

[7] Shutt, D. A., Smith, I. D. and Shearman, R. P. (1974). Oestrone, oestradiol-17beta and oestriol levels in human foetal plasma during gestation and at term. *J. Endocrinol*, *60*, 333-41.

[8] Keeling, J. W., Ozer, E., King, G. and Walker, F. (2000). Oestrogen receptor alpha in female fetal, infant, and child mammary tissue. *J. Pathol*, *191*, 449-51.

[9] Greco, T. L., Duello, T. M. and Gorski, J. (1993). Estrogen receptors, estradiol, and diethylstilbestrol in early development: the mouse as a model for the study of estrogen receptors and estrogen sensitivity in embryonic development of male and female reproductive tracts. *Endocr Rev*, *14*, 59-71.

[10] Mizejewski, G. J., Vonnegut, M. and Jacobson, H. I. (1983). Estradiol-activated alpha-fetoprotein suppresses the uterotropic response to estrogens. *Proc. Natl. Acad. Sci. U.S.A*, *80*, 2733-7.

[11] Jacobson, H. I., Bennett, J. A. and Mizejewski, G. J. (1990). Inhibition of estrogen-dependent breast cancer growth by a reaction product of alpha-fetoprotein and estradiol. *Cancer Res*, *50*, 415-20.

[12] Jacobson, H. I., Andersen, T. T. and Bennett, J. A. (2014). Development of an active site peptide analog of alpha-fetoprotein that prevents breast cancer. *Cancer Prev. Res. (Phila)*, *7*, 565-73.

[13] Abelev, G. I. (1968). Production of embryonal serum alpha-globulin by hepatomas: review of experimental and clinical data. *Cancer Res*, *28*, 1344-50.

[14] Nerad, V., Brzek, V., Skaunic, V. and Kopecny, J. (1969). Secondary amenorrhea, the first symptom of hepatoma. Sb Ved. Pr Lek. Fak. *Karlovy Univerzity Hradci Kralove*, *12*, 257-62.

[15] Couinaud, C., Schwarzmann, V., Ceoara, B., Orengo, P. and Fitterer, R. (1973). Malignant hepatoma with amenorrhea and galactorrhea. Disappearance of the endocrine syndrome following right hepatectomy. *Ann. Chir*, *27*, 151-6.

[16] Nerad, V. and Skaunic, V. (1969). Pregnancy following left-sided lobectomy of the liver during recurring hepatoma. *Cas Lek Cesk*, *108*, 228-9.

[17] Fujii, H., Ichikawa, K., Takagaki, T., Nakanishi, Y., Ikegami, M., Hirose, S. and Shimoda, T. (2003). Genetic evolution of alpha fetoprotein-producing gastric cancer. *J. Clin. Pathol*, *56*, 942-9.

[18] Trompetas, V., Varsamidakis, N., Frangia, K., Polimeropoulos, V. and Kalokairinos, E. (2003). Gastric hepatoid adenocarcinoma and familial investigation: does it always produce alpha-fetoprotein? *Eur J Gastroenterol Hepatol*, *15*, 1241-4.

[19] Soto, A. M. and Sonnenschein, C. (1980). Control of growth of estrogen-sensitive cells: role for alpha-fetoprotein. *Proc. Natl. Acad. Sci. U.S.A*, *77*, 2084-7.

[20] Crandall, B. F., Lebherz, T. B., Schroth, P. C. and Matsumoto, M. (1983). Alpha-fetoprotein concentrations in maternal serum: relation to race and body weight. *Clin. Chem*, *29*, 531-3.

[21] Sell, S., Nichols, M., Becker, F. F. and Leffert, H. L. (1974). Hepatocyte proliferation and alpha 1-fetoprotein in pregnant, neonatal, and partially hepatectomized rats. *Cancer Res*, *34*, 865-71.

[22] Pike, M. C., Spicer, D. V., Dahmoush, L. and Press, M. F. (1993). Estrogens, progestogens, normal breast cell proliferation, and breast cancer risk. *Epidemiol Rev*, *15*, 17-35.

[23] Sonnenschein, C. and Soto, A. M. (1979). Growth inhibition of estrogen-sensitive tumor cells in newborn rats. Probable role of alpha-fetoprotein. *J. Natl. Cancer Inst*, *63*, 835-41.

[24] Soto, A. M., Lee, H., Siiteri, P. K., Murai, J. T. and Sonnenschein, C. (1984). Estrogen induction of progestophilins in rat estrogen-sensitive cells grown in media supplemented with sera from castrated rats and from rats bearing an alpha-fetoprotein-secreting hepatoma. *Exp. Cell Res*, *150*, 390-9.

[25] Sonnenschein, C., Ucci, A. A. and Soto, A. M. (1980). Age-dependent growth inhibition of estrogen-sensitive rat mammary tumors. Probable role of alpha-fetoprotein. *J. Natl. Cancer Inst*, *64*, 1141-6.

[26] Sonnenschein, C., Ucci, A. A. and Soto, A. M. (1980). Growth inhibition of estrogen-sensitive rat mammary tumors. Effect of an alpha-fetoprotein-secreting hepatoma. *J. Natl. Cancer Inst*, *64*, 1147-52.

[27] Pool, T. B., Hagino, N. and Cameron, I. L. (1978). Relationship between functional castration and alpha-fetoprotein produced by hepatoma-bearing female rats. *J. Reprod. Fertil*, *53*, 39-44.

[28] Bennett, J. A., Zhu, S., Pagano-Mirarchi, A., Kellom, T. A. and Jacobson, H. I. (1998). Alpha-fetoprotein derived from a human hepatoma prevents growth of estrogen-dependent human breast cancer xenografts. *Clin. Cancer Res*, *4*, 2877-84.

[29] Allen, S. H., Bennett, J. A., Mizejewski, G. J., Andersen, T. T., Ferraris, S. and Jacobson, H. I. (1993). Purification of alpha-fetoprotein from human cord serum with demonstration of its antiestrogenic activity. *Biochim Biophys Acta*, *1202*, 135-42.

[30] Bennett, J. A., Semeniuk, D. J., Jacobson, H. I. and Murgita, R. A. (1997). Similarity between natural and recombinant human alpha-fetoprotein as inhibitors of estrogen-dependent breast cancer growth. *Breast Cancer Res. Treat*, *45*, 169-79.

[31] Festin, S. M., Bennett, J. A., Fletcher, P. W., Jacobson, H. I., Shaye, D. D. and Andersen, T. T. (1999). The recombinant third domain of human alpha-fetoprotein retains the antiestrotrophic activity found in the full-length molecule. *Biochim. Biophys. Acta*, *1427*, 307-14.

[32] Jacobson, H. I. and Janerich, D. T. (1989). Pregnancy-altered breast cancer risk: mediated by AFP? In: Mizejewski GJ, Jacobson HI, eds. Biological activities of alpha-fetoprotein. Boca Raton, FL: CRC Press, p. 93-100.

[33] Jacobson, H. I., Thompson, W. D. and Janerich, D. T. (1989). Multiple births and maternal risk of breast cancer. *Am. J. Epidemiol*, *129*, 865-73.

[34] Thompson, W. D., Jacobson, H. I., Negrini, B. and Janerich, D. T. (1989). Hypertension, pregnancy, and risk of breast cancer. *J. Natl. Cancer Inst*, *81*, 1571-4.

[35] Janerich, D. T., Mayne, S. T., Thompson, W. D., Stark, A. D., Fitzgerald, E. F. and Jacobson, H. I. (1990). Familial clustering of neural tube defects and gastric cancer. *Int. J. Epidemiol*, *19*, 516-21.

[36] Ekbom, A., Trichopoulos, D., Adami, H. O., Hsieh, C. C. and Lan, S. J. (1992). Evidence of prenatal influences on breast cancer risk. *Lancet*, *340*, 1015-8.

[37] Nasca, P. C., Weinstein, A., Baptiste, M. and Mahoney, M. (1992). The relation between multiple births and maternal risk of breast cancer. *Am. J. Epidemiol*, *136*, 1316-20.

[38] Dietz, A. T., Newcomb, P. A., Storer, B. E., Longnecker, M. P. and Mittendorf, R. (1995). Multiple births and risk of breast cancer. *Int. J. Cancer*, *62*, 162-4.

[39] Wyshak, G., Honeyman, M. S., Flannery, J. T. and Beck, A. S. (1983). Cancer in mothers of dizygotic twins. *J. Natl. Cancer Inst.*, *70*, 593-9.

[40] Albrektsen, G., Heuch, I. and Kvale, G. (1995). Multiple births, sex of children and subsequent breast-cancer risk for the mothers: a prospective study in Norway. *Int. J. Cancer*, *60*, 341-4.

[41] Brock, D. J. and Sutcliffe, R. G. (1972). Alpha-fetoprotein in the antenatal diagnosis of anencephaly and spina bifida. *Lancet*, *2*, 197-9.

[42] Kelsey, J. L. and Hill, D. L. (1983). Breast and gynecologic cancer epidemiology. In: Mizejewski GJ, Jacobson HI, editors. Biological activities of alpha-fetoprotein. Boca Raton: CRC Press.

[43] Gray, G. E., Henderson, B. E. and Pike, M. C. (1980). Changing ratio of breast cancer incidence rates with age of black females compared with white females in the United States. *J. Natl. Cancer Inst*, *64*, 461-3.

[44] Halmesmaki, E., Autti, I., Granstrom, M. L., Heikinheimo, M., Raivio, K. O. and Ylikorkala, O. (1986). Alpha-fetoprotein, human placental lactogen, and pregnancy-specific beta 1-glycoprotein in pregnant women who drink: relation to fetal alcohol syndrome. *Am J Obstet Gynecol*, *155*, 598-602.

[45] Howe, G., Rohan, T., Decarli, A., Iscovich, J., Kaldor, J., Katsouyanni, K., Marubini, E., Miller, A., Riboli, E., Toniolo, P., et al. (1991). The association between alcohol and breast cancer risk: evidence from the combined analysis of six dietary case-control studies. *Int. J. Cancer*, *47*, 707-10.

[46] Clayton-Hopkins, J. A., Oppezzo, P. and Blasé, A. P. (1982). Maternal serum AFP levels in pregnancy complicated by hypertension. *Prenatal Diagnosis*, *2*, 47-54.

[47] Wald, N., Barker, S., Peto, R., Brock, D. J. and Bonnar, J. (1975). Maternal serum alpha-fetoprotein levels in multiple pregnancy. *Br. Med.J.*, *1*, 651-2.

[48] Richardson, B. E., Hulka, B. S., Peck, J. L., Hughes, C. L., van den Berg, B. J., Christianson, R. E. and Calvin, J. A. (1998). Levels of maternal serum alpha-fetoprotein (AFP) in pregnant women and subsequent breast cancer risk. *Am. J. Epidemiol.*, *148*, 719-27.

[49] Melbye, M., Wohlfahrt, J., Lei, U., Norgaard-Pedersen, B., Mouridsen, H. T., Lambe, M. and Michels, K. B. (2000). Alpha-fetoprotein levels in maternal serum during pregnancy and maternal breast cancer incidence. *J. Natl. Cancer Inst.*, *92*, 1001-5.

[50] Grubbs, C. J., Juliana, M. M., Hill, D. L. and Whitaker, L. M. (1986). Suppression by pregnancy of chemically induced preneoplastic cells of the rat mammary gland. *Anticancer Res.*, *6*, 1395-400.

[51] Lemon, H. M. (1987). Antimammary carcinogenic activity of 17-alpha-ethinyl estriol. *Cancer*, *60*, 2873-81.

[52] Lemon, H. M. (1975). Estriol prevention of mammary carcinoma induced by 7, 12-dimethylbenzanthracene and procarbazine. *Cancer Res.*, *35*, 1341-53.

[53] Grubbs, C. J., Peckham, J. C. and McDonough, K. D. (1983). Effect of ovarian hormones on the induction of 1-methyl-1-nitrosourea-induced mammary cancer. *Carcinogenesis*, *4*, 495-7.

[54] Russo, I. H., Koszalka, M. and Russo, J. (1990). Human chorionic gonadotropin and rat mammary cancer prevention. *J. Natl. Cancer Inst*, *82*, 1286-9.

[55] Jacobson, H. I., Lemanski, N., Agarwal, A., Narendran, A., Turner, K. E., Bennett, J. A. and Andersen, T. T. (2010). A proposed unified mechanism for the reduction of human breast cancer risk by the hormones of pregnancy. *Cancer Prev. Res. (Phila Pa)*, *3*, 212-20.

[56] Mizejewski, G. J., Dias, J. A., Hauer, C. R., Henrikson, K. P. and Gierthy, J. (1996). Alpha-fetoprotein derived synthetic peptides: assay of an estrogen-modifying regulatory segment. *Mol. Cell Endocrinol.*, *118*, 15-23.

[57] Eisele, L. E., Mesfin, F. B., Bennett, J. A., Andersen, T. T., Jacobson, H. I., Vakharia, D. D., MacColl, R. and Mizejewski, G. J. (2001). Studies on analogs of a peptide derived from alpha-fetoprotein having antigrowth properties. *J. Pept. Res.*, *57*, 539-46.

[58] Eisele, L. E., Mesfin, F. B., Bennett, J. A., Andersen, T. T., Jacobson, H. I., Soldwedel, H., MacColl, R. and Mizejewski, G. J. (2001). Studies on a growth-inhibitory peptide derived from alpha-fetoprotein and some analogs. *J. Pept. Res*, *57*, 29-38.

[59] Mesfin, F. B., Bennett, J. A., Jacobson, H. I., Zhu, S. and Andersen, T. T. (2000). Alpha-fetoprotein-derived antiestrotrophic octapeptide. *Biochim. Biophys. Acta*, *1501*, 33-43.

[60] Kirschner, K. N., Lexa, K. W., Salisburg, A. M., Alser, K. A., Joseph, L., Andersen, T. T., Bennett, J. A., Jacobson, H. I. and Shields, G. C. (2007). Computational design and experimental discovery of an antiestrogenic peptide derived from alpha-fetoprotein. *J. Am. Chem. Soc*, *129*, 6263-8.

[61] Mesfin, F. B., Andersen, T. T., Jacobson, H. I., Zhu, S. and Bennett, J. A. (2001). Development of a synthetic cyclized peptide derived from alpha-fetoprotein that prevents the growth of human breast cancer. *J. Pept. Res.*, *58*, 246-56.

[62] DeFreest, L. A., Mesfin, F. B., Joseph, L., McLeod, D. J., Stallmer, A., Reddy, S., Balulad, S. S., Jacobson, H. I., Andersen, T. T. and Bennett, J. A. (2004). Synthetic peptide derived from alpha-fetoprotein inhibits growth of human breast cancer: investigation of the pharmacophore and synthesis optimization. *J. Pept. Res.*, *63*, 409-19.

[63] Joseph, L. C., Bennett, J. A., Kirschner, K. N., Shields, G. C., Hughes, J., Lostritto, N., Jacobson, H. I. and Andersen, T. T. (2009). Antiestrogenic and anticancer activities of peptides derived from the active site of alpha-fetoprotein. *J. Pept. Sci.*, *15*, 319-25.

[64] Andersen, T. T., Georgekutty, J., DeFreest, L. A., Amaratunga, G., Narendran, A., Lemanski, N., Jacobson, H. I. and Bennett, J. A. (2007). An alpha-fetoprotein-derived peptide reduces the uterine hyperplasia and increases the antitumour effect of tamoxifen. *Br. J. Cancer*, *97*, 327-33.

[65] Bennett, J. A., Mesfin, F. B., Andersen, T. T., Gierthy, J. F., Jacobson, H. I. (2002). A peptide derived from alpha-fetoprotein prevents the growth of estrogen-dependent human breast cancers sensitive and resistant to tamoxifen. *Proc. Natl. Acad. Sci. U.S.A*, *99*, 2211-5.

[66] Bennett, J. A., Defreest, L., Anaka, I., Saadati, H., Balulad, S., Jacobson, H. I. and Andersen, T. T. (2006). AFPep: an anti-breast cancer peptide that is orally active. *Breast Cancer Res. Treat.*, *98*, 133-41.

[67] Festin, S. M., Bennett, J. A., Fletcher, P. W., Jacobson, H. I., Shaye, D. D. and Andersen, T. T. (1999). The recombinant third domain of human alpha-fetoprotein retains the antiestrotrophic activity found in the full-length molecule. *Biochim. Biophys. Acta*, *1427*, 307-14.

[68] Parikh, R. R., Gildener-Leapman, N., Narendran, A., Lin, H. Y., Lemanski, N., Bennett, J. A., Jacobson, H. I. and Andersen, T. T. (2005). Prevention of N-methyl-N-nitrosourea-induced breast cancer by alpha-fetoprotein (AFP)-derived peptide, a peptide derived from the active site of AFP. *Clin. Cancer Res.*, *11*, 8512-20.

[69] Li M., Li, H., Li, C., Guo, L., Liu, H., Zhou, S., Liu, X., Chen, Z., Shi, S., Wei, J., et al. (2009). Cytoplasmic alpha-fetoprotein functions as a co-repressor in RA-RAR signaling to promote the growth of human hepatoma Bel 7402 cells. *Cancer Lett.*, *285*, 190-9.

[70] Murgita, R. A. and Tomasi, T. B. Jr. (1975). Suppression of the immune response by alpha-fetoprotein on the primary and secondary antibody response. *J. Exp. Med.*, *141*, 269-86.

[71] Sierralta, W. D., Epunan, M. J., Reyes, J. M., Valladares, L. E., Andersen, T. T., Bennett, J. A., Jacobson, H. I. and Pino, A. M. (2008). A peptide derived from alpha-

fetoprotein inhibits the proliferation induced by estradiol in mammary tumor cells in culture. *Oncol. Rep*, *19*, 229-35.

[72] Bryan, A., Joseph, L., Bennett, J. A., Jacobson, H. I. and Andersen, T. T. (2001). Design and synthesis of biologically active peptides: a 'tail' of amino acids can modulate activity of synthetic cyclic peptides. *Peptides*, *32*, 2504-10.

[73] Vlieghe, P., Lisowski, V., Martinez, J. and Khrestchatisky, M. (2010). Synthetic therapeutic peptides: science and market. *Drug Discov. Today*, *15*, 40-56.

[74] Vogel, V. G., Costantino, J. P., Wickerham, D. L., Cronin, W. M., Cecchini, R. S., Atkins, J. N., Bevers, T. B., Fehrenbacher, L., Pajon, E. R. Jr. and Wade, J. L., III, et al. (2006). Effects of tamoxifen vs raloxifene on the risk of developing invasive breast cancer and other disease outcomes: the NSABP Study of Tamoxifen and Raloxifene (STAR) P-2 trial. *JAMA*, *295*, 2727-41.

[75] Tamoxifen for early breast cancer: an overview of the randomised trials. Early Breast Cancer Trialists' Collaborative Group. (1998). *Lancet*, *351*, 1451-67.

[76] Kato, S., Endoh, H., Masuhiro, Y., Kitamoto, T., Uchiyama, S., Sasaki, H., Masushige, S., Gotoh, Y., Nishida, E., Kawashima, H., et al. (1995). Activation of the estrogen receptor through phosphorylation by mitogen-activated protein kinase. *Science*, *270*, 1491-4.

[77] Anzano, M. A., Byers, S. W., Smith, J. M., Peer, C. W., Mullen, L. T., Brown, C. C., Roberts, A. B. and Sporn, M. B. (1994). Prevention of breast cancer in the rat with 9-cis-retinoic acid as a single agent and in combination with tamoxifen. *Cancer Res.*, *54*, 4614-7.

[78] Li, J. J., Papa, D., Davis, M. F., Weroha, S. J., Aldaz, C. M., El-Bayoumy, K., Ballenger, J., Tawfik, O. and Li, S. A. (2002). Ploidy differences between hormone- and chemical carcinogen-induced rat mammary neoplasms: comparison to invasive human ductal breast cancer. *Mol. Carcinog.*, *33*, 56-65.

[79] Harvell, D. M., Buckles, L. K., Gould, K. A., Pennington, K. L., McComb, R. D. and Shull, J. D. (2003). Rat strain specific attenuation of estrogen action in the anterior pituitary gland by dietary energy restriction. *Endocrine*, *21*, 175-83.

[80] Harvell, D. M., Strecker, T. E., Xie, B., Pennington, K. L., McComb, R. D. and Shull, J. D. (2002). Dietary energy restriction inhibits estrogen-induced mammary, but not pituitary, tumorigenesis in the ACI rat. *Carcinogenesis*, *23*, 161-9.

[81] Harvell, D. M., Strecker, T. E., Tochacek, M., Xie, B., Pennington, K. L., McComb, R. D., Roy, S. K. and Shull, J. D. (2000). Rat strain-specific actions of 17beta-estradiol in the mammary gland: correlation between estrogen-induced lobuloalveolar hyperplasia and susceptibility to estrogen-induced mammary cancers. *Proc. Natl. Acad. Sci. U.S.A*, *97*, 2779-84.

[82] Spady, T. J., Harvell, D. M., Snyder, M. C., Pennington, K. L., McComb, R. D. and Shull, J. D. (1998). Estrogen-induced tumorigenesis in the Copenhagen rat: disparate susceptibilities to development of prolactin-producing pituitary tumors and mammary carcinomas. *Cancer Lett.*, *124*, 95-103.

[83] Li, S. A., Weroha, S. J., Tawfik, O. and Li, J. J. (2002). Prevention of solely estrogen-induced mammary tumors in female aci rats by tamoxifen: evidence for estrogen receptor mediation. *J. Endocrinol.*, *175*, 297-305.

[84] Weroha, S. J., Li, S. A., Tawfik, O. and Li, J. J. (2006). Overexpression of cyclins D1 and D3 during estrogen-induced breast oncogenesis in female ACI rats. *Carcinogenesis*, *27*, 491-8.

[85] Shull, J. D., Pennington, K. L., Reindl, T. M., Snyder, M. C., Strecker, T. E., Spady, T. J., Tochacek, M. and McComb, R. D. (2001). Susceptibility to estrogen-induced mammary cancer segregates as an incompletely dominant phenotype in reciprocal crosses between the ACI and Copenhagen rat strains. *Endocrinology*, *142*, 5124-30.

[86] Shull, J. D., Spady, T. J., Snyder, M. C., Johansson, S. L. and Pennington, K. L. (1997). Ovary-intact, but not ovariectomized female ACI rats treated with 17beta-estradiol rapidly develop mammary carcinoma. *Carcinogenesis*, *18*, 1595-601.

[87] Tower, A. M., Trinward, A., Lee, K., Joseph, L., Jacobson, H. I., Bennett, J. A. and Andersen, T. T. (2009). AFPep, a novel drug for the prevention and treatment of breast cancer, does not disrupt the estrous cycle or fertility in rats. *Oncol. Rep.*, *22*, 49-56.

[88] Aussel, C., Lafaurie, M., Masseyeff, R. and Stora, C. (1981). *In vivo* regulation of ovarian activity by alpha-fetoprotein. *Steroids*, *38*, 195-204.

[89] Sparks, R. L. and Grubbs, B. G. (1979). Suppression of estrogen action on mouse vagina by serum containing alpha-fetorprotein. *Anat. Rec*, *193*, 689-90.

In: Alpha-Fetoprotein: Functions and Clinical Applications ISBN: 978-1-63484-875-6
Editors: Nisha Lakhi and Michael Moretti

Chapter 17

RADIOLABELLING AND BIODISTRIBUTION STUDIES OF GIP PEPTIDES DERIVED FROM ALPHA-FETOPROTEIN

Michał Maurin[1], Piotr Garnuszek[1,*], Urszula Karczmarczyk[1], Piotr A. Baran[2], Ewa Laszuk[1] and Gerald J. Mizejewski[3]

[1]National Centre for Nuclear Research Radioisotope
Centre POLATOM, Otwock, Poland
[2]Institute of Biotechnology and Antibiotics, Warsaw, Poland
[3]Wadsworth Center, Albany, US

ABSTRACT

Human alpha-fetoprotein (hAFP), a 70 kDa fetal origin glycoprotein, is a major serum constituent throughout pregnancy and has long been employed as clinical marker for both fetal defect and tumor presence. The 34-amino acid fragment 447-480 of human AFP (P149) located on the third domain was derived and shown to retain the anti-estrogenic activity of parent AFP glycoprotein. P149 exhibited the anti-breast cancer activity and the growth inhibition in non-tumor mouse uterine assay. It probably plays a crucial role in binding AFP molecule to rAFP. This purified linear synthetic 34-mer segment known as the "growth inhibitory peptide" (GIP) has several oligomeric forms with diverse aggregation behavior. The dominant trimeric forms were found to be suppressive in assays of estrogen-induced growth. The shorter segment of the original 34-mer was found to display growth activities, with the middle segment (P149b) showed hydrophobic dye-binding properties.

The aim of present study was to develop 99mTc and 177Lu radiolabelling method of cyclic analogues of P149, i.e.: 11-aminoacid cGIP11 (cyclo-EKTPINPGVGQ) and 34-amino acid cGIP34 (cyclo-LSEDKLLACGEGAADIIIGHLCIRHRMTPVNPGV), which were conjugated with bifunctional chelators HYNIC and DOTA-NHS. The HYNIC conjugates were labelled with 99mTc using stannous chloride in presence of tricine as co-ligand. DOTA-cGIP34 was labelled with 177Lu in a presence of ascorbic acid. In addition, cGIP34 was radioiodinated (125I) using Iodo-gen method. Physiological

* Corresponding author: Piotr Garnuszek: Piotr.Garnuszek@polatom.pl.

distribution of the radioactive preparations was investigated in normal rats and in tumor-bearing mice. HPLC analysis revealed that 99mTc-HYNIC-cGIP11was obtained with 35% yield and 99mTc-HYNIC-cGIP34 with yield >90%. After the purification using AmiconUltra 3kDa centrifugal filters the complexes were radiochemically pure (>95%) and stable over 24h.

In vivo studies of the 99mTc- and 125I-labelled cGIP34 peptide have shown significant differences of biodistribution comparing to the previously studied linear P149-Q[125I]Y, 99mTc-P149-QY and 99mTc-HYNIC-P149-QY analogues. 125I-cGIP34 was rapidly excreted with urine (33%%ID and 90%ID after 2 and 24h, respectively) with no other critical organs. In comparison, 99mTc-HYNIC-cGIP34 was excreted with urine much slower (only 22%ID after 24h) and accumulated in kidneys (41% and 33%ID after 2 and 24h, respectively). 99mTc-cGIP11 was rapidly excreted with urine (over 2.5 and 3 times faster after 2 and 24 hours respectively than 99mTc-cGIP34) with much lover accumulation in kidney (15% and 10%ID after 2 and 24h respectively). However, both 99mTc-labelled peptides revealed low accumulation in tumor. Interestingly, for 99mTc-HYNIC-cGIP34 the increase of Tumor/Blood ratio has been observed at 24 h p.i.. This could be a proof for active targeting of 99mTc-HYNIC-cGIP34.

The results of our study still indicate the need for improvement in design of GIP analogues as radionuclide carriers for AFP receptor targeting.

INTRODUCTION

Human alpha-fetoprotein (hAFP), a 70 kDa fetal origin glycoprotein, is a major serum constituent throughout pregnancy and has long been employed as clinical marker for both fetal defect and tumour presence [1-2]. The precise biological role of AFP has not been yet determined but it has been demonstrated that it function as a transport protein for various ligands such as: retinoids, dyes, heavy metals, fatty acids and drugs [1-3]. It was postulated that growth of some tumour cells occurs, among others, through the interaction of alpha-fetoprotein with cell surface receptors. Later studies have documented that alpha-fetoprotein binds to a specific receptor (rAFP) on human breast cancer cells, and in parallel it was found that expression of rAFP on human unproliferating cells was very low [4-9]. In addition, the amount of rAFP molecules on lymphoma cells was about 10-times higher in comparison to normal proliferating T lymphocytes [10, 11]. The AFP binding proteins were detectable in both soluble and membrane bound forms revealing different molecular masses [11-13].

Alpha-fetoprotein has been reported to possess a growth inhibitory motif as an occult epitope in the compactly folded circulating form of protein. Intermediate unfolded forms of hAFP molecule induced by stress, shock, and high ligand concentrations have revealed the presence of an encrypted growth-suppressive segment on the third domain of hAFP. A purified linear synthetic 34-mer segment termed the “growth inhibitory peptide” (GIP) exhibits various oligomeric forms with complex aggregation behaviors, in with dominant trimeric forms were found to be suppressive in assays of estrogen-induced growth. The smaller segment of the original 34-mer was found to display growth activities of their own, with the middle segment (P149b) also showing hydrophobic dye-binding properties. Studies on amino acid sequence identity revealed that the GIP sequences display identity/similarity matches to both cytoplasmic and nucleus-cytosceleton-associated proteins. Experimental findings evidences that peptide is capable of modulating tubulin polymerization, cell shape, cell-surface aggregation phenomena reminiscent of a microtubule-associated protein. Because of involvement of GIP in experimental models of estrogen receptor/cytoskeleton, a mechanism of action is forwarded in which the linear GIP is proposed to be a G-coupled receptor binding ligand that is translocated across the plasma membrane via receptor-mediated endocytosis [14].

Recent studies on AFP have shown that in stress conditions conformational change of AFP take place, resulting in revealing the growth inhibition properties of the protein. The 34-aminoacid fragment 447-480 of human AFP (P149) located on the third domain was derived, and it was shown that the peptide retained the anti-estrogenic activity of parent AFP glycoprotein [15, 18]. It has been shown that P149 peptide possesses an anti-breast cancer activity and growth inhibition activity in non-tumour mouse uterine assay [15-19]. It probably plays a crucial role in binding AFP molecule to rAFP [16]. Our previous study [20] showed that the ^{99m}Tc-labelled P149-QY revealed preferable T/M (tumour/muscle) and T/B (tumour/blood) ratio in mammary adenocarcinoma bearing mice. Unfortunately, ^{99m}Tc-labelled P149 analogue accumulated in high levels in kidneys, liver and intestines.

The aim of present study was to develope the method for the ^{99m}Tc labelling of cyclic analogues of P149 named as cGIP34 and its fragment cGIP11. The cGIP34 was already evaluated as growth inhibitory peptide [21]. cGIP11 possesses the same active sequence. Both peptides are cyclic and may be less potent for formation of inactive aggregates that also accumulate in critical organs such as kidneys or liver. In the present study the diagnostic properties of the developed ^{99m}Tc-labelled preparations of cGIP peptides were additionally verified by biodistribution studies in normal and tumor-bearing animals and in-vitro using human breast cancer cell line MCF-7. In the present study the possibility of ^{177}Lu-labelling of the cGIP peptides was also investigated.

MATERIALS AND METHODS

Chemicals

The cyclic AFP fragments: 11 amino acid cGIP11 (cyclo-EKTPINPGVGQ) and 34 amino acid cGIP34 (cyclo-LSEDKLLACGEGAADIIIGHLCIRHRMTPVNPGV) were synthesized by Professor G.J. Mizejewski's team. HYNIC-NHS was synthesized on site at NCBJ OR POLATOM. DOTA-NHS was purchased from Macrocyclics, Inc. Technetium-99m in the form of eluate of sodium pertechnetate was obtained from ^{99}Mo/^{99m}Tc generator (POLGENTEC, NCBJ OR POLATOM, Świerk, Poland). Sodium iodide-125 was purchased from Perkin Elmer.

The other reagents used were commercially available and were of p.a. or HPLC grade.

^{99m}Tc-Labelling cGIP34 Peptide

A direct ^{99m}Tc-labelling of the cGIP34 conjugate was carried out in the presence of stannous chloride dihydrate used for ^{99m}Tc (VII) reduction to ^{99m}Tc (IV). To a solution containing ca. 50 μg of the peptide dissolved in water, 100 μL of 0.1M phosphate buffer pH 6.8 and 500 μL of $^{99m}TcO_4^-$ (ca. 700 MBq) was added followed by addition of $SnCl_2xH_2O$ in

0.05M HCl (30 μg). Reaction mixtures were incubated for 30min at room temperature. Labelling yield and radiochemical purity was checked by TLC and HPLC analysis.

Conjugation of cGIP11 and cGIP34 Peptides with HYNIC

Conjugation of the AFP fragments with HYNIC-NHS was conducted according to method described by Renen et al. [22]. Briefly the HYNIC derivatives of the GIP peptides were prepared by reaction of 2.5 molar excess of HYNIC-NHS dissolved in DMF with 0.2 mg of the GIP peptide dissolved in water. pH of reaction mixture was adjusted to 8.2 by addition of 50μl of 1M sodium carbonate. After 20 min incubation at RT, GIP-HYNIC conjugates were purified: cGIP34 conjugate - by 4-times filtration on AmiconUltra 3kDa centrifugal filters; cGIP11 by osmosis on GE MiniDialysis kit 1kDa. The HYNIC to peptide conjugation rate was checked by MS analysis.

^{99m}Tc-Labelling of HYNIC GIP11 and HYNIC cGIP34 Conjugates

^{99m}Tc-labellings of the HYNIC conjugates were carried out using stannous chloride dihydrate for ^{99m}Tc (VII) reduction to ^{99m}Tc (IV) in presence of tricine as co-ligand. The labelling procedure was optimized with regard to amounts of the peptide, stannous chloride and tricine. Generally, to a solution containing from 20 μg to 50 μg (up to 50μL) of the peptide dissolved in water, 100 μL of 0.1M phosphate buffer pH 6.8, tricine (from 2 to 4mg) and 300-500 μL of $^{99m}TcO_4^-$ (ca. 300-800 MBq) were added followed by addition of $SnCl_2$ in 0.05M HCl (from 25 to 40 μg). The reaction mixtures were incubated from 15 min to 30min at room temperature. ^{99m}Tc-labelled cGIP34 conjugate was purified by 4-times filtration on centrifugal filters (Amicon Ultra 3kDa). ^{99m}Tc-labelled cGIP11 conjugate was purified by HPLC (Jupiter Proteo 4u). Labelling yield and radiochemical purity were checked by TLC and HPLC analysis.

Conjugation of DOTA-NHS with cGIP34 Peptide

Conjugation of cGIP34 with DOTA-NHS was done by incubation of 200μg of the peptide with 10-times molar excess of DOTA-NHS. pH of the reaction mixture was adjusted to 8.2 by addition of 1M Na_2CO_3. After 30 min incubation at room temperature, the reaction mixture was then purified by 3-times centrifugation using AmiconUltra 3kDa filters.

Labelling of DOTA-cGIP34 Conjugate with ^{177}Lu

^{177}Lu-labelling of DOTA-cGIP34 conjugate was carried out by addition of 1.1 GBq of ^{177}Lu and 200 μL of ascorbic buffer (50 mg/mL pH 4.0) to 100μg (1mg/mL) of conjugate. The reaction mixture was incubated for 40 min at 100°C. After incubation the reaction mixture was then centrifuged 3-times on AmiconUltra 3 kDa filters. Determination of labelling yield and purification effectiveness was monitored by HPLC method followed by addition of 10mg of DTPA for chelation of the unbound ^{177}Lu.

Radioiodination of P149-QY Peptide

The radioiodination of hAFP and cGIP34 with I-125 was carried out using the Iodo-gen method according to the previously described procedure [23]. Radioiodinated peptides were purified by 4-time filtration on AmiconUltra 3 kDa centrifugal filters. The radiochemical purity (RCP) of radioiodinated peptides was determined by paper electrophoresis.

Radiochemical Analysis

The radiochemical purity of the obtained preparations was determined by RP-HPLC, TLC and paper electrophoresis methods. HPLC analysis was performed using 150 x 4.6 mm Jupiter Proteo 4μm column (Phenomenex) and Shimadzu system equipped with a dual wavelength spectrophotometer set at 240 nm and 280 nm and on-line radioactivity detector, connected in series. The chromatographic conditions were as follow: flow rate 0.7 mL/min; mobile phase A - 0.1% TFA in H_2O, mobile phase B - 0.1% TFA in ACN; gradient elution: from 5% to 80% B in 20min; oven temp. 30°C; injection 20 μL.

TLC analyses were performed on ITLC SG plates using two developing systems: MEK - for detection of ^{99m}Tc-pertechnetate (R_f = 1), and the mixture of EtOH, NH_4OH (25%) and H_2O (5:2:1 *v/v/v*) - for determination of ^{99m}Tc-Sn colloid (R_f = 0). The R_f values for the labelled peptides in these two systems were 0 and 1, respectively.

Electrophoresis was carried out on 15 cm strips of Whatman 1 paper developed in 0.025M $NaClO_4$; HV = 500 V; analysis time = 30 min.

The distributions of radioactivity on the radiochromatograms were determined by Radio-TLC Scanner MINI-SCAN (*BIOSCAN Washington, D.C.*) and Laura radio-chromatographic software (LabLogic System Limited, Sheffield, England).

MS Analysis

The mass spectra of the cGIP11 and cGIP34 were acquired using a 4800 MALDI TOF analyser obtained from Applied Biosystems (Foster city, CA). The mass spectrometer was fitted with a Nd: YAG laser (355 nm) to ionize samples. The laser-firing rate was 200 Hz. Before analysis the peptides were diluted in water (e.g., 1 μg/μL) and 1 μl of the sample was spotted on the MALDI target plate (stainless steel 192-well). The sample was allowed to air dry and 1 μl of matrix (α-cyano-4-hydroxycinnamic acid CHCA) was spotted (5 mg/mL of 50% ACN containing 0.1% TFA) allowed it to air dry. The MALDI plate was inserted in to the mass spectrometer to acquire the spectra.

Biodistribution Studies of Radiolabelled cGIP11 and cGIP34 Peptides in Normal Rats

Physiological distributions of the radioactive preparations were studied in normal rats (Wistar, male, weighing 180-220 g). The radiolabelled peptides were administered intravenously at doses of 2.0 - 10.0 MBq in 0.2 mL single injections (from 1.0 to 10 μg of the peptide per rat). During the experiments the animals were housed in metabolic cages, in which urine was collected. The animals were anaesthetised and sacrificed 2 h and 24h *p.i.v.*, and the radioactivity of the collected blood pool, urine, and samples of the weighted tissues were measured using a gamma counter supplied with an adapter for whole-body measurement. The results were calculated as a percentage of the dose in organs (%ID) or a percentage of the dose per gram of tissue (%ID/g).

All animal experiments were performed after approval and within the guidelines of The IVth Local Animal Ethics Committee in Warsaw, and were carried out in accordance with the principles of good laboratory practice (GLP).

Biodistribution Studies of ^{99m}Tc-Radiolabelled cGIP11 and cGIP34 Peptides in Tumour-Bearing Mice

For biodistribution study C3H female mice (25-30g) with subcutaneously implemented the breast cancer cells (derived from spontaneously developed mammary adenocarcinoma in C3H mice) were used. The diameter of inducted tumour was ca 1.5 – 2cm. The ^{99m}Tc-labelled cGIP11 and cGIP34 peptides were administered intravenously at doses of 5.5MBq (1.75 μg) and 10MBq (1.16μg) in 0.1 mL single injections per animal, respectively. The animals were anaesthetised and sacrificed at 0.5h, 2 h and 24h *p.i.v.*, and the radioactivity of the collected blood pool, urine, and samples of the weighted tissues were measured using a gamma counter supplied with an adapter for whole-body measurement. The results were calculated as a percentage of the dose in organs (%ID) or a percentage of the dose per gram of tissue (%ID/g).

All animal experiments were performed after approval and within the guidelines of The IVth Local Animal Ethics Committee in Warsaw, and were carried out in accordance with the principles of good laboratory practice (GLP).

In Vitro Study

The preliminary studies on bioactivity of the ^{99m}Tc-HYNIC-cGIP, ^{177}Lu-DOTA-cGIP34 conjugates and ^{125}I-cGIP34 were conducted on the human breast cancer cell line MCF-7. Binding of the radio-labelled peptides to the MCF-7 cells was studied by incubation of ca. 400 000 cells with 200 kcpm of radiolabelled peptides for 0.5, 2 and 4 hours. Radioactivity bound to the cells was measured using the automatic well gamma counter (LKB Wizard).

RESULTS AND DISCUSSION

Conjugation of cGIP11 and cGIP34 Peptides with HYNIC

The conducted MS study of GIP-HYNIC conjugates gave a piece of information about conjugation yield and enabled a tentative characteristic of the conjugates.

For cGIP11, the following monoisotopic signals were observed: m/z 1153.6 derived from the ion $[M + H]^+$, sodium adduct m/z 1175.6 $[M + Na]^+$, potassium adduct m/z 1191.6 $[M + K]^+$ and dimmer m/z 2306.2 $[2M + H]^+$, respectively (Figure 1). Relative peak area of above signals was 100:10:4:1.

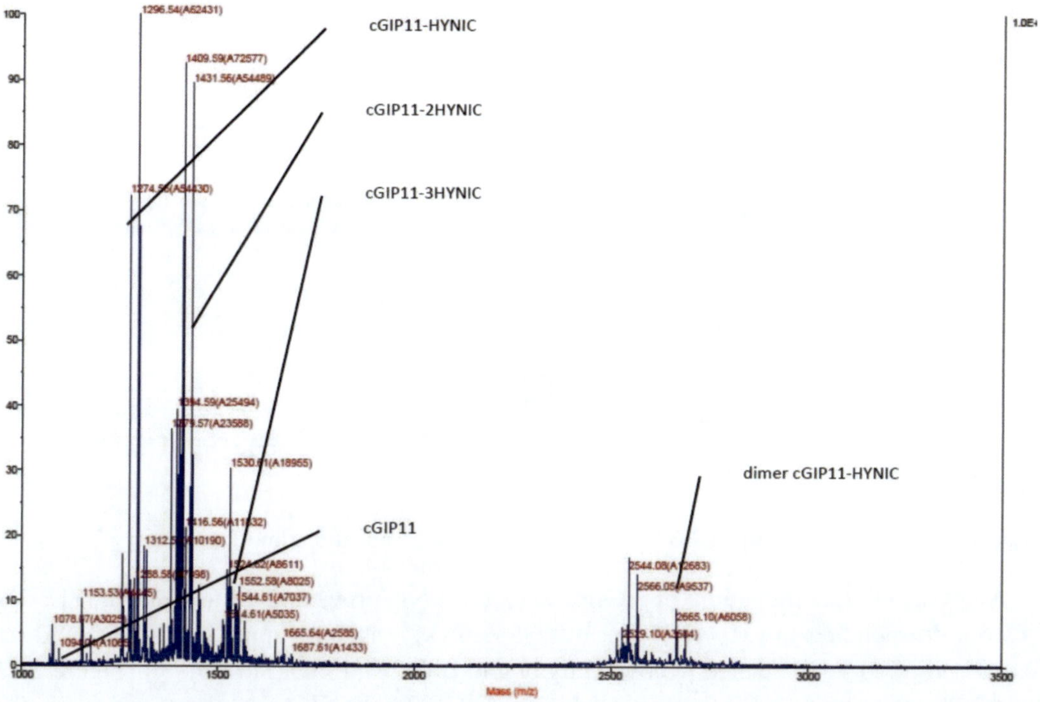

Figure 1. MS spectrum of cGIP11-HYNIC, conjugation 3h (after purification).

Spectrum derived from reaction products of cGIP11 with HYNIC represents signals corresponding to: covalently bound HYNIC (135Da) to cGIP11, m/z 1288.6 $[(M + 135) + H]^+$, its sodium adduct m/z 1310.5 $[(M +135) + Na]^+$, its product without water m/z 1270.6 $[(M + 135\text{-}H_2O) + H]^+$ and without ammonia m/z 1271.6 $[(M+135\text{-}NH_3)+H]^+$, and also product dimmer m/z 2441.1 $[(2M + 135) + H]^+$, its dimmer deprived of the water molecule m/z 2542.2 $[2(M + 135\text{-}NH_3) + H]^+$, and dimmer sodium adduct deprived of ammonia m/z 2564.2 $[2(M + 135\text{-}NH_3) + Na]^+$. On the basis of spectrum analysis it can be assumed that HYNIC-cGIP11 conjugate forms with high yield of 66%, however 23% of conjugate was in the form of dimmers. Increasing conjugation time from 20min to 3h resulted in formation of new aggregated forms of peptide visible on the spectrum as strong non-monoisotopic signals with average mass that could correspond to conjugates of cGIP11 with 2 and 3 bonded

HYNIC molecules. It can be assumed that percentage of cGIP11 conjugated with 2 HYNIC molecules was 43%, and 6% with 3 HYNIC molecules.

For cGIP34 peptide, only one monoisotopic signal was observed: m/z 3569.6 corresponding to $[M + H]^+$ ion (Figure 2).

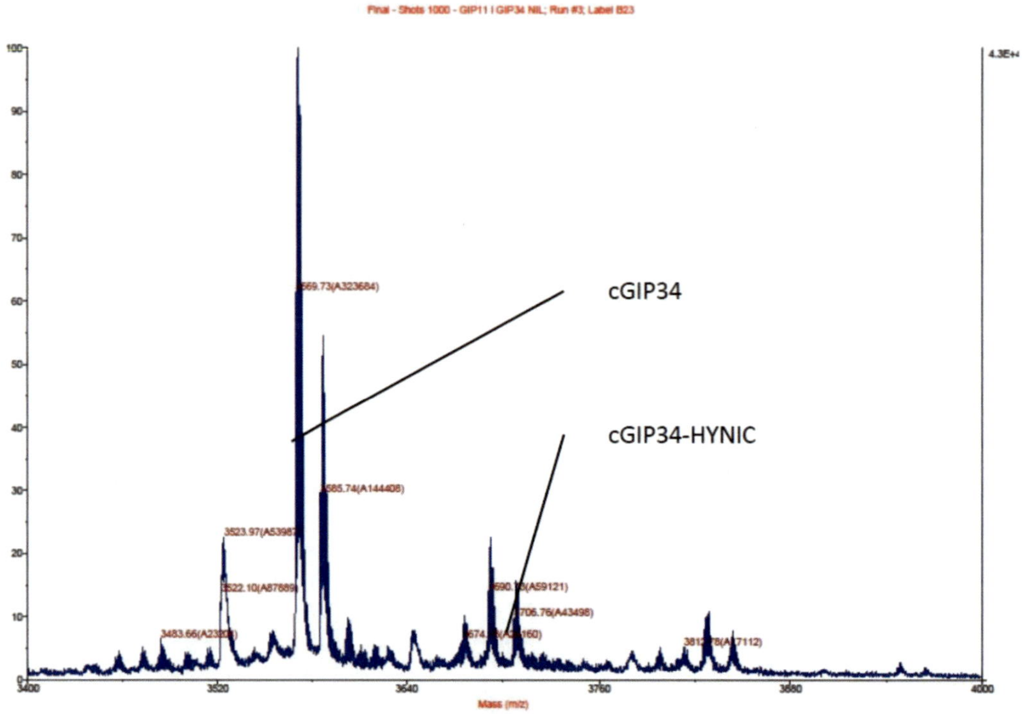

Figure 2. MS spectrum of cGIP34-HYNIC after purification, conjugation time 3h.

Analysis after conjugation of cGIP34 with HYNIC revealed following signals: m/z 3569.6 corresponding to $[M + H]^+$ ion, m/z 3585.6 $[M + NH_3]^+$ - ammonia adduct and m/z 3704.8 $[(M+135) + H]^+$ related to covalently bound HYNIC (135Da) to cGIP34 (Figure 2). In case of 20 min conjugation, the overall yield of cGIP34-HYNIC conjugate was only 10%. Unfortunately increase of conjugation time resulted in increase of yield to 15% only with simultaneous formation of some kind of aggregate (~5%) that was visible on mass spectrum as non-monoisotopic signal with an average mass corresponding to cGIP34 conjugated with 2 HYNIC molecules.

^{99m}Tc-Labelling HYNIC-Conjugates of cGIP11 and cGIP34

Study on the ^{99m}Tc radiolabelling of cGIP11 and cGIP34 peptides via HYNIC resulted in optimisation of the radiolabelling conditions. The optimum conditions for radiolabelling were 20-50 μg of the peptide-HYNIC conjugate, 2mg of tricine and 30 μg of $SnCl_2$ x $2H_2O$. For the both of peptides the optimum pH for radiolabelling was ranging from 6 to 7. HPLC analysis revealed that radiolabelling in this conditions resulted in ^{99m}Tc-HYNIC-cGIP11 with 35% yield and ^{99m}Tc-HYNIC-cGIP34 with yield >90%. TLC analyses of the radiolabelled peptides filtrated through 0.22μm membrane have not shown the existence of the Tc-Sn colloid impurities. Purification of the labelled conjugates resulted in obtainment of radiochemically pure (>95%) ^{99m}Tc-labelled peptides, which were stable over 24h (Figure 3). A low radiolabelling yield of HYNIC-cGIP11 might be due to existence of high percentage of multi HYNIC forms of the conjugate and observed dimeric forms of cGIP11-HYNIC-cGIP11, where HYNIC may be inaccessible for ^{99m}Tc ion. The furher changes of the labelling conditions resulted in a slight increase of the labelling yield while at the same time affected the stability of the radiolabelled conjugate.

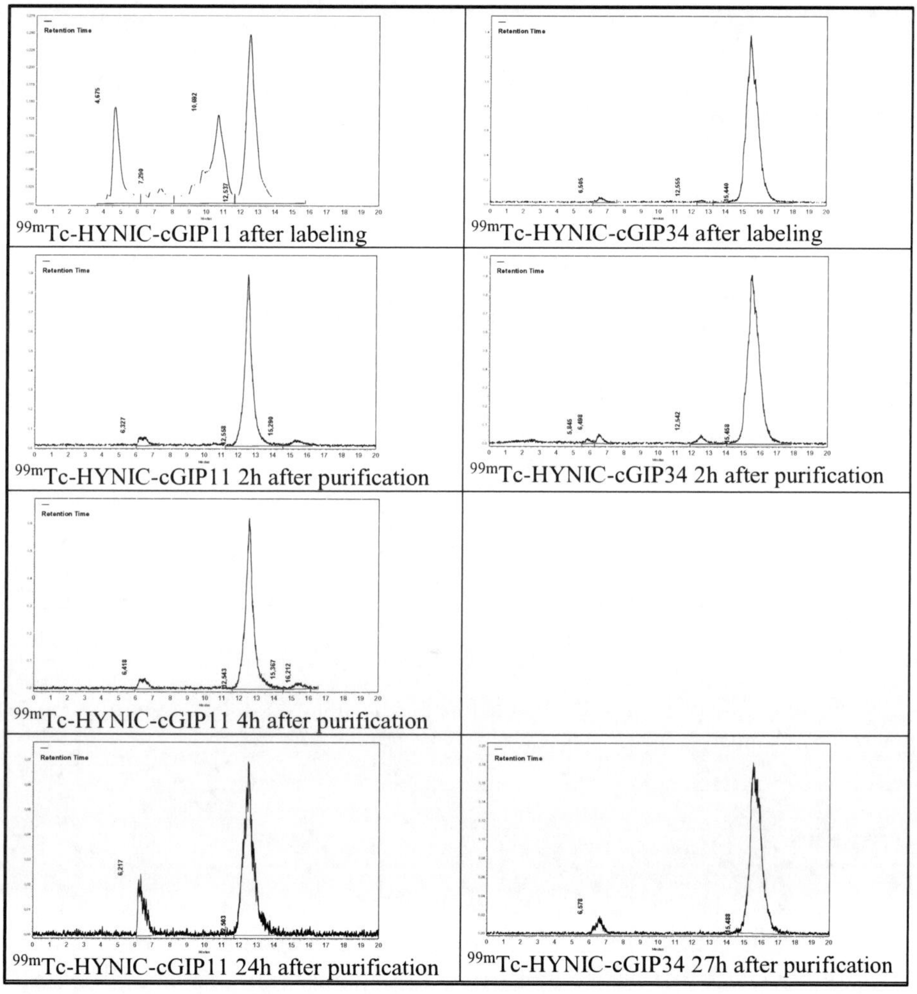

Figure 3. HPLC Radiochromatograms of ^{99m}Tc labelled cGIP-HYNIC. Jupiter Proteo 4μ, A: H_2O/0. 1%TFA, B: AcN/0.1%TFA gradient 5%B to 80%B in 20 min, 0.7ml/min.

The hypothesis of potential second mechanism of ^{99m}Tc chelation by sulfhydryl groups of cGIP34 has been verified by direct radiolabelling of cGIP34 peptide in various conditions. The results of this evidenced that a direct radiolabelling of this cyclised peptide did not occur.

Conjugation of cGIP34 with DOTA and Radiolabelling with 177Lu

This stage of study was focused on checking the possibility of conjugation of the peptide with DOTA and effectives of the ^{177}Lu radiolabelling. In case of the obtained conjugate (DOTA/peptide reaction molar ratio = 10), ^{177}Lu-DOTA-cGIP34 labelling yield monitored by HPLC was only 15%. Purification of the preparation resulted in the labelled peptide of 75% purity, which was stable within at least 27h. Probably the low labelling yield was due to insufficient amount of the bounded DOTA to cGIP34 peptide. Constant radiochemical purity (75%) observed, indicated that there is still possibility of additional purification of this radiopeptide.

Radioiodination of cGIP34

Radioiodination of cGIP34 was performed to obtain potentially unchanged radioactive peptide (with insignificant structural change comparing to the parent peptide molecule) that can be used for comparative study of the bio-activity of the radiolabelled DOTA and HYNIC conjugates. The yield of radio-iodination was 66%, and the purification by filtration resulted in the peptide of 98% radiochemical purity (Figure 4).

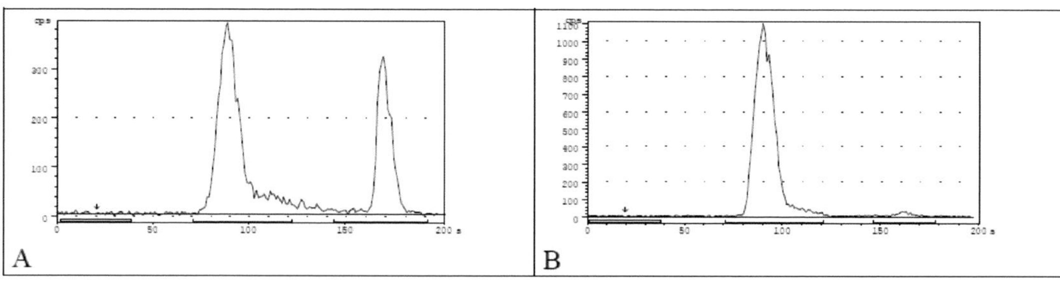

Figure 4. Radiochromatograms of 125I-cGIP34 before purification (A), and after purification (B). Paper electrophoresis on Whatman 1, 0.025 M $NaClO_4$, 500 V, 30 min.

In-Vivo Studies

In vivo studies of the ^{99m}Tc- and ^{125}I-labelled cGIP34 peptide have shown significant differences of biodistribution (Table 1) comparing to the previously studied P149-Q[^{125}I]Y [23], ^{99m}Tc-P149-QY and ^{99m}Tc-HYNIC-P149-QY analogues [20] (Table 2). ^{125}I-cGIP34 was rapidly excreted with urine (33%%ID and 90%ID after 2 and 24h, respectively) with no detectable critical organs. In comparison, ^{99m}Tc-HYNIC-cGIP34 was excreted with urine much slower (only 22%ID after 24h), and accumulated in kidneys (41% and 33%ID after 2 and 24h, respectively). Still, when compared with the previously studied linear P149QY peptide (analogue of linear GIP34) it exhibited much lower accumulation in organs such as kidney, liver and intestines, what may suggest a higher stability *in vivo*.

Table 1. Biodistribution of ^{99m}Tc-HYNIC-cGIP34 and ^{125}I-GIP34 in normal Wistar rats, 2h and 24h p.i.v. administration

Tissue	^{99m}Tc-HYNIC-cGIP34 2 h; n=8		^{99m}Tc-HYNIC-cGIP34 24 h; n=4		^{125}I-cGIP34 2 h; n=4		^{125}I-cGIP34 24 h; n=4	
	%ID/g	SD	%ID/g	SD	%ID/g	SD	%ID/g	SD
blood	0.06%	0.02%	0.00%	0.00%	0.37%	0.03%	0.04%	0.02%
thyroid gland	0.04%	0.01%	0.00%	0.00%	0.31%	0.07%	0.03%	0.01%
lungs	0.73%	0.77%	0.50%	0.26%	0.54%	0.09%	0.09%	0.05%
liver	1.02%	0.28%	0.69%	0.07%	0.39%	0.05%	0.05%	0.01%
spleen	0.30%	0.11%	0.12%	0.05%	0.34%	0.05%	0.04%	0.01%
pancreas	0.03%	0.01%	0.01%	0.01%	1.51%	0.28%	0.03%	0.01%
kidneys	41.96%	8.52%	32.28%	0.98%	2.41%	0.50%	0.34%	0.01%
intestines	0.09%	0.10%	0.08%	0.05%	0.40%	0.06%	0.15%	0.06%
stomach	0.07%	0.03%	0.01%	0.01%	1.07%	0.28%	0.09%	0.06%
muscle	0.02%	0.01%	0.00%	0.00%	0.49%	0.08%	0.09%	0.12%
urine %ID	11.64%	1.96%	20.44%	0.44%	33.30%	8.43%	90.09%	2.07%

Table 2. Biodistribution of the radiolabelled P149-QY peptide in tumour-bearing mice (2 h p.i.v.; %ID/g)

Tissue	P149-Q[^{131}I]Y $n = 6$	^{99m}Tc-P149-QY $n = 5$	^{99m}Tc-HYNIC- P149-QY (SH blocked) $n = 5$
blood 1ml	4.12 ± 0.43	2.25 ± 0,43	1.44 ± 0.12
thyroid gland	9.16 ± 3.84	1.19 ± 0.44	1.33 ± 0.26
lung	2.02 ± 0.85	1.75 ± 0.43	1.11 ± 0.08
liver	1.67 ± 0.15	3.36 ± 1.21	4.40 ± 0.24
kidney	13.83 ± 3.56	72.16 ± 3.35	80.76 ± 15.26
stomach	23.43 ± 7.13	3.83 ± 1.19	1.29 ± 0.32
intestines	3.62 ± 0.51	2.64 ± 0.54	2.55 ± 0.05
tumour	2.55 ± 0.52	0.66 ± 0.16	1.27 ± 0.26
muscle	1.22 ± 0.34	0.37 ± 0.16	0.58 ± 0.08

Biodistribution studies of ^{99m}Tc-labelled cGIP11 showed that the radioactive peptide was rapidly excreted with urine (over 2.5 and 3 times faster than the radioactive cGIP34 analogue after 2 and 24 hours, respectively) and exhibited much lover accumulation in kidneys (15% and 10%ID after 2 and 24h, respectively) (Table 3). The low accumulation in thyroid and gastrointestinal tract suggests its stability in vivo. The interesting differences were observed in physiological distribution of ^{99m}Tc labelled cGIP11 peptides purified in different ways. The peptide purified by HPLC exhibited lower accumulation in organs and faster excretion with urine than the peptide purified by filtration using centrifugal AmiconUltra filters. This observation indicates that in case of the 99m-Tc-labelled peptide after filtration, the aggregation or lose of monomeric form occur, due to concentration of peptide during centrifugation. The cGIP34 radiolabelled aggregates may reveal slower excretion from organism and higher accumulation in organs such as liver, lungs or intestines.

Table 3. Biodistribution of ^{99m}Tc-HYNIC-cGIP11 in normal Wistar rats, 2h and 24h p.i.v. administration

Tissue	Centrifugal purification				HPLC purification			
	^{99m}Tc-HYNIC-cGIP11 2 h; n = 4		^{99m}Tc-HYNIC-cGIP11 24 h; n = 4		^{99m}Tc-HYNIC-cGIP11 2 h; n = 4		^{99m}Tc-HYNIC-cGIP11 24 h; n = 3	
	%ID/g	SD	%ID/g	SD	%ID/g	SD	%ID/g	SD
blood	0.41%	0.07%	0.13%	0.02%	0.07%	0.01%	0.02%	0.00%
thyroid gland	0.21%	0.01%	0.15%	0.02%	0.05%	0.00%	0.06%	0.01%
lung	0.34%	0.05%	0.19%	0.03%	0.06%	0.01%	0.06%	0.02%
liver	0.79%	0.08%	0.52%	0.05%	0.06%	0.01%	0.05%	0.01%
spleen	0.41%	0.08%	0.33%	0.12%	0.04%	0.00%	0.11%	0.07%
pancreas	0.19%	0.03%	0.18%	0.08%	0.03%	0.01%	0.05%	0.02%
kidney	18.01%	1.93%	13.36%	0.76%	15.90%	2.20%	10.38%	1.67%
intestines	0.19%	0.03%	0.32%	0.18%	0.05%	0.01%	0.08%	0.02%
stomach	0.33%	0.04%	0.16%	0.05%	0.14%	0.03%	0.04%	0.00%
muscle	0.10%	0.01%	0.08%	0.01%	0.02%	0.00%	0.03%	0.02%
urine %ID	36.43%	7.69%	54.68%	1.74%	64.34%	4.30%	76.02%	4.55%

The study in the tumour model showed comparable biodistribution route (Table 4) as that observed in healthy rats (Table 1). Both peptides exhibited low accumulation in tumour tissue. The calculated T/M and T/B ratios are presented in Figure 5. Interesting result has been observed 24h pi in case of cGIP34 peptide for which the increase of T/B ratio has been observed. This observation could be a proof for active targeting of ^{99m}Tc-HYNIC-cGIP34 peptide. In case of ^{99m}Tc-labelled cGIP11 the increase of the T/B and T/M over time has not been observed.

Table 4. Biodistribution of ^{99m}Tc-HYNIC-cGIP11 and ^{99m}Tc-HYNIC-cGIP34 in C3H mice with inducted breast cancer, 0.5h, 2h and 24h p.i.v. administration (n = 4)

	^{99m}TcHYNIC-cGIP11					
	0.5h		2h		24h	
tissue	**%ID/g**	**SD**	**%ID/g**	**SD**	**%ID/g**	**SD**
blood	5.98%	2.51%	2.71%	0.15%	0.63%	0.14%
thyroid gland	4.50%	3.50%	3.01%	0.65%	2.92%	4.08%
lung	3.79%	2.37%	1.68%	0.19%	0.62%	0.12%
liver	2.94%	2.08%	1.56%	0.12%	0.99%	0.10%
spleen	1.58%	1.14%	0.79%	0.09%	0.45%	0.08%
pancreas	1.71%	1.22%	0.84%	0.06%	0.61%	0.10%
kidney	19.58%	15.72%	10.61%	2.11%	5.38%	1.14%
intestines	1.50%	0.93%	1.24%	0.06%	0.81%	0.02%
stomach	11.68%	9.88%	5.09%	0.68%	0.74%	0.11%
muscle	0.78%	0.38%	0.47%	0.08%	0.31%	0.46%
tumour	4.03%	3.52%	1.38%	0.09%	0.54%	0.23%
urine %ID	46.37%	30.95%	72.84%	1.92%	85.29%	1.59%

	^{99m}TcHYNIC-cGIP34					
	0.5h		2h		24h	
tissue	%ID/g	SD	%ID/g	SD	%ID/g	SD
blood	5.53%	2.79%	3.67%	3.32%	0.36%	0.04%
thyroid gland	2.18%	0.38%	3.38%	0.51%	0.33%	0.21%
lung	3.15%	0.83%	1.87%	0.21%	0.27%	0.06%
liver	12.02%	3.18%	11.06%	1.27%	7.00%	1.16%
spleen	3.59%	1.06%	2.87%	0.48%	1.54%	0.16%
pancreas	1.12%	0.12%	0.92%	0.26%	0.05%	0.01%
kidney	110.59%	54.88%	91.17%	17.90%	42.61%	6.00%
intestines	1.06%	0.21%	1.10%	0.12%	0.61%	0.13%
stomach	4.74%	0.59%	5.09%	1.30%	0.95%	0.10%
muscle	0.68%	0.18%	0.41%	0.09%	0.28%	0.11%
tumour	2.02%	0.79%	1.25%	0.42%	0.66%	0.12%
urine %ID	14.30%	11.19%	26.51%	0.71%	59.37%	1.49%

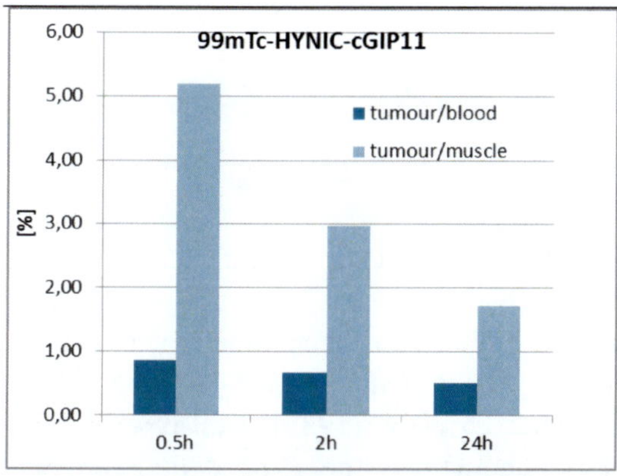

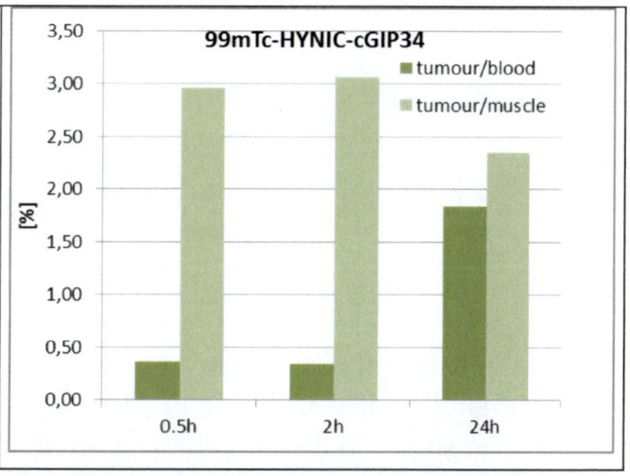

Figure 5. Tumour to blood and tumour to muscle ratio for ^{99m}Tc-labelled cGIP11 and cGIP34 peptides administered in breast cancer bearing C3H mice.

In-Vitro Study

Unfortunately performed preliminary in-vitro study on specific binding of ^{99m}Tc-labelled cGIP11 and cGIP34 peptides and ^{177}Lu-labelled cGIP34 peptide to receptors on human breast cancer cells MCF-7 have not indicated the specific binding. According to literature the MCF-7 cell line is one of the most susceptible on growth suppression effect of GIP peptides. The cell line has also documented biding affinity to ES receptor [14]. A lack of specific binding of the HYNIC and DOTA analogues of studied GIP peptides may be due to low receptor density on cell surface or, what is more probable, to the changed active structure of peptide by attachment of relatively big chelator to lysine in the peptide chain. Nevertheless, the conducted comparative study showed 85% greater relative binding of ^{125}I-labelled cGIP34 than ^{99m}Tc-labelled analogue, indicating that cGIP34 has a potential as a carrier of diagnostic or therapeutic radionuclides.

Conclusion

The conducted study on ^{99m}Tc-labelling of cGIP11 and cGIP34 peptides resulted in elaboration of an efficient radiolabelling method via HYNIC, which led to achievement of stable radiopeptides of high specific activities and high radiochemical purities. Biodistribution studies have shown that these peptides possess fast pharmacokinetics with high urinary excretion and a low accumulation in the soft tissues. Study on the animal tumour model showed that labelled peptides did not exhibit a high affinity to breast tumour (T/B ratios <1 at 0.5h and 24h p.i.). The results obtained at 24h p.i. may suggest that ^{99m}Tc labelled cGIP34 in opposition to cGIP11 was actively targeted to the mammary adenocarcinoma cells.

In-vitro study has shown that the peptides do not exhibit specific binding (times 0.5 – 4h) to MCF-7 cells. Within our study the preliminary experiments with conjugation of DOTA to cGIP34, and then ^{177}Lu-labelling were conducted, which have shown that cGIP34 peptide can be effectively labelled with therapeutic β–emitters such as ^{177}Lu.

This preliminary study showed that the ^{177}Lu-labelled peptide had comparative properties as ^{99m}Tc-labelled analogue. Nevertheless, much higher binding of the radioiodinated cGIP34 still shows that there is potential for GIP peptides as radionuclide carriers.

References

[1] Brock, D. J. H. and Sutcliffe, R. G. (1972). Alpha-fetoprotein in the antenatal diagnosis of anencephaly and spina bifida. *Lancet*, *2*, 197-199.

[2] Mizejewski, G. J. (1997). Alpha-fetoprotein as a biologic response modifier: relevance to domain and subdomain structure. *Proc Soc Exptl Biol Med*, *215*, 333-362.

[3] Naval, J., Villacampa, M. J., Goguel, A. F. and Uriel, J. (1985). Cell-type-specific receptors for alpha-fetoprotein in a mouse T-lymphoma cell line. *Proc. Natl. Acad. Sci, USA*, *82*, 330-5.

[4] Villacampa, M. J., Moro, R., Naval, J., Failly-Crepin, C., Lampreave, F. and Uriel, J. (1984). Alpha-fetoprotein receptors in a human breast cancer cell line. *Biochem. Biophys. Res. Commun*, *122*, 1322-7.

[5] Suzuki, Y., Zeng, C. Q. and Alpert, E. (1992). Isolation and partial characterization of a specific alpha-fetoprotein receptor on human monocytes. *J. Clin. Invest*, *90*, 1530-6.

[6] Torres, J. M., Geuskens, M. and Uriel, J. (1992). Activated human T lymphocytes express albumin binding proteins which cross-react with alpha-fetoprotein. *Eur. J. Cell Biol*, *57*, 222-8.

[7] Uriel, J., Laborda, J., Naval, J. and Geuskens, G. (1989). An overview in G.J. Mizejewski, H.I. *Jacobson; Boca Raton CRC*, vol *2*, 103.

[8] Geuskens, M., Dupressior, T. and Uriel, J. (1991). A study, by electron microscopy, of the specific uptake of alpha-fetoprotein by mouse embryonic fibroblasts in relation to in vitro aging, and by human mammary epithelial tumour cells in comparison with normal donors' cells. *Cytol. Pathol*, *23*, (1991), 59-66.

[9] Torres, J. M., Geuskens, M. and Uriel, J. (1991). Receptor-mediated endocytosis and recycling of alpha-fetoprotein in human B-lymphoma and T-leukemia cells. *Int. J. Cancer*, *47*, 110-7.

[10] Torres, J. M1., Laborda, J., Naval, J., Darracq, N., Calvo, M., Mishal, Z. and Uriel, J. (1989). Expression of alpha-fetoprotein receptors by human T-lymphocytes during blastic transformation. *Mol. Immunol*, *26*, 851-7.

[11] Kanevsky, V. Y., Pozdnyakova, L. P., Aksenova, O. A., Severin, S. E., Katukov, V. Y. and Severin, E. S. Biochem. (1997). Isolation and characterization of AFP-binding proteins from tumor and fetal human tissues. *Mol. Biol. Inter*, *41*, 1143-51.

[12] Mizejewski, G. J. (1995). Alpha-fetoprotein binding proteins: implications for transmembrane passage and subcellular localization. *Life Science*, *56*, 1-9.

[13] Mizejewski, G., Smith, G. and Butterstein, G. (2004). Review and proposed action of alpha-fetoprotein growth inhibitory peptides as estrogen and cytoskeleton-associated factors. *Cell Biol. Int*, *28*, 913-33.

[14] Mizejewski, G. J., Dias, J. A., Hauer, C. R., Henrikson, K. P. and Gierthy, J. (1996). Alpha-fetoprotein derived synthetic peptides: assay of an estrogen-modifying regulatory segment. *Molecular and Cellular Endocrinology*, *118*, 15-23.

[15] Eisele, L. E., Mesfin, F. B., Bennett, J. A., Andersen, T. T., Jacobson, H. I., Vakharia, D. D., MacColl, R. and Mizejewski, G. J. (2011). Studies on analogs of a peptide derived from alpha-fetoprotein having antigrowth properties. *J Pept Res*, *57*, 539-46.

[16] Eisele, L. E., Mesfin, F. B., Bennett, J. A., Andersen, T. T., Jacobson, H. I., Soldwedel, H., MacColl, R. and Mizejewski, G. J. (2001). Studies on a growth-inhibitory peptide derived from alpha-fetoprotein and some analogs. *J Pept Res*, *57*, 29-38.

[17] Vakharia, D. and Mizejewski, G. J. (2000). Human alpha-fetoprotein peptides bind estrogen receptor and estradiol, and suppress breast cancer. *Breast Cancer Res. Treatment*, *63*, 41-52.

[18] Mesfin, F. B., Bennett, J. A., Jacobson, H. I., Zhu, S. and Andersen, T. T. (2000). Alpha-fetoprotein-derived antiestrotrophic octapeptide. *Biochim. Biophys. Acta*, *1501*, 33-43.

[19] MacColl, R., Eisele, L. E., Stack, R. F., Hauer, C., Vakharia, D. D., Benno, A., Kelly, W. C. and Mizejewski, G. J. (2001). Interrelationships among biological activity, disulfide bonds, secondary structure, and metal ion binding for a chemically synthesized 34-amino-acid peptide derived from alpha-fetoprotein. *Biochim Biophys Acta*, *1528*, 127-34.

[20] Maurin, M., Garnuszek, P., Karczmarczyk, U. and Mirowski, M. (2008). Studies on 99mTc-labeling of the modified fragment of human alfa-fetoprotein (P149-QY). *J. Rad. Nucl. Chem*, *275*, 101-107.

[21] Muehlemann, M., Miller, K. D., Dauphinee, M. and Mizejewski, G. J. (2005). Review of Growth Inhibitory Peptide as a biotherapeutic agent for tumor growth, adhesion, and metastasis. *Canc Metas. Rev*, *24*, 441-467.

[22] Rennen, H. J. J. M., Boerman, O. C., Koenders, E. B., Oyen, W. J. G. and Corstens, F. H. M. (2000). Labeling proteins with Tc-99m via hydrazinonicotinamide (HYNIC): optimization of the conjugation reaction. *Nucl. Med. Biol*, *7*, 599-604.

[23] Garnuszek, P., Wiercioch, R., Sztajer, H., Karczmarczyk, U. and Mirowski, M. (2005). Uptake of radiolabelled modified fragment of human alfa-fetoptrotein by experimental mammary adenocarcinoma: *in vitro* and *in vivo* studies. *Nucl Med Rev Cent East Eur*, *8*, 6-10.

SECTION 5: ALPHA-FETOPROTEIN: REGULATORY ROLES, IMMUNOTOLERANCE, AND VACCINES

In: Alpha-Fetoprotein: Functions and Clinical Applications ISBN: 978-1-63484-875-6
Editors: Nisha Lakhi and Michael Moretti

Chapter 18

CRYPTIC EPITOPE P149 OF ALPHA-FETOPROTEIN: ROLE IN PREGNANCY AND BREAST CANCER

Dilip Vakharia[1,*], PhD, and Gerald Mizejewski[2], PhD
[1]University of South Carolina School of Medicine, Columbia, SC
[2]Wadsworth Center, New York State Department of Health, Albany, NY, US

ABSTRACT

The Chapter provides a brief summary of published work on the topic of a cryptic epitope (P149) peptide of alpha fetoprotein (AFP) and its role during pregnancy and breast cancer development since our earlier work was published in Breast Cancer Resch Treat 63: 41-52, 2000. It provides information on the potential utility of a diagnostic peptide in the management of pregnancies with complications. The plausible use of a synthetic peptide representing cryptic epitope P149 and its sub fragments in preventing insulin-induced fetotoxicity and its potential use as diagnostic marker for cystic fibrosis patients is discussed. An explanation for the reduced risk of breast cancer influenced by pregnancy hormonal levels and AFP is discussed in the context of breast development, breast stem cells, and the cryptic epitope peptide. Finally, the P149 peptide is presented as a possible anti-cancer agent based on its growth inhibition of multiple cancer types.

In the current chapter we are revisiting issues raised in our earlier well received publication [1]. The objective was to identify reports since our original publication that have touched upon issues presented by us. A further goal was to expand on a discussion on how best to utilize the new knowledge gained on the subject of alpha fetoprotein (AFP) epitopes for clinical and therapeutic application.

The 34 amino acid (AA) cryptic epitope P149 (growth inhibitory peptide) is located on the 3rd domain of AFP extending from AA# 464 to AA # 497. For a more detailed analysis of its function, P149 was synthesized as a 34 AA peptide and as 3 individual sub-fragments i.e., P149A, P149B and P149C (see Figure 1.)

---

* Corresponding author: Dilip Vakharia, University of South Carolina School of Medicine, Columbia, SC 29209. dilvakhar@gmail.com.

Domain I Domain 2 Domain 3

NH2------------||------------------------||------------|--------------|---------|-----------COOH

P149A P149B P149C

|_______P149_____|

P149	(464-497)	LSEDKLLACGEGAADIIIGHLCIRHEMTPVNPGV
P149A	(464-475)	LSEDKLLACGEG
P149B	(476-489)	AADIIIGHLCIRHE
P149C	(489-497)	EMTPVNPGV

Figure 1. Schematic representation of AFP-derived peptides. The amino acid sequences and their positions in native AFP are given in parentheses. The smaller peptide fragments of P149 (34AA) were synthesized as an amino terminal segment (P149A, 12AA), a middle segment (P149B, 14AA) and a carboxy terminal segment (P149C, 9AA).

Our study demonstrated that pretreatment of alpha fetoprotein (AFP) with excess molar amounts of estrogen receptor alpha (ERα)-ligand estradiol (E_2), facilitated a conformational change that exposed the cryptic epitope P149 of AFP detectable by an enzyme linked immunoassay (ELISA) using rabbit anti-P149 antibody. This P149- ELISA, but not a conventional AFP immunoassay, detected the cryptic epitope P149 on AFP. Sera of patients with AFP-secreting tumors or sera or amniotic fluid of pregnant women with fetal growth retardation showed high levels of the AFP cryptic epitope using a P-149-ELISA assay.

We demonstrated that in a recombinant ERα/E_2 competitive -binding assay, the 34 amino acid cryptic peptide P149 and its N-terminal, 12 AA fragment (P149A), bound ERα with low (1:6 x10^{-4} M) affinity compared to E_2 and tamoxifen (1:25 x 10^{-9}M), while the middle 14 AA fragment P149B, bound ^{3}H- E_2 with 10^5 fold less affinity compared to ERα. The 9 AA C-terminal fragment (P149C), the native AFP, and several other peptide controls bound neither ERα nor E_2. Thus, it may be of no mere coincidence that P149A share amino acid sequence identity/similarity to an estrogen (steroid co-activator) [2].

P149 and its fragments P149A and P149C were shown to significantly suppress estradiol-induced uterine growth whereas P149B did not [1, 2]. Both P149 and P149C suppressed human mammary tumor (6WI-101) growth in vivo and suppressed by more than the 75% growth of breast tumor cell lines MCF-7 (ER^+), MDA-MB-231 (ER^-), MDA-MB435 (ER^-) in vitro [1, 3, 4].

Regarding the observation that there is a low risk of breast cancer in women that displayed high AFP levels during pregnancy at an early maternal age of childbirth (age 20-24 years or less) [5, 6], we have hypothesized that continuous binding of low affinity ER-ligands (such as P149) with growth suppressive properties during much of the gestation early in life, may render breast tissue less susceptible to cancer later in life. Furthermore, presence of native AFP or AFP-P149 [transformed AFP (tAFP)] in serum and in breast effusions may influence the development and growth of the breast through its interaction with breast cells and stem cells present in the tissue and in the milk.

It is likely that in addition to excess molar amounts of estradiol, excess amounts of fatty acids or glucose during pregnancy may facilitate conformational changes in AFP exposing the cryptic epitope P149. A correction of 10-15% value is routinely applied to normalize the measured AFP level during human pregnancy in the New York prenatal screening program for insulin-dependent diabetes associated pregnancies [7]. This may be due to the presence of a conformational variant form of AFP that is not immunologically detected by commercial antibodies to AFP.

In the current chapter we will expand on the topics discussed above and provide the readers with the current standing on these issues.

Detection of AFP with cryptic epitope P149 (AFP-P149) in pregnancy: The results of three studies that have measured serum levels of normal AFP and AFP-P149 in women with normal or complications of pregnancies are discussed. Dr. Jose Bartha's laboratory in Spain have published three studies describing levels of AFP-P149 in normal maternal and fetal sera [8], in women with threatened preterm labor [9] and in women with intrauterine growth retardation [10]. The results of these and similar studies are summarized in Table 1. They have used a commercial kit (Serometrix, Syracuse, New York, USA) that employs a two-tier antibody competitive immunoadsorbent ELISA described by us [1]. This assay utilizes anti-P149 antibody (Ab-P149) on plates coated with P149. Standards with known concentrations of P149 or patient samples are added to the wells of the plate together with Ab-P149. The residual Ab-P149 not bound to P149 in solution binds to P149 coated to the plate. After washing the plates, the bound Ab-P149 is detected by appropriate second antibody conjugated to an enzyme that in the presence of appropriate substrate produces color. The term transformed AFP (tAFP) used by the authors is synonymous with AFP-P149 site exposed on the whole AFP polypeptide. The native AFP levels were determined by electro-chemiluminescence assay on the Elecsys 2010 system using the kit supplied by Roche Diagnostics, Indianapolis, USA.

Monitoring of cryptic epitope (P149) in normal and pregnancies with complications: For the study in normal maternal and fetal sera, two groups (Group I and II) of 25 women each, healthy Caucasian, non-smokers, without any pregnancy complications and delivering healthy babies without chromosomal abnormalities were studied [8]. For Group I, maternal venous blood was collected at 12, 20, 32 and 36 weeks of gestation. Group II represented 25 women who were admitted to the maternity ward to have an elective cesarean section (absence of labor). These included 15 cases of breech presentation, 5 cases of previous cephalopelvic disproportion, two cases of previous uterine scar (one case of myomectomy and two cases with two previous cesarean sections), one case of maternal extreme myopia and one case of vulvoperineal condylomata. In Group II, maternal blood was collected immediately before cesarean section. Fetal blood samples from the umbilical artery and vein were collected following birth after clamping the umbilical cord in both study groups. The native AFP median levels (ng/ml) in Group I increased from 15.70 (1st trimester) to 207.70 (first half of 3rd trimester, followed by drop to 145.32 ng/ml during second half of 3rd trimester. Whereas the corresponding AFP-P149 levels increased from 14.73 (1st trimester), 28.29 (2nd trimester) to 30.45 (first half of 3rd trimester, followed by drop to 8.06 during second half of 3rd trimester. The ratio of AFP-P149: AFP for 1st, 2nd, 3rd (first half) and 3rd (second half) trimesters was 0.94, 0.44, 0.15 and 0.05.

Table 1. Transformed alpha-fetoprotein (t-AFP or AFP-P149) maternal serum levels are shown for normal pregnancies versus those with congenital disorders and pregnancy complications*

A: Second Trimester t-AFP			
Type of Pregnancy	**Number of patients**	**Median value (ng/ml)**	**Range of values**
Normal pregnancy	25	28.3	12.9 - 123.7
Down syndrome	4	13.1 ↓	0.5 - 14.6
Neural tube defect	4	35.5 ↑	8.8 – 57.4
Smith- Lemli- Opitz syndrome (SLOS) with trisomy-18	4	22.8 ↓	7.8 – 38.3
Smith- Lemli- Opitz syndrome (SLOS) with growth retardation	11	31.5 ↑	9.39 – 79.9
B: Third Trimester t-AFP			
Type of Pregnancy	**Number of patients**	**Median value (ng/ml)**	**Range of values**
Early trimester normal pregnancy	20	30.5	10.5 – 168.2
Late trimester normal pregnancy	20	8.1	2.6 – 18.8
Term pregnancy	20	3.1	0.6 – 6.4
Threatened preterm labor (TPL) Control TPL	25 25	7.64 ↓ 33.38	1.78 – 29.1 13.8 - 190
Preterm delivery (PD) Control PD	10 40	5.52 ↓ 22.5	0.2 – 29.8 0.4 - 274
Intra-uterine growth retardation (IUGR) Control IUGR	30 30	15.4 ↑ 8.66	10.8 – 24.01 6.2 – 13.45

* Second trimester specimens were obtained as discarded samples from the Lenetix Medical Screening laboratory, Long Island, NY after all patient identification was removed.

Third trimester specimens were provided by Dr. Jose Bartha, Fetal Medicine Unit, University Hospital, Cadiz, Spain.

For comparison to native AFP levels see references 8-10.

The AFP levels in maternal serum (1st trimester to term) were 500-1000 fold lower than fetal levels in umbilical artery or vein. The AFP-P149 term maternal serum levels on the other hand were 10 fold higher than fetal umbilical artery or vein.

As discussed by the authors, the conformational change in native AFP is reflected as early as the first trimester of pregnancy with a corresponding increase in levels of both native and AFP with exposed P149 epitope up until early 3rd trimester of pregnancy and then dropping off at term. Secondly, the location and its associated chemical milieu, as in placenta, maternal tissues or the blood stream may be responsible for the conformational change in AFP leading to increased levels of AFP-P149 in maternal blood than in fetal blood. Further, the amount of AFP-P149 is not dependent on the concentration of AFP. The early increase in AFP-P149 or AFP during pregnancy may play a role in preventing immune-rejection of the feto-placental unit perhaps by affecting immune response properties of participating immune cells as suggested by recent studies [11, 12].

Threatened preterm labor (TPL) and AFP-P149: In a prospective case control study, maternal serum levels of native AFP and AFP-P149 (tAFP) were compared between two groups each of 25 gestational age matched pregnant women, one group with healthy pregnancies and the other group having pregnancies with TPL [9]. The TPL group represented women who were referred to the hospital for regular uterine contractions before 35 weeks of gestation. The TPL was based on evaluation of cervical effacement or dilation or both with a Bishop Score of ≥ 6 and the existence of regular and increasing uterine activity (4 uterine contractions/30 min). These women were treated with tocolytics and also received corticosteroids in order to improve the fetal lung maturation status in cases of preterm delivery. The control group was composed of women with matching gestational age to the TPL group. AFP and AFP-P149 levels were measured in venous blood samples collected before tocolytic administration and in a control group just before performing a normal third trimester ultrasound scan. As above, they have used a commercial ELISA kit (Serometrix, Syracuse, New York, USA) to measure AFP-P149 and measured native AFP using the kit supplied by Roche Diagnostics.

Compared to the control, the TPL group had significantly lower values for gestational age at delivery and for birth weight. Ten women in the TPL group had preterm deliveries. There were no significant differences in the native AFP levels between the two groups. However, the median AFP-P149 levels in TPL group [7.64 (1.78-29.06)] were significantly ($p = 0.006$) lower than the control group [33.38 (3.80 – 190.50)]. The median AFP-P149/AFP ratio was significantly lower in women with TPL [0.04 (0.004-0.12)] than in the control group [0.16 (0.05-0.80)]. Within the TPL group, there was no significant difference in AFP-P149 levels between those with ($n = 10$) or without ($n = 15$) preterm deliveries. Overall, the lower third trimester AFP-P149 levels in women with preterm deliveries [$n = 10$, (median 5.52 (0.19-29.8)] were significantly lower than without preterm deliveries [$n = 40$, 22.51 (0.42-273.91).

AFP-P149 levels and AFP-P149: AFP ratios were four fold lower in women with TPL than in normal pregnancies of the same gestational age. This was explained by the fact that during late third trimester of normal pregnancy, both AFP and AFP-P149 levels drop [8]. The authors have hypothesized that AFP-P149 may play an immunomodulatory role and the absence of sufficient amounts of AFP-P149 could cause a decrease in the maternal immunotolerance. This would promote generation of pro-inflammatory cytokines that may mediate pro-parturition effects via prostaglandins, endothelins and leukotrienes in both women delivering at term and in women with preterm delivery.

Intrauterine Growth Retardation (IUGR) and AFP-P149: In a prospective study, Bartha et al. [10] have measured serum AFP-P149 levels in age, gestational duration matched 30 healthy pregnant women and 30 women with intrauterine growth retardation (IUGR) using the ELISA assay. All women included in the control group showed normal fetal growth at 34 weeks of gestation and their infants had birth weights between the 10th and 90th percentiles for gestation. Ultrasound assessment showed anatomically normal fetuses. No multiple pregnancies and cases of chromosomal abnormalities were included in the study. IUGR was defined by fetal abdominal circumference below the 5th percentile for gestation and at delivery all infants had a weight below the 5th percentile for gestation. Maternal serum AFP-P149 levels were significantly higher in women with IUGR 15.39 (10.81-24.010 ng/ml) than in control group 8.66 (6.22-13.45 ng//ml). In women with IUGR no statistically significant differences were found between those with or without placental insufficiency; also, no significant correlation between AFP-P149 and difference values of the pulsatility index I in the umbilical artery was observed. Women with IUGR and hemodynamic redistribution (pulsatility index in the fetal middle cerebral artery below the 2.5th centile for gestational age) had increased maternal serum levels of AFP-P149 than those without hemodynamic redistribution. The authors discussed the concern of worsening hypoxia, perinatal death, and long-term neurodevelopmental complications if the delivery of the preterm growth–restricted fetuses is delayed. Allowing birth to a premature baby has its complications. In such a situation, in lieu of an optimal antepartum test to guide physicians about the timing of delivery for the preterm fetuses, authors speculate that AFP-P149 may be used as a marker of deterioration of fetal condition during management of fetuses with IUGR.

Correction of AFP levels during pregnancy: Effect of insulin and fatty acids It is now known that chemicals like estradiol (mostly in rodents), insulin, and fatty acids bind AFP and excess molar amounts of these compounds facilitate a conformational change in the AFP molecule [13]; this thereby affects true levels of AFP measured by standard immunoassays. The effect of exogenous insulin on the maternal serum level of AFP and fetal weight was studied by Andersen et al. in mice [14]. They observed that insulin given as a single injection on 13th day of pregnancy resulted in elevated maternal serum AFP levels, in greater number of fetuses, and with larger fetal mass per animal when observed on 18th day of pregnancy. In contrast, long-term treatment with insulin from the first day throughout pregnancy resulted in a lower level of maternal serum AFP and in significantly lower fetal mass per animal. Pregnant women with type 1 and type 2 diabetes show low levels of maternal serum AFP and both groups require a body weight correction and a diabetes correction factor of 10% when compared with nondiabetic AFP levels [15].

The fetotoxic effect of insulin as seen in mouse studies was tested in a chicken embryo assay by Butterstein et al. [16]. Injection of 2 U of insulin into the yolk sac resulted in death of 30% of fetuses after 7 days of egg incubation at 37°C. Fetal malformations were observed in 12% of dead fetuses. Injection of mixture of insulin and human AFP (pre-incubated for 1hr at room temperature before injection) resulted in a fetal death rate of only19%, a reduction of 63%. Injection of mixture of insulin and cryptic epitope peptide P149 resulted in a fetal death rate of 10%, a reduction of 70%. The presence of fetal malformations was reduced to 6% in the AFP/insulin group and to 5% in the P149/insulin group (a 50% reduction). In comparison, growth retardation in the AFP/insulin group occurred at a rate of 18% while the AFP-P149/insulin group exhibited a 30% rate, nearly a twofold increase over the insulin-alone rate. It was observed that AFP and P149 alone also induced some fetal demise (12-14%) and growth retardation (17-24%). These observations suggests that in in pregnant women with type 1 and type 2 diabetes or increased insulin levels, sufficient amounts of transformed AFP and AFP-P149 could play a role in reducing insulin-induced fetotoxicity at the expense of fetal growth retardation.

Fatty Acid, AFP and Cystic Fibrosis

The study of AFP levels in 23 infants found presumptive positive for cystic fibrosis (CF) in a pilot study of newborn CF screening in New York State, 18 of the 20 confirmed CF cases showed 60-80% lower values of AFP levels than age matched controls [17]. The other three infants unaffected by CF but showing false positive results for immunoreactive trypsin, did not exhibit decreased AFP levels. One of the explanation for the low AFP levels in CF given by the authors is the failure of the antibody in the AFP immunoassay to recognize AFP that is conformationally altered due to binding to polyunsaturated fatty acids. It has been determined in a mouse model that mimics the genetic disorder found in human CF [18], that arachidonic acid (AA)/docosahexaenoic acid (DHA) imbalance plays a role in the phenotypic expression of CF. DHA is a polyunsaturated fatty acid found during pregnancy and early infancy, and AFP plays an important role in binding and trafficking of AA and DHA into cells undergoing growth and differentiation. Polyunsaturated fatty acids have been demonstrated to induce conformational changes in the AFP molecule [19]. It is likely that binding of AA or DHA could induce a conformational change that exposes cryptic epitope P149 or another epitope not recognized by antibodies used in the conventional AFP immunoassay. It will be worthwhile to analyze these CF serum samples in the P149-ELISA assay used by others and confirm if AFP P149 could be used as a biomarker for CF in newborns. We have observed high levels of native serum AFP in CF infants at age of one year compared to healthy infants [20]. The physiological significance of the role of native AFP and AFP-P149 in the pathology of CF is not known.

AFP and Breast Cancer

Role of maternal serum and placental biomarkers including AFP in breast cancer during pregnancy has been discussed by Mizejewski [21]. Gestational breast cancer can occur in young and premenopausal women, however those with a high incidence in the cancer, range in age from 30-38 years. In population based studies, elevated levels of AFP during full term pregnancy have been shown to be associated with a reduced breast cancer risk in pre- and postmenopausal women [5, 6]. Similarly high AFP levels in pregnancies associated with multiple births [22], and pre-eclampsia [23, 24] have a reduced breast cancer risk. The ethnic difference in low breast cancer risk in Asian women in Shanghai, China, compared to Caucasian women in Boston, United States, was explained by the presence of high AFP levels [25]. However, in more recent reporting of this study population, the authors concluded that the observed higher difference in the levels of maternal serum AFP in Asian women compared to Caucasian women was reduced; this was after controlling for pre-pregnancy body mass index which is lower among Chinese women and inversely associated with AFP. The maternal serum AFP levels were unrelated to birth weight in both ethnic cohorts but were inversely associated with duration of gestation [26]. For pregnancies associated with high birthweight infants, a strong and independent association for breast cancer risk in women at later age has been reported [27]. In this study, an infant's birth weight had a strong positive relationship with the mother's serum estriol (E3)/AFP ratio during pregnancy. On the other hand, umbical cord blood AFP and not serum AFP in any gestation week was inversely correlated with birth weight [28].

Hormones of Pregnancy and Breast Cancer

The importance of AFP over other pregnancy hormones (estrogen and progesterone) to explain association of reduced breast cancer risk and parity was demonstrated in an animal model by Jacobson et al. [29, 30]. Their study showed that the cancer induced by nitrosomethylurea exposure in rats during pregnancy or following treatment with combination of estrogens (estradiol or estriol) and progesterone together with hCG showed that each hormone treatment reduced mammary cancer incidence and elevated serum AFP levels. In the in vitro system, treatment of HepG2 liver cells with each hormone regimen increased the levels of AFP in the culture medium. Medium containing AFP elicited by hCG inhibited the E2-stimulated proliferation of cultured human MCF7 breast cancer cells, whereas hCG alone did not inhibit their growth. Furthermore, antibodies to AFP neutralized the growth-inhibiting effect of AFP-containing HepG2 medium. From these studies the authors concluded that of the pregnancy hormones studied, AFP played a proximal role in inhibiting mammary gland cancer.

The results of two population based studies of serum levels of AFP during pregnancy and breast cancer incidence [5, 6] were discussed in our earlier report [1] to explain how increased levels of transformed AFP (AFP-P149) during pregnancy at early age may prevent subsequent risk of breast cancer development. We hypothesized that continuous binding of low affinity ER-ligands (such as P149) with growth suppressive properties during much of the gestation early in life, may render breast tissue cells (stem cells) less susceptible to breast cancer risk later in life.

AFP, Breast Milk

The question whether breast tissue synthesizes AFP is debatable. Malignant but not normal breast tissue produce [31] AFP and there is an AFP specific receptor present in the breast tissue and serum of tumor bearing cells [32, 33]. Early studies of Biddle and Sarcione showed that postmenopausal women with primary breast carcinoma had increased serum AFP levels and they suggested that the source of AFP was breast tissue [34]. Treatment of these serum samples with 0.4M KCl identified two forms of AFP, immunoreactive and nonimmunoreactive using a standard radioimmunoassay (RIA). The nonimmunoreactive form accounted for increase in the value of AFP measured by RIA by 44 IU/ml. This study suggested that a percentage of AFP in the serum was conformationally altered by AFP-binding ligands (removed by 0.4M KCl and not recognized by antibody in the assay).

The analysis of serum AFP using standard immunoassay, by assessing concanavalin A (ConA)- binding pattern and by immunohistochemical (IHC) AFP staining of breast and liver tissues showed high serum levels of AFP in a single male subject with breast cancer and hepatocellular carcinoma [35]. Seventy percent of the AFP did not bind ConA suggesting non-hepatocellular pattern whereas 30% ConA-bound AFP on elution showed pattern representative of hepatocellular carcinoma. However, LCA-lectin affinity chromatography predominantly showed a pattern representative of hepatocellular carcinoma. The AFP-immunohistochemical staining was positive for primary and metastatic cancers cells in the breast tissue whereas hepatocellular carcinoma cells were negative.

Similarly, small amounts [(0.1 ± 0.1), 0.0-0.7] of AFP (ng/ml) in breast cyst fluid compared to serum [(3.0 ± 1.2), 0.4-5.4] in breast macrocysts of women with fibrocystic breast disease have been determined using immune-bead-enzymatic assay. The serum AFP levels of women without breast macrocyst [(2.8 ± 1.7), 0.4-4.8] were comparable to those with macrocyst [36].

Thus it appears that breast tissue whether having benign or malignant disposition does produce AFP. Thus, it is likely that during pregnancy, breast milk and serum would also contain a percentage of AFP that is secreted by breast cells with the remaining contributed by the fetal liver. The role, if any, that non-secreted cytoplasmic AFP play specifically in liver tissue and perhaps in other organs has recently been discussed by Mizejewski [37].

Earlier study of measurement of AFP in body fluids using Ouchterlony agar immunodiffusion technique by Smith et al. [38] showed an absence of AFP in milk (colostrum) of women. Analysis of 21 breast milk samples collected from mothers feeding infants of ages 2-27 months for native AFP by standard immunoassay and transformed AFP (AFP-P149) by ELISA-P149 assay showed undetectable levels of native AFP and 12.1 ng/ml (2-42 ng/ml) of t-AFP (AFP-P149), respectively (unpublished observation, Kathleen Arcaro, University of Massachusetts).

Breast Development, Stem Cells and Cancer

No reports are available if AFP plays a role in the development and the growth of mammary gland. This AFP could be a non-secretory form similar to that found in liver [37] or the secretory form discussed above in male subject with breast and liver cancer [36] or native AFP found in pregnant women. It will be of interest to test the effect of AFP and AFP derived peptides on the breast tissue development. Also, their effect on various stem cell types present in the tissue and breast milk.

Russo and Russo have extensively studied the development, architecture and stem cell distribution in breast tissue from menarche to various stages of pregnancy in nulliparous women and parous women. During 1-2 years after the onset of the first menstrual period, the rudimentary mammary gland shows growth activity and starts forming ducts and alveolar buds leading to increasing numbers of lobular structures with each menstrual cycle [39]. Epithelial cells of premenopausal nulliparous women exhibit a significantly higher proliferative activity in Lobule 1 and high concentration of estrogen receptor than those in lobules 2, 3 and 4 with lobule 1 being a natural target of malignancy [39]. The breast architecture of nulliparious women exhibit predominant lobule 1 with lower percentage of lob 2 whereas the breast tissue of parous women contain lowest percentage of Lob 1, a slightly higher percentage of lob 2 and predominant lob 3 structure [39]. After menopause sets in, even when the preponderant structure in the breast is lobule 1, it is only the nulliparous women that are at high risk of developing breast cancer whereas the parous women remain protected [40]. Using cell-based and experimental parameter-based one dimensional and computer-generated 3D fractal modeling approach, Honeth et al. [41] have recently developed mammary lobule morphogenesis based on location and fate choices of stem and progenitor cells. This model has shown that ALDH1A1$^+$, CK14CK19$^+$stem cells are enriched in lobule 1 and are significantly lower in lobules 2 and 3. Comparison of representation of ALDH1A1$^+$ stem cells in mammary epithelium did not show expected significantly high numbers in nulliparous women when compared with samples from parous women observed by Russo et al. [40]. However, a significantly larger percentage of ALDH1A1$^+$ cells were seen in Lobule 1 from *BRCA*1/2 mutation carriers compared to non-carriers. Further, a considerable enrichment in ALDH1A1$^+$ cells was seen in Lobule 1 from nulliparous women compared to parous women when *BRCA*1/2 mutation carriers and non-carriers were analyzed separately [41].

Estrogen receptor and cytokeratin 19 double positive breast stem cell [with characteristic properties of cells described as 'side population (SP)] expressing several fold higher levels of genes- ERα, p21 (CIPI)) and Msi1 than non-SP cells are found localized intermediate in position between the luminal and basal myoepithelial layers in less developed or immature lobules, similar to lobule 1 [42, 43]. Russo Gabriela et al. [43] have called these stem cells as Stem Cell 1 found mostly in nulliparous women. Subsequent to first pregnancy, these Stem cell 1 differentiate to a new progenitor or Stem cell 2 which are refractory to carcinogenesis.

Hassiotou has discussed cells in the breast milk and breast tissue which include stem cells [44, 45]. According to Hassiotou, both CK19 ductal epithelial cells and CK14 myoepithelial cell lineages develop from common progenitor mammary stem cells (MaSC). In non-pregnant and non-lactating females MaSC has $CD49f^{high}/CD29^{+}/CD24^{low}$ phenotype and subsequent progenitors gradually lose the expression of the marker CD49f. These MaSC cells may be similar to one discussed by Honeth et al. [41] and enriched in lobule 1. In the breast milk a subpopulation of breast stem cells (hBSC, not found in resting breast) and distinct from MaSC, expressing pluripotency genes including OCT4, SOX2 and NANOG are found together with a small population of cells expressing hematopoietic stem markers (CD34, CD133), cells with mesenchymal stem cell (MSC) phenotype expressing CD90, CD105 and CD73 markers and stem cells with epithelial-to-mesenchymal transition (EMT) markers. These hBSC are compared with human embryonic stem cells (hESC) as in in vitro differentiation assays they differentiate into mammary lineage and cells with properties of neurons, cardiomyocytes, osteoblasts, adipocytes, hepatocytes or pancreatic beta cells [44]. However in contrast to hESC, hBSC do not form tumors in SCID mice in the teratoma assay.

As per the recent studies from Sell's laboratory [46], perhaps hBSC, MSC and stem cells with EMT markers present in breast milk discussed above may have originated from circulating bone marrow derived stem cells and MaSC cells would represent breast cancer stem cell perhaps distinct from so called transit-amplifying stem cells discussed by Sell [47, 48].

In the light of this discussion on the structural changes in the development of breast tissue, breast stem cells and role of pregnancy/lactation associated biomarkers, the studies on the direct effect, and on the mechanism of action of AFP and AFP-peptides in arresting development of breast tumor growth still remains to be elucidated.

Potential Diagnostic and Therapeutic Use of AFP-Peptides

The diagnostic utility of serum measurement of native AFP levels during pregnancy is well documented. The recent use of measurement of serum levels of AFP with exposed cryptic epitope P149 (AFP-P149) also called transformed AFP (t-AFP) in an AFP-P149 ELISA assay for monitoring and managing pregnancies with complications (intrauterine growth retardation, threatened preterm labor) provides additional clinical utility of these tests. We have proposed and discussed here the potential role AFP-149 peptides play in explaining the reduced risk of breast cancer associated with pregnancy.

More work is needed to establish how AFP-149 mediates its anti-tumor activity. We have shown that bioactivity of peptide P149 is represented by its shorter fragments P149A and P149C. Of these, P149 and P149A binds estrogen receptor whereas P149C does not [1]. However, irrespective of their ability to bind ER, all the three peptides inhibit growth of several breast tumor cell lines though not all the peptides inhibit all or the same cell lines [1]. This observation suggests that in addition to ER, other cellular effects including events involving cross-talk with ER [49] or others discussed by Leclercq et al. [50] may play a role.

AFP- Peptides –Mechanism of Action

Mizejewski has discussed in detail the potential mechanism of anti-tumor growth action of AFP peptides P149 and its fragment P149C [51]. Both linear and cyclic forms of the peptides have been studied. The biological activity is more potent in cyclic form (10^{-7} to 10^{-10}M) compared to linear form (10^{-4} to 10^{-6}M). The article reviews the binding characteristics of these peptides to cell surface and their cytoplasmic or subcellular targets explaining the cancer growth suppression property. In vivo recording of membrane potential and membrane resistance indicated that P149 at 10^{-6} M and lower concentrations serves as a cell membrane pore forming/cell penetrating peptide coincident with decreased cell membrane resistance; at high peptide concentrations (10^{-5} M and greater) P149 acts as a channel blocker coincident with increased cell membrane resistance. However, P149C acts only as a channel blocker [3]. Thus P149 can gain entrance into cancer cells by at least two pathways namely, (a) receptor-mediated endocytosis and (b) pore-forming/cell penetration. Once within the cytosol, P149 downregulates Cyclin-E, SKP2, and various cell cycle RNA transcripts of proteins which prevent Cyclin-E/Cdc2-mediated progression of cell cycle G1 to S-phase and it also blocks ubiquitin-initiated degradation of cyclin inhibitors such as p27 KIP and p21 CIP [51]. P149C activity initiates at the bi-lipid plasma cell membrane level followed by interaction with a receptor or an ion channel complex. It is suggested that Thr-Pro-Val-Asn (TPVN) sequence in P149C has a beta-turn structure known to bind and activate cell surface receptors linked to G-proteins and the Pro-Val-Asn-Pro (PVNP) sequence in P149C is involved in the activation of serine/threonine Src-kinases. For E_2-stimulated breast cancer, P149C blocks voltage-gated K^+ ion channels, interferes with EGFR or estradiol-stimulated MAPK-kinases activity, inhibit serine-118 human ER phosphorylation, interferes with p53 tumor suppressor phosphorylation, and increases p21 CIP levels thereby arresting cell cycle progression. For ER^- cells, P149C binds to the cell and enacts K+ ion channel blockage resulting in the cessation of cell cycle progression and subsequent mitosis [52]. P149C thus serves as a plasma membrane interactive agent that may block ion channels and interfere with cell surface-induced signal transduction pathways involved in tumor growth, progression, and metastasis.

The potential for the use of AFP-epitopes for hepatoma immunotherapy has further been extensively reviewed by Mizejewski GJ [53]. Pardee et al. [54] have recently studied dendritic cell-mediated antigen presentation of normal and hepatoma derived AFP and have identified mannose receptor (MR/CD206) on dendritic cells a major uptake pathway for both forms of AFP. In hepatocellular carcinoma patients, tumor-derived AFP is more immunogenic and its CD4+ T cell responses are not MR/CD206 dependent. In healthy controls, both forms of AFP are identically presented to CD8+ cells and its CD4+ T cell response is dependent on MR/CD206 and other scavenger receptors. Several of the scavenger receptors (mannose, low-density lipoprotein receptors, the asialoglycoprotein receptor, and the receptor for advanced glycation endproducts (RAGE) were found to bind to the cryptic or growth inhibitory peptide (GIP)-epitopes such as P149B and P149C especially the CIRHEMTPV AA sequence [37].

The potential for the use of AFP-peptides specifically P149 and P149C (referred to as GIP-34 and GIP-8 in the citation) as therapeutic agents to treat cancer including breast cancer was demonstrated in a multicenter study [3]. In this study P149 was effective against many epithelial cancers (adenocarcinomas) but not against tumors of sarcoma origin. Two different investigators in this multicenter study showed that P149 bound to 30kD or 15kD (possibly oligomers) protein moieties (AFP receptors or binding protein) in breast tumor cells. P149 appears to inhibit cancer cell growth by both S-phase cell cycle arrest and prevention of cyclin inhibitor p27 degradation by ubiquitinization and subsequent proteosomal processing. Also it may downregulate transcription of several ion channels, including both potassium and calcium types.

The shorter fragment of P149, P149C (8mer) conjugated to doxorubicin specifically targeted tumor cells (ovarian-SKOV3 cells, breast-MCF-7 cells) compared to peripheral blood lymphocytes with cytotoxic effects [3]. P149C thus serves as a plasma membrane interactive agent that may block ion channels and interfere with cell surface-induced signal transduction pathways involved in tumor growth, progression, and metastasis.

P149 and P149C do not bind to the AFP receptor which was previously described by Moro [4]. A family of proteins termed scavenger receptors has recently been proposed as candidate AFP receptors [55]. A more recent study of Pardee et al. [54] discussed above, has identified the mannose receptor on dendritic cells as a gateway into the cell for normal and tumor-derived AFP. Mannose receptors are indeed present on the surface of MCF-7 cells [56]. Identification of P149 and P149C specific mannose receptor (and perhaps mucins) on breast tumor cells could aid in developing a targeted therapeutic approach for cancer of the breast.

REFERENCES

[1] Vakharia, D. and Mizejewski, G. J. (2000). Human alpha-fetoprotein peptides bind estrogen receptor and estradiol, and suppress breast cancer. *Breast Cancer Res Treat*, *63*, 41-52.

[2] Mizejewski, G., Smith, G. and Butterstein, G. (2004). Review and proposed action of alpha-fetoprotein growth inhibitory peptides as estrogen and cytoskeleton-associated factors. *Cell biology International*, *28*, 913-33.

[3] Mizejewski, G. J., Mirowski, M., Garnuszek, P., Maurin, M., Cohen, B. D., Poiesz, B. J., et al. (2010). Targeted delivery of anti-cancer growth inhibitory peptides derived from human alpha-fetoprotein: review of an International Multi-Center Collaborative Study. *Journal of Drug Targeting.*, *18*, 575-88.

[4] Mizejewskia, G. J. and Butterstein, G. (2006). Survey of functional activities of alpha-fetoprotein derived growth inhibitory peptides: review and prospects. *Current protein and peptide science*, *7*, 73-100l.

[5] Richardson, B. E., Hulka, B. S., Peck, J. L., Hughes, C. L., van den Berg, B. J., Christianson, R. E., et al. (1998). Levels of maternal serum alpha-fetoprotein (AFP) in pregnant women and subsequent breast cancer risk. *Am J Epidemiol*, *148*, 719-27.

[6] Melbye, M., Wohlfahrt, J., Lei, U., Norgaard-Pedersen, B., Mouridsen, H. T., Lambe, M., et al. (2000). Alpha-fetoprotein levels in maternal serum during pregnancy and maternal breast cancer incidence. *J Natl Cancer Inst*, *92*, 1001-5.

[7] Martin, A. O., Dempsey, L. M., Minogue, J., Liu, K., Keller, J., Tamura, R., et al. (1990). Maternal serum alpha-fetoprotein levels in pregnancies complicated by diabetes: implications for screening programs. *American Journal of Obstetrics and Gynecology*, *163*, 1209-16.

[8] Gonzalez-Bugatto, F., Foncubierta, E., Bailen Mde, L., Illanes, S., Hervias-Vivancos, B. and Bartha, J. L. (2009). Maternal and fetal serum transformed alpha-fetoprotein levels in normal pregnancy. *The Journal of Obstetrics and Gynaecology Research*, *35*, 271-6.

[9] Gonzalez-Bugatto, F., Bailen Mde, L., Fernandez-Macias, R., Fernandez-Deudero, A., Hervias-Vivancos, B. and Bartha, J. L. (2009). Transformed alpha-fetoprotein (t-AFP) levels in women with threatened preterm labor. *Gynecologic and Obstetric Investigation*, *68*, 199-204.

[10] Bartha, J. L., Illanes, S., Gonzalez-Bugatto, F., Abdel-Fattah, S. A., Mizejewski, G. J. and Soothill, P. W. (2007). Maternal serum transformed alpha-fetoprotein levels in women with intrauterine growth retardation. *Fetal Diagnosis and Therapy*, *22*, 294-8.

[11] Dong, M., Ding, G., Zhou, J., Wang, H., Zhao, Y. and Huang, H. (2008). The effect of trophoblasts on T lymphocytes: possible regulatory effector molecules--a proteomic analysis. *Cellular physiology and Biochemistry: International Journal of Experimental Cellular Physiology, Biochemistry, and Pharmacology*, *21*, 463-72.

[12] Semeniuk, D. J., Boismenu, R., Tam, J., Weissenhofer, W. and Murgita, R. A. (1995). Evidence that immunosuppression is an intrinsic property of the alpha-fetoprotein molecule. *Adv Exp Med Biol*, *383*, 255-69.

[13] Mizejewski, G. J. (2001). Alpha-fetoprotein structure and function: relevance to isoforms, epitopes, and conformational variants. *Experimental biology and medicine*, *226*, 377-408.

[14] Andersen, L. L., Buschard, K. and Hau, J. (1991). Effect of exogenous insulin treatment on fetal growth and maternal alpha-fetoprotein levels in pregnant mice. *Biomedica biochimica acta.*, *50*, 901-6.

[15] Thornburg, L. L., Knight, K. M., Peterson, C. J., McCall, K. B., Mooney, R. A. and Pressman, E. K. (2008). Maternal serum alpha-fetoprotein values in type 1 and type 2 diabetic patients. *American journal of Obstetrics and Gynecology*, *199*, 135 e1-5.

[16] Butterstein, G., Morrison, J. and Mizejewski, G. J. (2003). Effect of alpha-fetoprotein and derived peptides on insulin- and estrogen-induced fetotoxicity. *Fetal diagnosis and therapy*, *18*, 360-9.

[17] Mizejewski, G. J. and Pass, K. A. (2001). Fatty acids, alpha-fetoprotein, and cystic fibrosis. *Pediatrics*, *108*, 1370-3.

[18] Freedman, S. D., Katz, M. H., Parker, E. M., Laposata, M., Urman, M. Y. and Alvarez, J. G. (1999). A membrane lipid imbalance plays a role in the phenotypic expression of cystic fibrosis in cftr(-/-) mice. *Proc Natl Acad Sci U S A*, *96*, 13995-4000.

[19] Vallette, G., Vranckx, R., Martin, M. E., Benassayag, C. and Nunez, E. A. (1989). Conformational changes in rodent and human alpha-fetoprotein: influence of fatty acids. *Biochim Biophys Acta*, *997*, 302-12.

[20] Mizejewski, G. J. (2003). Levels of alpha-fetoprotein during pregnancy and early infancy in normal and disease states. *Obstetrical and Gynecological Survey*, *58*, 804-26.

[21] Mizejewski, G. J. (2014). Cancer during pregnancy: what is the role of maternal serum and placental biomarkers? A review and commentary. *Tumori*, *100*, 581-9.

[22] Lambe, M., Hsieh, C., Tsaih, S., Ekbom, A., Adami, H. O. and Trichopoulos, D. (1996). Maternal risk of breast cancer following multiple births: A Nationwide Study in Sweden. *Cancer Causes Control*, *7*, 533-8.

[23] Vatten, L. J., Romundstad, P. R., Odegard, R. A., Nilsen, S. T., Trichopoulos, D. and Austgulen, R. (2002). Alpha-foetoprotein in umbilical cord in relation to severe pre-eclampsia, birth weight and future breast cancer risk. *Br J Cancer*, *86*, 728-31.

[24] Low, H. P., Tiwari, A., Janjanam, J., Qiu, L., Chang, C. I., Strohsnitter, W. C., et al. (2013). Screening preeclamptic cord plasma for proteins associated with decreased breast cancer susceptibility. *Genomics, proteomics and bioinformatics*, *11*, 335-44.

[25] Lambe, M., Trichopoulos, D., Hsieh, C. C., Wuu, J., Adami, H. O. and Wide, L. (2003). Ethnic differences in breast cancer risk: a possible role for pregnancy levels of alpha-fetoprotein? *Epidemiology*, *14*, 85-9.

[26] Lagiou, P., Samoli, E., Lagiou, A., Lambe, M., Trichopoulos, D., Adami, H. O., et al. (2007). Levels and correlates of alpha-fetoprotein in normal pregnancies among Caucasian and Chinese women. *European Journal of Cancer Prevention: The Official Journal of the European Cancer Prevention Organisation*, *16*, 178-83.

[27] Bukowski, R., Chlebowski, R. T., Thune, I., Furberg, A. S., Hankins, G. D., Malone, F. D., et al. (2006). Birth weight, breast cancer and the potential mediating hormonal environment. *PLoS One*, *7*, e40199.

[28] Nagata, C., Iwasa, S., Shiraki, M. and Shimizu, H. (2006). Estrogen and alpha-fetoprotein levels in maternal and umbilical cord blood samples in relation to birth weight. *Cancer Epidemiol Biomarkers Prev*, *15*, 1469-72.

[29] Jacobson, H. I., Lemanski, N., Narendran, A., Agarwal, A., Bennett, J. A. and Andersen, T. T. (2008). Hormones of pregnancy, alpha-fetoprotein, and reduction of breast cancer risk. *Adv Exp Med Biol*, *617*, 477-84.

[30] Jacobson, H. I., Lemanski, N., Agarwal, A., Narendran, A., Turner, K. E., 2nd, Bennett, J. A., et al. (2010). A proposed unified mechanism for the reduction of human breast cancer risk by the hormones of pregnancy. *Cancer Prev Res (Phila)*, *3*, 212-20.

[31] Sarcione, E. J. and Hart, D. (1985). Biosynthesis of alpha fetoprotein by MCF-7 human breast cancer cells. *Int J Cancer*, *35*, 315-8.

[32] Moro, R., Gulyaeva-Tcherkassova, J. and Stieber, P. (2012). Increased alpha-fetoprotein receptor in the serum of patients with early-stage breast cancer. *Current oncology*, *19*, e1-8.

[33] Biddle, W. and Sarcione, E. J. (1987). Specific cytoplasmic alpha-fetoprotein binding protein in MCF-7 human breast cancer cells and primary breast cancer tissue. *Breast Cancer Res Treat*, *10*, 279-86.

[34] Sarcione, E. J. and Biddle, W. (1987). Elevated serum alpha fetoprotein levels in postmenopausal women with primary breast carcinoma. *Disease markers*, *5*, 75-9.

[35] Sarui, H., Nakayama, T., Takeda, N., Ishizuka, T. and Yasuda, K. (2001). Alpha-fetoprotein-producing male breast cancer accompanied with hepatocellular carcinoma: assessment by lectin-affinity profile. *The American Journal of the Medical Sciences*, *322*, 369-72.

[36] Martinez, L., Castilla, J. A., Blanco, N., Peran, F. and Herruzo, A. (1995). CA 125, CA 15.3, CA 27.29, CEA, beta-hCG and alpha-fetoprotein levels in cyst fluid of breast macrocysts. International *Journal of Gynaecology and Obstetrics: The Official Publication of the International Federation of Gynaecology and Obstetrics*, *48*, 187-92.

[37] Mizejewski, G. J. (2015). Nonsecreted cytoplasmic alpha-fetoprotein: a newly discovered role in intracellular signaling and regulation. An update and commentary. *Tumour Biology: The Journal of the International Society for Oncodevelopmental Biology and Medicine*.

[38] Smith, J. A., Francis, T. I., Edington, G. M. and Williams, A. O. (1971). Human alpha fetoprotein in body fluids. *Br J Cancer*, *25*, 337-42.

[39] Russo, J. and Russo, I. H. (2004). Development of the human breast. *Maturitas*, *49*, 2-15.

[40] Russo, J., Rivera, R. and Russo, I. H. (1992). Influence of age and parity on the development of the human breast. *Breast Cancer Res Treat*, *23*, 211-8.

[41] Honeth, G., Schiavinotto, T., Vaggi, F., Marlow, R., Kanno, T., Shinomiya, I., et al. (2015). Models of breast morphogenesis based on localization of stem cells in the developing mammary lobule. *Stem cell reports*, *4*, 699-711.

[42] Clarke, R. B., Spence, K., Anderson, E., Howell, A., Okano, H. and Potten, C. S. (2005). A putative human breast stem cell population is enriched for steroid receptor-positive cells. *Developmental Biology*, *277*, 443-56.

[43] Russo, J., Balogh, G. A., Chen, J., Fernandez, S. V., Fernbaugh, R., Heulings, R., et al. (2006). The concept of stem cell in the mammary gland and its implication in morphogenesis, cancer and prevention. *Front Biosci*, *11*, 151-72.

[44] Hassiotou, F., Beltran, A., Chetwynd, E., Stuebe, A. M., Twigger, A. J., Metzger, P., et al. (2012). Breastmilk is a novel source of stem cells with multilineage differentiation potential. *Stem Cells*, *30*, 2164-74.

[45] Hassiotou, F., Geddes, D. T. and Hartmann, P. E. (2013). Cells in human milk: state of the science. *Journal of Human Lactation: Official Journal of International Lactation Consultant Association*, *29*, 171-82.

[46] Guest, I., Ilic, Z., Ma, J., Grant, D., Glinsky, G. and Sell, S. (2010). Direct and indirect contribution of bone marrow-derived cells to cancer. *Int J Cancer*, *126*, 2308-18.

[47] Sell, S. (2008). Alpha-fetoprotein, stem cells and cancer: how study of the production of alpha-fetoprotein during chemical hepatocarcinogenesis led to reaffirmation of the stem cell theory of cancer. *Tumour biology: The Journal of the International Society for Oncodevelopmental Biology and Medicine*, *29*, 161-80.

[48] Ma, J., Lanza, D. G., Guest, I., Uk-Lim, C., Glinskii, A., Glinsky, G., et al. (2012). Characterization of mammary cancer stem cells in the MMTV-PyMT mouse model. *Tumour biology: The Journal of the International Society for Oncodevelopmental Biology and Medicine*, *33*, 1983-96.

[49] O'Malley, B. W., Schrader, W. T., Mani, S., Smith, C., Weigel, N. L., Conneely, O. M., et al. (1995). An alternative ligand-independent pathway for activation of steroid receptors. *Recent Progress in Hormone Research*, *50*, 333-47.

[50] Leclercq, G., Lacroix, M., Laios, I. and Laurent, G. (2006). Estrogen receptor alpha: impact of ligands on intracellular shuttling and turnover rate in breast cancer cells. *Curr Cancer Drug Targets*, *6*, 39-64.

[51] Mizejewski, G. J. (2011). Mechanism of Cancer Growth Suppression of Alpha-Fetoprotein Derived Growth Inhibitory Peptides (GIP): Comparison of GIP-34 versus GIP-8 (AFPep). Updates and Prospects. *Cancers*, *3*, 2709-33.

[52] Sierralta, W. D., Epunan, M. J., Reyes, J. M., Valladares, L. E. and Pino, A. M. (2008). A synthetic peptide derived from alpha-fetoprotein inhibits the estradiol-induced proliferation of mammary tumor cells in culture through the modulation of p21. *Adv Exp Med Biol*, *617*, 463-8.

[53] Mizejewski, G. J. (2009). Alpha-fetoprotein (AFP)-derived peptides as epitopes for hepatoma immunotherapy: a commentary. *Cancer Immunol Immunother*, *58*, 159-70.

[54] Pardee, A. D., Yano, H., Weinstein, A. M., Ponce, A. A., Ethridge, A. D., Normolle, D. P., et al. (2015). Route of antigen delivery impacts the immunostimulatory activity of dendritic cell-based vaccines for hepatocellular carcinoma. *Journal for Immunotherapy of Cancer*, *3*, 32.

[55] Mizejewski, G. J. (2011). Review of the putative cell-surface receptors for alpha-fetoprotein: identification of a candidate receptor protein family. *Tumour Biology: The Journal of the International Society for Oncodevelopmental Biology and Medicine*, *32*, 241-58.

[56] Secret, E., Maynadier, M., Gallud, A., Chaix, A., Bouffard, E., Gary-Bobo, M., et al. (2014). Two-photon excitation of porphyrin-functionalized porous silicon nanoparticles for photodynamic therapy. *Advanced materials*, *26*, 7643-8.

In: Alpha-Fetoprotein: Functions and Clinical Applications
Editors: Nisha Lakhi and Michael Moretti
ISBN: 978-1-63484-875-6

Chapter 19

ALPHA-FETOPROTEIN-MEDIATED IMMUNE TOLERANCE AND ITS REVERSAL

Vladimir N. Pak[1,*], PhD, and Nikolai N. Belyaev[2], DrSci
[1]Freelance researcher, Toronto, Canada
[2]Head of Laboratory of Molecular Immunology and Immunobiotechnology Aitkhozhin's Institute of Molecular Biology and Biochemistry Almaty, Kazakhstan Republic

ABSTRACT

Alpha-fetoprotein (AFP) serves as a shuttle delivery protein to the AFP receptor-positive embryo and cancer cells. By delivering nutrients through receptor-mediated endocytosis, AFP stimulates the proliferation of receptor-expressing cells. On the other hand, AFP also has immune suppressive properties that also can be realized through nutrient delivery to suppressor cells. AFP receptor-positive early myeloid progenitors, myeloid-derived suppressor cells (MDSCs), are possibly the key regulatory cells in the immune cells hierarchy that suppress both innate and adaptive immunity. AFP-nutrient stimulated MDSCs induce local immune tolerance and can play an important role in embryogenesis, cancer, autoimmune disorders, inflammation and transplantation. However, instead of stimulating MDSCs with AFP-nutrient complex, one can deplete them with AFP receptor-targeted toxin, thus reducing the immune tolerance and allowing immune system to normalize the local immune balance and in particular, to attack cancer cells.

ABBREVIATIONS

AFP	alpha-fetoprotein
AFPR	AFP receptor
PUFA	polyunsaturated fatty acid
AFPT	AFP+toxin conjugate/complex
MDSCs	myeloid-derived suppressor cells

* Corresponding author: reducin@gmail.com.

NKs	natural killer cells
Tregs	T regulatory cells
DCs	dendritic cells
NSs	natural suppressor cells
CTLs	cytotoxic lymphocytes
ECs	Ehrlich carcinoma cells
GI	gastro-intestinal
GM-CSF	granulocyte-macrophage colony stimulating factor
TGF-β	transforming growth factor β
IL-10	interleukin-10
kDa	thousand Dalton
VEGF	vascular endothelial growth factor
TNFα	tumor necrosis factor alpha
IFNγ	interferon gamma
NF-κB	nuclear factor kappa-light-chain-enhancer of activated B cells
NO	nitrogen oxide
HLA	human leukocyte antigen
FITC	fluoresceinisothiocyanate
CCR5	C-C chemokine receptor type 5

Introduction

AFP is a shuttle delivery protein that can enter either embryo or cancer cells due to AFPR-mediated endocytosis. By delivering nutrients to the cells, AFP stimulates their growth and proliferation. A majority of cancers re-express AFPR, and that is why AFPT has been used as a targeted chemotherapy. AFP also has immune suppressive properties. By delivering nutrients to AFPR-positive immune suppressive cells, AFP can stimulate and generate immune tolerance in pregnancy and cancer. We have detected AFPR on myeloid progenitor subpopulations – MDSCs. Thus, in addition to using AFP for targeted cancer chemotherapy, AFPT can be used to deplete MDSCs, and therefore activate innate and adaptive immunity against cancer cells. AFPT was shown to generate a response to cancer treatments at very low doses compared to cytotoxic chemotherapy [1], and this can be explained by the AFPT immunotherapy mode of action.

AFP

AFP is an oncofetal protein of 68–72 kDa appearing during early embryonic development. It serves as a pregnancy and tumor marker [2]. AFP is an obligatory component of hematopoiesis. Its synthesis is initiated with embryonal hematopoiesis in the yolk sac visceral endoderm, which is the site of synthesis of all serum proteins. When embryonal hematopoiesis is replaced by fetal hematopoiesis, the liver becomes the main site of blood formation, and the synthesis of AFP is transferred to the liver. The AFP level in maternal serum during pregnancy ranges from 100-500 ng/ml. The level of AFP in the sera of newborn humans is about 0.1 mg/ml and it drops slowly over 3-4 months, at which time it reaches the background level of AFP that normally persists at 5-10 ng/ml in adults [3]. In adults, AFP stimulates tissue regeneration after injury or supports hematopoiesis. Bone marrow stem cells are at the root of the hematopoiesis and can be the source of AFP synthesis in adults.

AFP-related research papers were recently reviewed [4]. We will discuss two functions of AFP: nutrient delivery and immune suppression, and how they may be connected. The most important ligands transported by AFP are PUFAs, namely arachidonic (AA) and docosahexaenoic acid (DHA), and their metabolites that can be detected as complexes with the given protein circulating in blood. The presence of AFP significantly increases the flow of PUFAs into the cytoplasm of activated T-lymphocytes [5]. The increase in the concentration of PUFAs is of great importance because the said acids are not only necessary structural components of the cell membrane, but also serve as an additional source of energy for the cells. PUFAs metabolites, in particular those of AA, can act as secondary messengers, thus participating in the regulation of cellular growth and differentiation [6]. However, AFP is not a transport protein for only PUFAs, but also can form complexes with bilirubin, retinoids, steroids, heavy metals, flavonoids, phytoestrogens, dioxins, various organic drugs and other small molecules [7, 8].

About 30 AFP-binding proteins have been identified, including several classes of integral transmembrane proteins with a molecular mass of 18-250 kDa that have been reviewed by G.J. Mizejewski [9, 10]. The binding and uptake of AFP by the AFPR provides the basis for the concept of the autocrine AFP/AFPR system. The autocrine loop of the AFP growth stimulation has been confirmed in malignant cell lines, and is linked to the cell endocytotic delivery of ligands. AFP functions as a shuttle-transporter, bringing the ligand inside the cell, and then returning into circulation to repeat the cycle [11].

The second and most intriguing AFP function - immune suppression - has been studied for more than 30 years [12-15]. AFP has been shown to stimulate suppressor T lymphocytes [16, 17], inhibit B cell maturation and antibody production [18, 19], inhibit cytolytic T lymphocyte activity [20, 21] and cytolytic function of lymphokine activated NKs [22, 23], as well as modify the activity of antigen presenting cells [21, 24-26]. In this regard, AFP has been suggested to induce development of maternal immune tolerance to fetal antigens during pregnancy [7, 27]. AFP has shown high suppressive activity in experiments on transgenic animals [28, 29], and it is assumed that AFP plays a general role as an immunosuppressive molecule [30, 31]. The cellular interactions between the cells of the innate immune system and AFP have long been recognized [6, 32, 33]. AFP suppresses monocyte-derived DCs function [34, 35], as well as DCs activity, and stimulates the development of Tregs that produce TGF-β [36].

AFP Use in Treating Autoimmune Diseases

The ability of AFP to suppress the immune response was used to treat autoimmune diseases. Direct immunologic response observations on the role of AFP have been reported [37-39]. It was found that fetal AFP prevented the development of myasthenia gravis in rabbits. Furthermore, that clinical signs of the disease in effected animals disappeared when they were treated with AFP [37]. Experimental allergic encephalomyelitis, induced in guinea pigs, was successfully treated as well as partially prevented by administration of AFP [38]. However, AFP *per se* is essentially a growth-promoting protein and can induce undesired side effects, as the biologic response cannot be easily predicted. For example, natural AFP injected into the tumor-bearing mice can inhibit the immune response to cancer cells and therefore accelerate the tumor growth [40]. In cancer patients, AFP injections can induce an acute inflammatory immune response [41] and the therapeutic use of the human AFP can stimulate the growth of existing tumors [42].

Alfetin (75 mcg/ampoule), which is the natural human AFP (hAFP) extracted from aborted human tissues, was registered and used in Russia as an immune modulating drug [43], but it was later removed for ethical reasons. Alfetin injections (0.075-0.3 mg/day) led to a considerable relief from autoimmune diseases symptoms similar to those experienced naturally by pregnant women [7]. The recombinant hAFP has also demonstrated immune suppressive properties [44, 45], but failed to demonstrate efficacy in phase II trials. In our opinion, these clinical trials might have failed because of poor design. On the other hand, AFP cooperation with some drugs/ligands in the body was registered during Alfetin treatments, and should be taken into account in hAFP trials.

AFP And AFP-Ligand Complexes

The AFP hydrophobic pocket can bind PUFAs and many other ligands [46, 47]. With five days of half-life in the body, and requiring only about an hour for the ligand unloading [48], one molecule of AFP can shuttle dozens of molecules into receptor-positive cells. Thus, the hook effect can be observed in hepatocellular carcinoma: AFP is shuttling into the tumor and gives a low number in the automated immunoassay method as a blood tumor marker [49]. AFP delivers nutrients not only to embryo and cancer cells, but also to the key immune cells involved in immune regulation. For example, AFP serves as a chaperone protein for ligands, both endogenous and ubiquitous in nature, which functions cooperatively to impair DCs differentiation and function [50]. Thus it might be useful to investigate the substances that demonstrate binding affinity to AFP, together with their effect on the immune function, especially the immunosuppressive activity [51].

Natural AFP isolation and purification methods did not guarantee its separation from the non-covalently bound ligands. To purify AFP from the PUFAs and other ligands the delipidation procedure should be applied [52]. Before recombinant AFP became available, practically no data on the AFP immunosuppressor properties were obtained using the pure protein. Instead AFP-enriched sera or amniotic fluid was used during *in vitro* experiments. The immune-modulating properties of AFP were shown to be dependent on the origin of the preparation and the purification methods. There was a discussion about whether immune-modulating properties belong to AFP itself or to the ligand bound to it. Physiological levels of AFP can exhibit a dose-dependent growth-regulatory activity towards a sensitive tumor or developing cells. AFP could either stimulate or inhibit growth activity, depending on its relative concentration, exogenous or endogenous cytokines, and growth factors in the cell culture medium. The result is the combined effect of AFP and various other secreted growth factors [53].

AFP, depending on its origin and surrounding conditions, can perform different functions using different mechanisms. First, there is a regulatory effect on the concentration of the unbound form of its various ligands. In particular, PUFAs affect positively or negatively, many steps in the action chain of various steroids and many enzymes involved in the transduction of membrane-triggered signals. Second, different conformations and glycosylated isoforms of AFP [30], depending on the nature and concentration of the ligand(s) bound to it, might influence the binding of the protein to specific receptor(s) and as a consequence influence their biological properties. Third, the protein can influence effects associated with other signaling molecules, such as growth factors. In our opinion, though AFP molecule itself or its peptides have immunosuppressive effect, the biggest suppressive effect is provided by the delivery of dozens of ligands in a shuttle manner to specialized immune cells.

The importance of the ligand bound to AFP can be demonstrated by the two following examples. First, the reversible non-covalent complex of AFP extracted from human cord blood with 99% purity and containing no low molecular weight impurities and N-arachidonoyl aminoethylphosphate (N-AAP), a metabolite of AA, was used as an immunostimulating agent [54]. N-AAP itself does not exhibit immunogenic activity, AFP alone, administered in corresponding doses does not significantly affect the immunogenic characteristics of sheep erythrocytes in mice. However, administration of the AFP/N-AAP complex resulted in an increase of 87% in the relative amount of antibody-forming cells (AFC) in the spleen and an increase of 162% in the total AFC amount on the 5th day after injection in comparison with the amount of cells in animals immunized only with sheep erythrocytes.

In the second example, the recombinant hAFP was shown to synergistically potentiate the anti-cancer effects of apoptosis inducer 1’-S-1’ acetoxychavicol acetate when used as a non-covalent complex against human tumors [55]. Therefore, the ligands and their potency are most important for AFP action.

AFP-Mediated Immune Suppression in Pregnancy and Cancer

Pregnancy is a natural phenomenon that ensures survival of the species. Understanding how maternal immune system recognizes foreign antigens and how the mechanism behind gestational tolerance operates may inspire novel strategies that ensure immune tolerance in other conditions: regeneration, inflammation and transplantation. During pregnancy, the mother’s immune system is generating anti-embryo response, which is neutralized by the embryo-secreted factors leading to immune tolerance. This similar immune tolerance can be seen in cancer. Anti-cancer immune response exists, but it is suppressed by the factors secreted by cancer cells [56].

V. I. Govallo has studied the similarities between mother-fetus and host-tumor immune interaction systems [57]. He found that a protein extract of human placenta effectively blocks all expressions of cell immunity when added to a test tube of white blood cells. In cancer patients he found a quantitative reduction of the tumor mass after immunization with placental extract, and decreased production of immune blocking factors. It was not exactly clear how the placental extract works, but the statistics in his book show a 77.1% 5-year survival rate in the group of 35 patients with metastases of the breast, lung, hemangiopericytoma, synovial sarcoma, and other cancers [57].

Placenta extract injections lead to elimination of immune tolerance to the tumor only in uncompromised patients (with no prior chemotherapy or irradiation). This indicates that immune system should be healthy to generate anti-cancer response. More than 60% of the cancer patients who were considered incurable have survived for more than 10 years with immunity indicators showing a healthy immune system status. The de-blocking therapy of cancer through the use of products of embryonic origin can provide a new solution to the problem. Even a single subcutaneous immunization with a 20 ml of placental protein fraction extract achieved a regression of lung cancer metastases [57]. Fever, shivering, and slight weakness were sometimes observed following the injections. These symptoms did not require any medication, and disappeared within 24 hours. There were tactile sensations in the metastatic sites: patients previously unaware of the location of their tumors could now point to them.

Fetal placenta has a central role in promoting the homeostatic environment necessary for successful pregnancy. Decidual (uterine) macrophages and Tregs are key regulators of fetal tolerance, though it is not known how these tolerogenic leukocytes are induced [58]. The mother's immune tolerance to the embryo undoubtedly involves AFP, which only the embryo can secrete, unlike mother-secreted sex hormones that are also known to influence the tolerance pathways [58, 59]. Though the individual proteins of the extract were not identified, AFP obviously is the major one of them [60] and it is bound with unknown ligands. Hence, vaccination with the AFP-ligand complexes is likely responsible for the reversal of immune tolerance to tumor.

Cancer is an "embryo's evil twin"; its cells re-express many of the same genes and proteins, with AFP and AFPR among them. Although, AFP is found in high concentration in blood of patients with malignant tumors of the liver, gonads, and in cases of embryonic carcinoma [61, 62], the majority of cancers (breast, lung, prostate, ovarian, stomach, etc.) re-express AFPR [9, 63]. The dominance of the AFPR expressing cancers over AFP-expressing ones can suggest that the host origin AFP, that normally persists at 5-10 ng/ml in adults, and stimulates tissue regeneration or hematopoiesis, can be co-opted by cancers, thus locally feeding them and the AFPR positive immune suppression cells with stimulating ligands.

MDSCs

MDSCs are heterogeneous cell populations that possibly play a key role in the whole immune suppression network. MDSCs, being immature myeloid progenitors of granulocytic and monocytic branches of hematopoiesis, originate in the bone marrow from hematopoietic stem cells. In some circumstances they can leave their natural niche and spread throughout the body becoming immune cells with suppressor functions. Cancer uses the immune system to keep pumping out large numbers of mature and immature myeloid cells from the bone marrow to engage in the proliferation and activation at extramedullary locations [64]. Being normal host cells, MDSCs enhance cancer growth and suppress the specialized T cells that target and destroy cancer cells. In mouse tumor models, MDSCs infiltrate blood, normal peripheral organs, and tumors themselves, displaying persistent immature elements of the monocyte or granulocyte morphology. MDSCs and the Tregs are the major components of the immune-suppressive tumor microenvironment [65]. Induction of MDSCs is thought to play the major role in the tumor-mediated immune evasion [66]. The extreme level of the immunosuppressive activity of MDSCs is likely one of the reasons the immune system fails to inhibit cancer growth. MDSCs are more abundant in cancer patients than in healthy individuals. There exists an apparent correlation between the MDSCs levels in cancer patients' blood and the tumor progression [67]. Tumors were found to secrete a protein 10 – 100 kDa that stimulates MDSCs immune suppression activity (AFP with MW = 67,000 Da was not specifically tested) [68].

Historically, these cells appeared in the field of cancer immunology in the late 1980's under the name of "natural suppressor cells" (NSs) [69]. These cells have been described as myeloid progenitors, lacking linear differentiation markers which non-specifically inhibit various immune reactions [70, 71]. Their activity was augmented significantly in the bone marrow, spleen, and blood during the development of experimental tumors and cancer [72-74]. The term "natural" underlined the possibility of NSs to suppress different immune reactions without participation of MHC-dependent genetic restriction, by analogy with NKs. In those years several authors suggested that overwhelming majority of the AFP immunosuppressive activities both *in vitro* and *in vivo* was mediated by activation of special NSs [75-78].

We studied the influence of pure hAFP on the NSs activity of mice bone marrow cells [32, 40, and 79]. It was found that heavy fraction (buoyant density 1.09 g/ml) of bone marrow cells secreted some suppressor factors under influence of hAFP in dose-dependent manner. These factors inhibited proliferation of both myeloma cells and mitogen-activated lymphocytes. Moreover, these factors strongly inhibited NKs cytolytic activity. In fact, hAFP *per se* in the concentration used, induced some inhibition of NKs, but its level was minimal [79]. It was the first time when NSs stimulating activity of AFP was demonstrated directly. Anti-proliferative and NKs inhibiting activities of NSs correlated very closely. This allowed us to conclude that the major part of the anti-mitogenic and NKs-inhibiting activities was caused by the similar/same suppressor factors and AFP was a strong natural stimulator of NSs, which were able to produce the factors inhibiting NKs activity. So, AFP production in some cancer cases is not only a tumor marker, but also an important factor facilitating tumor growth. Such notion has been confirmed when a single injection of AFP into intact mice led to an increase of spontaneous NSs activity and to a 20% reduction of NKs activity [40]. Moreover, NSs activity was increased when ECs were subcutaneously inoculated three days after AFP injection. In the AFP-treated mice, the tumor mass at 14 days was 60% larger than that in the untreated mice.

Although convincing evidence for a negative role of NSs in tumor progression has been described [70-73], further studies in this field were limited by the absence of satisfactory methods to identify and measure NSs activity. The phenotype heterogeneity of NSs also causes additional difficulties. In 1998 a new approach to the study of NSs was described in the work published by V. Bronte and colleagues [80]. They described myeloid cell population with phenotype $Gr1^+CD11b^+$, which inhibited effector $CD8^+$ T cell activity. Over the next decade numerous studies on cells with such phenotype, named MDSCs [81], have been published.

Meanwhile studies in the field of cancer immunology have led to the conclusion that the immune system not only is unable to attack advanced cancers, but also promotes their growth. The novel view on the problem of relationship between tumor and immunity has been expressed in hypothesis of cancer immunoediting [82]. According to this hypothesis, development of interaction between malignant tumor and immune system passes through three sequential phases: elimination, equilibrium, and escape. Elimination represents implementation of immunological surveillance, when innate and adaptive immune systems work together to recognize growing nascent transformed cells and destroy them without noticeable clinical presentations. If few resistant transformed cells survive, elimination phase proceeds to equilibrium phase in which adaptive immune system prevents tumor growth but simultaneously edits tumor, making it less immunogenic [83]. NKs as a part of innate immune system can also edit tumor cells [84]. Tumor goes into a dormant state and the equilibrium phase may last over a period of many years. In the escape phase, cancer cells that have acquired the ability to circumvent immune recognition and destruction begin to rapidly proliferate, forming a visible tumor. Two related mechanisms are behind this event: the loss of immunogenicity due to immunoediting and changes in the immune system in response to immunosuppressive activity of growing tumor and its microenvironment.

Namely, the immunosuppressive microenvironment, formed in addition to the tumor by the presence of tumor-associated fibroblasts [85], suppressive endothelial cells [86], tumor-associated macrophages [87, 88], tolerogenic DCs [89], FoxP3$^+$ Tregs [90, 91] and MDSCs [91], provides functional anergy of cytolytic NKs, CD8$^+$ CTLs, and conversion of conventional CD4$^+$ T helper cells into suppressive Tregs. This results in an uncontrolled tumor growth.

Among the cells involved in suppression of antitumor immunity, MDSCs have a special place because they inevitably accompany the development of malignant tumor in both animal models [93] and in clinic [94].

To date several detailed reviews on the subject of MDSCs have been published [92-101]. Current ideas about MDSCs and their role in suppression of anti-tumor immunity are briefly described below.

MDSCs in various murine tumor models usually express integrin CD11b and linear differentiation antigen “granulocyte receptor-1” (Gr-1), which are markers of granulocytic and monocytic lines respectively. There are two subpopulations of MDSCs: monocytic (M-MDSCs) and granulocytic (G-MDSCs). A number of studies showed that phenotype of these cells in tumor-bearing mice is not identical to immature myeloid cells of intact animals. After tumor eradication or destruction of myeloid cells with monoclonal antibodies to CD11b, the immune system reverts to its original status and acquires lost anticancer activity.

Phenotype of human MDSCs is not completely clear. These cells in humans have been found not to express linear markers of mature cells (CD3, CD19, CD56, CD14, HLA-DR), but express myeloid markers CD31, CD33, CD11b, and CD15. MDSCs population with CD14$^+$ HLA-DR$^{-/low}$ phenotype has been found in prostate cancer and melanoma. Thus, different types of cancers can induce various MDSCs subpopulations.

Apparently, MDSCs accumulation is a general and dominant process in immunosuppression, observed in bone marrow, spleen, blood, liver of tumor-bearing animals, and blood of people with different types of cancer. Myeloid cells acquire suppressive capacity under the influence of tumor factors such as GM-CSF, M-CSF, IL-3, IL-10, TGF-β, VEGF,

IL-6, IL-1β, and SCF. Production of these factors by tumor is always associated with poor prognosis of the disease.

Tumor-associated cytokines and chemokines can also control the differentiation of neutrophils into MDSCs and enhance MDSCs homing to tumor. MDSCs express a number of chemokine and cytokine receptors, including CCR2, CXCR4, CXCR2, CD117 and VEGFR1. Among them CCR2 is considered the main marker of MDSCs. MDSCs have been shown to migrate into tumor in response to cellular damage, infection, or inflammatory mediators that belong to the family of S100 proteins binding calcium. These proteins mediate MDSCs accumulation in the locus of tumor microenvironment by blocking differentiation of myeloid precursors into mature DCs, and by chemoattraction of MDSCs into tumor (NF-κB-dependent signaling pathway).

The main property of MDSCs, isolated from tumors of animals, is the suppression of T- and NK-cell activation. The mechanism of this suppression varies depending on the subpopulation of MDSCs.

Suppression of T cells by MDSCs, is associated with the production of peroxynitrite (ONOO-), which is generated by arginase (ARG1) and inducible NO-synthase (iNOS). iNOS induced by TNFα and IFNγ, converts L-arginine into L-citrullin following by production of NO, which in turn leads to formation of peroxynitrite. iNOS inhibits T-cell activation and proliferation by blocking Jak 3 and STAT 5 activation. iNOS also inhibits expression of MHC II genes and induces apoptosis.

ARG1 converts L-arginine into L-ornithine, a substance needed for formation of polyamines (putrescin, spermidine, spermine) involved in cell transformation and tumor cell proliferation. ARG1 production by MDSCs induced by Th2-type cytokines (IL-4 and IL-13) leads to disruption of CD3 expression, a component of T-cell receptor. In addition, ARG1 increases production of superoxide anion, which after interaction with NO forms peroxynitrite and hydrogen peroxides causing suppression of T-cell functions in cancer.

iNOS can be induced in MDSCs by VEGF, GM-CSF and IL-6. Two mechanisms of participation of NO in suppressive activity of MDSCs have been suggested. The first one is connected with synergistic interaction of ARG1 and production of superoxide and NO stimulated by iNOS. The second one is likely dependent only on iNOS and NO production, which directly inhibits CTLs activity induced both by antigens and cytokines.

In addition to NO transforming growth, TGF-β is involved in suppressive activity of MDSCs. Production of this well-known cytokine is usually induced by IL-13 which is secreted, presumably, by NKs, which in turn, receive a signal for activation from antigen-presenting cells through a tumor antigen (glycolipid) associated with CD1b molecule. TGF-β directly inhibits antitumor activity of CTLs.

On top of their ability to directly suppress immune effector cells, MDSCs act as tolerogenic antigen-presenting cells and induces activation of $CD4^{+}CD25^{+}Foxp3^{+}$ Tregs that represent another family of tumor-associated suppressor cells [102]. This induction of Tregs requires ARG1 production by MDSCs and expression of CD80. Antigen-specific immunosuppression induced with G-MDSCs requires participation of IFNγ, but immunosuppression induced with M-MDSCs depends on participation of NO. Tumor associated MDSCs also inhibit other cells involved in the antitumor response such as DCs and macrophages.

Alpha-Fetoprotein-Mediated Immune Tolerance and Its Reversal

MDSCs, as one of the major factors of anticancer immunity suppression, can limit positive effect of cancer immunotherapy. Thereby development of new approaches that would effectively eliminate *in vivo* the number and/or function of MDSCs in tumor-bearing host is a critical problem of cancer immunology [103-105].

These approaches include: (1) blocking production of factors secreted by tumor and its microenvironment that attract and activate MDSCs (for instance inhibition of GM-CSF production by vitamin D3), (2) elimination of MDSCs with various cytotoxic agents (gemcitabin, 5-fluorouracil, docetaxel, doxorubicin, cyclophosphamide), (3) stimulation of MDSCs maturation into mature dendritic cells, macrophages and granulocytes (all-trans-retinoic acid), (4) blockage of immunosuppressive function of MDSCs (nitroaspirin, inhibitors of phosphodiesterase 5), (5) blockage of accumulation and suppressive function of MDSCs (sunitinib, anti-CD117 (c-Kit)). However, none of these agents (except 5-fluorouracil which has no significant effect on T cells, B cells, NKs, or DCs [106]) have selective effect on MDSCs. Moreover, many of these agents in *in vivo* tests on tumor-bearing mice showed antitumor effect only in combination with immunotherapy (immunization with tumor antigens, cell therapy with DCs) or other chemotherapy. Consequently, a search for more specific approaches that would allow eliminating MDSCs without damaging other cells of antitumor immunity is necessary.

We suggested depleting MDCSs by AFP-toxin conjugate that specifically recognizes MDSCs by AFPR [107]. AFPT can deplete MDSCs, break the immune tumor tolerance, and release the immune system to attack cancer cells. It is interesting to note that MDSCs additionally express CCR5 receptor that also directly binds AFP [67, 108, 109]. In our laboratory it has been shown that hAFP, obtained from umbilical cord blood [110], stimulated suppressor activity of NSs that are identical to MDSCs [32, 40, 79, 111-113]. Three populations of these cells expressed 13 AFP-binding proteins with molecular mass of 18-305 kDa, but only 4 of them bound serum albumin [114]. Seven specific AFP-binding membrane and cytoplasmic proteins with molecular masses of 42, 50, 65, 67, 74, 147 and 182 kDa with high constant of association (10^{10}-10^{11} M^{-1}) were found.

First we have studied AFP-FITC conjugate binding to different subsets (G-MDSCs and M-MDSCs) obtained by immunomagnetic separation from the spleen cell suspension of CBA mice bearing ECs (Figure 1, A, unpublished data). Approximately half of both MDSCs fractions possessed AFPRs. Cytotoxic conjugate, prepared from purified hAFP, chemically linked to anticancer drug daunorubicin by glutaraldehyde, has shown 50% killing efficacy in both MDSCs fractions *in vitro*. This conjugate was injected five times intraperitoneally over 2 days interval into CBA mice bearing subcutaneous ECs. Human AFP was shown to be immunogenic in a mouse; that is why the duration of our experiment was no more than 3 weeks. Furthermore, there are differences between human and mouse AFP and AFPR. Thus binding of hAFP to mouse MDSC might differ from that of mouse AFP to mouse AFPR. Apparently AFP receptors on the cells of different species recognize conservative epitopes of AFP, e.g., we have found virtually 100% cells of mouse ECs bounded to hAFP-FITC.

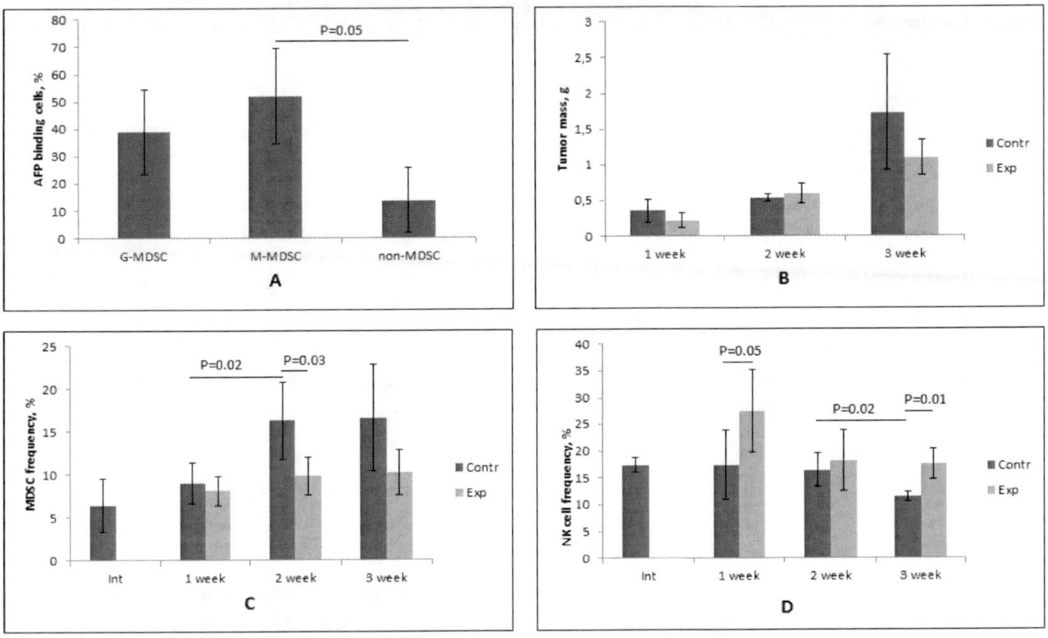

Figure 1. The effect of AFP conjugates on MDSCs and tumor growth. A – AFP binding to different MDSCs subsets. Purified hAFP obtained from the cord blood [109] was conjugated to FITC. MDSCs fractions (Gr-1^{high}CD11b^+ - G-MDSCs, Gr-1^{dim}CD11b^+ -M-MDSCs, the others cells – non-MDSCs) were obtained by immunomagnetic separation using MDSCs kit (Miteniy, Germany) from the spleen of CBA mice bearing subcutaneous ECs during two weeks. Each fraction was incubated with AFP-FITC conjugate and cell fluorescence was estimated on a FACSCalibur cytometer (DB Bioscience, USA). B, C, D – the influence of AFP-daunorubicin conjugate on tumor mass growth (B), MDSCs (C) and NKs (D) content in CBA mice spleen. ECs ($5x10^5$) were injected subcutaneously into CBA mice. After 3 days AFP-daunorubicin conjugate was injected intraperitoneally 5 times over 2 days interval. Every week after the ECs injections tumors were surgically resected and weighted. Simultaneously suspensions of the spleen cells were obtained incubated with fluorescent-labeled monoclonal antibodies (Gr-1-APC, CD11b-PerCP, and CD49b-APC) and percentage of MDSCs and NKs were estimated by flow cytometry. Abbreviations: Int – intact group, Contr – control group, Exp – experimental group.

Weekly analysis has shown delay of tumor mass growth in experimental group compared with control untreated animals (Figure 1, B unpublished data). Moreover, the inhibition of MDSCs (Gr-1^+CD11b^+) level increased, and NKs (CD49b^+) decreased level was observed during the study period (Figure 1, C, D unpublished data). Our data suggest that this approach can provide an effective and safe cancer therapy.

The AFPR presence on MDSCs provides a rational explanation for our previously published results regarding high anti-cancer efficacy of the AFPT with 6 out of 8 patients response [115]. In treating cancer patients (V. Pak, unpublished) we have found that:

1. The AFPT infusions [115] caused an acute phase immune response with fever and shivering.
2. The AFPT treatment decreased the population of monocytes in the blood.

3. The monocytes population that has dropped after AFPT treatments was successfully restored with GM-CSF.
4. The response lasted up to 3 months after 1 month of AFPT treatments, which can only be attributed to the continuing activity of the immune system. Sometimes there were tactile sensations in the metastatic sites: patients previously unaware of the location of their tumors could now point to them. We have assumed before, that main mechanism of AFPT efficacy is due to toxin targeted delivery to cancer cells, but the therapeutic doses needed to obtain the response in our experiments were surprisingly low.

AFPT can play the dual role in the cancer treatment: first, as a cancer immunotherapy by reducing the population of AFPR-positive MDSCs and second, as the targeted chemotherapy for AFPR-positive cancer cells.

AFPT can be a chemical conjugate or a non-covalent complex and comparison of the two was done in [1]. We used hAFP + amphotericin B complex for infusions [115], while the porcine AFP complexes with thapsigargin or atractyloside were used for oral administration [116]. Unlike oral formulations, infusions provided direct AFPT interaction with immune or cancer cells. Special drug formulations help AFPT non-covalent complexes to pass through acidic gastric environment without dissociation or being digested and to be released in the intestine. But the hypothetical transport of AFP or AFPT into the blood from the GI tract [1] was not reported. Nevertheless, the oral porcine AFP + atractyloside non-covalent complex formulation have shown response in 6 out of 12 metastatic colorectal cancer patients [117]. Taking into account the low AFPT dose used (0.6 mg AFP+0.012 mg atractyloside/day), this result was difficult to explain as targeted chemotherapy. Now we assume that the oral AFPT formulation serves as immunotherapy by depleting immune-suppressive cells with similar to MDSCs functions in Peyer's patches where AFPT are absorbed through the small pores in intestinal capillaries. The immune cells in the intestinal lining are the central components dedicated to the search of foreign antigens consumed with food. They receive information about microbes within the intestine, induce the production of both CTLs and the Tregs that control the immune response. The immune system of small intestine has a major impact on the immune response throughout the body. It collects antigens from the intestinal contents and controls the generation of T cells that suppress inflammation [118]. A subset of DCs within the intestinal lining transport the compounds absorbed from the intestine to the bloodstream. DCs collect and process both circulating and intestinal antigens, inducing the production of proper suppressor cells that in turn induce the tolerance to antigens in food. These intestinal suppressor cells secrete proteins that suppress the activation of helper cells, preventing intestinal inflammation.

We speculate that oral AFPT formulation can deplete the immune-suppressive cells in the GI tract's Peyer's patches, in response, the immune system will release the stimulating cytokines/cells. These cytokines/cells will travel in blood to the liver where the local immune tolerance to metastasis is reversed. This pathway can possibly explain the efficacy of the low oral AFPT doses in the metastatic colorectal cancer treatments.

DISCUSSION

Cancers do not "reinvent the wheel," rather they defend themselves from immune system attack using the same effective mechanism that embryos deploys during pregnancy. AFP-ligand complexes serve as an inducer of embryo and cancer tolerance suppressing the immune system and accelerating their growth. By substituting the ligand (e.g., PUFA) for the toxin in the AFP-ligand non-covalent complex [1] one can deplete immune suppressive cells and restore anti-cancer immunity.

For the first time we have demonstrated the AFPR presence on MDSCs and AFPT ability to specifically deplete MDSCs and consequently allow the immune system to attack cancer cells. Targeting MDSCs with toxins is a promising direction in immunotherapy [119]. As a result, the immune suppression will be reversed and the cancer cells will be targeted by NKs and the adaptive immune cells. AFPT immunotherapy can be used not only to treat AFPR-positive cancers but a majority of cancers.

The MDSCs depletion releases a stronger immune attack on cancer than the currently used immune checkpoint inhibitors [120] since it releases both the adaptive immunity and innate immunity, which is the main driver in cancer cells elimination. Thus MDSCs are an attractive target for developing safe and effective anticancer treatments, and the use of AFPT is a promising approach to solving this urgent problem.

A cancer can be characterized as an apoptosis failure at the cellular level and induced immune tolerance at the whole body level. Apoptosis can be addressed by targeted chemotherapy while the immune tolerance can be overcome with the immunotherapy. AFPT serves both of these purposes. The AFPT immunotherapy can be used alone or added to the existing chemotherapies. Low doses of the AFPT have demonstrated rapid regression of metastasis as well as improvements in survival. Thus, whether AFPT is an immunotherapy or a targeted chemotherapy or both, it is a promising anti-cancer drug.

REFERENCES

[1] Pak, V. (2014). The use of alpha-fetoprotein for the delivery of cytotoxic payloads to cancer cells. *Ther Deliv*, *5*, 885-92.

[2] Abelev, G. I. and Lazarevich, N. L. (1996). Alpha-Fetoprotein (AFP): Solved and Unsolved Problems. *Mc Gill J Med*, *2*, 127–34.

[3] Mizejewski, G. J. (2003). Levels of alpha-fetoprotein during pregnancy and early infancy in normal and disease states. *Obstetrical and Gynecological Survey*, *58*, 804-26.

[4] Terentiev, A. A. and Moldogazieva, N. T. (2013). Alpha-fetoprotein: a renaissance. *Tumor Biol*, *4*, 2075-91.

[5] Torres, J. M., Anel, A. and Uriel, J. (1992). Alpha-fetoprotein-mediated uptake of fatty acids by human T lymphocytes. *J. Cell. Physiol*, *150*, 456–62.

[6] Benassayag, C., Mignot, T. M., Haourigui, M., Civel, C., Hassid, J., Carbonne, B., Nunez, E. A. and Feme, F. (1997). High polyunsaturated fatty acid, thromboxane A, and alpha-fetoprotein concentrations at the human feto-maternal interface. *Journal of Lipid Research*, *38*, 276-86.

[7] Mizejewski, G. J. (2004). Biological roles of alpha-fetoprotein during pregnancy and perinatal development. *Exp Biol Med*, *229*, 439-63.
[8] Mizejewski, G. J. (2001). Alpha-fetoprotein structure and function: relevance to isoforms, epitopes, and conformational variants. *Exp Biol. Med*, *226*, 377-408.
[9] Mizejewski, G. J. (2010). Review of the putative cell-surface receptors for alpha-fetoprotein: identification of a candidate receptor protein family. *Tumor Biol*, *32*, 241-58.
[10] Mizejewski, G. J. (2014). The adenocarcinoma cell surface mucin receptor for alpha-fetoprotein: is the same receptor present on circulating monocytes and macrophages? A commentary. *Tumor Biol*, *35*, 7397-402.
[11] Esteban, C., Geuskens, M. and Uriel, J. (1991). Activation of an alpha-fetoprotein (AFP)/receptor autocrine loop in HT-29 human colon carcinoma cells. *International Journal of Cancer*, 49, 425-30.
[12] Murgita, R. A. and Tomasi, T. B. (1975). Suppression of the immune response by alpha-fetoprotein. *J. Exper. Med*, 141, 440-52.
[13] Chakraborty, M. and Mandal, C. (1993). Immuno-supressive effect of human alpha-fetoprotein: a cross species study. *Immunol. Invest.*, *22*, 329-39.
[14] Sheppard, H. W., Sell, S., Trefts, P. and Bahn, R. (1977). Effects of AFP on murine immune responses: I. Studies on mice. *J. Immunol*, 119, 91-7.
[15] Sell, S., Sheppard, H. W. and Poler, M. (1977). Effects of alpha-fetoprotein on murine immune responses. II. Studies on rats. *J. Immunol*, *119*, 98-103.
[16] Murgita, R. A., Goidl, E. A., Kontianen, S. and Wigzell, H. (1977). Alpha-fetoprotein induces suppressor T-cell *in vitro*. *Nature*, *19*, 257-69.
[17] Murgita, R. A., Goidl, E., Kontiainen, S., Beverley, P. C. and Wigzell, H. (1978). Adult murine T-cells activated in vitro by AFP and naturally occuring T-cells in newborn mice: identity in function and cell surface differentiation antigens. *Proc. Natl. Acad. Sci. U.S.A.*, *75*, 2897-901.
[18] Czokalo, M. and Wisniewski, L. (1979). Inhibition of blastic transformation and immunoglobulin release in lymphocytes cultured in the presence of alpha-fetoprotein (AFP). *Exp. Pathol*, 17, 176-80.
[19] Um, S. H., Mulhall, C., Alisa, A., Ives, A. R., Karani, J., Williams, R., Bertoletti, A. and Behboudi, S. (2004). α-Fetoprotein impairs APC function and induces their apoptosis. *J Immunol*, *173*, 1772-78.
[20] Mizejewski, G. J. (2002). Biological role of α-fetoprotein in cancer: prospects for anticancer therapy. *Expert Rev Anticancer Ther*, *2*, 709-35.
[21] Peck, A. B., Murgita, R. A. and Wigzell, H. (1978). Cellular and genetic restrictions in the immunoregulatory activity of alpha-fetoprotein. I. Selective inhibition of anti-Ia-associated proliferate reactions. *J. Exp. Med*, *147*, 667-83.
[22] Cohen, B. L., Orn, A., Gronvik, K. O., Gidlund, M., Wigzell, H. and Murgita, R. A. (1986). Suppression by alpha-fetoprotein of murine natural killer cell activity stimulated *in vitro* and *in vivo* by interferon and interleukin 2. *Scand. J. Immunol*, *23*, 211-23.
[23] Cardoso, E., Valdez, G., Comini, E. and Matera, L. (1991). Effect of human alpha-fetoprotein on native and *in vitro*-stimulated NK activity. *J. Clin. Lab. Immunol*, *34*, 183-88.

[24] Vetvicka, V., Holub, M., Kovcru, H., Siman, P. and Kovcru, F. (1988). Alpha-fetoprotein and phagocytosis in athymic nude mice. *Immunol. Lett*, *19*, 95-8.

[25] Aussel, C. and Fehlmann, M. (1986). Alpha-fetoprotein stimulates leukotriene synthesis in P388D1 macrophages. *Cell. Immunol*, *101*, 415-24.

[26] Laan-Potsep, K., Wigzell, H., Cotran, P. and Gidlund, M. (1991). Human alpha-fetoprotein (AFP) causes a selective down regulation of monocyte MHC class II molecules without altering other induced or noninduced monocyte markers or functions in monocytoid cell lines. *Cell. Immunol*, *133*, 506-8.

[27] Yin, Z. F. and Wang, C. H. (2003). Research advances on alpha-fetoprotein physiological function and clinical potential. *Ai Zheng*, *22*, 108-21.

[28] Yamashita, T., Nakane, A., Watanabe, T., Miyoshi, I. and Kasai, N. (1994). Evidence that alpha-fetoprotein suppresses the immunological function in transgenic mice. *Biochem. Biophys. Res. Commun*, *201*, 1154-9.

[29] Ogata, A., Yamashita, T., Koyama, Y., Sakai, M. and Nishi, S. (1995). Suppression of experimental antigen-induced arthritis in transgenic mice producing human alpha-fetoprotein. *Biochem. Biophys. Res. Commun*, *213*, 362-6.

[30] Oers, N. S. C., Cohen, B. L. and Murgita, R. A. (1989). Isolation and characterization of a distinct immunoregulatory isoform of alpha-fetoprotein produced by the normal fetus. *J Exp Med*, *170*, 811-25.

[31] Semeniuk, D. J., Boismenu, R., Tam, J., Weissenhofer, W. and Murgita, R. A. (1995). Evidence that immunosuppression is an intrinsic property of the alpha-fetoprotein molecule. *Adv Exp Med Biol*, *383*, 255-69.

[32] Belyaev, N. N., Bogdanov, A., Yu Savvulidi, Ph. G., Tleulieva, R. and Zakiryanova, G. K. (2005). Stimulation of natural suppressor cell activity with alpha-fetoprotein as possible mechanism of natural killer cell inhibition. *Health and Disease*, *43*, 14-19.

[33] Yamamoto, M., Tatsumi, T., Miyagi, T., Tsunematsu, H., Aketa, H., Hosui, A., Kanto, T., Hiramatsu, N., Hayashi, N. and Takehara, T. (2011). Alpha-fetoprotein impairs activation of natural killer cells by inhibiting the function of dendritic cells. *Clin. Exp. Immunol*, *165*, 211-9.

[34] Setiyono, A., Budiyati, A. D., Purwantomo, S., Anggelia, M. R., Fanany, I., Wibowo, G. A., Bachtiar, I., Utama, A. and Tai, S. (2011). Immunoregulatory effects of AFP domains on monocyte-derived dendritic cell function. *BMC Immunology*, *12*, 4-10.

[35] Budiyati, A. D., Setiyono, A., Tarigan, E. and Wibowo, H. (2012). The effect of alpha fetoprotein on NF-κB translocation in lipopolysaccharide induced monocyte-derived dendritic cell. *Med. J. Indones*, *21*, 97-101.

[36] Alisa, A., Boswell, S., Pathan, A. A., Ayaru, L., Williams, R. and Behboudi, S. (2008). Human CD4+ T cells recognize an epitope within α-fetoprotein sequence and develop into TGF-β-producing CD4+ T cell. *J. Immunol*, *180*, 5109–17.

[37] Brenner, T. and Abramsky, O. (1981). Immuno suppression of experimental autoimmune myasthenia gravis by alpha-fetoprotein rich formation. *Immunol. Lett*, 3, 163-7.

[38] Abramsky, O., Brenner, T., Mizrachi, R. and Soffer, D. (1982). Alpha-fetoprotein suppresses experimental allegric encephalomyelitis. *J. Neuroimmunol*, *2*, 1-7.

[39] Irony-Tur-Sinai, M., Grigoriadis, N., Lourbopoulos, A., Pinto-Maaravi, F, Abramsky, O. and Brenner, T. (2006). Amelioration of neuroinflammation by recombinant human alpha-fetoprotein. *Exp Neurol*, *198*, 136-44.

[40] Belyaev, N. N., Bogdanov, A. Yu., Savvulidi, P. G., Krasnoshtanov, V. K., Tleulieva, R. T., Alipov, G. K., Sekine, I., Bae, J. S., Lee, J. B., Young-Ki Min, Y. K. and Yang, H. M. (2008). The influence of alpha-fetoprotein on natural suppressor cell activity and Ehrlich carcinoma growth. *Korean J Physiol Pharmacol*, *12*, 193-7.

[41] Chereshnev, V. A., Rodionov, S., Yu, Vasil'ev, N. V., Orlov, O. A. and Cherkasov, V. A. (2011). Alpha-fetoprotein immunotherapy as a stage of combined treatment of cancer patients. *Vopr Onkol.*, 2005. *51*(1), 86-92. (in Russian).

[42] Mizejewski, G. J. Therapeutic use of human alpha-fetoprotein in clinical patients: is a cancer risk involved? *Int J Cancer*, *128*, 239-42.

[43] Chereshnev, V. A., Rodionov, S. Yu., Sherkasov, V. A., Malutina, N. N. and Orlov, O. A. (2004). Alpha-fetoprotein. Ekaterinburg: Uro RAN. (in Russian).

[44] Pollard, L. C., Murray, J., Moody, M., Stewart, E. J. and Choy, E. H. S. (2007). A randomised, double-blind, placebo-controlled trial of a recombinant version of human alpha-fetoprotein (MM-093) in patients with active rheumatoid arthritis. *Ann Rheum Dis*, *66*, 687-89.

[45] Dudich, E. (2007). MM-093, a recombinant human alpha-fetoprotein for the potential treatment of rheumatoid arthritis and other autoimmune diseases. *Curr Opin Mol Ther*, *9*, 603–10.

[46] Terentiev, A. A., Moldogazieva, N. T., Levtsova, O. V., Maximenko, D. M., Borozdenko, D. A. and Shaitan, K. V. (2012). Modeling of three dimensional structure of human alpha-fetoprotein complexed with diethylstilbestrol: docking and molecular dynamics simulation study. *J Bioinform Comput Biol.*, *10*, 1241012.

[47] Shen, J., Zhang, W., Fang, H., Perkins, R., Tong, W. and Hong, H. (2013). Homology modeling, molecular docking, and molecular dynamics simulations elucidated α-fetoprotein binding modes. *BMC Bioinformatics*, *14*, (Suppl 14), S6.

[48] Torres, J. M., Geuskens, M. and Uriel, J. (1991). Receptor-mediated endocytosis and recycling of alpha-fetoprotein in human B-lymphoma and T-leukemia cells. *Int J Cancer*, *47*, 110-7.

[49] Jassam, N., Jones, C. M., Briscoe, T. and Horner, J. H. (2006). The hook effect: a need for constant vigilance. *Ann. Clin. Biochem*, *43*, 314–7.

[50] Pardee, A. D., Shi, J. and Butterfield, L. H. (2014). Tumor–derived alpha-fetoprotein impairs the differentiation and T cell stimulatory activity of human dendritic cells. *The Journal of Immunology*, *193*, 5723-32.

[51] Hsia, J. C., Er, S. S., Estes, T. and Ruoslahti, E. (1980). Alpha-fetoprotein binding specificity for arachidonate, bilirubin, docosahexaenoate, and palmitate. A spin label study. *J Biol Chem*, *255*, 4224-7.

[52] Hirano, K., Watanabe, Y., Adachi, T., Ito, Y. and Sugiura, M. (1985). Drug binding properties of human alpha-fetoprotein. *Biochem. J*, 231, 189-91.

[53] Dudich, E., Semenkova, L., Gorbatova, E., Dudich, I., Khromykh, L., Tatulov, E., Grechko, G. and Sukhikh, G. (1998). Growth-regulative activity of human alpha-fetoprotein for different types of tumor and normal cells. *Tumor Biol*, *19*, 30–40.

[54] Strelchenok, O. "*Composition for the treatment of immune deficiencies and methods for its preparation and use,*" U.S. Patent 6, 599, 507 July 29, 2003.

[55] Arshad, N. M., In, L. L. A., Soh, T. L., Azmi, M. N., Ibrahim, H., Awang, K., Dudich, E., Tatulov, E. and Nagoor, N. H. (2015). Recombinant human alpha fetoprotein synergistically potentiates the anti-cancer effects of 1'-S-1'- acetoxychavicol acetate when used as a complex against human tumours harbouring AFP-receptors. *Oncotarget*, *6*, 1-17.

[56] Ochsenbein, A. F., Klenerman, P., Karrer, U., Ludewig, B., Pericin, M., Hengartner, H. and Zinkernagel, R. M. (1999). Immune surveillance against a solid tumor fails because of immunological ignorance. *Proc. Natl. Acad. Sci. U.S.A*, *96*, 2233–2238.

[57] Govallo, V. I. (1993). The immunology of pregnancy and cancer. New York: Nova Science Publishers, Commack.

[58] Svensson-Arvelund, J., Mehta, R. B., Lindau, R., Mirrasekhian, E., Rodriguez-Martinez, H., Berg, G., Lash, G. E., Jenmalm, M. C. and Ernerudh, J. (2015). The human fetal placenta promotes tolerance against the semiallogeneic fetus by inducing regulatory T cells and homeostatic M2 macrophages. J Immunol, *194*, 1534-44.

[59] Schumacher, A., Costa, S. D. and Zenclussen, A. C. (2014). Endocrine factors modulating immune responses in pregnancy. *Front Immunol*, *5*, 196.

[60] Newby, D., Dalgliesh, G., Lyall, F. and Aitken, D. A. (2005). Alphafetoprotein and alphafetoprotein receptor expression in the normal human placenta at term. *Placenta*, *26*, 190-200.

[61] Wu, A. H. B. and Sell, S. (1990). Markers of hepatocellular carcinoma and hepar cancer. In: Herberman RB, Mercer DW, editors. Immunodiagnosis of Cancer. New York: Marcel Dekker, 570-81.

[62] Jacobs, I. and Bast, R. C. (1990). Immunodiagnosis of germ cell tumors of the gonads. In: Herberman RB, Mercer DW, editors. Immunodiagnosis of Cancer. New York: Marcel Dekker, 323-38.

[63] Mizejewski, G. J. (2013). Review of the adenocarcinoma cell surface receptor for human alpha-fetoprotein, proposed identification of a widespread mucin as the tumor cell receptor. *Tumor Biol*, *34*, 1317-36.

[64] Umemura, N., Saio, M., Suwa, T., Kitoh, Y., Bai, J., Nonaka, K., Ouyang, G. F., Okada, M., Balazs, M., Adany, R., Shibata, T. and Takami, T. (2008). Tumor-infltrating myeloid-derived suppressor cells are pleiotropic-inflamed monocytes/macrophages that bear M1- and M2-type characteristics. *J Leukoc Biol*, *83*, 1136–44.

[65] Weiss, J. M., Subleski, J. J., Back, T., Chen, X., Watkins, S. K., Yagita, H., Sayers, T. J., Murphy, W. J. and Wiltrout, R. H. (2014). Regulatory T cells and myeloid-derived suppressor cells in the tumor microenvironment undergo Fas-dependent cell death during IL-2/αCD40 therapy. *J Immunol*, *192*, 5821-9.

[66] Najjar, Y. G. and Finke, J. H. (2013). Clinical perspectives on targeting of myeloid derived suppressor cells in the treatment of cancer. *Frontiers in Oncology*, *3*, 1-9.

[67] Finke, J. H., Rayman, P. A., Ko, J. S., Bradley, J. M., Gendler, S. J. and Cohen, P. A. (2013). Modification of the Tumor Microenvironment as a Novel Target of Renal Cell Carcinoma Therapeutics. *Cancer J*, *19*, 353-64.

[68] Kano, A. (2015). Tumor cell secretion of soluble factor(s) for specific immunosuppression. *Sci Rep.*, *9*, 8913.

[69] Forghani, P., Khorramizadeh, M. R. and Waller, E. K. (2012). Natural suppressor cells, past, present and future. *Frontiers in Bioscience*, *4*, 1237-45.

[70] Maier, T., Holda, J. H. and Claman, H. N. (1989). Natural suppressor cells. *Prog Clin Biol Res*, *288*, 235–44.

[71] Subiza, J. L., Vinuela, J. E., Rodriguez, R., Gil, J., Figueredo, M. A. and de la Concha, E. G. (1989). Development of splenic natural suppressor (NS) cells in Ehrlich tumor-bearing mice. *Int. J. Cancer*, *44*, 307-14.

[72] Wright, M. A., Wiers, K., Vellody, K., Djordjevic, D. and Young, M. R. (1998). Stimulation of immune suppressive CD34+ cells from normal bone marrow by Lewis lung carcinoma tumours. *Cancer Immunol. Immunother*, *46*, 253-60.

[73] Young, M. R. I., Wright, M. A., Lozano, Y., Prechel, M. M., Benefield, J., Leonetti, J., Colluns, Sh. L. and Petruzzelli, G. J. (1997). Increased recurrence and metastasis in patients whose primary head and neck squamous cell carcinomas secreted granulocyte-macrophage colony-stimulating factor and contained CD34+ natural suppressor cells. *Int. J. Cancer*, *74*, 69-74.

[74] Yamaguchi, V., Takashima, I., Funakoshi, M., Kawami, H. and Toge, T. (1994). Defective natural killer activity in gastric cancer patients: possible involvement of suppressor factor receptor. *In Vivo*, *8*, 279-83.

[75] Alpert, E., Dienstag, J. L., Sepersky, S., Littman, B. and Rocklin, R. (1978). Immunosuppressive characteristics of human AFP: effect on tests of cell mediated immunity and induction of human suppressor cells. *Immunol. Commun*, *7*, 163-85.

[76] Hooper, D. C. and Murgita, R. A. (1981). Regulation on murine T cell responses to autologous antigens by alpha-fetoprotein. *Cell. Immunol*, *15*, 417-25.

[77] Yamada, A. and Hayami, M. (1983). Suppression of natural killer cell activity by chicken alpha-fetoprotein in Japanese quails. *J. Natl. Cancer Inst*, *70*, 735-38.

[78] Hoskin, D. W. and Murgita, R. A. (1989). Specific maternal anti-fetal lymphoproliferative responses and their regulation by natural immunosuppressive factors. *Clin. Exp. Immunol*, *76*, 262-267.

[79] Belyaev, N. N., Bogdanov, A. Yu., Savvulidi, Ph. G. and Tleulieva, R. (2003). Alpha-fetoprotein as an inductor of natural suppressor cells, in: *IX International Symposium on Biology and Clinical Usefulness of Tumor Markers.* Abstracts, 12-15 Feb. 2003. Spain, Barcelona, 89.

[80] Bronte, V., Wang, M., Overwijk, W. W., Surman, D. R., Pericle, F., Rosenberg, S. A. and Restifo, N. P. (1998). Apoptotic death of CD8+ T lymphocytes after immunization: induction of a suppressive population of Mac-1+/Gr-1+ cells. *J. Immunol*, *161*, 5313–20.

[81] Gabrilovich, D. I., Bronte, V., Chen, S. H., Colombo, M. P., Ochoa, A. and Ostrand-Rosenberg, S. (2007). The terminology issue for myeloid-derived suppressor cells. *Cancer Res*, *67*, 425.

[82] Schreiber, R. D., Old, L. J. and Smyth, M. J. (2011). Cancer immunoedditing: Integrating immunity's roles in cancer suppression and promotion. *Science*, *331*, 1565-70.

[83] DuPage, M., Mazumdar, C., Schmidt, L. M., Cheung, A. F. and Jacks, T. (2012). Expression of tumour-specific antigens underlies cancer immunoediting. *Nature*, *482*, 405–09.

[84] Balsamo, M., Pietra, G., Vermi, W., Moretta, L., Mingari, M. C. and Vitale, M. (2012). Melanoma immunoediting by NK cells. *OncoImmunology*, *1*, 1–3.

[85] Cirri, P. and Chiarugi, P. (2012). Cancer-associated-fibroblasts and tumour cells: a diabolic liaison driving cancer progression. *Cancer Metastasis Rev*, *31*, 195-208.

[86] Mulligan, K. J. and Young, M. R. I. (2010). Tumors induce the formation of suppressor endothelial cells *in vivo*. *Cancer Immunol. Immunother*, *59*, 267–77.

[87] Quatromoni, J. G. and Eruslanov, E. (2012). Tumor-associated macrophages: function, phenotype, and link to prognosis in human lung cancer. *Am J Transl Res, 4*, 376-89.

[88] Hao, N. B., Lü, M. H., Fan, Y. H., Cao, Y. L., Zhang, Zh. R. and Yang, Sh. M. (2012). Macrophages in tumor microenvironments and the progression of tumors. *Clin. Develop. Immunol.*, 2012. Article ID 948098, 11 pages.

[89] Ma, Y., Shurin, G. V., Peiyuan, Z. H. and Shurin, M. R. (2013). Dendritic cells in the cancer microenvironment. *J. Cancer*, *4*, 36-44.

[90] Nishikawa, H. and Sakaguchi, S. H. (2010). Regulatory T cells in tumor immunity. *Int. J. Cancer*, *127*, 759–67.

[91] Byrne, W. L., Mills, K. H. G., Lederer, J. A. and O'Sullivan, G. C. (2011). Targeting regulatory T cells in cancer. *Cancer Res*, *71*, 6915-20.

[92] Gabrilovich, D. I., Ostrand-Rosenberg, S. and Bronte, V. (2012). Coordinated regulation of myeloid cells by tumours. *Nature Rev. Immunol*, *12*, 253-68.

[93] Youn, J. I., Nagaraj, S., Collazo, M. and Gabrilovich, D. I. (2008). Subsets of myeloid-derived suppressor cells in tumor-bearing mice. *J. Immunol*, *181*, 5791-802.

[94] Greten, T. F., Manns, M. P. and Korangy, F. (2011). Myeloid derived suppressor cells in human diseases. *International Immunopharmacology*, *11*, 802–7.

[95] Marigo, I., Dolcetti, L., Serafini, P., Zanovello, P. and Bronte, V. (2008). Tumor-induced tolerance and immune suppression by myeloid derived suppressor cells. *Immunol. Rev*, *222*, 162-79.

[96] Gabrilovich, D. I. and Nagaraj, S. (2009). Myeloid-derived suppressor cells as regulators of the immune system. *Nature Rev. Immunol*, *9*, 162-74.

[97] Ostrand-Rosenberg, S. and Sinha, P. (2009). Myeloid-derived suppressor cells: Linking inflammation and cancer. *J. Immunol*, *182*, 4499–506.

[98] Peranzoni, E., Zilio, S., Marigo, I., Dolcetti, L., Zanovello, P., Mandruzzato, S. and Bronte, V. (2010). Myeloid-derived suppressor cell heterogeneity and subset definition. *Curr. Opin. Immunol*, *22*, 238–44.

[99] Ye, X. Z., Yu, S. C. and Bian, X. W. (2010). Contribution of myeloid-derived suppressor cells to tumor-induced immune suppression, angiogenesis, invasion and metastasis. *J. Genet. Genomics*, *37*, 423−30.

[100] Khaled, Y. S., Ammori, B. J. and Elkord, E. (2013). Myeloid-derived suppressor cells in cancer: recent progress and prospects. *Immunology and Cell Biology*, 1–10.

[101] Serafini, P. and Bronte, V. (2014). Myeloid-Derived Suppressor Cells in Tumor-Induced T Cell Suppression and Tolerance. In: Tumor-Induced Immune Suppression. Mechanisms and Therapeutic Reversal. Gabrilovich, DI, Hurwitz, AA, editors. New York, Heidelberg, Dordrecht, London: Springer, 99-150.

[102] Huang, B., Pan, P. Y., Li, Q., Sato, A. I., Levy, D. E., Bromberg, J., Divino, C. M. and Chen, S. H. (2006). Gr-1+CD115+ immature myeloid suppressor cells mediate the development of tumor-induced regulatory cells and T cell anergy in tumor-bearing host. *Cancer Res*, *66*, 1123–31.

[103] Kao, J., Ko, E. C., Eisenstein, S., Sikora, A. G., Fu, S. and Chen, S. H. (2011). Targeting immune suppressing myeloid-derived suppressor cells in oncology. *Crit. Rev. Oncol. Hematol*, *77*, 12-9.

[104] Martin, F., Apetoh, L. and Ghiringhelli, F. (2012). Role of myeloid-derived suppressor cells in tumor immunotherapy. *Immunotherapy*, *4*, 43–57.

[105] Lindau, D., Gielen, P., Kroesen, M., Wesseling, P. and Adema, G. J. (2013). The immunosuppressive tumor network: myeloid-derived suppressor cells, regulatory T cells and natural killer T cells. *Immunology*, *138*, 105-15.

[106] Vincent, J., Mignot, G., Chalmin, F., Ladoire, S., Bruchard, M., Chevriaux, A., Martin, F., Apetoh, L., Rebe, C. and Ghiringhelli, F. (2010). 5-Fluorouracil selectively kills tumor-associated myeloid-derived suppressor cells resulting in enhanced T cell-dependent antitumor immunity. *Cancer Res*, *70*, 3052–61.

[107] Belyaev, N. N. (2014). Myeloid-derived suppressor cells (MDSC) as a main tumor induced negative regulators of cancer immunity and possible ways for their elimination. *KAZNU Bulletin Biology series*, *1*, 79-83.

[108] Atemezem, A., Mbemba, E., Marfaing, R., Vaysse, J., Pontet, M., Saffar, L., Charnaux, N. and Gattegno, L. (2002). Human alpha-fetoprotein binds to primary macrophages. *Biochem Biophys Res Commun*, *296*, 507-14.

[109] Zubkova, E., Semenkova, L., Dudich, E., Dudich, I., Parfyonova, Y. and Menshikov, M. (2013). Alpha-fetoprotein contributes to THP-1 cell invasion and chemotaxis via protein kinase and Gi-protein-dependent pathways. *Molecular and Cellular Biochemistry*, *379*, 283-93.

[110] Sinenko, S. A., Belyaev, N. N., Kuzovlev, V. D. and Khakimzhanov, A. A. (199). The immunochemical distinctions of human α-fetoprotein preparations as revealed with novel monoclonal antibodies. *Dokladi MS-AS RK*, *1*, 72-84.

[111] Bogdanov, A. Yu., Savvulidi, Ph. G., Tleulieva, R. T. and Belyaev, N. N. (2004). Alpha-fetoprotein as inducer of natural suppressor (NS) cells of bone marrow. I. Isopycnic fractions of NS cells and their suppressor effects. *Biothechnology. Theory and Practice*, *3*, 83-9. (in Russian).

[112] Bogdanov, A. Yu. and Belyaev, N. N. (2004). Alpha-fetoprotein as an inductor of bone marrow natural suppressor (NS) cells. II. Molecular nature of suppressor factors. *Biothechnology. Theory and Practice*, *3*, 90-8. (in Russian).

[113] Belyaev, N., Bogdanov, A. Yu., Savvulidi, Ph. G., Zakiryanova, G. K. and Tleulieva, R. T. (2003). NS1 and NS2 subpopulations of bone marrow natural suppressor cells. *Biology of blood and marrow transplantation*, *9*, 127-28.

[114] Bogdanov, A. Yu. (2005). Molecular mechanisms of natural suppressor cells activation with alpha-fetoprotein. PhD Thesis on Molecular Biology. *Almaty.*, 22 pages. (in Russian).

[115] Pak, V. N., Pak, N. A., Reshetnikov, S. S., Nikonov, S. D. and Ogirenko, A, P. (2005). "*Method for treatment of malignant neoplasms and a complex preparation having antineoplastic activity for use in such treatment*" U.S. Patent 6,878,688, April 12, 2005.

[116] Pak, V. (2011). "*Compositions of alpha-fetoprotein and inducers of apoptosis for the treatment of cancer,*" U.S. Patent 8,071,547, December 6, 2011.

[117] Pak, V., Molchanov, O. and Vincent, M. (2007). Treatment of metastatic colorectal cancer with aimpila, a glycoside/alpha-fetoprotein complex. *J of Clin Oncol. ASCO Annual Meeting Proceedings (Post-Meeting Edition)*, *25*, (Suppl.18S), 3589.

[118] Dasgupta, S., Erturk-Hasdemir, D., Ochoa-Reparaz, J., Reinecker, H. C. and Kasper, D. L. (2014). Plasmacytoid dendritic cells mediate anti-inflammatory responses to a gut commensal molecule via both innate and adaptive mechanisms. *Cell Host Microbe*, *15*, 413-23.

[119] Wesolowski, R., Markowitz, J. and Carson III, W. E. (2013). Myeloid derived suppressor cells – a new therapeutic target in the treatment of cancer. *Journal for ImmunoTherapy of Cancer*, *1*, 1-10.

[120] Tumeh, P. C., Harview, C. L., Yearley, J. H., Shintaku, I. P., Taylor, E. J., Robert Chmielowski, L. B., Spasic, M., Henry, G., Ciobanu, V., West, A. N., Carmona, M., Kivork, C., Seja, E., Cherry, G., Gutierrez, A. J., Grogan, T. R., Mateus, C., Tomasic, G., Glaspy, J. A., Emerson, R. O., Robins, H., Pierce, R. H., Elashoff, D. A. C., Robert, C. and Ribas, A. (2014). PD-1 blockade induces responses by inhibiting adaptive immune resistance. *Nature*, *515*, 568–71.

In: Alpha-Fetoprotein: Functions and Clinical Applications
Editors: Nisha Lakhi and Michael Moretti
ISBN: 978-1-63484-875-6

Chapter 20

ENGINEERING ALPHA-FETOPROTEIN TO DEVELOP VACCINES AND T CELL IMMUNOTHERAPIES FOR HEPATOCELLULAR CARCINOMA: FROM PRECLINICAL STUDIES TO FUTURE PERSPECTIVES

Yukai He**, MD, PhD, and Sha Wu, PhD
Augusta University Cancer Center,
Cancer Immunology, Inflammation, and Tolerance Program,
Augusta University, Augusta, GA, US

ABSTRACT

AFP has been serving as a biomarker for the diagnosis of HCC for many years. Now, research has been conducted to use it as an immunological target for HCC immunotherapy. However, because of the self-nature of AFP, the immune activation and antitumor effects of the current AFP-based vaccines are weak. T cell epitope engineering was employed to create immunogenic AFP. A proof of principle was obtained in an animal model, demonstrating that the epitope-engineered-AFP gene vaccine indeed potently activated AFP-specific CD8 T cells, which critically could cross-recognize native-AFP peptides and kill AFP+ tumor cells. Importantly, immunization with engineered AFP increased liver infiltration of AFP-specific CD8 T cells and prevented the carcinogen-induced autochthonous HCC. In addition, prime-boost immunization with engineered AFP gene vaccine generated CD8 memory cells that could detect and respond to evolving HCC tumor cells, enhancing the prevention of autochthonous HCC. This study provides a blueprint for us to develop effective human AFP-based HCC vaccines that may be used to prevent HCC relapse after liver resection and *de novo* tumor development in high-risk populations. In addition, the new AFP-based HCC vaccines may allow us to identify high-quality TCR genes for engineering human T cells for adoptive immunotherapy of HCC as our effort.

* Corresponding author: Yukai He, MD/PhD, Augusta University Cancer Center, Cancer Immunology, Inflammation, and Tolerance Program, Augusta, GA 30912, email: yhe@gru.edu.

INTRODUCTION

Liver cancer is the sixth most common cancer in the world with ~782,000 new cases in 2012 (http://www.cdc.gov/cancer/international/statistics). In the US, liver cancer, especially primary hepatocellular carcinoma (HCC), is one of the most rapidly growing malignancies. The number of new HCC cases in the US has doubled within the last 10 years. According to The American Cancer Society, the estimated number of new cases of liver cancer in 2015 is 35,660, approximately three-fourths of which are primary HCC. The worldwide HBV vaccination program has significantly reduced the incidence of new HBV infections and thus will decrease HBV-associated HCC in the long run [1]. However, the large pool of existing HBV (350 million) [2] and HCV (170 million) [3] infected patients and the recognition that obesity/diabetes has become the major risk factor of HCC in the developed countries [4, 5] suggest that the HCC incidence will continue to increase over the next few decades. To make matters worse, the treatment options for HCC are limited and ineffective [6]. Although liver resection and transplantation are curative, the lack of post-surgery adjuvant therapies is a critical barrier to the success of liver resection, which results in a 5-year recurrence rate of ~70% [7]. With a dismal 3-year survival rate of 17% [8], the mortality rate of HCC remains high due to the lack of adjuvant therapies and high recurrence rate [7, 8]. The 745,000 annual deaths make liver cancer the 2nd most common cause of cancer death worldwide. The lack of effective treatments, the high recurrence rate, and the poor prognosis of HCC urgently demand new intervention approaches for HCC management.

Recent progress in immunotherapy demonstrates that activation of the patient's immune system has a great potential to control tumor growth [9, 10]. The human body's immune system is not only important to monitor, respond to, and eliminate invading microbes, but is also critical to conduct immune surveillance [11, 12] to detect and eliminate mutated cells, thus preventing or at least delaying tumor formation. These arguments favor the suggestion that developing immunological approaches, such as cancer vaccines and T cell immunotherapies, will have a great potential to control HCC tumor growth by filling the gap of HCC management after liver resection and improve the prognosis of HCC patients [13]. Approximately 84% of HCC tumors (but not normal hepatocytes) express the oncofetal proteins alpha-fetoprotein (AFP), glypican 3 (GPC3), or both [14, 15], making them excellent immune targets for HCC immunotherapy. However, these proteins are self/tumor antigens and weakly immunogenic. Modifications of the self/tumor antigens by antigen engineering will be required to create immunogenic tumor antigens to effectively activate antigen-specific immune effector cells to generate a significant antitumor effect. In this chapter, we will focus on reviewing the recent development of utilizing and engineering AFP to prevent and treat HCC in mouse models and in human clinical trials.

CANCER VACCINES ARE IMPORTANT COMPONENTS OF CANCER IMMUNOTHERAPIES

Over the last few years, significant progress in immunotherapy has been made in the form of using checkpoint blockade antibodies to unleash the function of immune effectors [10] to achieve a potent antitumor effect. Despite the risk of inducing autoimmune destruction, this antigen-undefined and nonspecific approach generates much needed excitement for the field of cancer immunotherapy. However, it also raised the question of whether antigen-specific cancer vaccines are effective [16] because of the unsuccessful clinical trials of many cancer vaccines [17]. Recent clinical studies have found that the efficacy of immune checkpoint blockade antibodies was correlated to the pre-existence of tumor-specific immunity [18] and to the number of tumor-infiltrating CD8 T cells [19, 20]. We have found in an animal model that the number of tumor-infiltrating lymphocytes can be significantly increased by effective antigen-specific vaccination [21, 22]. In addition, in many cases, cancer vaccines are needed to identify tumor antigen-specific T cells and TCR genes that can be used for T cell engineering [23] and adoptive immunotherapy [24]. Furthermore, cancer vaccines, like those for microbial infections, should have great preventive effect against cancer development *de novo* and relapse [25]. This can be especially important for preventing HCC since the high-risk population of HCC is easy to identify, and targeted immunization is plausible. Immunization in high-risk populations such as people with chronic HBV and HCV infections or with obesity may greatly reduce the HCC incidence if effective HCC vaccines are available. Thus, we believe that effective cancer vaccines are still important components of the immunological arsenals for tumor immunotherapy and tumor immunoprevention. However, current cancer vaccines are obviously not potent enough to activate sufficient high-quality tumor antigen-specific immune effector cells to generate an antitumor effect and thus will need further improvement to enhance their efficacy.

AFP Serves as the HCC-Associated Tumor Antigen for Developing HCC Vaccines

To develop effective cancer vaccines, as a rule of thumb, it is desirable to target the tumor antigens specifically expressed in tumor cells to generate tumor-specific immunity. In addition, to avoid immune escape, ideal tumor-specific antigens should play an indispensable role for tumor cell survival and growth. Unfortunately, in the case of HCC, as in many other cancers, no such tumor-specific and critical molecules can be found to serve as the ideal tumor antigens. Most of the cancer antigens are tumor-associated antigens that are over-expressed but not strictly expressed in tumor cells and may not be essential for tumor cell growth. Several such HCC-associated antigens have been identified over the last few decades, including the oncofetal proteins of AFP [26, 27] and GPC3 [28], cancer/testis antigens of MAGE-A [29] and NY-ESO [30], SSX-2 (synovial sarcoma X (SSX) breakpoint proteins [31]), and the widely over-expressed antigens such as telomerase reverse transcriptase [32]. Although a number of antigen-undefined HCC vaccines that consist of the entire HCC tumor cells have been tested in clinical trials for treating HCC patients (previous review [33]), the most promising and exciting immunotherapy approaches involve targeting the above mentioned HCC-associated antigens. HCC vaccines based on these HCC-associated antigens have been tested in animal models and clinical trials [26, 27, 32, 34] in the form of peptides, proteins, or genes with and without dendritic cells (DCs). Here, we focus on work using the native wild-type AFP for HCC vaccine development, on the recent effort of engineering AFP to create more potent HCC vaccines, and on the future perspectives of using AFP-specific T cells for HCC adoptive immunotherapy.

1. Native AFP as Self/Tumor Antigen for HCC Vaccine Development and Immunotherapy

AFP is expressed in fetal hepatocytes and its gene expression is gradually turned off or significantly reduced after birth. However, AFP is highly re-expressed in 70-80% of HCC [35]. Thus, AFP has been used as a biomarker for diagnosis for many decades. Because of its high level of expression and secretion, AFP has not been considered a good candidate of tumor antigen because any high avidity T cells may have been deleted during T cell development in the thymus or rendered tolerant because of its high level in HCC patients. However, evidence has been documented that AFP-specific T cells can be found in HCC patients even though the immune responses are insufficient to generate an antitumor effect in patients. Over the last 20 years, scientists and physicians have been trying to develop AFP-based cancer vaccines to activate AFP-specific immune responses in order to generate an antitumor effect in animal studies [36, 37] and clinical investigations [26, 34, 38].

A. **Preclinical Studies**: In animal models, plasmid DNA encoding native mouse AFP [39-41], DCs transduced with AFP viral vectors [37, 42], or DCs pulsed with peptides [43] have been tested as HCC cancer vaccines to induce AFP-specific immune responses. Despite those efforts, only low levels of immune activation and weak antitumor effects have been documented in the immunized mice. For example, using the ELISPOT assay, only 10 spots/1X10^6 splenocytes were detected after two intramuscular immunizations with mouse AFP plasmid DNA [39]. In another study, multiple immunizations with bone marrow derived DCs expressing mouse AFP only protected 1 out of 4 mice from AFP+ tumor challenge [37]. In a more recent study, mouse AFP DNA formulated with cationic polymers activated moderate T cell responses, but generated a more pronounced antitumor overall effect [40]. To increase the AFP-specific immune responses, xenogenic AFP peptides [43] or proteins [44] have also been used. Furthermore, a combination of AFP with heat shock proteins [45, 46] or stimulatory cytokines [36, 47] was employed to enhance the AFP-specific responses. However, such efforts only resulted in marginal improvement of immune activation in mice. The prime-boost immunization strategy, such as DNA prime followed by vaccinia vector [48] or adenoviral vector [49] boost, generated a higher immune response to 100-200 spots/1X10^6 splenocytes. However, the overall antitumor effect was modest and inconsistent. In general, the HCC vaccines based on wild-type native AFP demonstrated insufficient immune activation and limited antitumor responses in animal models. This induction of weak immune activation by native AFP in mice may also explain why no H-2b- restricted AFP epitopes have been identified until recently [50].

B. **Clinical Studies**: Butterfield and her group pioneered AFP-based human HCC vaccine studies [51]. They first identified 4 HLA-A2 restricted AFP epitopes using HLA-A2 transgenic mice and cultured human CD8 T cells [52, 53]. Then, Butterfield et al. [38] conducted the first clinical trial in heavily pre-treated stage IV patients with 3 biweekly vaccinations of 4 different AFP peptides emulsified in incomplete Freund's adjuvant. They found that all 6 patients generated T cell responses to most or all of the 4 peptides as measured by ELISPOT and tetramer assays. Thus, the human T cell repertoire is capable of recognizing AFP in the context of MHC class I despite a tolerating environment of high circulating serum levels of AFP. In a follow-up phase I/II trial, DCs pulsed with the AFP peptides were further used [34]. Although low level AFP-specific responses were detected, no significant antitumor effect was observed. It appeared that the AFP-specific immune responses were insufficient or irrelevant to generate an antitumor effect. Recently, an additional clinical trial was initiated to include AFP DNA vaccines followed by an adenoviral vector boost to generate a better antitumor effect [54]. In addition, cytokine GM-CSF was also incorporated into a second new clinical trial. The hope is that these added measures will generate stronger immune responses and improve antitumor effects.

In conclusion, the HCC cancer vaccines based on native AFP protein only induced a low level of AFP-specific immune responses and did not generate a consistent antitumor effect [27]. Therefore, new measures are required to increase the immune responses and improve the antitumor effect.

2. T-Cell Epitope Engineering to Create Effective AFP-Based HCC Vaccines

In order to effectively activate T cells with low avidity for self-antigens, it is necessary to increase the immunogenicity of self-antigens by antigen engineering to a degree that the self-reactive T cells will view the modified antigen as foreign. The term of antigen engineering refers to modifications of protein antigen sequences and structures to make the antigen more immunogenic to effectively induce antibody [55] or T cell responses [56]. The latter case is also called T-cell epitope engineering [56], which can be done in three different ways: 1) improving the antigen epitope binding affinity for MHC; 2) improving the antigen binding affinity for TCR; and 3) improving peptide stability. Here, we only focus on the first approach of T-cell epitope engineering to generate engineered epitopes that can effectively activate the low avidity T cells specific for self/tumor antigen. We will discuss the rationale of utilizing T-cell epitope engineering to create effective cancer vaccines (A) and the recent development of using this technology to develop AFP-based HCC vaccines (B).

A. The rationale of utilizing T-cell epitope engineering to develop cancer vaccines

a. The mechanism of central tolerance creates the need for T-cell epitope engineering in order to effectively activate self-reactive T cells. During the process of T cell development and maturation in the thymus, when immature T cells meet with self-antigens, they undergo clonal deletion, anergy, or receptor-editing so that high avidity T cells toward self-antigens will be removed from the body (negative selection) [57]. The mature T cells that survive the thymus development are T cells with low avidity for self-antigens, but can recognize foreign antigens with high avidity to protect the body from hazardous foreign invasion. While this mechanism of central tolerance (deletion of self-reactive high-avidity T cells) is in place to protect organisms from autoimmune attack by their own immune system, it creates a major obstacle for generating effective antitumor immunity because most tumor-associated antigens are self-antigens. The native self/tumor antigens are unlikely to efficiently activate potent antitumor immunity because no high-avidity T cells against self/tumor antigens are available in the T cell repertoire. Indeed, previous studies demonstrated that immunization with native self/tumor antigens was usually unable to activate the host T cells [21, 58, 59]. Consistent with this argument, immunization with vaccinia vector expressing native, full-length mouse AFP was incapable of inducing AFP-specific CD8 responses in mice [60]. In our own study, we also cannot detect any AFP epitope-specific immune responses after immunization with lentivector expressing native mouse AFP [50], in agreement with the concept that native self/tumor antigens are unable to activate self T cells (Figure 1A).

b. **Mutated self/tumor antigens are able to generate a stronger antitumor effect.** On the other hand, according to the theory of central tolerance, the high-avidity T cells recognizing the mutated self/tumor antigens should not be deleted during T cell development in the thymus and thus can be activated by corresponding mutated self/tumor antigens (Figure 1B). Indeed, it was recently found that self/tumor antigens may acquire mutations during cancer development and such mutated self/tumor antigens could effectively activate the T cells that can generate a better antitumor effect [61-63]. However, one big hurdle for using mutated self/tumor antigens to activate T cells specific for the mutated antigen is that every single tumor has its own unique mutations. Thus, mutanome sequencing of every single tumor is required to identify the relevant and appropriate mutated epitopes. Furthermore, in the prophylactic setting, there is no such tumor associated mutation yet for developing mutation- based vaccines. Thus, it is impractical if not impossible to develop cancer vaccines based on the unique mutations in each individual tumor, but it does provide us the proof of concept that engineering the antigen may be the way to create more effective cancer vaccines.

c. **The TCR cross-reactivity makes it possible to utilize engineered antigens to activate T cells that can cross-recognize native tumor antigens.** The T cells activated by engineered T cell epitopes must be able to cross-react with native epitopes on tumor cells in order to generate an antitumor effect. The cross-reactivity of TCR [64, 65], which refers to the fact that one T cell can react productively with many (as many as a million) different MHC/peptide complex, is the theoretical basis for using T cell epitope engineering to generate effective cancer vaccines. Each T cell can recognize numerous peptides, i.e., be cross-reactive, suggesting that the chance of any native peptide-MHC ligand being an optimal fit for its cognate TCR is small, creating the room for engineering more potent agonist peptides to better activate the T cells. More importantly, the TCR cross-reactivity also allows the T cells activated by optimized peptides to recognize the native peptides presented by tumor cells. Based on this theory, the mutated (also called heteroclitic) epitope peptides were designed to potently activate T cells that are capable of cross-recognizing native epitopes in the tumor-associated antigen (Figure 1C). Although this is not the best scenario where high-avidity T cells recognize mutated tumor antigen, the T cell cross-reactivity should allow T cells that are activated by mutated epitopes to be able to recognize and kill tumor cells. If there are enough cross-reacting T cells, the emerging novice tumors can be controlled and the growth of established tumors may be slowed down.

B. **Development of engineered AFP-Based HCC vaccines**

a. **Altered peptide ligands induce more robust immune responses:** The altered peptide ligand (APL) or heteroclitic peptide approach has been extensively studied [59, 66-68] as a means of activating self/tumor antigen-reactive T cells. The principle of APL design is to increase the stability of the APL and MHC complexes [69], which can then optimally engage with TCR to activate cognate T cells. In the design of APLs, the changes in the peptides must be significant enough to activate the T cell clones that cannot be activated by a native antigen, but at the same time, the change must be subtle enough to allow the activated T cells to cross-recognize the native self/tumor antigen. However, even with careful design, the T cell repertoire activated by APLs may be skewed [70] and the quality of clonotypic T cells may be suboptimal [71] in generating an antitumor effect. Furthermore, some of the synthetic short peptide epitopes predicted by computer may not indeed be processed and presented by target tumor cells [53]. These drawbacks of APLs may explain why recent APL clinical trials failed to show significant antitumor effects [17]. To our best knowledge, there is only one report demonstrating that the APL technology was used to activate AFP-specific T cells [72]. Unfortunately, no antitumor effect was reported in the study.

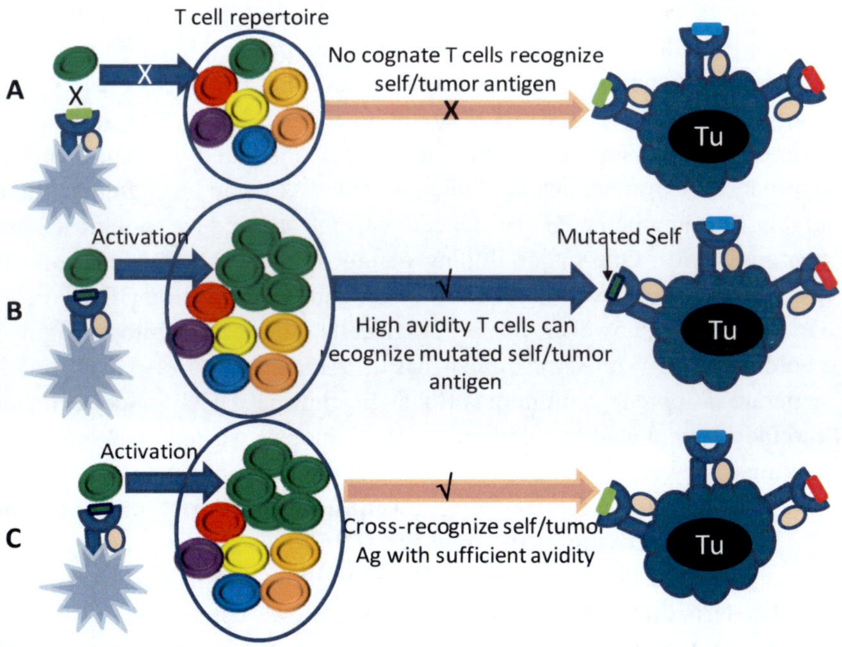

Figure 1. Different scenarios of immune activation and tumor recognition: A. Native self/tumor antigen is unlikely to activate T cells to recognize and kill tumor cells because of the mechanism of central tolerance. B. Mutated self/tumor antigens can activate high affinity T cells to recognize and kill tumor cells with cognate mutated antigen. C. Rationally designed antigen activate high affinity T cells that can cross-recognize and kill tumor cells expressing native self/tumor antigens.

b. **Gene vaccines:** One important caveat in APL design is that the epitopes identified by computer algorithm are not necessarily presented by target tumor cells. If they are not presented by tumor cells, no matter how potent T cell responses to these epitopes are, they are not able to control tumor growth. On the other hand, compared to APLs, gene vaccines encoding the epitope-engineered tumor antigens may have a better chance of success of inducing tumor antigen-specific CD8 responses that are correlated to the antitumor effect. Unlike exogenously APL short peptides, gene vaccines in the format of recombinant virus may better recapitulate natural antigen processing and presentation. The tumor antigen ectopically expressed within antigen-presenting cells following gene transduction will likely traverse the same processing and presentation pathways as the native antigen expressed in target tumor cells. Thus, any immune cell activated by a gene vaccine will likely recognize target tumor cells. Secondly, the full-length antigen offers a much broader immune response than shorter APL. Thirdly, one single immunization will allow us to examine immune responses against multiple epitopes, facilitating identification of antigen-specific immune responses. Consistent with this reasoning, immunization with plasmid DNA or recombinant lentivector encoding the full-length melanoma antigen of tyrosinase related protein 1 (TRP1), which contained multiple engineered CD8 epitopes, stimulated potent TRP1-specific CD8 responses [21, 58]. One single immunization with recombinant lentivector encoding the engineered-TRP1 full-length antigen induced ~10% of circulating CD8 T cells to secrete IFNγ in response to native TRP1 antigen [21]. Thus, epitope-engineering of self/tumor antigen together with genetic immunization presents an effective approach to induce potent immune responses against self/tumor antigens [73].

c. **Engineered AFP-based HCC gene vaccines:** Using this strategy of T cell epitope-engineering and the lentivector-mediated genetic immunization approach, we systematically studied whether AFP can be engineered to become a highly immunogenic HCC cancer vaccine to induce potent CD8 immune responses [50]. One major finding of our study is that the creation of epitope-engineered mouse AFP is required to activate AFP-specific CD8 T cells in mice. Once activated, however, these CD8 T cells induced by epitope-engineered AFP could efficiently cross-recognize native AFP epitopes on HCC tumor cells to generate a potent antitumor effect. In murine HCC models, including the carcinogen- induced autochthonous HCC model, we reported several important findings.

   i. **T cell epitope engineering is required to create immunogenic mouse AFP to activate AFP-specific CD8 T cells in mice.** A number of algorithmic programs are available to allow scientists to search for different MHC I-restricted epitopes. In our study, we used the algorithmic program developed by Guevara-Patino and colleagues [74] and identified 10 putative mouse H-2b restricted CD8 epitopes in the mouse AFP protein. Then, we optimized the epitopes to increase their binding affinity to the mouse H-2b molecule. Based on the data from this algorithmic program, we synthesized the entire AFP gene containing the 10 optimized epitopes. To study whether the epitope-engineered AFP could induce more potent CD8 responses than native AFP, we utilized the approach of lentivector-mediated genetic immunization because it has been proven to be one of the most effective approaches to activate CD8 and CD4 T cells in our previous studies [75-80]. The AFP-specific CD8 responses were detected by utilizing the technique of intracellular cytokine staining after *ex vivo* stimulation with native mouse AFP peptides. The native AFP peptides were used to make sure that any immune responses detected will have a better chance of recognizing tumor cells because they express native mouse AFP. During this process, we were able to identify CD8 T cells that could recognize 3 native mouse AFP epitopes: AFP_{212}, AFP_{419}, and AFP_{499}. In contrast, immunization with lentivector expressing native AFP did not activate CD8 T cells that can recognize any of the 10 epitopes. These data indicate that epitope-engineering of AFP is required to effectively activate AFP-specific CD8 T cells, which are capable of cross-recognizing native-AFP epitopes. We also noticed that the 3 native-AFP epitopes identified in the study had an intermediate affinity for H-2b molecules; this agrees with the theory that T cells with the highest and lowest affinity to self-antigens are deleted through negative selection and negligence during thymus development [81].

ii. **The CD8 T cells activated by epitope-engineered AFP recognize and kill AFP+ tumor cells.** One of the major drawbacks of APL technology is that CD8 T cells activated by APL may not recognize antigen-expressing tumor cells because the native epitopes corresponding to the APL may not be necessarily processed and presented by target tumor cells. However, as we previously reasoned that the CD8 T cells activated by gene vaccines will have a better chance of recognizing tumor cells because the epitopes presented by professional antigen-presenting cells such as DCs and tumor cells will likely be similar. This prompted us to study whether CD8 T cells activated by full-length engineered AFP gene vaccines could recognize native AFP epitopes on AFP-expressing tumor cells. We found that the CD8 T cells activated by the engineered-AFP gene vaccine could effectively recognize AFP+ EL4 tumor cells but not the parental EL4 tumor cells. Importantly, the CD8 T cells could also effectively kill AFP+ EL4 target tumor cells. These data strongly suggest that CD8 T cells activated by engineered-AFP gene vaccine can recognize and kill tumor cells expressing native AFP. The data also supports the hypothesis that the AFP CD8 epitopes identified in this study are presented by AFP+ tumor cells, making them recognizable by the activated CD8 T cells. The antitumor effect of AFP vaccines was then examined *in vivo*. Compared to naïve control mice, native AFP gene vaccine did not protect mice from tumor challenge although tumor growth was slowed, a limited antitumor effect was found similar to a previous report [37]. In contrast, one single immunization of the engineered AFP gene vaccine completely protected mice from a lethal dose of AFP+ EL4 tumor challenge. Furthermore, the antitumor effect of the engineered AFP gene immunization was mediated by the CD8 T cells because CD8 T cell depletion completely abolished the antitumor effect, which is consistent with previous findings [37]. The role of CD4 T cells in the antitumor effect of AFP vaccines has not been established.

iii. **Immunization with the engineered AFP gene vaccine significantly increases liver infiltration of the AFP-specific CD8 T cells and reduces tumor load of carcinogen-induced autochthonous HCC.** To study the antitumor effect of epitope-engineered AFP gene vaccines, we also utilized a more rigorous HCC model, the well-established autochthonous HCC model induced by the carcinogen, diethylnitrosamine (DEN). Our data showed that immunization with the engineered AFP but not native AFP gene vaccines significantly reduced the number of DEN-induced autochthonous HCC tumor nodules. Furthermore, immunological analysis demonstrated that the engineered AFP gene vaccine significantly increased the liver infiltration of functional AFP-specific CD8 T cell.

iv. **Prime-boost with engineered AFP gene vaccine increases the AFP-specific CD8 response, liver infiltration of AFP-specific CD8 T cells, and antitumor effect in the autochthonous HCC model.** Prime-boost immunization strategy is a well-established approach to increase antigen-specific immune responses [82]. We recently demonstrated that such a prime-boost strategy could better prevent mice from developing autochthonous melanoma [79]. Thus, in the AFP vaccination study, we also utilized a prime-boost strategy to increase the magnitude of AFP-specific CD8 responses. We found that priming the mice with recombinant lentivector expressing engineered AFP gene vaccine and then boosting with vaccinia vector expressing the same engineered AFP significantly enhanced the systemic AFP-specific CD8 effector responses. The mean fluorescent intensity of IFNγ in the prime-boosted group was significantly higher compared to that of the prime-alone group, suggesting that the amount of IFNγ produced is significantly greater. Importantly, the AFP-specific CD8 T cell responses were markedly increased again at the age of 6 months, corresponding to the emerging DEN-induced HCC tumor and AFP expression. The spike of AFP-specific CD8 responses suggests that the AFP-specific memory CD8 T cells induced by vaccines may detect the emerging AFP antigen in the liver and respond by expansion, which would then control autochthonous HCC growth. The mouse liver from the prime-boosted mice was infiltrated with AFP-specific CD8 T cells. In addition, compared to the spleen, the percentage of AFP-specific CD8 T cells in the tumor was significantly higher, suggesting either that AFP-specific CD8 T cells are preferentially recruited into liver tissue or may simply reflect the fact that the AFP-specific CD8 memory T cells respond to the emerging HCC tumor cells in the liver by expansion; thus, they may be retained in liver tissue because of the constant emerging AFP antigens. On the other hand, no AFP-specific CD8 T cells were found in the livers of native AFP gene vaccine prime-boosted mice. Thus, we concluded that the creation of highly immunogenic AFP by epitope engineering is absolutely required to activate and to increase liver infiltration of functional AFP-specific CD8 T cells to prevent autochthonous HCC in the DEN-treated mice.

FUTURE PERSPECTIVES

As research is progressing and new technology becomes available, we predict that the future development of AFP-based cancer vaccines and immunotherapy for HCC will proceed at the following areas.

1. **Develop immunogenic human HCC vaccines:** The novel findings of T cell epitope engineering to create immunogenic HCC-associated AFP antigen made in animal studies provide the evidence and proof of principle that it is possible to induce more potent immune responses against self/tumor antigen and generate much stronger antitumor effect through vaccine design and construction of effective genetic vaccines. We expect that this will be translated into human studies to develop novel human AFP-based HCC vaccines.
2. **Generation of re-programmed autologous T cells that can specifically target tumor cells:** Recent studies demonstrated that the delivery of an antigen-specific TCR gene into autologous T cells to create TCR re-programmed T cells that are specific for particular tumor antigen is a potent approach to achieve an antitumor effect [83]. We predict that novel human AFP-based HCC vaccines will allow us to identify and clone high-affinity TCR genes that can be used to re-program human T cells for adoptive immunotherapy of HCC.
3. **Combination therapy with checkpoint blockade:** In the last few years, checkpoint blockade of targeting CTLA-4 and PD1 have generated impressing antitumor effects in cancer patients [84]. Our own previous study found that PD1 blockade could increase efficacy of lentivector immunization and rescue the effector function of tumor infiltrating T cells [22]. It is envisioned that such a non-specific approach will generate a synergistic effect when combined with AFP-based HCC cancer vaccines not only by increasing the magnitude of antigen-specific immune responses but also by removing the immune blockade at the effector phase in tumor lesions [85].
4. **Immune prevention in a high-risk population:** Vaccines are most effective when they are used to prevent diseases. However, it has been a challenge to employ cancer vaccines to prevent cancer development because of the difficulty of gauging the risk/benefit ratio of cancer vaccination. HCC may offer an excellent model for testing the idea of immune prevention by cancer vaccines. HBV and HCV patients have a~146-fold higher chance of developing HCC [86]. Even worse, ~24% of male HBV patients infected by vertical transmission develop HCC by age 70 [87]. These patients have a >10,000 times higher risk than HBV-negative people. There is no intervention to stop the progression from chronic HBV infection to HCC. These high-risk patients, living in the fear of HCC, will likely accept HCC vaccines and should benefit from them. In addition, the lack of treatment options and high recurrence rate of HCC after liver resection will expedite translation of research bench findings to become a form of adjuvant therapy to prevent HCC relapse.

Conclusion

Oncofetal protein AFP is not only a biomarker for the diagnosis of HCC, but also can serve as an immunological target of therapy. However, the self-nature of AFP-based vaccine makes it impossible to induce potent immune responses and significant antitumor effects. In an animal model, we recently obtained the data demonstrating that it is possible to improve the AFP-specific immune responses via T cell epitope engineering. Immunization with recombinant lentivector expressing the epitope-optimized AFP generated potent AFP-specific CD8 responses, allowing us to identify three novel H-2b restricted CD8 epitopes. Importantly, the CD8 T cells activated by the engineered AFP gene vaccine could cross-recognize native AFP peptides and kill AFP+ tumor cells; this generated a potent antitumor effect. Immunization with engineered AFP increased liver infiltration of AFP-specific CD8 T cells and prevented the carcinogen-induced autochthonous HCC. This study provides a proof of principle and blueprint for creating effective human AFP-based HCC vaccines that may produce a better antitumor effect than the current HCC vaccines to prevent HCC relapse after liver resection and *de novo* tumor development in high-risk populations. It also gives us the tool to identify and clone high-quality TCR genes for engineering human T cells for adoptive immunotherapy of HCC as our future research endeavor.

Acknowledgment

The related research has been supported by NIH grants, Georgia Research Alliance, and Augusta University Cancer Center startup fund.

References

[1] Chang, M. H., You, S. L., Chen, C. J., et al. (2009). Decreased incidence of hepatocellular carcinoma in hepatitis B vaccinees: a 20-year follow-up study. *J Natl Cancer Inst*, *101*, 1348-1355.

[2] Wright, T. L. (2006). Introduction to chronic hepatitis B infection. *Am J Gastroenterol*, *101* Suppl 1, S1-6.

[3] Sy, T. and Jamal, M. M. (2006). Epidemiology of hepatitis C virus (HCV) infection. *Int J Med Sci*, *3*, 41-46.

[4] Welzel, T. M., Graubard, B. I., Quraishi, S., et al. (2013). Population-Attributable Fractions of Risk Factors for Hepatocellular Carcinoma in the United States. *Am J Gastroenterol*, *108*, 1314–1321.

[5] El-Serag, H. B., Tran, T. and Everhart, J. E. (2004). Diabetes increases the risk of chronic liver disease and hepatocellular carcinoma. *Gastroenterology*, *126*, 460-468.

[6] Corey, K. E. and Pratt, D. S. (2009). Current status of therapy for hepatocellular carcinoma. *Therap Adv Gastroenterol*, *2*, 45-57.

[7] Kao, W. Y., Su, C. W., Chau, G. Y., Lui, W. Y., Wu, C. W. and Wu, J. C. (2011). A comparison of prognosis between patients with hepatitis B and C virus-related hepatocellular carcinoma undergoing resection surgery. *World J Surg*, *35*, 858-867.

[8] Ferlay, J., Shin, H. R., Bray, F., Forman, D., Mathers, C. and Parkin, D. M. (2010). Estimates of worldwide burden of cancer in 2008: GLOBOCAN 2008. *Int J Cancer*, *127*, 2893-2917.

[9] Yang, Y. (2015). Cancer immunotherapy: harnessing the immune system to battle cancer. *J Clin Invest*, *125*, 3335-3337.

[10] Pardoll, D. M. (2012). The blockade of immune checkpoints in cancer immunotherapy. *Nat Rev Cancer*, *12*, 252-264.

[11] Dunn, G. P., Koebel, C. M. and Schreiber, R. D. (2006). Interferons, immunity and cancer immunoediting. *Nat Rev Immunol*, *6*, 836-848.

[12] Swann, J. B. and Smyth, M. J. (2007). Immune surveillance of tumors. *J Clin Invest*, *117*, 1137-1146.

[13] Greten, T. F., Wang, X. W. and Korangy, F. (2015). Current concepts of immune based treatments for patients with HCC: from basic science to novel treatment approaches. *Gut*, *64*, 842-848.

[14] Mizejewski, G. J. (2009). Alpha-fetoprotein (AFP)-derived peptides as epitopes for hepatoma immunotherapy: a commentary. *Cancer Immunol Immunother*, *58*, 159-170.

[15] Capurro, M., Wanless, I. R., Sherman, M., et al. (2003). Glypican-3: a novel serum and histochemical marker for hepatocellular carcinoma. *Gastroenterology*, *125*, 89-97.

[16] Hodi, F. S., O'day, S. J. and Mcdermott, D. F. (2010). Improved survival with ipilimumab in patients with metastatic melanoma. *N Engl J Med*, *363*, 711-723.

[17] Filipazzi, P., Pilla, L., Mariani, L., et al. (2012). Limited induction of tumor cross-reactive T cells without a measurable clinical benefit in early melanoma patients vaccinated with human leukocyte antigen class I-modified peptides. *Clin Cancer Res*, *18*, 6485-6496.

[18] Yuan, J., Adamow, M. and Ginsberg, B. A. (2011). Integrated NY-ESO-1 antibody and CD8+ T-cell responses correlate with clinical benefit in advanced melanoma patients treated with ipilimumab. *Proc Natl Acad Sci U S A*, *108*, 16723-16728.

[19] Tumeh, P. C., Harview, C. L., Yearley, J. H., et al. (2014). PD-1 blockade induces responses by inhibiting adaptive immune resistance. *Nature*, *515*, 568-571.

[20] Pico De Coana, Y., Choudhury, A. and Kiessling, R. (2015). Checkpoint blockade for cancer therapy: revitalizing a suppressed immune system. *Trends Mol Med*, *21*, 482-491.

[21] Liu, Y., Peng, Y., Mi, M., et al. (2009). Lentivector immunization stimulates potent CD8 T cell responses against melanoma self-antigen tyrosinase-related protein 1 and generates antitumor immunity in mice. *J Immunol*, *182*, 5960-5969.

[22] Zhou, Q., Xiao, H., Liu, Y., et al. (2010). Blockade of programmed death-1 pathway rescues the effector function of tumor-infiltrating T cells and enhances the antitumor efficacy of lentivector immunization. *J Immunol*, *185*, 5082-5092.

[23] Schumacher, T. N. (2002). T-cell-receptor gene therapy. *Nat Rev Immunol*, *2*, 512-519.

[24] Restifo, N. P., Dudley, M. E. and Rosenberg, S. A. (2012). Adoptive immunotherapy for cancer: harnessing the T cell response. *Nat Rev Immunol*, *12*, 269-281.

[25] Lollini, P. L., Cavallo, F., Nanni, P. and Quaglino, E. (2015). The Promise of Preventive Cancer Vaccines. *Vaccines (Basel)*, *3*, 467-489.

[26] Bei, R. and Mizejewski, G. J. (2011). Alpha fetoprotein is more than a hepatocellular cancer biomarker: from spontaneous immune response in cancer patients to the development of an AFP-based cancer vaccine. *Curr Mol Med*, *11*, 564-581.

[27] Buonaguro, L., Petrizzo, A., Tagliamonte, M., Tornesello, M. L. and Buonaguro, F. M. (2013). Challenges in cancer vaccine development for hepatocellular carcinoma. *J Hepatol*, *59*, 897-903.

[28] Chen, M., Li, G., Yan, J., et al. (2013). Reevaluation of glypican-3 as a serological marker for hepatocellular carcinoma. *Clin Chim Acta*, *423*, 105-111.

[29] Tada, F., Abe, M., Hirooka, M., et al. (2012). Phase I/II study of immunotherapy using tumor antigen-pulsed dendritic cells in patients with hepatocellular carcinoma. *Int J Oncol*, *41*, 1601-1609.

[30] Flecken, T., Schmidt, N., Hild, S., et al. (2013). Immunodominance and functional alterations of tumor-associated antigen-specific CD8 T-cell responses in hepatocellular carcinoma. *Hepatology*, *59*, 1415-26.

[31] Gehring, A. J., Ho, Z. Z., Tan, A. T., et al. (2009). Profile of tumor antigen-specific CD8 T cells in patients with hepatitis B virus-related hepatocellular carcinoma. *Gastroenterology*, *137*, 682-690.

[32] Greten, T. F., Forner, A., Korangy, F., et al. (2010). A phase II open label trial evaluating safety and efficacy of a telomerase peptide vaccination in patients with advanced hepatocellular carcinoma. *BMC Cancer*, *10*, 209.

[33] He, Y., Hong, Y. and Mizejewski, G. J. (2014). Engineering alpha-fetoprotein-based gene vaccines to prevent and treat hepatocellular carcinoma: review and future prospects. *Immunotherapy*, *6*, 725-736.

[34] Butterfield, L. H., Ribas, A., Dissette, V. B., et al. (2006). A phase I/II trial testing immunization of hepatocellular carcinoma patients with dendritic cells pulsed with four alpha-fetoprotein peptides. *Clin Cancer Res*, *12*, 2817-2825.

[35] Liu, C., Xiao, G. Q., Yan, L. N., et al. (2013). Value of alpha-fetoprotein in association with clinicopathological features of hepatocellular carcinoma. *World J Gastroenterol*, *19*, 1811-1819.

[36] Grimm, C. F., Ortmann, D., Mohr, L., et al. (2000). Mouse alpha-fetoprotein-specific DNA-based immunotherapy of hepatocellular carcinoma leads to tumor regression in mice. *Gastroenterology*, *119*, 1104-1112.

[37] Vollmer, C. M., Jr. Eilber, F. C., Butterfield, L. H., et al. (1999). Alpha-fetoprotein-specific genetic immunotherapy for hepatocellular carcinoma. *Cancer Res*, *59*, 3064-3067.

[38] Butterfield, L. H., Ribas, A., Meng, W. S., et al. (2003). T-cell responses to HLA-A*0201 immunodominant peptides derived from alpha-fetoprotein in patients with hepatocellular cancer. *Clin Cancer Res*, *9*, 5902-5908.

[39] Tian, G., Yi, J. L. and Xiong, P. (2004). Specific cellular immunity and antitumor responses in C57BL/6 mice induced by DNA vaccine encoding murine AFP. *Hepatobiliary Pancreat Dis Int*, *3*, 440-443.

[40] Cany, J., Barteau, B., Tran, L., et al. (2011). AFP-specific immunotherapy impairs growth of autochthonous hepatocellular carcinoma in mice. *J Hepatol*, *54*, 115-121.

[41] Hanke, P., Serwe, M., Dombrowski, F., Sauerbruch, T. and Caselmann, W. H. (2002). DNA vaccination with AFP-encoding plasmid DNA prevents growth of subcutaneous AFP-expressing tumors and does not interfere with liver regeneration in mice. *Cancer Gene Ther*, *9*, 346-355.

[42] Tan, X. H., Zhu, Q., Liu, C., Liu, X. L., Shao, X. T. and Wei, B. (2006). [Immunization with dendritic cells infected with human AFP adenovirus vector effectively elicits

immunity against mouse hepatocellular carcinomas]. *Zhonghua Zhong Liu Za Zhi*, *28*, 13-16.

[43] Pang, X. H., Chen, M. S., Jia, W. H. and Zhou, X. X. (2008). [Inhibitory effects of human AFP-derived peptide-pulsed dendritic cells on mouse hepatocellular carcinoma]. *Ai Zheng*, *27*, 1233-1238.

[44] Zhang, W., Liu, J., Wu, Y., et al. (2008). Immunotherapy of hepatocellular carcinoma with a vaccine based on xenogeneic homologous alpha fetoprotein in mice. *Biochem Biophys Res Commun*, *376*, 10-14.

[45] Lan, Y. H., Li, Y. G., Liang, Z. W., et al. (2007). A DNA vaccine against chimeric AFP enhanced by HSP70 suppresses growth of hepatocellular carcinoma. *Cancer Immunol Immunother*, *56*, 1009-1016.

[46] Wang, X., Lin, H. and Wang, Q. (2012). Specific genetic immunotherapy induced by recombinant vaccine alpha-fetoprotein-heat shock protein 70 complex. *Physics Procedia*, *33*, 738-742.

[47] Rodriguez, M. M., Ryu, S. M., Qian, C., et al. (2008). Immunotherapy of murine hepatocellular carcinoma by alpha-fetoprotein DNA vaccination combined with adenovirus-mediated chemokine and cytokine expression. *Hum Gene Ther*, *19*, 753-759.

[48] Meng, W. S., Butterfield, L. H., Ribas, A., et al. (2001). Alpha-Fetoprotein-specific tumor immunity induced by plasmid prime-adenovirus boost genetic vaccination. *Cancer Res*, *61*(24), 8782-8786.

[49] Saeki, A., Nakao, K., Nagayama, Y., et al. (2004). Diverse efficacy of vaccination therapy using the alpha-fetoprotein gene against mouse hepatocellular carcinoma. *Int J Mol Med*, *13*, 111-116.

[50] Hong, Y., Peng, Y., Guo, Z. S., et al. (2014). Epitope-optimized alpha-fetoprotein genetic vaccines prevent carcinogen-induced murine autochthonous hepatocellular carcinoma. *Hepatology*, *59*, 1448-1458.

[51] Evdokimova, V. N. and Butterfield, L. H. (2008). Alpha-fetoprotein and other tumour-associated antigens for immunotherapy of hepatocellular cancer. *Expert Opin Biol Ther*, *8*, 325-336.

[52] Butterfield, L. H., Koh, A., Meng, W., et al. (1999). Generation of human T-cell responses to an HLA-A2.1-restricted peptide epitope derived from alpha-fetoprotein. *Cancer Res*, *59*, 3134-3142.

[53] Butterfield, L. H., Meng, W. S., Koh, A., et al. (2001). T cell responses to HLA-A*0201-restricted peptides derived from human alpha fetoprotein. *J Immunol*, *166*, 5300-5308.

[54] Butterfield, L. H., Economou, J. S., Gamblin, T. C. and Geller, D. A. (2014). Alpha fetoprotein DNA prime and adenovirus boost immunization of two hepatocellular cancer patients. *J Transl Med*, *12*, 86.

[55] Pantophlet, R. and Burton, D. R. (2003). Immunofocusing: antigen engineering to promote the induction of HIV-neutralizing antibodies. *Trends Mol Med*, *9*, 468-473.

[56] Pentier, J. M., Sewell, A. K. and Miles, J. J. (2013). Advances in T-cell epitope engineering. *Front Immunol*, *4*, 133.

[57] Hogquist, K. A., Baldwin, T. A. and Jameson, S. C. (2005). Central tolerance: learning self-control in the thymus. *Nat Rev Immunol*, *5*, 772-782.

[58] Guevara-Patino, J. A., Engelhorn, M. E., Turk, M. J., et al. (2006). Optimization of a self antigen for presentation of multiple epitopes in cancer immunity. *J Clin Invest*, *116*, 1382-1390.

[59] Overwijk, W. W., Tsung, A., Irvine, K. R., et al. (1998). gp100/pmel 17 is a murine tumor rejection antigen: induction of "self"-reactive, tumoricidal T cells using high-affinity, altered peptide ligand. *J Exp Med*, *188*, 277-286.

[60] Tran, L., Judor, J. P., Gauttier, V., et al. (2012). The immunogenicity of the tumor-associated antigen alpha-fetoprotein is enhanced by a fusion with a transmembrane domain. *J Biomed Biotechnol*, 2012, 878657.

[61] Castle, J. C., Kreiter, S., Diekmann, J., et al. (2012). Exploiting the mutanome for tumor vaccination. *Cancer Res*, *72*, 1081-1091.

[62] Matsushita, H., Vesely, M. D., Koboldt, D. C., et al. (2012). Cancer exome analysis reveals a T-cell-dependent mechanism of cancer immunoediting. *Nature*, *482*, 400-404.

[63] Robbins, P. F., Lu, Y. C., El-Gamil, M., et al. (2012). Mining exomic sequencing data to identify mutated antigens recognized by adoptively transferred tumor-reactive T cells. *Nat Med*, *19*, 747-752.

[64] Sewell, A. K. (2012). Why must T cells be cross-reactive? *Nat Rev Immunol*, *12*, 669-677.

[65] Mason, D. (1998). A very high level of crossreactivity is an essential feature of the T-cell receptor. *Immunol Today*, *19*, 395-404.

[66] Dyall, R., Bowne, W. B., Weber, L. W., et al. (1998). Heteroclitic immunization induces tumor immunity. *J Exp Med*, *188*, 1553-1561.

[67] Slansky, J. E., Rattis, F. M., Boyd, L. F., et al. (2000). Enhanced antigen-specific antitumor immunity with altered peptide ligands that stabilize the MHC-peptide-TCR complex. *Immunity*, *13*, 529-538.

[68] Van Stipdonk, M. J., Badia-Martinez, D., Sluijter, M., Offringa, R., Van Hall, T. and Achour, A. (2009). Design of agonistic altered peptides for the robust induction of CTL directed towards H-2Db in complex with the melanoma-associated epitope gp100. *Cancer Res*, *69*, 7784-7792.

[69] Van Der Burg, S. H., Visseren, M. J., Brandt, R. M., Kast, W. M. and Melief, C. J. (1996). Immunogenicity of peptides bound to MHC class I molecules depends on the MHC-peptide complex stability. *J Immunol*, *156*, 3308-3314.

[70] Ekeruche-Makinde, J., Clement, M., Cole, D. K., et al. (2012). T-cell receptor-optimized peptide skewing of the T-cell repertoire can enhance antigen targeting. *The Journal of Biological Chemistry*, *287*, 37269-37281.

[71] Speiser, D. E., Baumgaertner, P., Voelter, V., et al. (2008). Unmodified self antigen triggers human CD8 T cells with stronger tumor reactivity than altered antigen. *Proc Natl Acad Sci U S A*, *105*, 3849-3854.

[72] Meng, W. S., Butterfield, L. H., Ribas, A., et al. (2000). Fine specificity analysis of an HLA-A2.1-restricted immunodominant T cell epitope derived from human alpha-fetoprotein. *Mol Immunol*, *37*, 943-950.

[73] He, Y., Munn, D. and Falo, L. D. Jr. (2007). Recombinant lentivector as a genetic immunization vehicle for antitumor immunity. *Expert Rev Vaccines*, *6*, 913-924.

[74] Houghton, C. S., Engelhorn, M. E., Liu, C., et al. (2007). Immunological validation of the EpitOptimizer program for streamlined design of heteroclitic epitopes. *Vaccine*, *25*, 5330-5342.

[75] He, Y., Zhang, J., Donahue, C. and Falo, L. D. Jr. (2006). Skin-derived dendritic cells induce potent CD8(+) T cell immunity in recombinant lentivector-mediated genetic immunization. *Immunity*, *24*, 643-656.

[76] He, Y. and Falo, L. D. Jr. (2007). Lentivirus as a potent and mechanistically distinct vector for genetic immunization. *Curr Opin Mol Ther*, *9*, 439-446.

[77] Esslinger, C., Chapatte, L., Finke, D., et al. (2003). *In vivo* administration of a lentiviral vaccine targets DCs and induces efficient CD8(+) T cell responses. *J Clin Invest*, *111*, 1673-1681.

[78] Arce, F., Breckpot, K., Collins, M. and Escors, D. (2011). Targeting lentiviral vectors for cancer immunotherapy. *Curr Cancer Ther Rev*, *7*, 248-260.

[79] Xiao, H., Peng, Y., Hong, Y., et al. (2011). Lentivector prime and vaccinia virus vector boost generate high-quality CD8 memory T cells and prevent autochthonous mouse melanoma. *J Immunol*, *187*, 1788-1796.

[80] Hong, Y., Peng, Y., Mi, M., et al. (2011). Lentivector expressing HBsAg and immunoglobulin Fc fusion antigen induces potent immune responses and results in seroconversion in HBsAg transgenic mice. *Vaccine*, *29*, 3909-3916.

[81] Starr, T. K., Jameson, S. C. and Hogquist, K. A. (2003). Positive and negative selection of T cells. *Annu Rev Immunol*, *21*, 139-176.

[82] Woodland, D. L. (2004). Jump-starting the immune system: prime-boosting comes of age. *Trends Immunol*, *25*, 98-104.

[83] Park, T. S., Rosenberg, S. A. and Morgan, R. A. (2011). Treating cancer with genetically engineered T cells. *Trends Biotechnol*, *29*, 550-557.

[84] Callahan, M. K. and Wolchok, J. D. (2013). At the bedside: CTLA-4- and PD-1-blocking antibodies in cancer immunotherapy. *J Leukoc Biol*, *94*, 41-53.

[85] Hato, T., Goyal, L., Greten, T. F., Duda, D. G. and Zhu, A. X. (2014). Immune checkpoint blockade in hepatocellular carcinoma: current progress and future directions. *Hepatology*, *60*, 1776-1782.

[86] Ulcickas Yood, M., Quesenberry, C. P., Jr. Guo, D., et al. (2008). Incidence of hepatocellular carcinoma among individuals with hepatitis B virus infection identified using an automated data algorithm. *J Viral Hepat*, *15*, 28-36.

[87] Yu, M. W., Chang, H. C., Liaw, Y. F., et al. (2000). Familial risk of hepatocellular carcinoma among chronic hepatitis B carriers and their relatives. *J Natl Cancer Inst*, *92*, 1159-1164.

ABOUT THE EDITORS

Nisha Lakhi, MD

Clinical Professor of Obstetrics and Gynecology
Director of Research
Co-Director of Minimally Invasive Surgery
Richmond University Medical Center
Staten Island, New York, US
Email: nlakhi@yahoo.com

Nisha A. Lakhi, MD is a board certified Obstetrician and Gynecologist with extensive background in both basic science and clinical research. As an undergraduate and medical student, she worked in a research laboratory at The Ohio State University, Columbus, Ohio, developing and synthesizing compounds to selectively target tumor cells for use in Boron Neutron Capture Therapy.

Dr. Lakhi completed her residency in Obstetrics and Gynecology at the New York Medical College/Richmond Program, followed by a fellowship, with specialized training in minimally invasive surgery, at Atlanta Gynecologic Oncology, Northside Hospital. During fellowship, she worked extensively with the Da Vinci surgical robotic system, and conducted clinical studies on the use of the Da Vinci robot for the staging and treatment of ovarian cancer and advanced stage uterine malignancies.

Following completion of her fellowship, she joined the faculty of Richmond University Medical Center in Staten Island, New York, where she is the Director of Research, as well as a practicing clinician and an educator. She has served as the principal investigator of multiple studies within the institution. She regularly presents her data in peer reviewed journals and at several national and international conferences. Some of her research interests include oxidative stress in pregnancy, pregnancy complications, treatment of gynecological malignancies, and minimally invasive surgery.

Michael Moretti, MD

Chairman
Department of Obstetrics and Gynecology
Maternal Fetal Medicine Specialist
Richmond University Medical Center
Staten Island, New York, US
Email: mmoretti@rumcsi.org

Michael L. Moretti, MD is the Chairman of the Department of Obstetrics and Gynecology at Richmond University Medical Center. After receiving his M.D. from Universidad del Noreste, Tampico, Mexico, Dr. Moretti graduated from the Fifth Pathway Program at New York Medical College/St. Vincent's Hospital, Staten Island, NY in 1982. Dr. Moretti received his residency training at Staten Island University Hospital and then went on to complete a Fellowship in Maternal-Fetal Medicine at the University of Tennessee, Memphis, followed by two years of Advanced Obstetrical Ultrasound Training at Yale University, New Haven Hospital, New Haven, Connecticut.

Dr. Moretti is board-certified in both Obstetrics and Gynecology and Maternal-Fetal Medicine. He maintains an active involvement in both clinical practice in Maternal-Fetal Medicine as well as clinical research, and has numerous publications in peer-reviewed journals as well as contributions to scholarly texts in the field of Maternal-Fetal Medicine. He continues to be directly involved in Graduate Medical Education and conducts an active schedule of departmental and inter-departmental teaching rounds, conferences, and clinical instruction.

INDEX

#

9

A

B

C

D

E

F

G

H

I

K

L

M

N

O

S

T

U

V

W

X

Y